www.wadsworth.com

www.wadsworth.com is the World Wide Web site for Thomson Wadsworth and is your direct source to dozens of online resources.

At *www.wadsworth.com* you can find out about supplements, demonstration software, and student resources. You can also send email to many of our authors and preview new publications and exciting new technologies.

www.wadsworth.com
Changing the way the world learns®

SIXTH EDITION

Essentials of Sociology

David B. Brinkerhoff
University of Nebraska, Lincoln

Lynn K. White
University of Nebraska, Lincoln

Suzanne T. Ortega
University of Missouri, Columbia

Rose Weitz
Arizona State University

THOMSON

WADSWORTH

Australia • Canada • Mexico • Singapore • Spain
United Kingdom • United States

THOMSON

WADSWORTH

Acquisitions Editor: *Jay Whitney*

Development Editor: *Shelley Murphy*

Assistant Editor: *Stephanie Monzon*

Editorial Assistant: *Jennifer Walsh*

Technology Project Manager: *Dee Dee Zobian*

Marketing Manager: *Matthew Wright*

Marketing Assistant: *Tara Pierson*

Advertising Project Manager: *Linda Yip*

Project Manager, Editorial Production: *Matt Ballantyne*

Art Director: *Rob Hugel*

Print/Media Buyer: *Karen Hunt*

Permissions Editor: *Sarah Harkrader*

Production: *Mary Douglas, Rogue Valley Publications*

Compositor: *Graphic World, Inc.*

Text Designer: *Roy Neuhaus*

Photo Researcher: *Roberta Broyer*

Copy Editor: *Susan Defosset*

Illustrator: *Electronic Publishing Services, Inc., NYC, Graphic World Publishing Services*

Cover Designer: *Yvo*

Cover Image: *Copyright © Alberto Incrocci/Getty Images*

Cover Printer: *Phoenix Color Corp*

Printer: *RR Donnelley—Willard*

For more information about our products, contact us at:
Thomson Learning Academic Resource Center
1-800-423-0563

For permission to use material from this text or product, submit a request online at http://www.thomsonrights.com.

Any additional questions about permissions can be submitted by email to thomsonrights@thomson.com.

Library of Congress Control Number: 2004102003

Student Edition: ISBN 0-534-62676-9

Instructor's Edition: ISBN 0-534-62677-7

Thomson Wadsworth
10 Davis Drive
Belmont, CA 94002-3098
USA

Asia
Thomson Learning
5 Shenton Way #01-01
UIC Building
Singapore 068808

Australia/New Zealand
Thomson Learning
102 Dodds Street
Southbank, Victoria 3006
Australia

Canada
Nelson
1120 Birchmount Road
Toronto, Ontario M1K 5G4
Canada

Europe/Middle East/Africa
Thomson Learning
High Holborn House
50/51 Bedford Row
London WC1R 4LR
United Kingdom

Latin America
Thomson Learning
Seneca, 53
Colonia Polanco
11560 Mexico D.F.
Mexico

Spain/Portugal
Paraninfo
Calle Magallanes, 25
28015 Madrid, Spain

Brief Contents

Contents

CHAPTER 9   *Racial and Ethnic Inequality   208*

CHAPTER 11 *Age and Health Inequalities*   256

CHAPTER 12 *Family*   284

Preface

Sociology deals with the major issues—both macro and micro—that confront our planet, our nation, and our lives. At the micro level, sociology explores the substance of ordinary life—getting a job, getting married (and staying that way), caring for children, and caring for aging parents. At the macro level, sociologists grapple with the crucial national and international problems of our times: homelessness, health care reform, environmental degradation, poverty, and international conflict and dependence. An introductory textbook offers the opportunity to help students understand both of these levels and the connections between them. In this textbook, we hope to give students a sociological framework that will help them better understand how social structures shape both their personal lives and the larger social world that surrounds them.

This sixth edition of *Essentials of Sociology* has the same goal as the preceding editions: to provide a concise, balanced introduction to the field of sociology. Like the first five editions, the sixth edition provides a careful blend of theory and the latest research combined with a series of examples, case studies, and applications that will help students develop the sociological imagination. The sixth edition of *Essentials of Sociology* will be particularly helpful for those who wish to supplement their text with readers, literature, or activities or for instructors who teach short academic terms and have difficulty covering the material contained in a more comprehensive text. In condensing the field, we have retained all the central concepts, theories, and research findings. We also offer many examples to help students grasp this basic information and to keep the book engaging and readable.

Changes in the Sixth Edition

The book has been revised to take into account recent developments in the field as well as "hot" new topics such as terrorism, "school choice," felon disenfranchisement, and the rise of American consumer culture. We have carefully reviewed all the major journals and many of the specialty journals to provide students with the most current findings. Figures and graphs have been revised to incorporate the very latest data available.

Major changes in this edition include expanded coverage of globalization, sexuality, and technology. Chapter 8, now titled "Global Inequality and Globalization," includes sections on the sources of globalization; the cultural, economic, and political impacts of globalization; and the links between globalization, global inequality, and the terrorists attacks of 9/11. Where relevant, those attacks and related topics dealing with terrorism have also been woven into the text in other chapters.

Chapter 10, "Sex, Gender, and Sexuality," now includes new sections on sexuality, the social construction of gender, and the feminist movement. An expanded discussion of technology now appears in Chapter 16, which includes sections on reproductive technologies, the technological imperative, and "normal accidents."

Coverage of **diversity issues** has been continued in this edition, both in the body of the chapters and in the **Focus On** boxes and **Intersections.** In addition, Chapter 9 now includes a new section on Arab Americans and an expanded discussion of Hispanic Americans.

New Features of the Sixth Edition

New **Connections** boxes have been added to each chapter to clarify concepts and to help students see how the concepts apply to their own lives. Four types of connections are presented in the book: examples, personal applications for students, historical notes, and social policy applications.

New end-of-chapter sections titled **Where This Leaves Us** tie the chapter concepts together and show how the material relates to the larger themes, problems, and issues discussed in the chapter. These sections will help students improve their grasp of the main points in the chapter.

Professors may now choose to have their students conduct research in Wadsworth's **Opposing Viewpoints Resource Center**. As in the previous edition, each chapter ends with a **Sociology on the Net** section, which now includes suggested topics that students can search for in the Opposing Viewpoints database.

The **Focus On** boxes—included in each chapter and continued from the last edition—now include Internet and **InfoTrac College Edition** links for students to use. These links give students access to rich data sources.

Other Features of the Sixth Edition

The sixth edition of *Essentials of Sociology* retains the pedagogical features that have made it successful and adds new features. Each chapter contains at least one high-interest boxed insert. Clearly identified concepts, **Concept Summaries**, and chapter summaries continue to aid students in mastering the material. As in the fifth edition, discussion of the three major theoretical perspectives in sociology appears near the beginning of each chapter, and **Critical Thinking** questions appear at the end of each chapter. These questions challenge students to apply sociological concepts and theory to problems relevant to their immediate lives and can be used for group discussion or individual writing assignments.

This edition also continues our tradition of incorporating cross-national examples throughout the text. Chapter 11, for instance, includes a case study of declining life expectancies in Eastern Europe and Russia. In Chapter 4, the **Focus on a Global Perspective** box features an analysis of impression management among Australian aborigines and the Focus On box in Chapter 5 assesses the strength of weak ties in China and Singapore.

Focus On Boxes

A boxed insert in each chapter introduces provocative and interesting issues. These Focus On boxes fall into four different formats: Focus on American Diversity, Focus on a Global Perspective, Focus on Technology, and the all new Focus on the Environment feature. To demonstrate to students the importance of understanding the increasing diversity of American society, the **Focus on American Diversity** boxes examine issues such as the measurement of IQ, gay and lesbian families, and gender differences in mathematics. The **Focus on a Global Perspective** series introduces students to a comparative approach to social issues and social science research, and it deals with topics such as women and development, international migration, and conducting survey research in Nigeria. **The Focus on Technology** series links technological innovation to social change by analyzing the implications of technology for social relationships, social control, and social institutions. Finally, the **Focus on the Environment** boxes address important social issues and policy concerns such as criminally negligent forms of environmental contamination, environmental racism, and the environmental movement. With this enhanced focus on the interconnection between society and ecology, *Essentials of Sociology,* Sixth Edition, has better coverage of today's environmental issues than any other intro text on the market.

Intersections

This popular feature continues in the current edition. Located between chapters throughout the text, essays are designed to help students understand the connections between concepts and topics that have been treated separately in preceding chapters. The Intersections demonstrate how institutions, social processes, culture, and systems of inequality often reinforce each other. After the first four chapters, for instance, Intersections asks students to think about the relationships between culture and social structure by exploring how they simultaneously impact the natural environment. After Chapter 14, students are challenged to think further about the relationships between values and institutions by considering how the political process fosters the appearance of what some scholars have called the *culture wars*. This effective feature helps students understand the connections between important sociological concepts.

Intersections Exercises Using MicroCase Online

So that students can see actual data that reveals the relationships explored in the Intersections essays, a corresponding exercise exists on the **Companion Web Site**. For each Intersections in the book, the Web exercise asks students to utilize the **MicroCase Data Analysis** program. Here, students are instructed to select data sets and variables, analyze and compare the statistical data that results, and examine possible relationships between the ideas introduced in the Intersections. Students then answer a series of questions about the analysis, which guides them to think critically about the interrelated ideas. An answer key is available for instructors, because students can email their responses to these questions to the instructor of their course. These Intersections Exercises are available at the book's Companion Web Site: http://sociology.wadsworth.com/brinkerhoff/essentials6e/.

Concept Learning Aids

Learning new concepts and new vocabulary is vital to developing a new perspective. In *Essentials of Sociology*, this learning is facilitated in four ways: (1) When new terms and concepts first appear in the text, they are printed in boldface type, and complete definitions are set out clearly in the margin. (2) When several related concepts are introduced (for example, pluralist, power elite, and state autonomy models of American government), a Concept Summary is included to summarize the definitions, give examples, and clarify differences. (3) A glossary is included in the back of the book for handy reference. (4) Critical Thinking questions encourage students to make concepts and terms a working part of their vocabulary by using them to discuss a problem of personal or social relevance.

Chapter Summaries

A short point-by-point summary lists the chief points made in each chapter. These summaries will aid beginning students as they study the text and help them distinguish the central concepts from the supporting points.

Sociology on the Net

The Sociology on the Net feature, appearing at the end of each chapter, emphasizes the importance of understanding technology and its impact on society as a whole by providing key URLs and InfoTrac® College Edition search terms for each chapter.

Sociology on the Net identifies several of the most relevant sites on the Web for finding information related to issues discussed in *Essentials of Sociology*. For example, browsers are directed to sites that specialize in multicultural issues (Chapter 2), world poverty (Chapter 8), gender discrimination (Chapter 10), and the sociology of cyberspace (Chapter 16).

Supplements for the Sixth Edition

For Students and Instructors

WebTutor™ ToolBox for WebCT or Blackboard

Preloaded with content and available free via pincode when packaged with this text, *WebTutor* ToolBox pairs all the content of this text's rich Book Companion Web Site with all the sophisticated course management functionality of a WebCT or Blackboard product. You can assign materials (including online quizzes) and have the results flow automatically to your gradebook. ToolBox is ready to use as soon as you log on—or you can customize its preloaded content by uploading images and other resources, adding Web links, or creating your own practice materials. Students only have access to student resources on the Web site. Instructors can enter a pincode for access to password-protected Instructor Resources.

InfoTrac® College Edition

Each purchase of a new copy of the text includes a free four-month passcode to InfoTrac College Edition, the online library that gives students anytime, anywhere access to reliable resources. This fully searchable database offers 20 years' worth of full-text articles from almost 5,000 diverse sources, such as academic journals, newsletters, and up-to-the-minute periodicals including *Time, Newsweek, Science, Forbes,* and *USA Today.* This incredible depth and breadth of material—available 24 hours a day from any computer with Internet access—makes conducting research so easy that your students will want to use it to enhance their work in every course! Through InfoTrac's InfoWrite, students now also have instant access to critical-thinking and paper-writing tools. Both adopters and their students receive unlimited access for four months.

Opposing Viewpoints Resource Center (OVRC)

Newly available from Wadsworth, this online center presents varying perspectives on today's most compelling issues. OVRC draws on Greenhaven Press's acclaimed Social Issues Series, as well as core reference content from other Gale and Macmillan Reference USA sources. The result is a dynamic online library of current events topics—the facts as well as the arguments of each topic's proponents and detractors. Special sections focus on critical thinking, which walk students through the steps involved in critically evaluating point-counterpoint arguments, and researching and writing papers. OVRC is also available through Wadsworth's Sociology Online Resources and Writing Companion.

For Instructors

Instructor's Edition

The Instructor's Edition contains a visual walk-through of the text that provides an overview of the key features, themes, and available supplements.

Instructor's Resource Manual with Test Bank (with Multimedia Manager CD-ROM)

This supplement offers the instructor chapter outlines, class projects and assignments, discussion and lecture topics, Internet and InfoTrac Activities, as well as personal applications and journal topics for each chapter. Unique to this supplement is the **Student Data Set** (SDS), which is based on a class questionnaire that enables instructors to directly link students to the data and relationships examined in the main text. The manual includes suggestions for lectures, discussions, and class activities involving the Student Data Set. The test bank contains 100 multiple-choice questions, 15 true/false questions and 10 short-answer questions with answers and page references along with 10 essay questions for each chapter. A table of contents for the *CNN Today* Sociology Video Series and concise user guides for both InfoTrac and WebTutor are also included.

An all NEW **Multimedia Manager Instructor Resource CD-ROM** is now packaged with the IRM/TB. This new instructor resource includes book-specific PowerPoint® Lecture Slides, graphics from the book itself, the IRM/TB word docs, CNN Video Clips, and links to many of Wadsworth's important sociology resources. All of your media teaching resources in one place!

ExamView® Computerized Testing

Create, deliver, and customize tests and study guides (both print and online) in minutes with this easy-to-use assessment and tutorial system. *ExamView* offers both a *Quick Test Wizard* and an *Online Test Wizard* that guide you step by step through the process of creating tests. Each test appears on screen exactly as it will print or display online.

Using *ExamView*'s complete word processing capabilities, you can enter an unlimited number of new questions or edit existing questions.

Wadsworth's Introduction to Sociology 2004 Transparency Acetates

A set of four-color acetates consisting of tables and figures from Wadsworth's introductory sociology texts is available to help prepare lecture presentations. Free to qualified adopters.

CNN Today: Sociology Videos, Volumes I–VII

Integrate the up-to-the-minute programming power of CNN and its affiliate networks right into your classroom! Updated yearly, *CNN Today Sociology Videos* are course-specific videos that can help you launch a lecture, spark a discussion, or demonstrate an application using the top-notch business, science, consumer, and political reporting of the CNN networks. Produced by Turner Learning, Inc., these 45-minute videos show your students how the principles they learn in the classroom apply to the stories they see on television. Special adoption conditions apply.

Wadsworth's Lecture Launchers for Introductory Sociology

An exclusive offering jointly created by Wadsworth/Thomson Learning and DALLAS TeleLearning, this video contains a collection of video highlights taken from the *Exploring Society: An Introduction to Sociology* Telecourse (formerly the *Sociological Imagination*). Each 3–6 minute video segment has been especially chosen to enhance and enliven class lectures and discussions of 20 key topics covered in any introductory sociology text. Accompanying the video is a brief written description of each clip, along with suggested discussion questions to help effectively incorporate the material into the classroom. Available both on VHS and DVD.

Sociology: Core Concepts

An exclusive offering jointly created by Wadsworth/Thomson Learning and DALLAS TeleLearning, this video contains a collection of video highlights taken from the telecourse *Exploring Society: An Introduction to Sociology* (formerly the *Sociological Imagination*). Each 15–20 minute video segment will enhance student learning of the essential concepts in the introductory course and can be used to initiate class lectures, discussion, and review. The video covers topics such as the sociological imagination, stratification, race and ethnic relations, social change, and more. Available both on VHS and DVD.

Wadsworth Sociology Video Library

Bring sociological concepts to life with videos from Wadsworth's *Sociology Video Library*, which includes thought-provoking offerings from Films for the Humanities as well as other excellent educational video sources. This extensive collection illustrates the important sociological concepts covered in many sociology courses. Certain adoption conditions apply.

For Students

Study Guide

Each chapter of the *Study Guide* includes learning objectives, detailed chapter outlines, matching exercises, Internet exercises, 30–35 multiple-choice questions and 10 true/false questions with answers and page references, 8–10 short answer questions, and 5 essay questions to enhance student understanding of chapter concepts.

Practice Tests

Each chapter of the *Practice Tests* contains 20–25 multiple-choice questions and 10–15 true/false questions, all with corresponding rejoinders and page references to the main text, as well as 5 short-answer questions with page references.

Wadsworth's Sociology Online Resources and Writing Companion, First Edition

New! This valuable guide shows students how they can use Wadsworth's exclusive online resources—*InfoTrac College Edition,* the *Opposing Viewpoints Resource Center (OVRC),* and *MicroCase Online*—to assist them in their study of sociology and build essential research and writing skills. Part One provides informative user guides that introduce each of these powerful research tools. Part Two contains directed exercises designed to develop research and critical thinking proficiency for each of the core topics in sociology. Part Three provides an overview of some of the research and writing tools available online, such as *InfoWrite* and the *OVRC Research Guide,* and shows students how they can effectively integrate their research findings into class assignments.

Researching Sociology on the Internet, Third Edition

Prepared by D. R. Wilson of Houston Baptist University, this guide is designed to assist sociology students with doing research on the Internet. Part One contains the general information necessary to get started and answers questions about security, the type of sociology material available on the Internet, the information that is reliable and the sites that are not, the best ways to find research, and the best links to take students where they want to go. Part Two looks at each main topic in the area of sociology and refers students to sites where they can obtain the most enlightening research and information.

Acknowledgments

As with the earlier editions, in preparing this edition we have accumulated many debts. We are especially grateful for the good-natured and generous advice of our colleagues at the University of Nebraska at Lincoln, Arizona State University, and the University of Missouri. Special thanks go to Robert Benford, Miguel Carranza, Jay Corzine, Julie Harms Cannon, Kurt Johnson, Jennifer Lehmann, Helen Moore, Wayne Osgood, Keith Parker, and Al Williams. They were always willing to share their expert knowledge and to comment and advise on our own forays into their substantive areas. Thanks also to Tim Pippert and Sharon Larson who contributed InfoTrac exercises and a number of innovative classroom activities to the Instructor's Manual. In addition, we are particularly grateful for the research assistance so generously provided by Lisa Tichavsky for this edition.

Special thanks go to the people at West and Wadsworth Publishing, including Clyde Perlee, who first prompted us to become authors and Denise Simon, who was generous with encouragement and advice. Eve Howard, Stephanie Monzon, Lisa Weber, and Maryalice Ditzler have continued to be helpful over the years; Shelley Murphy and Jay Whitney played crucial roles in helping us craft this edition. At all levels, the people at Wadsworth have been a pleasure to work with and helped us make our book the best possible, while leaving the substance and direction of the book in our hands.

We would like to express our gratitude to those people who reviewed the manuscript for us:

Tim Brezina, *Tulane University*

Lisa Linares, *Madison Area Technical College*

Ronald Matson, *Wichita State University*

Peter Venturelli, *Valparaiso University*

Allison Vetter, *University of Central Arkansas*

Leslie Wang, *University of Toledo*

Once again, we thank those people who reviewed the manuscript for previous editions of *Sociology* and *Essentials of Sociology*. Their suggestions and comments made a substantial contribution to the project:

Margaret Abraham, *Hofstra University, New York*

Paul J. Baker, *Illinois State University*

Robert Benford, *University of Nebraska*

Susan Blackwell, *Delgado Community College*

Marie Butler, *Oxnard College*

John K. Cochran, *Wichita State University, Kansas*

Carolie Coffey, *Cabrillo College, California*

Paul Colomy, *University of Akron, Ohio*

Ed Crenshaw, *University of Oklahoma*

Raymonda P. Dennis, *Delgado Community College, New Orleans;*

Lynda Dodgen, *North Harris County College, Texas*

David A. Edwards, *San Antonio College, Texas*

Laura Eells, *Wichita State University, Kansas*

William Egelman, *Iona College, New York*

Constance Elsberg, *Northern Virginia Community College*

Christopher Ezell, *Vincennes University, Indiana*

Joseph Faltmeier, *South Dakota State University*

Daniel E. Ferritor, *University of Arkansas*

Charles E. Garrison, *East Carolina University, North Carolina*

James R. George, *Kutztown State College, Pennsylvania*

Harold C. Guy, *Prince George Community College, Maryland*

Rose Hall, *Diablo Valley College, California*

Sharon E. Hogan, *Blue River Community College*

Michael G. Horton, *Pensacola Junior College, Florida*

Cornelius G. Hughes, *University of Southern Colorado*

Jon Ianitti, *SUNY Morrisville*

Carol Jenkins, *Glendale Community College*

William C. Jenné, *Oregon State University*

Dennis L. Kalob, *Loyola University, New Orleans*

Sidney J. Kaplan, *University of Toledo, Ohio*

Florence Karlstrom, *Northern Arizona University*

Diane Kayongo-Male, *South Dakota State University*

William Kelly, *University of Texas*

James A. Kithens, *North Texas State University*

Phillip R. Kunz, *Brigham Young University, Utah*

Billie J. Laney, *Central Texas College*

Charles Langford, *Oregon State University*

Mary N. Legg, *Valencia Community College, Florida*

John Leib, *Georgia State University*

Joseph J. Leon, *California State Polytechnic University, Pomona*

J. Robert Lilly, *Northern Kentucky University*

Jan Lin, *University of Houston*

James Lindberg, *Montgomery College*

Richard L. Loper, *Seminole Community College, Florida*

Carol May, *Illinois Central College*

Rodney C. Metzger, *Lane Community College, Oregon*

Vera L. Milam, *Northeastern Illinois University*

Purna C. Mohanty, *Paine College*

James S. Munro, *Macomb College, Michigan*

Lynn D. Nelson, *Virginia Commonwealth University*

J. Christopher O'Brien, *Northern Virginia Community College*

Charles O'Connor, *Bemidji State University, Minnesota*

Jane Ollenberger, *University of Minnesota-Duluth*

Robert L. Petty, *San Diego Mesa College, California*

Ruth A. Pigott, *Kearney State College, Nebraska*

John W. Prehn, *Gustavus Adolphus College, St. Peter, Minnesota*

Adrian Rapp, *North Harris County College, Texas*

Mike Robinson, *Elizabethtown Community College, Kentucky*

Joe Rogers, *Big Bend Community College*

Will Rushton, *Del Mar College, Texas*

Rita P. Sakitt, *Suffolk Community College, New York*

Martin Scheffer, *Boise State University, Idaho*

Richard Scott, *University of Central Arkansas*

Ida Harper Simpson, *Duke University, North Carolina*

James B. Skellenger, *Kent State Unversity, Ohio*

Ricky L. Slavings, *Radford University, Virginia*

John M. Smith, Jr., *Augusta College, Georgia*

Evelyn Spiers, *College of the Canyons*

James Steele, *James Madison University, Virginia*

Michael Stein, *University of Missouri–St. Louis*

Barbara Stenross, *University of North Carolina*

Jack Stirton, *San Joaquin Delta College, California*
Deidre Tyler, *Salt Lake Commuity College*
Emil Vajda, *Northern Michigan University*
Henry Vandenburgh, *SUNY, Oswego*
Steven L. Vassar, *Mankato State University, Minnesota*
Jane B. Wedemeyer, *Sante Fe Community College, Florida*
Dorether M. Welch, *Penn Valley Community College*
Thomas J. Yacovone, *Los Angeles Valley College, California*
David L. Zierath, *University of Wisconsin.*

CHAPTER 1

The Study of Society

© CORBIS

What Is Sociology?

Each of us starts the study of society with the study of individuals. We wonder why Theresa keeps getting involved with men who treat her badly, why Mike never learns to quit drinking before he gets sick, why our aunt puts up with our uncle, and why anybody ever liked the Spice Girls. We wonder why people we've known for years seem to change drastically when they get married or change jobs.

If Theresa were the only woman with bad taste in men or if Mike were the only man who ever drank too much, then we might try to understand their behavior by peering into their personalities. We know, however, that there are thousands, maybe millions, of men and women who have disappointing romances and who drink too much. To understand Mike and Theresa, then, we must place them in a larger context and examine the forces that seem to compel so many people to behave in a similar way.

Sociologists tend to view these common human situations as if they were plays. They might, for example, title a common human drama *Boy Meets Girl*. Just as *Hamlet* has been performed all around the world for 400 years with different actors and different interpretations, *Boy Meets Girl* has also been performed countless times. Of course, the drama is acted out a little bit differently each time, depending on the scenery, the people in the lead roles, and the century—but the essentials are the same. Thus, we can read nineteenth- or even sixteenth-century love stories and understand why those people did what they did. They were playing roles in a play that is still performed daily.

More formal definitions will be introduced later, but the metaphor of the theater can be used now to introduce two of the most basic concepts in sociology: role and social structure. By *role*, we mean the expected performance of someone who occupies a specific position. Mothers have roles, teachers have roles, students have roles, and lovers have roles. Each position has an established script that suggests appropriate lines, gestures, and relationships with others. Discovering what each society offers as a stock set of roles is one of the major themes in sociology. Sociologists try to find the common roles that appear in society and to determine why some people play one role rather than another.

The second major sociological concept is *social structure*, which is concerned with the larger structure of the play in which the roles appear. What is the whole set of roles that appears in this play, and how are the roles interrelated? Thus, the role of mother is understood in the context of the social structure we call the family. The role of student is understood in the context of the social structure we call education. Through these two major ideas, role and social structure, sociologists try to understand the human drama.

The Sociological Imagination

The **sociological imagination** is the ability to see the intimate realities of our own lives in the context of common social structures; it is the ability to see personal troubles as public issues.

The ability to see the intimate realities of our own lives in the context of common social structures has been called the **sociological imagination** (Mills 1959, 15). Sociologist C. Wright Mills suggests that the sociological imagination is developed when we can place personal troubles (such as poverty, divorce, or loss of faith) into a larger social context and, as a result, recognize them as common public issues. He suggests that many of the things we experience as individuals are

AFP/Wide World Photos

■ In any given year, about 3.5 million Americans—almost one-third of whom are children—experience homelessness. Although some homeless individuals suffer from mental illness or substance abuse, the extent of homelessness largely reflects the lack of affordable housing and adequately paying jobs in the United States.

really beyond our control. They have to do with society as a whole, its historical development, and the way it is organized. Mills gives us some examples of the differences between a personal trouble and a public issue:

> When, in a city of 100,000, only one man is unemployed, that is his personal trouble, and for its relief we properly look to the character of the man, his skills, and his immediate opportunities. But when in a nation of 50 million employees, 15 million men are unemployed, that is an issue, and we may not hope to find its solution within the range of opportunities open to any one individual. The very structure of opportunities has collapsed. Both the correct statement of the problem and the range of possible solutions require us to consider the economic and political institutions of the society, and not merely the personal situation and character of a scatter of individuals. . . .
>
> Consider marriage. Inside a marriage a man and a woman may experience personal troubles, but when the divorce rate during the first four years of marriage is 250 out of every 1000 attempts, this is an indication of a structural issue having to do with the institutions of marriage and the family and other institutions that bear on them. (Mills 1959, 9)

In everyday life, we do not define personal experiences in these terms. We rarely consider the impact of history and social structures on our own experiences. If a child becomes a drug addict, parents tend to blame themselves; if spouses divorce, their friends usually focus on their personality problems; if a student does poorly in school, everyone will be likely to blame the student alone. To develop the sociological imagination is to understand how outcomes such as these are, in part, a product of society and not fully within the control of the individual.

Some people do poorly in school, for example, not because they are stupid or lazy but because they are faced with conflicting role expectations. The "this is the best time of your life" play calls for very different roles from the "education is the key to success" play. Other people may do poorly because they come from a family that does not give them the financial or psychological support that they need. In fact, their family may need them to earn an income to help support their younger brothers and sisters. These students may be working 25 hours a week in addition to going to school; they may be going to school despite their family's lack of understanding of why college is important, or why college students need quiet and privacy for studying. In contrast, other students

Connections
Social Policy

How can we help the unemployed? How we answer this question will depend on what we think caused the problem. Psychologists and social workers tend to focus on individual-level solutions such as offering workshops in job interviewing and short-term economic assistance to laid-off workers. Sociologists instead focus on understanding the broader social forces that cause widespread unemployment. They raise questions such as "How can we halt the movement of American jobs overseas?" and "How can we ensure that children from poorer families get the education needed to obtain good jobs in a shrinking job market?"

may find it difficult to fail: Their parents provide tuition, living expenses, a personal computer, a car, and moral support. As we will discuss in more detail in Chapter 13, parents' social class is one of the best predictors of who will fail and who will graduate. Success or failure is thus not entirely an individual matter; it is socially structured.

The sociological imagination—the ability to see our own lives and those of others as part of a larger social structure and a larger human drama—is central to sociology. Once we develop this imagination, we will be less likely to explain others' behavior through their personality and will increasingly look to the roles and social structures that determine behavior. We will also recognize that the solutions to many social problems lie not in changing individuals but in changing the social structures and roles that are available to them. Although poverty, divorce, illegitimacy, and racism are experienced as intensely personal hardships, they are unlikely to be reduced effectively through massive personal therapy. To solve these and many other social problems, we need to change social structures; we need to rewrite the play. The sociological imagination offers a new way to look at—and a new way to solve—common troubles and dilemmas that face individuals.

Sociology as a Social Science

Sociology is concerned with people and with the rules of behavior that structure the ways in which people interact. As one of the social sciences, sociology has much in common with political science, economics, psychology, and anthropology. All these fields share an interest in human social behavior and, to some extent, an interest in society. In addition, they all share an emphasis on the scientific method as the best approach to knowledge. This means that they rely on critical and systematic examination of the evidence before reaching any conclusions, and their practitioners approach each research question from a position of moral neutrality—that is, they try to be objective observers. This scientific approach is what distinguishes the social sciences from journalism and other fields that comment on the human condition.

Sociology is the systematic study of human social interaction.

Sociology is a social science whose unique province is the systematic study of human social interaction. Its emphasis is on relationships and patterns of interaction—how these patterns develop, how they are maintained, and how they change.

The Emergence of Sociology

Sociology emerged as a field of inquiry during the political, economic, and intellectual upheavals of the eighteenth and nineteenth centuries. Rationalism and science replaced tradition as methods of understanding the world, leading to changes in government, education, economic production, and even religion and family life. The clearest symbol of this turmoil is the French Revolution (1789), with its bloody uprising and rejection of the past.

Although less dramatic, the Industrial Revolution had an even greater impact. Within a few generations, traditional rural societies were replaced by industrialized urban societies. The rapidity and scope of the change resulted in substantial social disorganization. It was as if society had changed the play without bothering to tell the actors, who were still trying to read from old scripts. Although a few people prospered mightily, millions struggled desperately to make the adjustment from rural peasantry to urban working class.

The picture of urban life during these years—in London, Chicago, or Hamburg—was one of disorganization, poverty, and dynamic and exciting change. This turmoil and tragedy provided the inspiration for much of the intellectual effort of the nineteenth century: Charles Dickens's novels, Jane Addams's reform work, Karl Marx's revolutionary theory. It also inspired the scientific study of society. These were the years in which science was a new enterprise and nothing seemed too much to hope for. After electricity, the telegraph, and the X-ray, who was to say that science could not discover how to turn stones into gold or how to eliminate poverty or war? Many hoped that the tools of science could help in understanding and controlling a rapidly changing society.

The Founders: Comte, Spencer, Marx, Durkheim, and Weber

The upheavals of the nineteenth century in Europe and Great Britain stimulated the development of sociology as a scientific discipline. We will look at five theorists who may be considered the founders of sociology.

August Comte (1798–1857)

The first major figure to be concerned with the science of society was the French philosopher August Comte. He coined the term *sociology* in 1839 and is generally considered the founder of this field.

Comte was among the first to suggest that the scientific method could be applied to social events (Konig 1968). The philosophy of positivism, which he developed, suggests that the social world can be studied with the same scientific accuracy and assurance as the natural world. Once the laws of social behavior were learned, he believed, scientists could accurately predict and control events. Although thoughtful people wonder whether we will ever be able to predict human behavior as accurately as we can predict the behavior of molecules, the scientific method remains central to sociology.

Auguste Comte, 1798–1857

Another of Comte's lasting contributions was his recognition that an understanding of society requires a concern for both the sources of order and continuity and the sources of change. Comte called these divisions the theory of statics and the theory of dynamics. Although sociologists no longer use his terms, Comte's basic divisions of sociology continue under the labels of social structure (statics) and social process (dynamics).

Herbert Spencer (1820–1903)

Another pioneer in sociology was the British philosopher-scientist Herbert Spencer, who advanced the thesis that evolution accounts for the development of social, as well as natural, life. Spencer viewed society as similar to a giant organism: Just as the heart and lungs work together to sustain the life of the organism, so the parts of society work together to maintain society. Spencer's analogy led him to some conclusions that seem foolish by modern standards, but they also led him to some basic principles that still guide the study of sociology.

One of Spencer's guiding principles was that society must be understood as an adaptation to its environment. This principle of adaptation implies that to understand society, we must focus on processes of growth and change. It also implies that there is no "right" way for a society to be organized. Instead, societies will change as circumstances change.

Herbert Spencer, 1820–1903

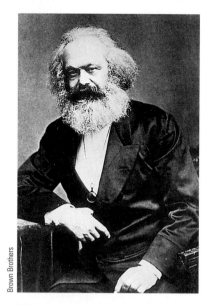

■ Karl Marx, 1818–1883

Spencer's second major contribution was his concern with the scientific method. More than many scholars of his day, Spencer was aware of the importance of objectivity and moral neutrality in investigation. In essays on the bias of class, the bias of patriotism, and the bias of theology, he warned sociologists that they must suspend their own opinions and wishes when studying the facts of society (Turner & Beeghley 1981).

Karl Marx (1818–1883)

A philosopher, economist, and social activist, Karl Marx was born in Germany to middle-class parents. Marx received his doctorate in philosophy at the age of 23, but because of his radical views he was unable to obtain a university appointment and spent most of his adult life in exile and poverty (Rubel 1968).

Marx was repulsed by the poverty and inequality that characterized the nineteenth century. Unlike other scholars of his day, he was unwilling to see poverty as either a natural or a God-given condition of the human species. Instead, he viewed poverty and inequality as human-made conditions fostered by private property and capitalism. As a result, he devoted his intellectual efforts to understanding—and eliminating—capitalism. Many of Marx's ideas are of more interest to political scientists and economists than to sociologists, but he left two enduring legacies to sociology: the theories of economic determinism and the dialectic.

Economic Determinism   Marx began his analysis of society by assuming that the most basic task of any human society is to provide food and shelter to sustain itself. Marx argued that the ways in which society does this—its modes of production—provide the foundations on which all other social and political arrangements are built. Thus, he believed that family, law, and religion all develop after and adapt to the economic structure; in short, they are determined by economic relationships. This idea is called **economic determinism.**

A good illustration of economic determinism is the influence of economic conditions on marriage choices. In traditional agricultural societies, young people often remain economically dependent upon their parents until well into adulthood because the only economic resource, land, is controlled by the older generation. In order to support themselves now and in the future, they must remain in their parents' good graces; this means they cannot marry without their parents' approval. In societies where young people can earn a living without their parents' help, however, they can marry when they please and whomever they please. Marx would argue that this shift in mate selection practices is the result of changing economic relationships. Because Marx saw all human relations as stemming ultimately from the economic systems, he suggested that the major goal of a social scientist is to understand economic relationships: Who owns what, and how does this pattern of ownership affect human relationships?

Economic determinism means that economic relationships provide the foundation on which all other social and political arrangements are built.

The Dialectic   Marx's other major contribution was a theory of social change. Many nineteenth-century scholars applied Darwin's theories of biological evolution to society; they believed that social change was the result of a natural process of adaptation. Marx, however, argued that the basis of change was conflict, not adaptation. He argued that conflicts between opposing economic interests lead to change.

Marx's thinking on conflict was influenced by the German philosopher George Hegel, who suggested that for every idea (thesis), a counter idea (antithesis) develops to challenge it. As a result of conflict between the two ideas, a new idea (synthesis) is produced. This process of change is called the **dialectic** (Figure 1.1).

Dialectic philosophy views change as a product of contradictions and conflict between the parts of society.

Marx's contribution was to apply this model of ideological change to change in economic and material systems. Within capitalism, Marx suggested, the capitalist class was the thesis and the working class was the antithesis. He predicted that conflicts between them would lead to a new synthesis, a new economic system that would be socialism. Indeed, in his role as social activist, Marx hoped to encourage conflict and ignite the revolution that would bring about the desired change. The workers, he declared, "have nothing to lose but their chains" (Marx & Engels 1967, 7:258).

Although few sociologists are revolutionaries, many accept Marx's ideas on the importance of economic relationships and economic conflicts. Much more controversial is Marx's argument that the social scientist should also be a social activist, a person who not only tries to understand social relationships but also tries to change them.

Emile Durkheim (1858–1917)

Like Marx, Durkheim was born into a middle-class family. While Marx was starving as an exile in England, however, Durkheim spent most of his career occupying a prestigious professorship at the Sorbonne in France. Far from rejecting society, Durkheim embraced it, and much of his outstanding scholarly energy was devoted to understanding the stability of society and the importance of social participation for individual happiness. Whereas the lasting legacy of Marx is a theory that looks for the conflict-laden and ever-changing aspects of social practices, the lasting legacy of Durkheim is a theory that examines the positive contributions and stability of social patterns. Together they allow us to see both order and change.

Durkheim's major works are still considered essential reading in sociology. These include his studies of suicide, education, divorce, crime, and social change. Two enduring contributions are his ideas about the relationship between individuals and society and the development of a method for social science.

One of Durkheim's major concerns was the balance between social regulation and personal freedom. He argued that community standards of morality, which he called the collective conscience, not only confine our behavior but also give us a sense of belonging and integration. For example, many people complain about having to dress up; they complain about having to shave their faces or their legs or having to wear a tie or pantyhose. "What's wrong with my jeans?" they want to know. At the same time, most of us feel a sense of satisfaction when we appear in public in our best clothes. We know that we will be considered attractive and successful. Although we may complain about having to meet what appear to be arbitrary standards, we often feel a sense of satisfaction in being able to meet those standards successfully. In Durkheim's words, "institutions may impose themselves upon us, but we cling to them; they compel us, and we love them" (Durkheim [1895] 1938, 3). This beneficial regulation, however, must not rob the individual of all freedom of choice.

In his classic study, *Suicide,* Durkheim identified two types of suicide that stem from an imbalance between social regulation and personal freedom. Fatalistic suicide occurs when society overregulates and allows too little freedom; when our behavior is so confined by social institutions that we cannot exercise our independence ([1897] 1951, 276). Durkheim gave as an example of fatalistic suicide the very young husband who feels overburdened by the demands of work, household, and family. Anomic suicide, on the other hand, occurs when there is too much freedom and too little regulation, when society's influence does not check individual passions ([1897] 1951, 258). Durkheim said that this kind of suicide is most likely to occur in times of rapid social change. When established ways of doing things have lost their meaning, but no clear alternatives have developed, individuals feel lost. The high suicide

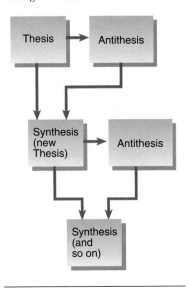

FIGURE 1.1
The Dialectic
The dialectic model of change suggests that change occurs through conflict and resolution rather than through evolution.

Emile Durkheim, 1858–1917

Bettmann Archive

rate of Native Americans (approximately twice that of white Americans) is generally attributed to the weakening of traditional social regulation.

Durkheim was among the first to stress the importance of using reliable statistics to logically rule out incorrect theories of social life and to identify more promising theories. Each of his works illustrates his ideal social scientist: an objective observer who only seeks the facts. As sociology became an established discipline, this ideal of objective observation replaced Marx's social activism as the standard model for social science.

Max Weber (1864–1920)

A German economist, historian, and philosopher, Max Weber (vay-ber) provided the theoretical base for half a dozen areas of sociological inquiry. He wrote on religion, bureaucracy, method, and politics. In all these areas, his work is still valuable and insightful; it is covered in detail in later chapters. Three of Weber's more general contributions were an emphasis on the subjective meanings of social actions, on social as opposed to material causes, and on the need for objectivity in studying social issues.

Weber believed that knowing patterns of behavior was less important than knowing the meanings people attach to behavior. For example, Weber would argue that it is relatively meaningless to compile statistics such as one-half of all marriages contracted today may end in divorce compared to only 10 percent in 1890 (Cherlin 1992). More critical, he would argue, is understanding how the meaning of divorce has changed in the past hundred years. Weber's emphasis on the subjective meanings of human actions has been the foundation of scholarly work on topics as varied as religion and immigration.

Weber trained as an economist, and much of his work concerned the interplay of things material and things social. He rejected Marx's idea that economic factors were the determinants of all other social relationships. In a classic study, *The Protestant Ethic and the Spirit of Capitalism* ([1904–05] 1958), Weber tried to show how social and religious values may be the foundation of economic systems. This argument is developed more fully in Chapter 13, but its major thesis is that the religious values

Brown Brothers

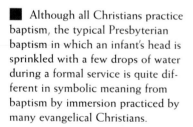

Max Weber, 1864–1920

Although all Christians practice baptism, the typical Presbyterian baptism in which an infant's head is sprinkled with a few drops of water during a formal service is quite different in symbolic meaning from baptism by immersion practiced by many evangelical Christians.

© Ed Kashi

of early Protestantism (self-discipline, thrift, and individualism) were the foundation for capitalism.

One of Weber's more influential ideas was his declaration that sociology must be **value-free.** Weber argued that sociology should be concerned with establishing what is and not what ought to be. Weber's dictum is at the heart of the standard scientific approach that is generally advocated by modern sociologists. Thus, although one may study poverty or racial inequality because of a sense of moral outrage, such feelings must be set aside to achieve an objective grasp of the facts. This position of neutrality is directly contradictory to the Marxist emphasis on social activism, and sociologists who adhere to Marxist principles generally reject the notion of value-free sociology. Most modern sociologists, however, try to be value-free in their scholarly work.

Value-free sociology concerns itself with establishing what is, not what ought to be.

Sociology in the United States

Sociology in the United States developed somewhat differently than it did in Europe. Although U.S. sociology has the same intellectual roots as European sociology, it has some distinctive characteristics. Three features that have characterized U.S. sociology from its beginning are a concern with social problems, a reforming rather than a radical approach to these problems, and an emphasis on the scientific method.

One reason that U.S. sociology developed differently from European sociology is that our social problems differ. Slavery, the Civil War, and high immigration rates, for example, made racism and ethnic discrimination much more salient issues in the United States. One of the first sociologists to study these issues was W. E. B. DuBois. DuBois, who received his doctorate in 1895 from Harvard University, devoted his career to developing empirical data about African Americans and to using those data to combat racism. The work of Jane Addams, another early sociologist and recipient

W. E. B. DuBois, 1868–1963

Jane Addams, 1860–1935

MAP 1.1
Country of Residence of International Sociological Association Members, 2002

Sociology is a global discipline, although the vast majority of members in the International Sociological Association are citizens of wealthy, Western nations. Many of these European and U.S. sociologists study developing regions of Asia, South America, or Africa. But indigenous peoples increasingly bring their own experiences to analyzing their societies, especially in issues concerning population and development.
SOURCE: Data supplied by the International Sociological Association.

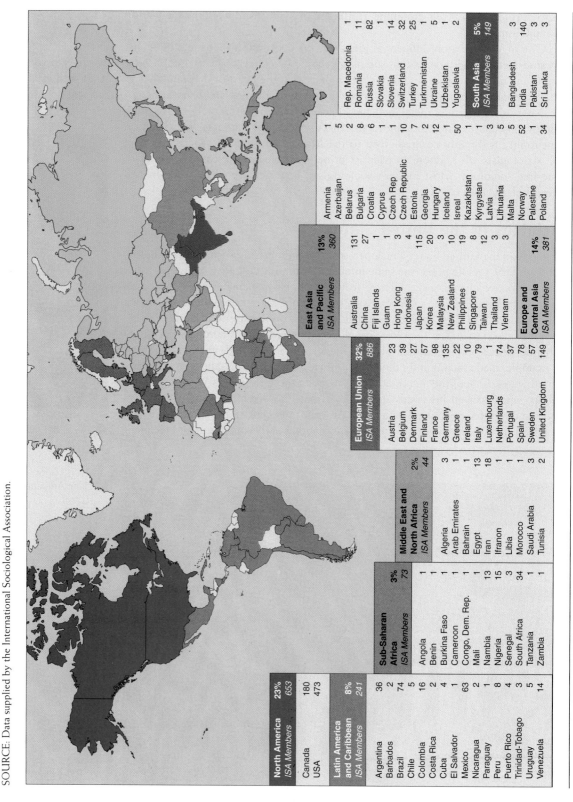

North America	23%	
ISA Members		653
Canada		180
USA		473

Latin America and Caribbean	8%	
ISA Members		241
Argentina		36
Barbados		2
Brazil		74
Chile		5
Colombia		16
Costa Rica		4
Cuba		2
El Salvador		1
Mexico		63
Nicaragua		2
Paraguay		1
Peru		8
Puerto Rico		4
Trinidad-Tobago		3
Uruguay		5
Venezuela		14

Sub-Saharan Africa	3%	
ISA Members		73
Angola		1
Benin		1
Burkina Faso		1
Cameroon		1
Congo, Dem. Rep.		1
Mali		1
Namibia		13
Nigeria		15
Senegal		3
South Africa		34
Tanzania		1
Zambia		1

Middle East and North Africa	2%	
ISA Members		44
Algeria		1
Arab Emirates		3
Bahrain		1
Egypt		13
Iran		18
Ifranon		1
Libia		1
Morocco		1
Saudi Arabia		3
Tunisia		2

European Union	32%	
ISA Members		886
Austria		23
Belgium		39
Denmark		27
Finland		57
France		98
Germany		135
Greece		22
Ireland		10
Italy		79
Luxembourg		1
Netherlands		74
Portugal		37
Spain		78
Sweden		57
United Kingdom		149

East Asia and Pacific	13%	
ISA Members		360
Australia		131
China		27
Fiji Islands		1
Guam		1
Hong Kong		3
Indonesia		4
Japan		115
Korea		20
Malaysia		3
New Zealand		10
Philippines		19
Singapore		8
Taiwan		12
Thailand		1
Vietnam		3

Europe and Central Asia	14%	
ISA Members		381
Armenia		1
Azerbaijan		5
Belarus		2
Bulgaria		8
Croatia		6
Cyprus		1
Czech Rep		10
Czech Republic		3
Estonia		7
Georgia		12
Hungary		3
Iceland		10
Isreal		19
Kazakhstan		8
Kyrgystan		1
Latvia		3
Lithuania		3
Malta		5
Norway		52
Palestine		1
Poland		34
Rep. Macedonia		1
Romania		11
Russia		82
Slovakia		1
Slovenia		14
Switzerland		32
Turkey		25
Turkmenistan		1
Ukraine		5
Uzbekistan		1
Yugoslavia		2

South Asia	5%	
ISA Members		149
Bangladesh		3
India		140
Pakistan		3
Sri Lanka		3

of the 1931 Nobel Peace Prize, illustrates the reformist approach of much U.S. sociology. Addams was the founder of Hull House, a famous center for social services and community activism located in a Chicago slum. She and her colleagues used quantitative social science data to lobby successfully for legislation mandating safer working conditions, a better juvenile justice system, better public sanitation, and services for the poor (Deegan 1987).

As sociology became more established, it also became more conservative. In the years between the two World Wars, a new generation of sociology professors became convinced that social activism was incompatible with academic respectability. By the 1950s, however, and into the 1960s, sociologists such as C. Wright Mills and Ralf Dahrendorf turned renewed attention to social problems and social conflict. These days, most sociologists are interested in these concerns, although few are activists.

The first sociology course in the United States was taught at Yale University in 1876. By 1910, most colleges and universities in the United States offered sociology courses, although separate departments were slower to develop. Most of the courses were offered jointly with other departments, most often with economics but frequently with history, political science, philosophy, or general social science departments.

These days, almost all colleges and universities offer undergraduate degrees in sociology. Most universities offer master's degrees in sociology and approximately 125 offer doctoral degree programs. Graduate sociology programs are more popular in the United States than in any other country in the world. This is partly because sociology in the United States has always been oriented toward the practical as well as the theoretical. The focus has consistently been on finding solutions to social issues and problems, with the result that U.S. sociologists not only teach sociology but also work in government and industry.

As recognition has grown that the solutions to problems such as AIDS, environmental degradation, poverty, and ethnic conflict require international effort, U.S. sociologists are developing closer working relationships with their counterparts in agencies and universities throughout the world. Map 1.1 shows the distribution of sociologists around the globe.

Current Perspectives in Sociology

As this brief review of the history of sociological thought has demonstrated, there are many ways of approaching the study of human social interaction. The ideas of Marx, Weber, Durkheim, and others have given rise to dozens of theories about human behavior. In this section, we bring together and summarize ideas that are the foundation of the three dominant theoretical perspectives in sociology today: structural functionalism, conflict theory, and symbolic interactionism.

Structural-Functional Theory

Structural-functional theory addresses the question of social organization and how it is maintained. This theoretical perspective has its roots in natural science and the analogy between society and an organism. In the analysis of a living organism, the scientist's task is to identify the various parts (structures) and to determine how they work (function). In the study of society, a sociologist with this perspective tries to identify the structures of society and how they function, hence the name *structural functionalism*.

Structural-functional theory addresses the question of social organization (structure) and how it is maintained (function).

Concept Summary

Major Theoretical Perspectives in Sociology

	Structural Functionalism	Conflict Theory	Symbolic Interactionism
Nature of society	Interrelated social structures that fit together to form an integrated whole	Competing interests, each seeking to secure its own ends	Interacting individuals and groups
Basis of interaction	Consensus and shared values	Constraint, power, and competition	Shared symbolic meanings
Major question	What are social structures? Do they contribute to social stability?	Who benefits? How are these benefits maintained?	How do social structures relate to individual subjective experiences?
Level of analysis	Social structure	Social structure	Interpersonal interaction

The Assumptions Behind Structural Functionalism

In the sense that any study of society must begin with an identification of the parts of society and how they work, structural functionalism is basic to all perspectives. Scholars who use this perspective, however, are distinguished from other social analysts by their reliance on three major assumptions:

1. *Stability*. The chief evaluative criterion for any social pattern is whether it contributes to the maintenance of society.
2. *Harmony*. As the parts of an organism work together for the good of the whole, so the parts of society are also characterized by harmony.
3. *Evolution*. Change occurs through evolution—the adaptation of social structures to new needs and demands and the elimination of unnecessary or outmoded structures.

Structural-Functional Analysis

A structural-functional analysis asks two basic questions: What is the nature of this social structure (what patterns exist)? What are the consequences of this social structure (does it promote stability and harmony)? In this analysis, positive consequences are called **functions** and negative consequences are called **dysfunctions.** A distinction is also drawn between **manifest** (recognized and intended) consequences and **latent** (unrecognized and unintended) consequences.

The basic strategy of looking for structures along with their manifest and latent functions and dysfunctions is common to nearly all sociological analysis. Scholars from widely different theoretical perspectives use this framework for examining society. What sets structural-functional theorists apart from others who use this language are their assumptions about harmony and stability.

Consider, for example, new laws now under consideration in many states that would allow battered women to use the "battered women's syndrome" as a legal defense if they assault or kill their abuser. Such laws would explicitly recognize the right of women who assault or kill an abusive partner to plead not guilty by reason of temporary insanity. What would be the consequences of this new social structure? Its manifest function (intended positive outcome) is, of course, to give legal recognition to the devastating long-term psychological consequences of domestic violence. The manifest dysfunction is that some offenders might use the battered women's syndrome defense as an excuse for a malicious, premeditated assault on a significant other. A latent dysfunction may be that women who are acquitted of

Functions are consequences of social structures that have positive effects on the stability of society.

Dysfunctions are consequences of social structures that have negative effects on the stability of society.

Manifest functions or dysfunctions are consequences of social structures that are intended or recognized.

Latent functions or dysfunctions are consequences of social structures that are neither intended nor recognized.

■ Volleyball games offer a graphic metaphor of social structure. Each person on the team occupies a different status and each plays a relatively unique role. Structural-functionalists focus on the benefits that these statuses, roles, and the institution of sports itself provide to society.

AP / Wide World Photos

legal charges on the basis of a temporary insanity plea could find it difficult to retain custody of their children, given the stigma often attached to individuals with any diagnosis of mental disorder. Another latent outcome may be the perpetuation of the view that women are irrational: that they stay with men who beat them because they are incapable of logically thinking through their options, and that they only leave when they "snap" mentally. Is this latent outcome a function or a dysfunction? Remember that structural-functional analysis typically starts from the assumption that any social action or structure that contributes to the maintenance of society and preserves the status quo is functional and that any action or structure that challenges the status quo is dysfunctional. Because perpetuating the view that women are irrational would reinforce existing gender roles, this would be judged a latent function (Table 1.1).

   As this example suggests, a social pattern that contributes to the maintenance of society may benefit some groups more than others. A pattern may be functional—that is, help maintain the status quo—without being either desirable or equitable. In general, however, structural-functionalists emphasize how social structures work together to create a smooth-running society.

Connections

Historical Note

Until 1962, American husbands legally could beat their wives. Under laws inherited from the English system, men were assumed to be intellectually and morally superior to women. They were legally held responsible for their wives' actions and so were given the right to beat their wives if needed to control the women's behavior. The term "rule of thumb" comes from the English court ruling under which a man could beat his wife so long as the stick he used was no wider than his thumb.

TABLE 1.1

A Structural-Functional Analysis of the "Battered Women's Syndrome" as a Legal Defense

Structural-functional analysis examines the intended and unintended consequences of social structures. It also assesses whether the consequences are positive (functional) or negative (dysfunctional). There is no moral dimension to the assessment that an outcome is positive; it merely means that the outcome contributes to the stability of society.

	Manifest	Latent
Function	Gives legal recognition to the psychological consequences of domestic violence.	Encourages the view that women are irrational.
Dysfunction	May serve as an excuse for violence against abusers.	Makes it more difficult for victims of domestic violence to retain custody of children.

Conflict Theory

Whereas structural-functional theory sees the world in terms of consensus and stability, conflict theory sees the world in terms of conflict and change. Conflict theorists contend that a full understanding of society requires a critical examination of the competition and conflict in society, especially of the processes by which some people are winners and others losers. As a result, **conflict theory** addresses the points of stress and conflict in society and the ways in which they contribute to social change.

Assumptions Underlying Conflict Theory

Conflict theory is derived from Marx's ideas. The following are three primary assumptions of modern conflict theory:

1. *Competition.* Competition over scarce resources (money, leisure, sexual partners, and so on) is at the heart of all social relationships. Competition rather than consensus is characteristic of human relationships.

Conflict theory addresses the points of stress and conflict in society and the ways in which they contribute to social change.

■ Conflict theory can help us understand why modern-day Latina immigrants, like Jewish immigrant women a century ago, so often are poorly paid and work in garment-industry sweatshops.

© Michael Grecco/Stock Boston

2. *Structural inequality.* Inequalities in power and reward are built into all social structures. Individuals and groups that benefit from any particular structure strive to
   see it maintained.
3. *Social change.* Change occurs as a result of conflict between competing interests
   rather than through adaptation. It is often abrupt and revolutionary rather than
   evolutionary and is usually helpful rather than harmful.

Conflict Analysis

Like structural functionalists, conflict theorists are interested in social structures. The
two questions they ask, however, are different. Conflict theorists ask: Who benefits
from those social structures? And how do those who benefit maintain their advantages?

A conflict analysis of domestic violence, for example, would begin by noting
that women are battered far more often and far more severely than are men, and that
the popular term "domestic violence" hides this reality. Conflict theorists' answer to
the question "Who benefits?" is that battering helps men to retain their dominance
over women. These theorists go on to ask how this situation developed and how
it is maintained. Their answers would focus on issues such as how some religions
traditionally have taught women to submit to their husbands' wishes and to accept
violence within marriage; how until recently the law did not regard woman battering
as a crime; and how some police officers still consider battering merely an unimportant family matter.

Symbolic Interaction Theory

Both structural-functional and conflict theories focus on social structures and the relationships between them. What about the relationship between individuals and social
structures? Sociologists who focus on the ways that individuals relate to and are affected
by social structures generally use symbolic interaction theory. **Symbolic interaction
theory** addresses the subjective meanings of human acts and the processes through
which we come to develop and share these subjective meanings. The name of this
theory comes from the fact that it studies the symbolic (or subjective) meaning of
human interaction.

Symbolic interaction theory
addresses the subjective meanings
of human acts and the processes
through which people come to
develop and communicate shared
meanings.

Assumptions Underlying Symbolic Interaction Theory

When symbolic interactionists study human behavior, they begin with three major
premises (Blumer 1969):

1. Symbolic meanings are important. Any behavior, gesture, or word can have multiple interpretations (can symbolize many things). In order to understand human
   behavior, we must learn what it means to the participants.
2. Meanings grow out of relationships. When relationships change, so do meanings.
3. Meanings are negotiated. We do not accept others' meanings uncritically. Each of
   us plays an active role in negotiating the meaning that things will have for us.

Symbolic Interaction Analysis

These premises direct symbolic interactionists to the study of how individuals are
shaped by relationships and social structures. Symbolic interactionists who study
violence against women have researched how boys are socialized to consider
aggression a natural part of being male through football and hockey games, dads
who tell their sons to fight back when someone makes fun of them, older brothers
who physically push them around, and the like. Symbolic interactionists also have
explored how teachers unintentionally reinforce the idea that girls are inferior

when teachers allow boys to take over schoolyards and to make fun of girls in the classroom. All of these experiences, some researchers believe, set the stage for later violence against women.

Symbolic interactionists are also interested in the active role individuals play in modifying and negotiating their way in relationships. Why do two children raised in the same family turn out differently? In part, because each child experiences subtly different relationships and situations; the meanings that the youngest child derives from the family experience may be different from those the oldest child derives.

Most generally, symbolic interaction is concerned with how individuals are shaped by relationships. This question leads first to a concern with childhood and the initial steps we take to learn and interpret our social worlds. It is also concerned with later relationships with lovers, friends, employers, teachers, and others. The strength of symbolic interactionism is that it focuses attention on how larger social structures affect our everyday lives, sense of self, and interpersonal relationships and encounters.

Interchangeable Lenses

Neither symbolic interaction theory, conflict theory, nor structural-functional theory is complete in itself. Together, however, they provide a valuable set of tools for understanding the relationship between the individual and society. These three sociological theories can be regarded as interchangeable lenses through which society may be viewed. Just as a telephoto lens is not always superior to a wide-angle lens, one sociological theory will not always be superior to another.

Occasionally, the same subject can be viewed through any of these perspectives. One will generally get better pictures, however, by selecting the theoretical perspective that is best suited to the particular subject. In general, structural functionalism and conflict theory are well suited to the study of social structures or **macrosociology.** Symbolic interactionism is well suited to the study of the relationship between individual meanings and social structures, or **microsociology.** Following are the three "snapshots" of female prostitution taken through the theoretical lens of structural-functional, conflict, and symbolic interaction theory.

Macrosociology focuses on social structures and organizations and the relationships between them.

Microsociology focuses on interactions among individuals.

The Functions of Prostitution

The functional analysis of female prostitution begins by examining its social structure. It identifies the recurrent patterns of relationships among pimps, prostitutes, and customers. Then it examines the consequences of this social structure. In 1961 Kingsley Davis listed the following outcomes of prostitution:

- It provides a sexual outlet for men who cannot compete on the marriage market—the physically or mentally handicapped or the very poor.
- It provides a sexual outlet for men who are away from home a lot, such as salesmen and sailors.
- It provides a sexual outlet for the kinky.

Provision of these services is the manifest or intended function of prostitution. Davis goes on to note that, by providing these services, prostitution has the latent function of protecting the institution of marriage from malcontents who, for one reason or another, do not receive adequate sexual service through marriage. Prostitution is the safety valve that makes it possible to restrict respectable sexual relationships (and hence childbearing and child rearing) to marital relationships while still allowing for the variability of human sexual appetites.

Some sociologists view prostitution as an outgrowth of poverty and sexism; others consider it voluntary and functional for society. Still others are interested in understanding how prostitutes maintain a positive identity in a stigmatized occupation.

© Joel Gordon

Prostitution: Marketing a Scarce Resource

Conflict theorists analyze prostitution as part of the larger problem of unequal allocation of scarce resources. Women, they argue, have not had equal access to economic opportunity. In some societies, they are forbidden to own property; in others, they suffer substantial discrimination in opportunities to work and earn. Because of this inability to support themselves, women have had to rely on economic support from men. They get this support by exchanging the one scarce resource they have to offer: sexual availability. To a conflict theorist, it makes little difference whether a woman barters her sexual availability through prostitution or through marriage; the underlying cause is the same.

Although most analyses of prostitution focus on adult women, the conflict perspective helps explain the growing problem of prostitution among runaway and homeless boys and girls. These young people have few realistic opportunities to support themselves by regular jobs: many are not old enough to work legally and, in any case, would be unable to support themselves adequately on the minimum wage. Their young bodies are their most marketable resource.

Prostitution: Learning the Trade

Symbolic interactionists who examine prostitution take an entirely different perspective. They want to know, for example, how prostitutes learn the trade and how they manage their self-concept so that they continue to think positively of themselves in spite of engaging in a socially disapproved profession. One such study was done by Wendy Chapkis (1997), who interviewed more than fifty women "sex workers"—prostitutes, call girls, pornographic actresses, and others. Many of the women she interviewed felt proud of their work. They felt that the services they offered were not substantially different from those offered by day-care workers or psychotherapists, who are also expected to provide services while acting as if they like and care for their clients. Chapkis found that as long as prostitutes are able to keep a healthy distance between their emotions and their work, they can maintain their self-esteem and mental health. As one woman described: "Sex work hasn't all been a bed of roses and I've learned some painful things. But I also feel strong in what I do. I'm good at it and I know how to maintain my emotional distance. Just like if you are a fire fighter or a brain surgeon or a psychiatrist, you have to deal with some heavy stuff and that means divorcing yourself from your feelings on a certain level. You just have to be able to do that to do your job" (Chapkis, 79).

As these examples illustrate, many topics can be fruitfully studied with any of the three theoretical perspectives. Consequently, each sociologist decides which perspective will work best for a given research project.

The Science of Society

The things that sociologists study—for example, deviance, marital happiness, and poverty—have probably interested you for a long time. You may have developed your own opinions about why some people have good marriages and some have bad marriages or why some people break the law and others do not. Sociology is an academic discipline that uses the procedures of science to critically examine common-sense explanations of human social behavior. Science is not divorced from common sense but is an extension of it.

Defining Science

The ultimate aim of science is to better understand the world. Science directs us to find this understanding by observing and measuring what actually happens. This is not the only means of acquiring knowledge. Some people get their perceptions from the Bible or the Koran or the Book of Mormon. Others get their answers from their mothers or their husbands or their girlfriends. When you ask such people, "But how do you know that that is true?" their answer is simple: "My mother told me" or "I read it in *Reader's Digest*."

Science is a way of knowing based on empirical evidence.

Science differs from these other ways of knowing in that it requires empirical evidence as a basis for knowledge; that is, it requires evidence that can be confirmed by the normal human senses. We must be able to see, hear, smell, or feel it. Before social scientists would agree that they "knew" religious differences increased the likelihood of divorce, for example, they would want to see statistical evidence.

Science has two major goals: accurate description and accurate explanation. In sociology, we are concerned with accurate descriptions of human interaction (how many people marry, how many people abuse their children, how many people flunk out of school). After we know the patterns, we hope to be able to explain them, to say why people marry, abuse their children, or flunk out of school.

The Research Process

At each stage of the scientific process, researchers use certain conventional procedures to ensure that their findings will be accepted as scientific knowledge. The procedures used in sociological research are covered in depth in classes on research methods, statistics, and theory construction. At this point, we merely want to introduce a few ideas that you must understand if you are to be an educated consumer of research results. We look at the four steps of the general research process, and in doing so review three concepts central to research: variables, operational definitions, and sampling.

Step One: Stating the Problem

The first step in the research process is carefully stating the issue to be investigated. We may select a topic because of a personal experience or out of commonsense observation. For example, we may have observed that African Americans appear more likely to experience unemployment and poverty than do white Americans. Alternatively, we might begin with a theory that predicts, for instance, that African Americans will have higher unemployment and poverty rates than white Americans because they have been discriminated against in schools and the workplace. In either case, we begin by reviewing the research of other scholars to help us specify exactly what it is that we want to know. If a good deal of research has already been conducted on the issue and good theoretical explanations have been advanced for some of the patterns, then a problem may be stated in the form of a **hypothesis**—a statement about relationships that we expect to observe if our theory is correct. A hypothesis must be testable; that is, there must be some way in which data can help weed out a wrong conclusion and identify a correct one. For example, the *belief* that whites deserve better jobs than African Americans cannot be tested, but the *hypothesis* that whites receive better job offers than African Americans can be tested.

A **hypothesis** is a statement about relationships that we expect to find if our theory is correct.

Step Two: Gathering Data

To narrow the scope of a problem to manageable size, researchers focus on variables rather than on people.

Variables   **Variables** are measured characteristics that vary from one individual, situation, or group to the next (Babbie 2004). If we wish to analyze differences in rates of African American/white unemployment, we need information on two variables: race and unemployment. The individuals included in our study would be complex and interesting human beings, but for our purposes, we would be interested only in these two aspects of each person's life.

Variables are measured characteristics that vary from one individual or group to the next.

When we hypothesize a cause-and-effect relationship between two variables, the cause is called the **independent variable,** and the effect is called the **dependent variable.** In our example, race is the independent variable, and unemployment is the dependent variable; that is, we hypothesize that unemployment *depends on* one's race.

The independent variable is the cause in cause-and-effect relationships.

The **dependent variable** is the effect in cause-and-effect relationships. It is dependent on the actions of the independent variable.

Operational Definitions   In order to describe a pattern or test a hypothesis, each variable must be precisely defined. Before we can describe racial differences in unemployment rates, for instance, we need to be able to decide whether an individual is unemployed. The exact procedure by which a variable is measured is called an **operational definition.** Reaching general agreement about these definitions may pose a problem. For instance, people are typically considered to be unemployed if they are actively seeking work but cannot find it. This definition ignores all the people who became so discouraged in their search for work that they simply gave up. Obviously, including discouraged workers in our definition of the unemployed might lead to a different description of patterns of unemployment. We always need to carefully check what operational definitions were used when we evaluate research results.

An **operational definition** describes the exact procedure by which a variable is measured.

Sampling   It would be time consuming, expensive, and probably nearly impossible to get information on race and employment status for all adults. It is also unnecessary. The process of **sampling**—taking a systematic selection of representative cases from a larger population—allows us to get accurate empirical data at a fraction of the cost that examining all possible cases would involve.

Sampling is a systematic selection of representative cases from the larger population.

Sampling involves two processes: (1) obtaining a list of the population you want to study and (2) selecting a representative subset or sample from the list. Selecting from the list is easy; choosing a relatively large number by a random procedure like tossing a coin generally assures that the sample will be unbiased. The more difficult task is getting a list. A central principle of sampling is that a sample is only representative of the list from which it is drawn. If we draw a list of people from the telephone directory, then our sample can only be said to describe households listed in the directory; it will omit those with unlisted numbers, those with no telephones, those who use only cell phones, and those who have moved since the directory was issued. The best surveys begin with a list of all the households, individuals, or telephone numbers in a target region or group.

Step Three: Finding Patterns

Correlation occurs when there is an empirical relationship between two variables.

The third step in the research process is to analyze the data, looking for patterns. If we study unemployment, for example, we will find that African Americans are more than twice as likely as white Americans to experience unemployment (U.S. Bureau of the Census 2002). This generalization notes a **correlation,** an empirical relationship between two variables—in this case, between race and employment.

Step Four: Generating Theories

A **theory** is an interrelated set of assumptions that explains observed patterns.

After a pattern is found, the next step in the research process is to explain it. As we will discuss in the section entitled "The Survey," finding a correlation between two variables does not necessarily mean that one variable causes the other. For example, even though there is a correlation between race and unemployment, not all African Americans are unemployed, and being African American is not the only cause of unemployment. Nevertheless, if we have good empirical evidence that being black increases the *probability* of unemployment, the next task is to explain why that should be so. Explanations are usually embodied in a **theory,** an interrelated set of assumptions that explains observed patterns. Theory always goes beyond the facts at hand; it includes untested assumptions that explain the empirical evidence.

In our unemployment example, we might theorize that the reason African Americans face more unemployment than whites is because many of today's African American adults grew up in a time when the racial difference in educational opportunity was much greater than it is now. This simple explanation goes beyond the facts at hand to include some assumptions about how education is related to race and unemployment. Although theory rests on an empirical generalization, the theory itself is not empirical; it is . . . well, theoretical.

It should be noted that many different theories can be compatible with a given empirical generalization. We have proposed that education explains the correlation between race and unemployment. An alternative theory might argue that the correlation arises because of discrimination. Because there are often many plausible explanations for any correlation, theory development is not the end of the research process. We must go on to test the assumption of the theory by gathering new data.

The scientific process can be viewed as a wheel that continuously moves us from theory to data and back again (Figure 1.2). Two examples illustrate how theory leads to the need for new data and how data can lead to the development of new theory.

As we have noted, data show that unemployment rates are higher among African Americans than among white Americans. One theoretical explanation for this pattern links higher African American unemployment to educational deficits. From this theory, we can deduce the hypothesis that African Americans and whites of equal education will experience equal unemployment. To test this

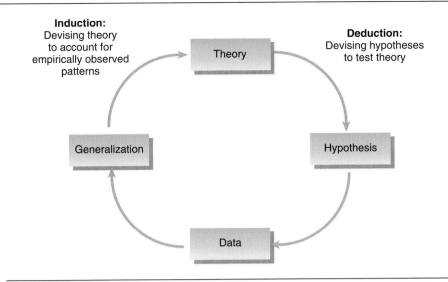

Induction:
Devising theory to account for empirically observed patterns

Deduction:
Devising hypotheses to test theory

Theory

Generalization

Hypothesis

Data

FIGURE 1.2
The Wheel of Science
The process of science can be viewed as a continuously turning wheel that moves us from data to theory and back again.
SOURCE: Adapted from Wallace, Walter. 1969. Sociological Theory. Chicago: Aldine.

hypothesis, we need more data, this time about education and its relationship to race and unemployment.

A study by Lori Reid (2002) tests this hypothesis for black women. Reid asked whether educational deficits explained why African American women are more likely to lose their jobs than are whites. She found that education does play a small role. However, other factors played larger roles in explaining black unemployment. Those factors, including black women's segregation in vulnerable occupations and residence in regions where unemployment was rising, were considerably more important predictors of unemployment.

Reid's findings can be the basis for revised theories. These new theories will again be subject to empirical testing, and the process will begin anew. In the language of

■ High rates of unemployment among African Americans have led sociologists to develop and test two competing hypotheses. The first is that African Americans are more likely to be unemployed because they lack education. The second is that African Americans are more likely to be unemployed *even if they get an education* because they more often work in declining industries and live in regions with high unemployment.

© Chuck Savage/CORBIS

Induction is the process of moving from data to theory by devising theories that account for empirically observed patterns.

Deduction is the process of moving from theory to data by testing hypotheses drawn from theory.

science, the process of moving from data to theory is called **induction,** and the process of moving from theory to data is called **deduction.** These two processes and their interrelationships are also illustrated in Figure 1.2.

Strategies for Gathering Data

The theories and findings reported in this book represent a variety of research strategies. Three of the more general strategies are outlined here: experiments, survey research, and participant observation. In this section, we review each method and illustrate its advantages and disadvantages by showing how it would approach the test of a common hypothesis, namely, that alcohol use reduces grades in college.

The Experiment

The **experiment** is a method in which the researcher manipulates independent variables to test theories of cause and effect.

An **experimental group** is the group in an experiment that experiences the independent variable. Results for this group are compared with those for the control group.

A **control group** is the group in an experiment that does not receive the independent variable.

The **experiment** is a research method in which the researcher manipulates the independent variable to test theories of cause and effect. In the classic experiment, a group that experiences the independent variable, an **experimental group,** is compared with a **control group** that does not. If the groups are equal as far as everything else, a comparison between them will show whether experience with the independent variable is associated with unique change in the dependent variable.

An experiment designed to assess whether alcohol use affects grades, for example, would need to compare an experimental group that drank alcohol with a control group that did not. A hypothetical experiment might begin by observing student grades to assess students' typical performance levels. Then the class would be divided randomly into two groups. If the initial pool is large enough, we can assume that the two groups are probably equal on nearly everything. For example, we assume that both groups probably contain an equal mix of good and poor students and of lazy and ambitious students. The control group might be requested to abstain from alcohol use for five weeks, and the experimental group might be requested to drink three times a week during the same period. At the end of the five weeks, we would compare the grades of the two groups. Both groups might have experienced a drop in grades because of normal factors such as fatigue, burnout, and overwork. The existence of the control group, however, allows us to determine whether alcohol use causes a reduction in grades over and above that which normally occurs.

Experiments are excellent devices for testing hypotheses about cause and effect. They have three drawbacks, however. First, experiments are often unethical because they expose subjects to the possibility of harm. The study on alcohol use, for example, might damage student grades, introduce students to bad habits, or otherwise injure them. Because of such ethical issues, many areas of sociological interest cannot be studied with the experimental method.

A second drawback to experiments is that subjects often behave differently when they are under scientific observation than they would in their normal environment. For example, although alcohol consumption might normally have the effect of lowering student grades, the participants in our study might find the research so interesting that their grades would actually improve. In this case, the subjects' knowledge that they are participating in an experiment affects their response to the independent variable. This response is called the **guinea-pig effect.**

The **guinea-pig effect** occurs when subjects' knowledge that they are participating in an experiment affects their response to the independent variable.

A final drawback to the experimental method is that laboratory experiments are often highly artificial. When researchers try to set up social situations in laboratories, they often must omit many of the factors that would influence the same behavior in a real-life situation. The result is often a very unnatural situation. Like the guinea-pig effect, this artificiality has the effect of reducing our confidence that the results that appear in the experiment can be generalized to the more complex conditions of the real world.

Because of these problems, few sociologists use experiments.

The Survey

The most common research strategy in sociology is the survey. In **survey research,** the investigator asks a relatively large number of people the same set of standardized questions. These questions may be asked in a personal interview, over the telephone, online, or in a paper-and-pencil format. Because it asks the same questions of a large number of people, it is an ideal method for furnishing evidence on incidence, trends, and differentials. Thus, survey data on alcohol use may allow us to say such things as the following: 80 percent of the undergraduates of Midwestern State currently use alcohol (**incidence**); the proportion using alcohol has remained about the same over the last 10 years (**trend**); and the proportion using alcohol is higher for males than for females (**differential**). Survey research is extremely versatile; it can be used to study attitudes, behavior, ideals, and values. If you can think of some way to ask a question about such matters, then you can study it with survey research.

Most surveys use what is called a **cross-sectional design;** they take a sample (or cross section) of the population at a single point in time and expect it to show some variability on the independent variable. Thus, in our study of alcohol use we would take a sample of students, expecting to find that some of them drink and some do not. We could then compare these two groups to see which gets the best grades.

In 1997 we conducted such a survey of our undergraduate students. The results are displayed in Table 1.2. The table shows that students who drink alcohol report getting worse grades: Although 51 percent of the nondrinkers have grade point averages over 3.5, only 28 percent of those who admit drinking do. At the other end of the grade distribution, drinkers were more likely than abstainers to report a grade point average of less than 2.5.

The difficulty with the cross-sectional design is that we cannot reach any firm conclusions about cause and effect. We cannot tell whether alcohol use causes bad grades or whether bad grades lead to alcohol use. A more striking problem is that we cannot be sure there is a cause-and-effect relationship at all. Because the drinkers and

Survey research is a method that involves asking a relatively large number of people the same set of standardized questions.

Incidence is the frequency with which an attitude or behavior occurs.

A **trend** is a change in a variable over time.

A **differential** is a difference in the incidence of a phenomenon across subcategories of the population.

A **cross-sectional design** uses a sample (or cross section) of the population at a single point in time.

TABLE 1.2
Cross-Tabulation of College Grades by Use of Alcohol
556 undergraduate students at a midwestern state university, 1997.

College Grades	Use Alcohol	
	No	Yes
B+ or A	51%	28%
C+ or B	42	60
Below C+	7	12
Total	100%	100%
Number	114	452

A Global Perspective

focus on

Survey Research in Nigeria

"**H**ow do you feel about the current government?" "Do you think men and women ought to be treated equally?" "How many televisions do you own?" Without much effort, we can imagine places in the world where questions like these that are so commonplace to us would appear foolish and perhaps dangerous to ask.

In the United States and much of the developed world, we are accustomed to questions about the most intimate details of our lives. It is common for bureaucratic agencies to record data pertaining to our height, weight, IQ, taxes, fertility, and credit rating. Our acceptance of intrusive questioning is based on the assumption that this information is somehow necessary to good governance and the trust that our privacy will be protected. Because we are familiar with routine bureaucratic data gathering, it is relatively easy for survey re-

searchers to enlist our cooperation. When asked personally, most Americans will respond to survey researchers' inquiries on topics ranging from politics to religion. (The only question Americans

routinely refuse to answer is, "What is your income?")

When survey research is conducted in developing nations, many of these conditions do not hold. Often people

■ Although mothers and mothering are universal, the meanings attached to children and to childbearing are not. Survey research on fertility and family planning may give seriously misleading results in nations such as Nigeria where conception is viewed as being "up to God," not "up to the individual," and answering questions about personal topics is considered indecent, if not downright dangerous.

---

the abstainers were not randomly assigned to the two categories, the two groups differ on many other variables besides drinking. For example, the drinkers may have less conventional families, come from worse neighborhoods, or simply hate reading. It could be that one of these factors is causing both the poor grades and the drinking.

The **panel design** follows a sample over a period of time.

A different strategy used in survey research is the **panel design,** in which a sample is followed over a period of time. During this period of time, some sample members will experience the independent variable, and we can observe how they differ from those who do not both before and after this experience. To use this design for examining the effect of alcohol use on grades would require surveying a group of young people at several points in time, say, from when they were 12 until they were 25. This design would not alter the fact that some people choose to drink and others do not, but it would let us look at the same people before and after their decision. It would allow us to see whether students' grades actually fell after they started to drink alcohol or whether they were always poor students. The major disadvantage of a panel design is, of course, that it is expensive and time consuming.

Social-desirability bias is the tendency of people to color the truth so that they sound nicer, richer, and more desirable than they really are.

Another important drawback of survey research in general is that respondents may misrepresent the truth. Both those who drink heavily *and* those who don't drink at all may lie about their habits for fear that others will think they aren't "cool." Such misrepresentation is known as **social-desirability bias**—the tendency for people to

there have good reason not to trust their governments. More generally, they simply are not accustomed to opening their private lives to the probing of bureaucratic agents.

One of the most common areas in which Western survey research methods meet resistance from people in developing nations is research on fertility (birthrates) and family planning. Agencies such as the United Nations and the U.S. Agency for International Development wish to know how many children women in Kenya, Nigeria, and other high-fertility nations want so that they can learn whether the women might be interested in contraception. Agnes Riedmann's (1993) analysis of a survey research project among the Yoruba of Nigeria highlights several cultural clashes that can impede such research efforts.

- Among the Yorubans, as in many non-Western cultures, fertility is "up to God." It would be unthinkable for individuals to put their own opinions forward. A question such as, "If you had more money, would you rather have a new car or another child?" presumes that individuals have a choice about fertility and, moreover, that dollar values can be assigned to children.
- Asking a woman's opinion is often considered indecent (unless her husband is present) or simply a waste of time, because women's opinions do not count.
- Yorubans hold a profound distrust of strangers who ask personal questions. After submitting to an ill-understood interview on fertility, for example, one Yoruban respondent asked whether the police were now coming to take her away.

In many cases Yoruban respondents mocked, yelled at, and ran from interviewers. The persistent inquiry of strangers into their private lives seemed to some to be just one more instance of the crazy behavior of *oyinbos,* or white people. Other Yorubans speculated that the researchers were asking these questions simply because they didn't have enough to do! If badgered into participation, Yorubans politely told interviewers whatever they thought the interviewers wanted to hear.

 For more information, look up the following subjects in InfoTrac College Edition:
Interviewing in sociology
Yoruba
Family planning

 Or visit the following Web site:
Art and Life in Africa: Yoruba Information
http://www.uiowa.edu/~africart/toc/people/Yoruba.html

color the truth so that they appear to be nicer, richer, and generally more desirable than they really are. The consequences of this bias vary in seriousness depending on the research aim and topic. Obviously, it is a greater problem for sensitive topics such as alcohol use or prejudice. In the study just described, students who care strongly how others view them will be more likely both to work to get good grades and to under-report their drinking.

Survey research is designed to obtain standard answers to standard questions. It is not the best strategy for studying deviant or undesirable behaviors or for examining ideas and feelings that cannot easily be reduced to questionnaire form. An additional drawback of survey research is that it is designed to study individuals rather than contexts. Thus, it focuses on the individual alcohol user or abstainer rather than on the setting and relationships in which drinking takes place. For these kinds of answers, we must turn to participant observation.

Participant Observation

Under the label **participant observation,** we classify a variety of research strategies—participating, interviewing, observing—that examine the context and meanings of human behavior. Instead of sending forth an army of interviewers, participant observers

Participant observation includes a variety of research strategies—participating, interviewing, observing—that examine the context and meanings of human behavior.

Research Methods

Concept Summary

Controlled Experiments	
Procedure	Dividing subjects into two equivalent groups, applying the independent variable to one group only, and observing the differences between the two groups on the dependent variable
Advantages	Excellent for analysis of cause-and-effect relationships; can stimulate events and behaviors that do not occur outside the laboratory in any regular way
Disadvantages	Based on small, nonrepresentative samples examined under highly artificial circumstances; unclear that people would behave the same way outside the laboratory; unethical to experiment in many areas
Survey Research	
Procedure	Asking the same set of standard questions of a relatively large, systematically selected sample
Advantages	Very versatile—can study anything that we can ask about; can be done with large, random samples so that results represent many people; good for incidence, trends, and differentials
Disadvantages	Shallow—does not get at depth and shades of meaning; affected by social-desirability bias; better for studying people than situations
Participant Observation	
Procedure	Observing people's behavior in its normal context; experiencing others' social settings as a participant; in-depth interviewing
Advantages	Seeing behavior in context; getting at meanings associated with behavior; seeing what people do rather than what they say they do
Disadvantages	Limited to small, nonrepresentative samples; dependent on interpretation of single investigator

go out into the field themselves to see firsthand what is going on. These strategies are used more often by sociologists interested in symbolic interaction theory—that is, researchers who want to understand subjective meanings, personal relationships, and the process of social life. The goals of this research method are to discover patterns of interaction and to find the meaning of the patterns for the individuals involved.

The three major tasks involved in participant observation are interviewing, participating, and observing. A researcher goes to the scene of the action, where she may interview people informally in the normal course of conversation, participate in whatever they are doing, and observe the activities of other participants. Not every participant observation study includes all three dimensions equally. A participant observer studying alcohol use on campus, for example, would not need to get smashed every night. She would, however, probably do long, informal interviews with both users and nonusers, attend student parties and activities, and attempt to get a feel for how alcohol use fits in with certain student subcultures.

The data produced by participant observation are often based on small numbers of individuals who have not been selected according to random-sampling tech-

■ To understand the beliefs and motives of those involved in stigmatized activities, such as this Ku Klux Klan march, researchers usually rely on participant observation as well as surveys.

© AP/Wide World Photos

niques. The data tend to be unsystematic and the samples not very representative; however, we do learn a great deal about the few individuals involved. This detail is often useful for generating ideas that can then be examined more systematically with other techniques. For this reason, participant observation may be viewed as a form of initial exploration of a research topic.

In some situations, however, participant observation is the only reasonable way to approach a subject. This is especially likely when we are examining (1) stigmatized behaviors or populations, (2) real behavior rather than attitudes, or (3) alienated populations. In the first instance, social-desirability bias makes it difficult to get good information. Nevertheless, a researcher who takes the time to get to know a group, conduct in-depth interviews, or collect long-term observations is more likely to obtain accurate information than a researcher who simply distributes a survey. Thus, what we know about the lives of "Deadheads" who followed the Grateful Dead band (Adams and Sardiello 2000), sex workers (Chapkis 1997), and other marginalized groups rests largely on the reports of participant observers.

In the second case, participant observation is well suited to studies of behavior—what people do rather than what they say they do. Behaviors are sometimes misrepresented in surveys simply because people are unaware of their actions or don't remember them very well. For example, individuals may believe they are not prejudiced; yet observational research might demonstrate that these same people systematically choose not to sit next to persons of another race on the bus or in public places. Sometimes, actions speak louder than words.

In the third case, participant observation is often the only reliable way to obtain information about alienated populations. Groups that are alienated because they lack the skills or resources to participate in mainstream society, such as those who are illiterate, are also likely to lack the skills needed to accurately answer surveys. Groups that are alienated because they have chosen to reject mainstream society—gang members, skinheads, or rioters, for example—are not likely to agree to answer surveys.

A major disadvantage of participant observation is that the observations and generalizations rely on the interpretation of one investigator. Because researchers are

Replication is the repetition of empirical studies with another investigator or a different sample to see if the same results occur.

not robots, it seems likely that their findings reflect some of their own world view. This is a greater problem with participant observation than with survey or experimental work, but all social science suffers to some extent from this phenomenon. The answer to this dilemma is **replication,** redoing the same study with another investigator to see if the same results occur.

Alternative Strategies

The bulk of sociological research uses these three strategies. There are, however, a dozen or more other imaginative and useful methods sociologists use to do research, many of them involving the analysis of social artifacts rather than people. For example, Peter Conrad used news articles on the human genome project to investigate how American ideas about individual responsibility for illness have changed over time (Conrad 1999), and Celesta Albonetti (1998) used court records to demonstrate that white collar criminals received lighter sentences if they were educated, wealthy, or female. Studies of other government records and statistics have demonstrated incidence, trends, and differentials in many areas of sociological interest.

Sociologists: What Do They Do?

A concern with social problems has been a continuing focus of U.S. sociology. This is evident both in the kinds of courses that sociology departments offer (social problems, race and ethnic relations, crime and delinquency, for example) and in the kinds of research sociologists do. These days, though, that research can take place in a wide variety of settings.

Working in Colleges and Universities

About three-quarters of U.S. sociologists are employed in colleges and universities, where they are required both to teach and to do research. Much of this research is basic sociology, which has no immediate practical application and is motivated simply by a desire to describe or explain some aspect of human social behavior more fully. Even basic research, however, often has implications for social policy. In addition, many sociology professors work closely with government, businesses, or nonprofit organizations to provide immediate practical answers to problems of concern to these institutions.

Working in Government

A long tradition of sociological work in government has to do with measuring and forecasting population trends. This work is vital for decisions about where to put roads and schools and when to stop building schools and start building nursing homes. In addition, sociologists have been employed to design and evaluate public policies in a wide variety of areas. In World War II, sociologists designed policies to increase the morale and fighting efficiency of the armed forces, and for decades, sociologists have helped plan and evaluate programs to reduce poverty.

Sociologists work in nearly every branch of government. For example, sociologists are employed by the Centers for Disease Control and Prevention (CDC), where they examine how social relationships are related to the transmission of AIDS, how intravenous drug users share needles, and why AIDS is transmitted along chains of sexual partners. While the physicians and biologists of the CDC examine the medical aspects of AIDS, sociologists work to understand the social aspects.

Working in Business

Sociologists are employed by General Motors and Pillsbury as well as by advertisers and management consulting firms. Some work in internal affairs (bureaucratic structures and labor relations), but many others conduct market research. Business and industry employ sociologists so that they can use their knowledge of society to predict which way consumer demand is likely to jump. For example, the greater incidence today of single-person households has important implications for life insurance companies, food packagers, and the construction industry. To stay profitable, companies need to be able to predict and plan for such trends. Sociologists are also extensively involved in the preparation of environmental impact statements, in which they try to assess the likely impact of, say, a new coal mine on the social and economic fabric of a community.

Working in Nonprofit Organizations

Nonprofit organizations range from hospitals and clinics to social-activist organizations and private think tanks; sociologists are employed in all of them. Sociologists at Planned Parenthood, for example, are interested in determining the causes and consequences of teenage sexual activity, evaluating communication strategies that can be used to prevent teenage pregnancy, and devising effective strategies to pursue some of that organization's more controversial goals, such as the preservation of legal access to abortion on demand.

The training that sociologists receive has a strong research orientation and is very different from the therapy-oriented training received by psychologists and social workers. Nevertheless, a thorough understanding of the ways that social structures impinge on individuals can be useful in helping individuals cope with personal troubles. Consequently, some sociologists also do marriage counseling, family counseling, and rehabilitation counseling.

Working to Serve the Public

Although most sociologists are committed to a value-free approach to their work as scholars, many are equally committed to changing society for the better, whether they work in government, business, nonprofit organizations, or academia. They see sociology as a "calling"—work that is inseparable from the rest of one's life and driven by a sense of moral responsibility for people's welfare (Yamane 1994). As a result, sociologists have served on a wide variety of public commissions and in public offices in order to effect social change. They work for change independently, too. For example, one sociologist, Claire Gilbert, publishes an environmental newsletter, *Blazing Tattles*, that reports adverse effects of pollution (Alesci 1994).

Perhaps the clearest example of sociology in the public service is the award of the 1982 Nobel Peace Prize to Swedish sociologist Alva Myrdal for her unflagging efforts to increase awareness of the dangers of nuclear armaments. Value-free scholarship does not mean value-free citizenship.

Where This Leaves Us

Sociology is a diverse and exciting field. From its beginnings in the nineteenth century, it has grown into a core social science that plays a central role in university education. Its three major perspectives—structural-functionalism, conflict theory, and symbolic interactionism—provide a complementary set of lenses for viewing the world, while its varied methodological approaches provide the tools needed to study social life in all its complexity. These lenses and tools position sociologists not only to understand the world, but to help change it for the better.

Summary

1. Sociology is the systematic study of social behavior. Sociologists use the concepts of role and social structure to analyze common human dramas. Learning to understand how individual behavior is affected by social structures is the process of developing the "sociological imagination."

2. The rapid social change that followed the industrial revolution was an important inspiration for the development of sociology. Problems caused by rapid social change stimulated the demand for accurate information about social processes. This social-problems orientation remains an important aspect of sociology.

3. Sociology has three major theoretical perspectives: structural-functional theory, conflict theory, and symbolic interaction theory. The three can be seen as alternative lenses through which to view society, with each having value as a tool for understanding how social structures shape human behavior.

4. Structural functionalism has its roots in evolutionary theory. It identifies social structures and analyzes their consequences for social harmony and stability. Identification of manifest and latent functions and dysfunctions is part of its analytic framework.

5. Conflict theory developed from Karl Marx's ideas about the importance of conflict and competition in structuring human behavior. It analyzes social structures by asking who benefits and how are these benefits maintained. It assumes that competition is more important than consensus and that change is a positive result of conflict.

6. Symbolic interaction theory examines the subjective meanings of human interaction and the processes through which people come to develop and communicate shared symbolic meanings. Although structural functionalism and conflict theory study macrosociology, symbolic interactionism is a form of microsociology.

7. Sociology is a social science. This means it relies on critical and systematic examination of the evidence before reaching any conclusions and that it approaches each research question from a position of neutrality. This is called value-free sociology.

8. The four steps in the research process are stating the problem, gathering the data, finding patterns, and generating theory. These steps form a continuous loop called the "wheel of science." The movement from data to theory is called induction, and the movement from theory to hypothesis to data is called deduction.

9. A design for gathering data depends on identifying the variables under study, agreeing on precise operational definitions of these variables, and obtaining a representative sample of cases in which to study relationships among the variables.

10. The experiment is a method designed to test cause-and-effect hypotheses. Although it is the best method for this purpose, it has three disadvantages: ethical problems, the guinea-pig effect, and highly artificial conditions. It is most often used for small-group research and for simulation of situations not often found in real life.

11. Survey research is a method that asks a large number of people a set of standard questions. It is good for describing incidence, trends, and differentials for random samples, but it is not as good for describing the contexts of human behavior or for establishing causal relationships.

12. Participant observation is a method in which a small number of individuals who are not randomly chosen

are observed or interviewed in depth. The strength of this method is the detail about the contexts of human behavior and its subjective meanings; its weaknesses are poor samples and lack of verification by independent observers.

13. Most sociologists teach and do research in academic settings. A growing minority are employed in government, nonprofit organizations, and business, where they do applied research. Regardless of the setting, sociological theory and research have implications for social policy.

Thinking Critically

1. Which of your own personal troubles might reasonably be reframed as a public issue? Does such a reframing change the nature of the solutions you can see?
2. Consider how a structural-functional analysis of gender roles might differ from a conflict analysis. Are men or women more likely to choose conflict theory?
3. Can you think of situations in which a change of friends, living arrangements, or jobs has caused you to have new interpretations of the events surrounding you?

4. Consider what study design you could ethically use to determine whether drinking alcohol, living in a sorority, or growing up with a single parent reduces academic performance.

Sociology on the Net

The Wadsworth Sociology Resource Center: Virtual Society
http://sociology.wadsworth.com/

The companion Web site for this book includes a range of enrichment material. Further your understanding of the chapter by accessing Online Practice Quizzes, Internet Exercises, InfoTrac College Edition Exercises, and many more compelling learning tools.

Chapter-Related Suggested Web Sites

American Sociological Association
http://www.asanet.org

Dead Sociologists Society
http://www2.pfeiffer.edu/~lridener/DSS/DEADSOC .HTML

SocioSite
http://www.pscw.uva.nl/sociosite/index.html

SocioWeb
http://www.socioweb.com/~markbl/socioweb/

Professor Kearl's Tour Through Cybersociety
http://www.trinity.edu/mkearl/

InfoTrac College Edition
http://www.infotrac-college.com/wadsworth

Access the latest news and research articles online—updated daily and spanning four years. InfoTrac College Edition is an easy-to-use online database of reliable, full-length articles from hundreds of top academic journals and popular sources. Conduct an electronic search using the following key search terms:

Sociology

Sociological theory

Sociological research

For more information related to this chapter, visit the Opposing Viewpoints Resource Center. Be sure you check all the options on the toolbar—viewpoints, references, statistics, and so on—and search under the following subject:

Sociology

Suggested Readings

Babbie, Earl. 1994. *The Sociological Spirit*. Belmont, Calif.: Wadsworth. From a dedicated sociologist who believes that the world and national problems that concern us most have their solutions in the realm addressed by sociology; a book of essays that not only introduces the sociological imagination but also motivates the reader to develop it.

Berger, Peter L. 1963. *Invitation to Sociology: A Humanistic Perspective*. Garden City, N.Y.: Doubleday Anchor. Forty years old and a classic. This is a delightful, well-written introduction to what sociology is and how it differs from other social sciences. It blends a serious exploration of basic sociological understandings with scenes from everyday life—encounters that are easy to relate to and that make sociology both interesting and relevant.

Mills, C. Wright. 1959. *The Sociological Imagination*. New York: Oxford University Press. A penetrating account of how the study of sociology expands understanding of common experiences.

Neuman, W. Lawrence. 2002. *Social Research Methods: Qualitative and Quantitative Approaches*. 5th ed. Boston: Allyn & Bacon. A textbook for undergraduates that covers the major research techniques in sociology. Coverage is up-to-date, thorough, and readable.

CHAPTER 2

Culture

AP/World Wide Photos

Outline

Introduction to Culture

In Chapter 1 we said that sociology is concerned with analyzing the contexts of human behavior and how these contexts affect our behavior. Our neighborhood, our family, and our social class provide part of that context, but the broadest context of all is our culture. **Culture** is the total way of life shared by members of a community.

In some places, a culture cuts across national boundaries and takes in people who live in two, three, or four nations. In other places, two distinct cultures (English and French in Canada, for example) may coexist within a single national boundary. For this reason, we distinguish between cultures and societies. A **society** is the population that shares the same territory and is bound together by economic and political ties. Often the members share a common culture, but not always.

Culture resides essentially in nontangible forms such as language, values, and symbolic meanings, but it also includes technology and material objects. A common image is that culture is a "tool kit" that provides us with the equipment necessary to deal with the common problems of everyday life (Swidler 1986). Consider how culture provides patterned activities of eating and drinking. People living in the United States share a common set of tools and technologies in the form of refrigerators, ovens, toasters, microwaves, and coffeepots. As the advertisers suggest, we share similar feelings of psychological release and satisfaction when, after a hard day of working or playing, we take a break with a cup of coffee or a cold beer. The beverages we choose and the meanings attached to them are part of our culture. Despite many shared meanings and values, however, this example also illustrates some of the difficulties inherent in any discussion of a single common culture: Although Mormon Americans and Muslim Americans share our American culture, the former do not drink coffee and neither group drinks alcohol.

Culture can be roughly divided into two categories: material and nonmaterial. *Nonmaterial culture* consists of language, values, rules, knowledge, and meanings shared by the members of a society. *Material culture* includes the physical objects that a society produces—tools, streets, sculptures, and toys, to name but a few. These material objects depend on the nonmaterial culture for meaning. For example, Barbie dolls and figurines of fertility goddesses share many common physical features; their meaning depends on nonmaterial culture.

Theoretical Perspectives on Culture

Within sociology, there are two approaches to the study of culture (Wuthnow & Witten 1988). The first approach treats culture as the underlying basis of interaction. It accepts culture as a given and is more interested in how culture shapes us than in how culture itself is shaped. Scholars taking this approach have concentrated on illustrating how norms, values, and language guide our behavior. The second approach focuses on culture as a social product. It asks why particular aspects of culture develop. These scholars would be interested, for example, in why the content of commercial television is so different from the content of public television. How does the economic structure of television affect its products? How does the way arts are funded in the United States affect what art becomes accepted by the public? Generally, the first perspective on culture is characteristic of structural-functional

---

Culture is the total way of life shared by members of a community. It includes not only language, values, and symbolic meanings but also technology and material objects.

A society is the population that shares the same territory and is bound together by economic and political ties.

theory, while conflict theorists are more interested in the determinants of culture. Because both the content of culture and the determinants of culture are of interest to us, we will use both perspectives.

Bases of Human Behavior: Culture and Biology

Why do people behave as they do? What determines human behavior? To answer these questions, we must be able to explain both the varieties and the similarities in human behavior. Generally, we will argue that biological factors help explain what is common to humankind across societies, whereas culture explains why people and societies differ from one another.

Cultural Perspective

Regardless of whether they are structural functionalists or conflict theorists, sociologists share some common orientations toward culture: Nearly all hold that culture is *problem solving*, culture is *relative*, and culture is a *social product*.

Culture Is Problem Solving

Regardless of whether people live in tropical forests or in the crowded cities of New York, London, or Tokyo, they confront some common problems. They all must eat, they all need shelter from the elements (and often from each other), and they all need to raise children to take their place and continue their way of life. Although

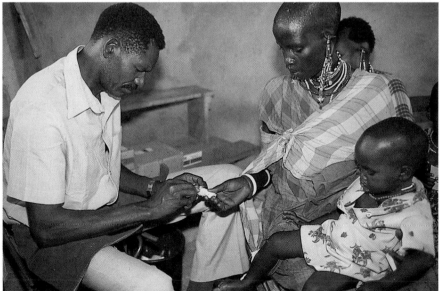

■ Despite the obvious differences between this traditional Brazilian medicine man and the modern Kenyan doctor, a close look shows many underlying similarities in their means of dealing with illness and injury.

these problems are universal, the solutions are highly varied. For example, responsibility for child rearing may be assigned to the mother's brother, as is done in the Trobriand Islands; to the natural mother and father, as is done in the United States; or to communal nurseries, as was done in early Israeli kibbutzim.

Whenever people face a recurrent problem, cultural patterns will evolve to provide a ready-made answer. This does not mean that the answer provided is the best answer or the only answer or the fairest answer, but merely that culture provides a standard pattern for dealing with this common dilemma. One of the issues that divides conflict and functional theorists is how these answers develop. Functionalists argue that the solutions we use today have evolved over generations of trial and error and that they have survived because they work, because they help us meet basic needs. A conflict theorist would add that these solutions work better for some people than others. Conflict theorists argue that elites manipulate culture in order to rationalize and maintain solutions that work to their advantage. Scholars from both perspectives agree that culture provides ready-made answers for most of the recurrent situations that we face in daily life; they disagree in their answer to the question of who benefits from a particular solution.

Culture Is Relative

The solutions that each culture devises may be startlingly different. Among the Wodaabe of Niger, for example, mothers are not allowed to speak directly to their first- or second-born children and, except for nursing, are not even allowed to touch them. The babies' grandmothers and aunts, however, lavish affection and attention on them (Beckwith 1983). The effect of this pattern of child rearing is to emphasize loyalties and affections throughout the entire kinship group rather than just with one's own children or spouse. This practice helps ensure that each new entrant will be loyal to the group as a whole. Is it a good or a bad practice? That is a question we can answer only by seeing how it fits in with the rest of the Wodaabe culture and by taking the viewpoint of one or another social group. Does it help the people meet recurrent problems and maintain a stable society? If so, then, structural functionalists would say it works; it is functional. Conflict theorists, on the other hand, would want to know which groups are hurt and which are helped by the practice. Both sets of theorists, however, believe that each cultural trait should be evaluated in the context of its own culture. This belief is called **cultural relativity.** A corollary of cultural relativity is that no practice is universally good or universally bad; goodness and badness are relative, not absolute.

This type of evaluation is sometimes a difficult intellectual feat. For example, no matter how objective we try to be, most of us believe that infanticide, human sacrifice, and cannibalism are absolutely and universally wrong. Such an attitude reflects **ethnocentrism**—the tendency to use the norms and values of our own culture as standards against which to judge the practices of others. Ethnocentrism usually means that we see our way as the right way and everybody else's way as the wrong way. When American missionaries came to the South Sea Islands, they found that many things were done differently in Polynesian culture. The missionaries, however, were unable to view Polynesian practices as simply different. If those practices were not like American practices, then they must be wrong and were probably wicked. As a result, the missionaries taught the islanders that the only acceptable way (the American way) to have sexual intercourse was in a face-to-face position with the man on top, the now-famous "missionary position." They taught the Polynesians that women must cover their breasts, that they should have clocks and come on time to

Cultural relativity requires that each cultural trait be evaluated in the context of its own culture.

Ethnocentrism is the tendency to view the norms and values of our own culture as standards against which to judge the practices of other cultures.

appointments, and a variety of other Americanisms that the missionaries maintained to be morally right behavior.

Ethnocentrism is often a barrier to interaction among people from different cultures, leading to much confusion and misinterpretation. It is not, however, altogether bad. In the sense that it is pride in our own culture and confidence in our own way of life, ethnocentrism is essential for social integration. In other words, we learn to follow the ways of our culture because we believe that they are the right ways; if we did not share that belief, there would be little conformity in society. Ethnocentrism, then, is a natural and even desirable product of growing up in a culture. An undesirable consequence, however, is that we simultaneously discredit or diminish the value of other ways of thinking and feeling.

Culture Is a Social Product

A final assumption sociologists make about culture is that culture is a social, not a biological, product. The immense cultural diversity that characterizes human societies is not the product of isolated gene pools, but of cultural evolution.

Some aspects of culture are produced deliberately. Shakespeare picked up paper and pen in order to write *Hamlet*; some advertising executive worked to invent the Taco Bell chihuahua. Governments, bankers, and homeowners deliberately commission the designing of homes, offices, and public buildings, and people buy publishing empires so that they can spread their own version of the truth. Other aspects of culture—such as our culture's ideas about right and wrong, its dress patterns, and its language—develop gradually out of social interaction. But all of these aspects of culture are human products; none of them is instinctive. People *learn* culture and, as they use it, they modify it and change it.

Culture depends on a unique human attribute: the capacity for language. Only after language is invented can bits of practical knowledge (such as "fire is good" and "don't use electricity in the bathtub") or ideas (such as "God exists") be transmitted from one generation to the next. Inventions, discoveries, and forms of social organization are socially bestowed and intentionally passed on so that each new generation potentially elaborates on and modifies the accumulated knowledge of the previous generations. In short, culture is cumulative only because of language.

Because of language, human beings are not limited to the slow process of genetic evolution in adapting to their circumstances. Cultural evolution is a uniquely human way for a species to adapt to its environment. Whereas biological evolution may require literally hundreds of generations to adapt the organism fully to new circumstances, cultural evolution allows the changes to be made within a short period of time. In this sense, cultural evolution is an extension of biological evolution, one that speeds up the processes of change and adjustment to new circumstances in the environment (van den Berghe 1978).

Biological Perspective

As the continued popularity of *National Geographic* attests, the wide diversity of human cultures is a continuous source of fascination. Costumes, eating habits, and living arrangements vary dramatically. It is tempting to focus on the exotic variety of human behavior and to conclude that there are no limits to what humankind can devise. A closer look, however, suggests that there are some basic similarities in cultures around the world the universal existence of the family, religion, aggression, and warfare. When we focus on these universals, cultural explanations are likely to be supplemented with biological explanations.

Sociobiology is the study of the biological basis of all forms of human (and nonhuman) behavior.

Sociobiology is the study of the biological basis of all forms of human (and nonhuman) behavior (Alcock 2001; Wilson 1978). Sociobiologists believe that humans and all other life-forms developed through evolution and natural selection. According to this perspective, change in a species occurs primarily through one mechanism: Some individuals are more successful at reproduction than others. As the offspring of successful reproducers increase in number relative to those of the less successful reproducers, the species comes to be characterized by the traits that mark successful reproducers.

Who are the successful reproducers? They are those who have more children and raise more of them until they are old enough to reproduce themselves (Alcock 2001; Daly & Wilson 1983). For example, sociobiologists suggest that parents who are willing to make sacrifices for their children, occasionally even giving their life for them, are more successful reproducers; by ensuring their children's survival, they increase the likelihood that their own genes will contribute to succeeding generations. Thus, sociobiologists argue that we have evolved biological predispositions toward altruism (an unselfish concern for others), *but only insofar as our own kin are concerned.*

Sociobiology provides an interesting theory about how the human species has evolved over tens of thousands of years. Most of the scholars who study the effect of biology on human behavior, however, are concerned with more contemporary questions such as "How do hormones, genes, and chromosomes affect human behavior today?" Joint work by biologists and social scientists is helping us to understand how biological and social factors work together to determine human behavior. For example, Booth and Osgood (1993) found that men were statistically more likely to engage in deviant behavior if they had *both* high levels of testosterone *and* low levels of social integration. Research such as this suggests that only by recognizing and taking into account the joint effects of culture and biology can we fully understand human behavior.

The Carriers of Culture

In the following sections, we review three vital aspects of nonmaterial culture—language, values, and norms—and show how they shape both societies and individuals.

Language

The essence of culture is the sharing of meanings among members of a society. The chief mechanism for this sharing is a common language. Language is the ability to communicate in symbols—orally, by manual sign, or in writing.

What does *communicate with symbols* mean? It means, for example, that when you see the combinations of circles and lines that appear on your textbook page as the word *orally,* you are able to understand that it means "speaking aloud." On a different level, it means that the noise we use to symbolize "dog" brings to your mind a four-legged domestic canine. Almost all communication is done through symbols. Even the meanings of physical gestures such as touching or pointing are learned as part of culture.

Scholars of sociolinguistics (the relationship of language to society) agree that language has three distinct relationships to culture: Language embodies culture, language is a symbol of culture, and language creates a framework for culture (Romaine 2000; Trudgill 2000).

© AFP/Corbis

■ When languages die, cultures often die too. This is why immigrant parents often send their children to special classes to make sure they learn their traditional language.

Language as an Embodiment of Culture

Language is the carrier of culture; it embodies the values and meanings of a society as well as its rituals, ceremonies, stories, and prayers. Until you share the language of a culture, you cannot fully participate in it (Romaine 2000; Trudgill 2000).

A corollary is that loss of language may mean loss of a culture. Of the approximately 300 to 400 Native American languages once spoken in the United States, only about 20 are expected to survive much longer (Dalby 2003, 147–148). When these languages die, important aspects of these Native American cultures will be lost. This vital link between language and culture is why many Jewish and Chinese parents in the United States send their children to special classes after school or on weekends to learn Hebrew or Chinese. This is also why U.S. law requires that people must be able to speak English before they can be naturalized as U.S. citizens. To participate fully in Jewish or Chinese culture requires some knowledge of these languages; to participate in U.S. culture requires some knowledge of English.

Language as a Symbol

A common language is often the most obvious outward sign that people share a common culture. This is true of national cultures such as French and Italian and subcultures such as youth. A distinctive language symbolizes a group's separation from others while it simultaneously symbolizes unity within the group of speakers (Joseph et al. 2003; Romaine 2000; Trudgill 2000). For this reason, groups seeking to mobilize their members often insist on their own distinct language. For example, Jewish pioneers who moved in the early 1900s from the ghettos of Europe to what was then Palestine declared that everyone within their communities must speak Hebrew. Yet no one had spoken Hebrew except in prayers for hundreds of years. Nevertheless, within a few decades, Hebrew became the national language of Israel.

Similarly, in the last two decades some Americans have opposed bilingual education and pushed to declare English the national language of the United States, while French Canadians have fought to make French the official language of Quebec (Dalby

© AP/Wide World Photos

■ Because language is such a potent symbol of culture, some Americans oppose any use of languages other than English in the United States. Arizona Proposition 203 (defeated in 2000) would have banned bilingual education in public schools.

2003). Meanwhile, government organizations in Mexico fight to keep English words from creeping into Spanish. All of these efforts are largely symbolic; in any country, both immigrants and native-born citizens will continue to use or will quickly adopt whichever language has the most social status and social utility (Ricento & Burnaby 1998). As a result, grandchildren of Mexican immigrants to the United States rarely speak Spanish, and most French-speaking Canadians are also fluent in English (Dalby 2003).

Language as Framework

According to some linguists, languages not only symbolize our culture but also help to create a framework in which culture develops. The **linguistic relativity hypothesis** associated with Whorf (1956) argues that the grammar, structure, and categories embodied in each language influence how its speakers see reality. According to this hypothesis, for example, because Hopi grammar does not have past, present, and future grammatical tenses (e.g., "I had," "I have," "I will have"), Hopi speakers think differently about time than do English speakers. This theory has come under attack in recent years, and most linguists now believe that although differences among languages influence thought in small ways, the universal qualities of all language and human thought far overshadow those differences. The difficulties of translating from one language to another illustrate the conceptual differences among languages; that translation is possible illustrates that, despite those differences, people in all cultures have essentially the same linguistic capabilities (Trudgill 2000).

Values

After language, the most central and distinguishing aspect of culture is **values,** shared ideas about desirable goals. Values are typically couched in terms of whether a thing is good or bad—desirable or undesirable. For example, many people in the United States believe that a happy marriage is desirable. In this case and many others, values may be very general. They do not, for example, specify what a happy marriage consists of.

Some cultures value tenderness, others toughness; some value cooperation, others competition. Nevertheless, because all human populations face common dilemmas, certain values tend to be universal. For example, nearly every culture values stability and security, a strong family, and good health. There are, however, dramatic differences in the guidelines that cultures offer for pursuing these goals. In societies like ours, an individual may try to ensure security by putting money in the bank or investing in an education. In many traditional societies, security is maximized by having a large number of relatives. In societies such as that of the Kwakiutl of the Pacific Northwest, security is achieved, not by saving your wealth, but by giving it away. The reasoning is that all of the people who accept your goods are now under an obligation to you. If you should ever need help, you would feel free to call on them and they would feel obliged to help. Thus, although many cultures place a value on establishing security against uncertainty and old age, the specific guidelines for reaching this goal vary. The guidelines are called norms.

Norms

Shared rules of conduct are **norms.** They specify what people *ought* to or *ought not* to do. The list of things we ought to do sometimes seems endless. We begin the day with an "I'm awfully tired, but I ought to get up," and many of us end the

---

The **linguistic relativity hypothesis** argues that the grammar, structure, and categories embodied in each language affect how its speakers see reality.

Values are shared ideas about desirable goals.

Norms are shared rules of conduct that specify how people ought to think and act.

Connections

Personal Application

As you sit in your college classroom, you are following a long list of norms. Your very presence in the classroom reflects your acknowledgment that higher education is useful. No matter how bored you might be, you sit reasonably still and try not to fidget. If you are falling asleep, you pull your cap brim down to hide your droopy eyelids. You raise your hand rather than call out to demonstrate your respect for the teacher. And you write down whatever the teacher says, or at least write something down so it looks like you are taking notes.

day with "This is an awfully good show, but I ought to go to bed." In between, we ought to brush our teeth, eat our vegetables, work hard, love our neighbors, and on and on. The list is so extensive that we may occasionally feel that we have too many obligations and too few choices. Of course, some pursuits are optional and allow us to make choices, but the whole idea of culture is that it provides a blueprint for living, a pattern to follow.

Norms vary enormously in their importance both to individuals and to society. Some, such as fashions, are powerful while they last but are not central to society's values. Others, such as those supporting monogamy and democracy, are central to our culture. Generally, we distinguish between two kinds of norms: folkways and mores.

Folkways

The word **folkways** describes norms that are simply the customary, normal, habitual ways a group does things. Folkways is a broad concept that covers relatively permanent traditions (such as fireworks on the Fourth of July) as well as short-lived fads and fashions (such as wearing baseball caps backwards or sneakers with "pumps").

A key feature of all folkways is that no strong feeling of right or wrong is attached to them. They are simply the way people usually do things. For example, if you choose to have hamburgers for breakfast and oatmeal for dinner, or to sleep on the floor and dye your hair purple, you will be violating U.S. folkways. Doing these things may lead others to brand you as eccentric, weird, or crazy, but you will not be regarded as immoral or criminal.

Norms that govern daily life are usually not as clear as the rules in this Tennessee barbershop. Nevertheless, most of us pick up a pretty good understanding of these everyday rules by observing how others behave and how they react to our behavior.

Mores

Some norms are associated with strong feelings of right and wrong. These norms are called **mores** (more-ays). Whereas eating oatmeal for dinner may only cause you to be considered crazy (or lazy), other things that you can do will really offend your neighbors. If you eat your dog or spend your last dollar on liquor when your child needs shoes, you will be violating mores. At this point, your friends and neighbors may decide that they have to do something about you. They may turn you in to the police or to a child protection association; they may cut off all interaction with you or even chase you out of the neighborhood. Not all violations of mores result in legal punishment, but all result in such informal reprisals as ostracism, shunning, or reprimand. These punishments, formal and informal, reduce the likelihood that people will violate mores.

Laws

Rules that are enforced and sanctioned by the authority of government are **laws.** Very often the important mores of society become laws and are enforced by agencies of the government. If the laws cease to be supported by norms and values, they are either stricken from the record or become dead-letter laws, no longer considered important enough to enforce. Not all laws, of course, are supported by public sentiment; in fact, many have come into existence as the result of lobbying by powerful interest groups. Laws regulating marijuana use in the United States, for example, owe their origins to lobbying by the liquor industry. Similarly, laws requiring the wearing of seat belts are not a response to social norms. In these cases, laws are trying to *create* norms rather than respond to them.

Folkways are norms that are the customary, normal, habitual ways a group does things.

Mores are norms associated with fairly strong ideas of right or wrong; they carry a moral connotation.

Laws are rules that are enforced and sanctioned by the authority of government. They may or may not be norms.

American Diversity

focus on

How Feminists Research Culture

Many of the classic field studies and much of the current survey research on culture simply do not capture the experience of women. Many aspects of women's lives are unexplored, under-researched, or misinterpreted. For this reason, feminist scholars are developing a variety of new methods and new concepts that will help in gaining a clearer understanding of the values and norms that shape men and women's behavior and relationships.

Kristin Yount (1991) noted, for instance, that despite many studies of coal miners and coal-mining communities, very little was known about the condition of women miners. By living in mining communities over a period of several months, Yount had an opportunity to go to parties, ball games, and company picnics and observe work in the mines. In doing so, she discovered a culture of sexual harassment among miners, which the women miners adapted to by playing one of three roles: Lady, Flirt, or Tomboy.

By focusing on traditionally masculine values and behaviors—politics, war, or wage labor, for example—researchers often have overlooked the everyday behaviors of most importance to women. Consequently, some feminist researchers have begun to examine mundane cultural objects like Girl Scout manuals and cookbooks to understand how girls learn values and norms regarding expected women's behavior. Other researchers examine mundane everyday social interactions. Nancy Henley (1985) studied interrupting behavior in conversations. She taped same-sex and mixed-sex pairs in conversation and found that an amazing proportion (96 percent) of interruptions were made by male speakers. Thus, Henley uncovered one of many important gender-related norms in our culture.

In addition to exploring areas of life that other researchers have ignored, feminist researchers also bring to their work a different notion of how research should be conducted. Feminist researchers are more likely than others to regard research as a collaborative effort between respondents and researchers. This collaboration implies that respondents will somehow benefit from the research they take part in. For this reason, feminist research often actively seeks to influence public policy.

By concentrating on women in a culture, researching behaviors that sociologists traditionally overlooked, developing new research techniques, and emphasizing collaboration, political activism, *and* scientific rigor, feminist research methods are changing the ways in which sociology studies culture.

For more information, look up the following subjects in InfoTrac College Edition:
Feminist research
Nonverbal communication—analysis

Or visit the following Web site:
Feminist Majority Foundation
http://www.feminist.org/research/1_public.html

Social Control

From our earliest childhood, we are taught to observe norms, first within our families and later within peer groups, at school, and in the larger society. After a period of time, following the norms becomes so habitual that we can hardly imagine living any other way—they are so much a part of our lives that we may not even be aware of them as constraints. We do not think, "I ought to brush my teeth or else my friends and family will shun me"; instead we think, "It would be disgusting not to brush my teeth, and I'll hate myself if I don't brush them." For thousands of generations, no human considered it disgusting to go around with unbrushed teeth. For most people in the United States, however, brushing their teeth is so much a part of their feeling about themselves, about who they are and the kind of person they are, that they would disgust themselves by not observing the norm.

Through indoctrination, learning, and experience, many of society's norms come to seem so natural that we cannot imagine acting differently. No society relies com-

Values, Norms, and Laws

Concept	Definition	Example from Marriage	Relationship to Values
Values	Shared ideas about desirable goals	It is desirable that marriage include physical love between wife and husband	
Norms	Shared rules of conduct	Have sexual intercourse regularly with each other, but not with anyone else	Generally accepted means to achieve value
Folkways	Norms that are customary or usual	Share a bedroom and a bed; kids sleep in a different room	Optional but usual means to achieve value
Mores	Norms with strong feelings of right and wrong	Thou shalt not commit adultery	Morally required means to achieve value
Laws	Formal standards of conduct, enforced by public agencies	Illegal for husband to rape wife; sexual relations must be voluntary	Legally required means; may or may not be supported by norms

pletely on this voluntary compliance, however, and all encourage conformity by the use of **sanctions**—rewards for conformity and punishments for nonconformity. Some sanctions are formal, in the sense that the legal codes identify specific penalties, fines, and punishments that are to be meted out to individuals found guilty of violating formal laws. Formal sanctions are also built into most large organizations to control absenteeism and productivity. Some of the most effective sanctions, however, are informal. Positive sanctions such as affection, approval, and inclusion encourage normative behavior, whereas negative sanctions such as a cold shoulder, disapproval, and exclusion discourage norm violations.

Despite these sanctions, norms are not always a good guide to what people actually do, and it is important to distinguish between normative behavior (what we should do) and actual behavior. For example, our own society has powerful mores supporting marital fidelity. Yet research has shown that nearly half of all married men and women in our society have committed adultery (Laumann et al. 1994). In this instance, culture expresses expectations that differ significantly from actual behavior. This does not mean the norm is unimportant. Even norms that a large minority, or even a majority, fail to live up to are still important guides to behavior. The discrepancy between actual behavior and normative behavior—termed *deviance*—is a major area of sociological research and inquiry (see Chapter 6).

Sanctions are rewards for conformity and punishments for nonconformity.

Subcultures are groups that share in the overall culture of society but also maintain a distinctive set of values, norms, lifestyles, and even language.

Cultural Diversity and Change

By definition, members of a community share a culture. But that culture is never completely homogeneous. In the following sections, we will look at two expressions of diversity within cultures—subcultures and countercultures—and at the processes by which cultures change.

Subcultures and Countercultures

When segments of society face substantially different social environments, subcultures grow up to help them adapt to these unique problems. **Subcultures** share in the overall culture of society, but also maintain a distinctive set of values, norms, lifestyles, and even language. In the United States, for example, some of the most prominent subcultures are based on region. Empirical studies show that people who live in the South are more likely than others to buy gospel music and to watch television frequently and are less likely to listen to heavy metal music or to read books (Weiss 1994). Although regional differences are weaker among African Americans than among whites, in general, Southerners are also more likely to approve of school prayer and neighborhood segregation and to disapprove of homosexuality and women politicians (Weakliem & Biggert 1999).

Countercultures are groups having values, interests, beliefs, and lifestyles that are opposed to those of the larger culture.

Countercultures are groups that have values, interests, beliefs, and lifestyles that conflict with those of the larger culture. This theme of conflict is clear among one recent U.S. countercultural group—punkers. Some punkers are just part-timers who dye their hair purple and listen to death rock but nevertheless manage to go to school or hold a job. Hardcore punkers, however, angrily reject straight society. They refuse to work or to receive charity; they live angry and sometimes hungry lives on the streets. As one young girl explained to a (formerly punk) sociologist, "Punks are saying 'Fuck this, we're pissed off'. . . When you're really young, living with your parents in their all-white rich suburb [and you] shave your head and dye [the remainder of] your hair green, then it's saying, 'Well, fuck society, fuck the corporate world, I'm going to be against all this shit'" (Leblanc 1999, 87).

An Example: Deafness as Culture

Most of us view deafness as undesirable, even catastrophic (Dolnick 1993). At best, we see being deaf as a medical condition to be remedied. However, some deaf people maintain that deafness is not a disability but a culture (Dolnick 1993). To these individuals, the essence of deafness is not the inability to hear but a culture based on their shared language, American Sign Language (ASL). ASL is not just a way to "speak" English with one's hands but is a language of its own, complete with its own rules of grammar, puns, and poetry. Furthermore, it is a language that is learned and shared. Whereas hearing babies begin to jabber nonsense syllables,

▪ These deaf students believe that they share a common culture and should have rights like those given to any minority culture.

deaf babies of parents who sign begin to "babble" nonsense signs with their fingers (Dolnick 1993). This shared language encourages, in turn, shared values and a positive group identity. Studies show, for instance, that many deaf people would not choose to join the "hearing" culture even if they could.

Thinking of deafness as culture illustrates many of the points made earlier. For instance, culture is problem solving, and deaf culture embodies a way to solve the human problem of communication. Using ASL shapes deaf people's experiences, reminding them of their common values, norms, and cultural identity. For this reason, many deaf individuals have reacted with outrage to the increasing use of cochlear implants (Arana-Ward 1997). These devices, when surgically implanted in the ear, help some otherwise-deaf persons to hear sounds. Hearing sounds, however, is not the same as understanding what one hears: many implant recipients—especially older children who were born deaf—are frustrated by a cacophony of sounds that they cannot interpret, even after months or years of training. Deaf activists argue that most children who receive implants waste their formative years in an often futile struggle to assimilate into the hearing world, when they could instead have become native speakers of ASL and valued members of the deaf community. These activists, therefore, view cochlear implants not as a neutral medical technology but as an example of the ethnocentrism of hearing persons.

Acculturation or Multiculturalism?

Until very recently, most Americans believed it would be best if the various ethnic and religious subcultures within American society would become acculturated to a single majority culture. **Acculturation** refers to the process through which individuals learn and adopt the values and social practices of the dominant group. As discussed in more detail in Chapter 13, acculturation was, and to some extent still is, one of the major goals of our educational institutions (Spring 1997). In schools, immigrant children learn not only to read and write English but also to consider "American" foods, ideas, and social practices preferable to those of their own native culture or subculture. Children named Juan or Hans may be encouraged to go by John or Harry. School curricula focus on the history, art, literature, and scientific contributions of Europeans and European Americans while downplaying the contributions of U.S. minority groups and non-Western cultures.

In the last quarter century, however, more and more Americans have concluded that America has always been more of a "salad bowl" of cultures than a melting pot. Moreover, many have come to believe that this "salad bowl" is one of Americans' greatest strengths and that it should be cherished rather than eliminated. These beliefs are often referred to as *multiculturalism*. Reflecting these beliefs, many schools and universities now incorporate materials that more accurately reflect American cultural diversity.

Acculturation is the process of acquiring the values and social practices of the dominant group and surrounding society.

Sources of Cultural Diversity and Change

Culture provides solutions to common and not-so-common problems. The solutions devised are immensely variable. Among the reasons for this variability are environment, isolation, cultural diffusion, technology, and dominant cultural themes.

Environment

Why are the French different from Australian aborigines, the Finns different from the Navajo? One obvious reason is the very different environmental conditions to which they must adapt. These different environmental conditions determine which kinds of economies can flourish and, to a significant extent, the degree of scarcity or abundance.

Isolation

When a culture is cut off from interaction with other cultures, it is likely to develop unique norms and values. Where conditions of isolation preclude contact with others, a culture continues on its own course, unaltered and uncontaminated by others. Since the nineteenth century, however, almost no cultures have been able to maintain their isolation from other cultures.

Cultural Diffusion

If isolation is a major reason why cultures remain both stable and different from each other, then cultural diffusion is a major reason why cultures change and become more similar over time. **Cultural diffusion** is the process by which aspects of one culture or subculture enter and are incorporated into another culture. For example, not only have many residents of Mexico City become regular consumers of McDonald's hamburgers, but belief in the value of "fast food" is gradually replacing Mexicans' traditional belief in the value of long, family-centered meals. Meanwhile, salsa now outsells ketchup in the United States, and Heinz now offers a green ketchup specifically to compete with salsa.

At its broadest level, cultural diffusion becomes the **globalization of culture,** in which certain cultural elements are shared by persons around the world. Nowadays, taxi-drivers in Bombay, Senegal, and Peru blare U.S. popular music from their radios, while businesspeople around the globe travel to Japan to study Japanese organizational styles. The globalization of culture is likely to proceed even more rapidly in the future due to growth in access to the Internet.

Technology

The tools available to a culture will affect its norms and values and its economic and social relationships. The rise of the automobile, for example, allowed young dating couples to escape the watchful eyes of concerned parents, contributing to changes in both courtship patterns and the roles of women (Scharff 1991). The automobile also encouraged individuals to live farther from their places of employment, to shop in malls rather than downtown, and to live in houses with three-car garages rather than front porches. As a result, among the unintended consequences of the automobile was a decline of inner cities, locally owned businesses, and a sense of neighborliness (Kunstler 1994).

Dominant Cultural Themes

Cultures generally contain dominant themes that give a distinct character and direction to the culture; they also create, in part, a closed system. New ideas, values, and inventions are usually accepted only when they fit into the existing culture or rep-

---

Cultural diffusion is the process by which aspects of one culture or subculture enter and are incorporated into another.

Globalization of culture is the process through which cultural elements (including musical styles, fashion trends, and cultural values) spread around the globe.

■ Diffusion of modern technology is particularly rapid when new tools enhance a society's ability to meet basic human needs at the same time that they are consistent with existing cultural patterns. Leaders, regardless of time, place, or the cultural bases of their authority, share a common need to communicate effectively with large numbers of followers.

resent changes that can be absorbed without too greatly distorting existing patterns. The Native American hunter, for example, was pleased to adopt the rifle as an aid to the established cultural theme of hunting. Western types of housing and legal customs regarding land ownership, however, were rejected because they were alien to a nomadic and communal way of life.

A Case Study: American Consumer Culture

U.S. culture is a unique blend of complex elements. It is a product of the United States' environment, its immigrants, its technology, and its place in history. These days, one of the ways in which U.S. culture diverges most strongly from other cultures is in its exceptionally strong emphasis on consumerism.

Consumerism is a philosophy that says "buying is good." In turn, this philosophy is based on the belief that "we are what we buy," and that through buying certain goods we can assert or improve our social status. In American consumer culture, children attempt to improve their social status by buying certain brands of breakfast cereals, teenagers by buying certain brands of clothing, and adults by buying certain models of cars. Ironically, consumers also believe they are asserting their individuality through their purchases, rarely noticing that millions of others are buying the same goods for the same reasons.

How did this consumer culture develop? The simple answer is that more consumer goods are available and affordable than ever before. But this is only a partial answer. Research suggests that the most important cause is a change in the comparisons we use in deciding whether to make a purchase (Schor 1998). Advertising now permeates our lives more than ever before—billboards adorn public buses and sports stadiums, movie theaters show advertisements before the films, ads pop up at popular Internet sites, schools broadcast television programs laced with commercials in the classroom, and so on—instilling in children and adults the belief that they need certain products in order to be a certain sort of person (Quart 2003). Similarly, as the number of hours Americans watch television per week has soared, so has Americans' desire for the goods they see on television. Instead of deciding what kind of shoes to wear or what kind of kitchen appliances to buy by looking at what their classmates or neighbors own, Americans now seek out consumer goods like those used by their favorite television characters. In fact, for every hour of television watched each week, individuals' annual spending on consumer goods increases by more than $200 (Schor 1998). Finally, in the past, women (who do most family shopping) typically compared their belongings to those of their neighbors, whose family incomes were usually similar to their own. Now that a majority of women work outside the home, most compare their belongings to those of their fellow workers, including supervisors with much higher incomes. As a result, families now spend higher percentages of their income on consumer goods, both big and small. For example, the median house size has increased from 1500 square feet in 1970 to 2300 square feet today—with prices to match (U.S. Bureau of the Census 2001).

Consumer culture affects our lives in many ways. Shopping and window-shopping have become major forms of recreation, and shopping malls have replaced parks, athletic fields, church basements, and backyards as popular gathering spots. College students work extra hours and put their grades at risk to buy the latest gadgets or fashions. And adults carry heavy debts and risk bankruptcy to buy expensive cars and houses as a way to "prove" their success and improve their social status. Figure 2.1 illustrates the rising debts of families between 1989 and 2001.

Consumerism is the philosophy that says "buying is good" because "we are what we buy."

FIGURE 2.1
Median Value of Debt, Among American Families That Hold Debts*
The rise in debts reflects both economic hard times and growing consumer desires.
*All amounts given in $2001 dollars.

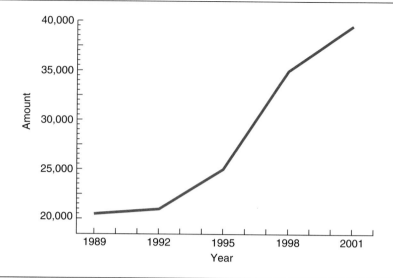

Consequences of Cultural Diversity and Change

Culture Shock

Culture shock refers to the discomfort that arises from exposure to a different culture.

In the long run, cultural diversity and cultural change often result in improvements in quality of life. In the short run, however, people often find both diversity and change unsettling. **Culture shock** refers to the disconcerting and unpleasant experiences that can accompany exposure to a different culture. For example, U.S. citizens who travel or work in Greece are often taken aback by the Greek customs of hugging acquaintances and standing very close (by American standards) to anyone they are speaking with. Greeks who travel or work in the United States are similarly confused by American customs that limit greetings to simple handshakes and dictate maintaining considerable physical distance during conversations. As a result, Americans sometimes conclude that Greeks are pushy or even sexually aggressive, while Greeks sometimes conclude that Americans are elitist or emotionally cold.

Cultural Lag

Cultural lag occurs when one part of culture changes more rapidly than another.

Cultural change can also foster social problems when one part of a culture changes more rapidly than another part. This situation is known as **cultural lag** (Ogburn 1922). Most often, cultural lags occur when social practices and values do not keep up with technological changes. The recent rise in genetic testing for breast cancer illustrates some of the problems that arise when law, values, and social practices lag behind technological change (Lewin 2000). Although most cases of breast cancer are not hereditary, women who carry the BRCA1 genetic mutation have a 50 percent lifetime risk of breast cancer. For this reason, some women who have a family history of breast cancer now opt to be tested for the mutated gene. If they do carry the gene, the only way they can reduce their chances of breast cancer is to have both

Connections

Example

American education is imbued with the value of competition. Children are encouraged to best their peers in the classroom and in the schoolyard—to strive for the highest test scores or the most home runs. In many African countries, children are chastised by teachers as well as other students if they behave competitively. Instead, children are encouraged to help each other learn and to work together for the good of the group. These differences can create culture shock for children who emigrate from Africa to the United States.

breasts surgically removed. Yet half of those who test positive will never develop breast cancer and some who have their breasts removed will nevertheless die of cancer. Because of these uncertainties, some women with family histories of breast cancer choose not to be tested. But what happens if a woman decides against getting tested and her mother (or sister or daughter) decides in favor of testing? If the mother subsequently has her breasts removed, the daughter will know immediately that her mother tested positive and will know that she, herself, is more likely to carry the mutated gene. Thus, the mother's decisions force the daughter to confront information that she may have chosen not to seek. At the same time, genetic counselors worry that both they and their clients might face legal liability if the client does *not* tell her relatives they too are at risk.

All these problems reflect cultural lag. Individuals forced to decide about genetic testing must do so in the absence of agreed-upon social values, clear legal decisions, or standard social practices for how persons in their situation should act.

Uses of Culture

Current sociological research on culture often focuses on one of two issues: the political uses of culture—how culture can create symbolic boundaries or collective memories that increase one social group's power or influence over another (Berezin 1997)—and the relationship between culture and mass media.

Symbolic Boundaries

One way that culture can affect power relationships between groups is through the establishment of symbolic boundaries (Bourdieu 1984; Lamont & Fournier 1992). *Symbolic boundaries* refer to perceived cultural differences that make it difficult for individuals to move from one social group to another. Different musical tastes, for instance, or a preference for different types of leisure activities often form symbolic boundaries. In the United States, opera, ballet, "serious fiction," and modern art are elements of culture often associated with persons of high social status (Halle 1993). These cultural preferences are often referred to as **high culture.** Because of them, upper-class persons may believe that they have little in common with individuals who prefer heavy metal or gospel music (Bryson 1996); as a result, they may be less likely to marry, admit to their social clubs, or hire individuals from different class and cultural backgrounds.

Collective Memory

Culture, politics, and social influence are also interwoven in the creation of **collective memories**—a culture's ideas about its own past. Collective memories appear in history textbooks and traditional folk songs and are displayed in museums and monuments such as the Vietnam War Memorial. As cultures change, so do collective memories. During World War II, for instance, Americans widely revered Abraham Lincoln as a symbol of the moral need to wage war. In post-Vietnam America, however, as our culture has increasingly rejected both authority and nationalism, Lincoln's prestige has declined considerably (Schwartz 1996, 1998).

High culture refers to the cultural preferences associated with the upper class.

Collective memories are a culture's ideas about its own past.

© Charlyn Zlotnik/Woodfin Camp & Associates

■ Monuments such as the Vietnam War Memorial symbolize a society's image about its past. Cultural change is often accompanied by competing versions of such collective memories.

focus on *Technology*

Models, Magazines, and Self-Esteem

Over the last several decades, the average American has gotten older and heavier. During the same time period, however, Miss America has gotten taller and slimmer, and America's favorite doll—Barbie—has lost her curves. Fashion magazines, movies, and television increasingly celebrate and glamorize an extremely rare body type, far slimmer than the typical American woman's (Milkie 1999). The net result is that the gap between media images of ideal female body shape and actual female body shapes has increased substantially (Wiseman et al. 1992). What effect has this had on girls' self-concepts? And what is the additional impact on young African American, Asian American, Native American, and Latina women when the media still primarily showcase fair-skinned, long-haired, and European-featured girls (Collins 1991)?

Many scholars believe that unrealistic beauty images in the media damage girls' self-esteem and have contributed to epidemic levels of eating disorders (Evans et al. 1991). Others argue that girls critically evaluate what they see and read in the media and recognize the media's unrealistic standards of beauty as myths. Sociologist Melissa Milkie (1999) studied this issue through in-depth interviews with 60 urban and rural, African American and white high-school girls. Here is what she found.

Most of the girls in Milkie's study, both African American and white, believed that the images of female beauty shown on the pages of girls' magazines were unrealistic. As one respondent noted, even the so-called "problem" bodies shown in magazines are perfect:

> These magazines are trying to tell you "Do this and do that" . . . if you have a problem body, . . . a big butt, big chest . . . And these girls that they are showing don't have that problem. I mean you can tell they don't, and that makes me mad. . . . They say if you got a stick figure, wear a one-piece [swimsuit] . . . and I'm looking at the girl and she doesn't have a stick figure. . . . (Milkie 1999, 198–199)

Despite their critiques of unrealistic, idealized media images of beauty, Milkie found that the self-concepts of white girls in her study were harmed by those images. Even though the girls rejected these images, they believed that their friends, boyfriends, and potential boyfriends judged them by how well they matched those images, and so they judged themselves accordingly. Interestingly, minority girls were less affected by media images because they rejected them as white-only ideals and because they believed their friends, boyfriends, and potential boyfriends also rejected those ideals.

Taken together, Milkie's results suggest that (1) individuals are active consumers of media messages, (2) different audiences interpret the same media messages differently, and (3) media do shape culture and individual experiences, at least in part because we judge ourselves through the "media-filled" eyes of others who matter to us.

© Joel Gordon

Unrealistic images of female beauty in magazines, on TV, and in fashion advertisements contribute to poor body image and low self-esteem among young American women.

For more information, look up the following subjects in InfoTrac College Edition:
Body image
Women's magazines
Girls and self-esteem

Or visit the following Web site:
National Women's Health Network
http://www.womenshealthnetwork.org/clearinghouse/eatd.htm

Cultural change is often accompanied by competing versions of collective memory (Rhea 1997; Sturken 1997). To some extent, current debates over multicultural education are really debates about which version of collective memory is most correct: For example, should American history textbooks describe the Civil War primarily as a fight against slavery or as a fight for state sovereignty? And should discussion of the fight against slavery emphasize military battles or the day-to-day resistance of slaves against their masters?

Mass Media and Audiences

Popular culture refers to aspects of culture that are widely accessible and broadly shared, especially among "ordinary" folks. Most of the research on popular culture focuses on the mass media: television; genre fiction such as romances, mysteries, or science fiction; and popular music styles like rock and roll or hip-hop. One important question such research seeks to answer is the extent to which mass media simply reflect American values and popular culture and the extent to which they actually change them. The answer is that media probably do both.

For much of the past century, movies and television portrayed African Americans as lazy or foolish and unmarried women as evil, disturbed, or unhappy (Entman & Rojecki 2000; Levy 1990). These depictions were certainly a reflection of American cultural beliefs of the time. By studying how media representations have changed, scholars have one important indication of the extent to which American values about U.S. racial and ethnic groups and women have or have not changed.

At the same time, media also affect our ideas about the world (Schudson 1995). For instance, even though crime rates have dropped dramatically over the last decade, Americans are more afraid of crime than ever. This is probably at least in part because news broadcasters continue to highlight crime stories (Glassner 1999). Of course, audiences do not simply absorb the messages sent by the media, but rather actively interpret and use the news, television shows, movies, music, and literature they consume. In fact, different audiences may respond quite differently to identical media messages. Darnell Hunt (1997) showed, for instance, that African American, Hispanic, and white viewers watched and interpreted the same news report on the rioting that followed the acquittal of the police officers charged in the Rodney King beating in very different ways. African American viewers were much more likely to question the credibility of the news report and to view the incident in racial terms, whereas white viewers were much more likely to listen respectfully and think of the issue in strictly judicial terms. Hispanic viewers responded in a manner midway between the other two. Results such as these suggest that the ability of the media to change entire cultures will be sharply limited, especially in societies as multicultural as that of the United States.

Popular culture refers to aspects of culture that are widely accessible and commonly shared by most members of a society, especially those in the middle, working, and lower classes.

Where This Leaves Us

Most of the time, we think of culture simply as something that we have, in the same way that we have two arms or a home. As this chapter has shown, though, culture is dynamic: constantly changing as the world changes around us. Languages, eating habits, clothing fashions, and the rest evolve, spread, or die. Ask your parents about the clothing they wore as children, the slang they spoke as teenagers, or the first time they ate a bagel or a tortilla.

Culture is also active, a force that changes us as it changes the world in which we live. The rise of American consumer culture is only one example of the way culture changes and the effects cultural changes have on all aspects of our lives, from how many hours we work each day to how we define ourselves as individuals.

Summary

1. Culture is a design for living that provides ready-made solutions to the basic problems of a society. It can be conceived of as a tool kit of material and nonmaterial components that help people adapt to their circumstances. Culture is normative, learned, and relative. Sociobiologists believe that human culture and behavior also have biological roots.

2. Language, or symbolic communication, is a central component of culture. Language embodies culture, serves as a framework for perceiving the world, and symbolizes common bonds among a social group.

3. Values spell out the goals that a culture finds worth pursuing, and norms specify the appropriate means to reach them.

4. The cultures of large and complex societies are not homogeneous. Subcultures and countercultures with distinct lifestyles and folkways develop to meet unique regional, class, and ethnic needs.

5. Consumer culture—the philosophy that buying is good, and we are what we buy—now plays a major role in American culture. As a result, shopping has become an important form of recreation and many Americans are both overworked and in debt.

6. The most important factors accounting for cultural diversity and change are the physical and natural environment, isolation from other cultures, cultural diffusion, level of technological development, and dominant cultural themes.

7. Cultural diversity and change can lead to culture shock and cultural lag. Culture shock refers to the disconcerting experiences that accompany rapid cultural change or exposure to a different culture. Cultural lag occurs when changes in one part of the culture do not keep up with changes in another part.

8. High culture refers to the cultural preferences associated with persons of high social status; popular culture refers to cultural elements associated with "ordinary" individuals. These cultural differences often create symbolic boundaries.

9. Mass media and collective memories both reflect and help to change culture. Audiences do not passively accept messages from the media, and groups sometimes compete over differing versions of collective memory.

Thinking Critically

1. What features of U.S. society might explain why children are raised in small nuclear families rather than in extended kin groups?

2. Can you think of an example from U.S. culture for which values, norms, and laws are not consistent with each other? What consequences, if any, are these inconsistencies likely to have?

3. How do environment, isolation, technology, and dominant cultural themes contribute to the maintenance and diffusion of the youth subculture?

4. Will the Internet increase or decrease the symbolic boundaries between high and popular culture? Explain how and why you think this will happen.

5. How would a structural functionalist analyze U.S. collective memories of the Vietnam era? A conflict theorist? A symbolic interactionist?

Sociology on the Net

The Wadsworth Sociology Resource Center: Virtual Society

http://sociology.wadsworth.com/

The companion Web site for this book includes a range of enrichment material. Further your understanding of the chapter by accessing Online Practice Quizzes, Internet Exercises, InfoTrac College Edition Exercises, and many more compelling learning tools.

Chapter-Related Suggested Web Sites

What Is Culture?
http://www.culturalstudies.net

Cultural Studies Central
http://www.culturalstudies.net

The Popular Culture Association/American Culture Association
http://www2.h-net.msu.edu/~pcaaca/

Cultural Exchange
http://www.centerforculturalexchange.org/

InfoTrac College Edition

http://www.infotrac-college.com/wadsworth

Access the latest news and research articles online—updated daily and spanning four years. InfoTrac College Edition is an easy-to-use online database of reliable, full-length articles from hundreds of top academic journals and popular sources. Conduct an electronic search using the following key search terms:

Culture

Subculture

Counterculture

Pop culture

Cultural diversity/multiculturalism

Language and culture

For more information related to this chapter, visit the Opposing Viewpoints Resource Center. Be sure you check all the options on the toolbar—viewpoints, references, statistics, and so on—and search under the following subjects:

Deafness

Sociobiology

English language

Bilingual education

Suggested Readings

Fadiman, Anne. 1997. *The Spirit Catches You and You Fall Down: A Hmong Child, Her American Doctors, and the Collision of Two Cultures.* New York: Farrar, Straus, and Giroux. A heart-rending true story of the culture clash between a Hmong immigrant family and their American doctors, illustrating some of the consequences of cultural diversity, culture shock, and ethnocentrism.

Glassner, Barry. 1999. *The Culture of Fear: Why Americans Worry about the Wrong Things.* New York: Basic Books. A fascinating account of how and why Americans have come to greatly overestimate the dangers of drugs, teenage motherhood, and "road rage" among other things. The author uses a broad range of data to explore the role of the media, scientists, politicians, corporations, and others in fostering these fears.

Kephart, William M., and Zellner, William W. 1994. *Extraordinary Groups: The Sociology of Unconventional Life-Styles* (5th ed.). New York: St. Martin's Press. A fascinating tour of some of the most interesting subcultures and countercultures in the United States, both past and present: the Amish, gypsies, Father Divines, and Jehovah's Witnesses. This is painless sociology—it applies basic concepts and theory to truly extraordinary groups.

Quart, Alissa. 2003. *Branded: The Buying and Selling of Teenagers.* Cambridge, MA: Perseus. Trenchant description of how advertisers aim directly at teenagers, from product placement in video games to soda machines in school cafeterias, and how teenagers use consumer goods, from designer jeans to designer breasts, to create their self-images.

Socialization

© David Young-Wolff/Photo Edit

What Is Socialization?

At the heart of sociology is a concern with *people*. Sociology is interesting and useful to the extent that it helps us explain why people do what they do. It should let us see ourselves, our family, and our acquaintances in a new light.

In this chapter we deal directly with individuals, focusing on the process whereby people learn the roles, statuses, and values necessary for participation in social institutions. This process is called **socialization.** Socialization is a lifelong process. It begins with learning the norms and roles of our family and our subculture and making these part of our self-concept. As we grow older and join new groups and assume new roles, we learn new norms and redefine our self-concept.

Socialization is the process of learning the roles, statuses, and values necessary for participation in social institutions.

The Self and Self-Concept

From the small lump of flesh that is the newborn infant develops a complex and fascinating human being, a human being who is simultaneously much like every other human being and at the same time exactly like no other.

Each individual **self** may be thought of as a combination of unique attributes and normative, or socially expected, responses. Within sociology, these two parts of the self are called the *I* and the *me* (Mead 1934).

The **I** is the spontaneous, creative part of the self; the **me** is the self as social object, the part of the self that responds to others' expectations. In English grammar, *me* is used when we speak of ourselves as the object of others' actions (She sent me to the office); *I* is used when we speak of ourselves as the actor (I acted up). Sociological use follows this convention.

As this description of the self implies, the two parts may pull us in different directions. For example, many people face a daily conflict between their I and their me when the alarm clock goes off in the morning—the I wants to roll over and go back to sleep, but the me knows it is supposed to get up and go to class. Some of these conflicts are resolved in favor of the me and some in favor of the I. Daily behavior, however, is the result of an ongoing internal dialogue between the I and the me.

The self is enormously complex, and we are often not fully aware of our own motives, capabilities, and characteristics. The self that we are aware of is our **self-concept.** It consists of our thoughts about our personality and social roles. For example, a young man's self-concept might include such qualities as young, male, Methodist, good athlete, student, shy, awkward with girls, responsible, American. His self-concept includes all the images he has of himself in the dozens of different settings in which he interacts with others.

The **self** is a complex whole that includes unique attributes and normative responses. In sociology, these two parts are called the *I* and the *me.*

The **I** is the spontaneous, creative part of the self.

The **me** represents the self as social object.

The **self-concept** is the self we are aware of. It is our thoughts about our personality and social roles.

Learning to Be Human

What is human nature? Are we born with a tendency to be cooperative and sharing or with a tendency to be selfish and aggressive? The question of the basic nature of humankind has been a staple of philosophical debate for thousands of years. It continues to be a topic of debate because it is so difficult (some would say impossible) to separate the part of human behavior that arises from our genetic heritage from the part that is developed after birth. The one thing we are sure of is that nature is never enough.

The Necessity of Nurture

Each of us begins life with a set of human potentials: the potential to walk, to communicate, to love, and to learn. By themselves, however, these natural capacities are not enough to enable us to join the human family. Without nurture—without love and attention and hugging—the human infant is unlikely to survive, much less prosper. The effects of neglect are sometimes fatal and, depending on severity and length, almost always result in retarded intellectual and social development.

How can we determine the importance of nurture? One way is to look at what happens when children are seriously neglected. Some of the clearest evidence comes from studies of children raised in low-quality orphanages, where their physical needs were met, but where staff were too overworked to give individual attention or to truly nurture each child.

Children who spend the first years of their lives in this type of institutional environment are devastated by the experience. Because of limited personal attention, such children withdraw from the social world; they seldom cry and often are indifferent to everything around them. Even if they are later adopted into good homes, they are significantly more likely to experience difficulties in thinking or learning. They are also more likely to experience problems in social relationships—either engaging in indiscriminate friendliness or withdrawing into autistic or near-autistic behaviors. These effects are illustrated by a study that compared children adopted by British parents from high-quality British orphanages and from low-quality Romanian institutions. The researchers found that 12 percent of the Romanian-born children exhibited autistic or quasi-autistic patterns, but none of the children from the British orphanages did (Rutter et al. 1999).

Deprivation can also occur in homes in which parents fail to provide adequate social and emotional stimulation. Children who have their physical needs met but are otherwise ignored by their parents have been found to exhibit many of the same symptoms as institutionalized infants. The bottom line is that children need intensive interaction with others to survive and develop normally. Much of the evidence for this conclusion, however, is derived from atypical situations in which unfortunate children have been subjected to extreme and unusual circumstances. To test the limits of these findings and to examine the reversibility of deprivation effects, researchers have turned to experiments with monkeys.

Monkeying with Isolation and Deprivation

In a classic series of experiments, Harry Harlow and his associates studied what happened when monkeys were raised in total isolation. The infants lived in individual cages with a mechanical mother figure that provided milk. Although the infant monkeys' nutritional needs were met, their social needs were not. As a result, both their physical and social growth suffered. They exhibited bizarre behavior such as biting themselves and hiding in corners. As adults, these monkeys refused to mate; if artificially impregnated, the females would not nurse or care for their babies (Harlow & Harlow 1966). These experiments provide dramatic evidence about the importance of social contact; even apparently innate behaviors such as sexuality and maternal

Connections

Example

The true story of Genie illustrates the consequences of severe deprivation. Until the age of 13, Genie's abusive father kept her locked in a small room, tied to a chair by a sort of harness. Her mother—blind, disabled, and cowed—could do nothing to help her. Genie was never spoken to or socialized in any way. When Genie's mother finally ran away with her, Genie could not talk, walk, or even use a toilet. After years of therapy, her abilities improved. However, her experiences left her with subnormal intelligence.

behavior must be developed through interaction. On the bright side, the monkey experiments affirm that some of the ill effects of isolation and deprivation are reversible. The young monkeys experienced almost total recovery when placed in a supportive social environment (Suomi, Harlow, & McKinney 1972).

Although it is dangerous to generalize from monkeys to humans, the evidence from the monkey experiments supports the observations about human infants: Physical and social development depend on interaction with others. Even being a monkey does not come naturally. Walking, talking, loving, and laughing all depend on sustained and intimate interaction with others. Clearly, our identities, even our lives, are socially bestowed and socially sustained (Berger 1963, 98).

Theories of Socialization

Socialization refers broadly to the processes through which we come to share roughly the same norms and values as other members of our culture. But how does this learning take place? What are the processes through which, as children and adults, we translate our social experiences into part of our self-concept? In the following pages we look at some psychological and sociological theories of socialization.

Freudian Theory

We begin by looking at the theory of socialization proposed by psychoanalyst Sigmund Freud at the beginning of the twentieth century. Although largely discredited at this point, Freud's theory is still worth understanding because of its considerable influence on American culture and social science: The theory is still used (in a much revamped form) by some psychologists and even some sociologists (e.g., Chodorow 1999).

This infant monkey and others studied by psychologist Harry Harlow grew up locked in individual cages without social contact. The monkeys suffered physical and emotional stunting and had difficulty learning how to have sexual intercourse or raise their own babies. These experiments suggest that even apparently innate behaviors such as sexuality must be developed through interaction.

© Martin Rogers/Stock Boston

Freud's theory of socialization links social development to biological cues. According to Freud, to become mentally healthy adults, children must develop a proper balance between their **id**—natural biological drives, such as hunger and sexual urges—and their **superego**—internalized social ideas about right and wrong. To find that balance, children must respond successfully to a series of developmental issues, each occurring at a particular stage and linked to biological changes in the body. For example, during the oral stage, infants and toddlers derive their greatest satisfaction from sucking a breast or bottle. Those who do not learn how to signal and fulfill those needs, Freud concluded, would later develop traits such as dependency and narcissism.

The phallic stage (between about ages 3 and 6) plays an especially important role in socialization in Freud's theory because that is when the superego is hypothesized to develop. According to Freud, during the phallic stage, children first start noticing their genitalia. At this time, they experience sexual attraction toward their opposite-sex parent, viewing their same-sex parent as a rival. When boys first learn that girls lack

The **id** is the natural, unsocialized, biological portion of self, including hunger and sexual urges.

The **superego** is composed of internalized social ideas about right and wrong.

American Diversity

Gender Differences in Mathematics

Despite the many changes in gender roles over the last 50 years, boys still outperform girls on standardized math tests such as the SAT. By high school, boys substantially outnumber girls in advanced mathematics courses, and in adulthood women remain substantially underrepresented in occupations like engineering that depend heavily on mathematics. Are males actually better than females at math and, if so, is this difference based on nature or nurture?

Neuroscientists interested in this question have begun exploring the relationship between fetal exposure to sex hormones and characteristic differences in the brains of adult men and women. For instance, higher levels of fetal exposure to testosterone (a "male" hormone) are associated with right-brain dominance, while lower exposure levels are associated with left-brain dominance. This association may help explain why, compared to the opposite sex, men more often are left-handed with good visual-spatial skills (a "right-brain" trait) and women more often are right-handed with good verbal skills (a "left-brain" trait).

From findings such as these, some researchers reason that gender differ-

On the average, males outperform females on mathematical tasks that require the kind of spatial skill measured by mental rotation tests. Although this difference may have a genetic basis, many scholars believe that the difference stems primarily from socialization. Changes in teaching and parenting are helping to reduce the gender gap in mathematical performance.

© Seth Resnick/Stock Boston

---

penises, however, they conclude that girls have been castrated by their fathers as punishment for some wrongdoing. Fearing the same fate, boys abandon their attraction to their similarly castrated mothers and identify with their fathers and try to obtain their love by adopting their values. Through this process, boys develop a strong superego. In contrast, Freud argued that because girls lack fear of castration, they can never develop a strong superego. Instead, girls' realization that they lack a penis drives them to jealousy, narcissism, and a sense of inferiority (Freud 1925 [1971:241–260]).

Freud based his theory on his personal interpretations of the lives and dreams of his patients. He did not test his ideas scientifically, and did not test them on a broad sample of the population. Nevertheless, Freud's conception of human nature and socialization continues to permeate American culture.

Piaget and Cognitive Development

Another influential psychological theory of socialization is cognitive development theory. This theory has its roots in the work of Swiss psychologist Jean Piaget (1954). Piaget developed his theory through intensive observations of normal young

ences in mathematical ability are at least partially a result of hormonal differences. But just because hormonal differences are *associated* with mathematical abilities does not mean that the hormonal differences *caused* the differences in abilities. At any rate, the gender differences in mathematical abilities are small. Because the differences within each sex are so much larger than the differences between them, critics of the biological perspective argue that hormones can explain only a very small part of the overall variation in mathematical aptitude. This leaves a great deal of room for the influence of social factors. Evidence for this point of view comes from two lines of research.

First, sociologists Richard Felson and Lisa Trudeau (1991) have shown that, with the exception of the SAT, girls actually outperform boys on most tests of mathematical performance. Similarly, although boys are more likely to choose advanced mathematics courses in high school, girls earn higher grades in mathematics and are more likely to be placed in

advanced tracks for required mathematics courses. Based on these results, Felson and Trudeau argue that males do not have a general advantage in mathematical skills. Males only appear to be better at math than females because of gender bias in questions asked on the SAT and because males do indeed do better on one specific mathematical skill, the mental rotation of objects.

Second, studies clearly show that the male advantage in mathematical performance has been declining steadily (Hyde, Fennema, & Lamon 1990). One possible explanation for this pattern is that boys and girls are now being

socialized more similarly, thereby reducing the traditional male advantage in math. Biopsychologist Janice Juraska has demonstrated that female rats have fewer nerve connections than males into the hippocampus, a brain region associated with spatial relations and memory. But, when the cages of female rats were "enriched" with stimulating toys, the females developed more neural connections. As Juraska says, "Hormones do affect things—it's crazy to deny that—but there is no telling which way sex differences might go if we completely changed the environment" (as quoted in Gorman 1992).

 For more information, look up the following subjects in InfoTrac College Edition:
Girls and mathematics
Sex differences
Education of women

 Or visit the following Web site:
Educational Issues for Girls and Women in Mathematics
http://camel.math.ca/Women/EDU/Education.html

children. His goal was to identify the stages that children go through in the process of learning to think about the world.

Piaget's observations led him to conclude that there are four stages of cognitive development. The **sensorimotor stage** occurs from birth until about the age of 2. During this stage, children's understanding about the world is limited to what they can actually see, touch, feel, smell, or hear at a given moment. As a result, they do not understand cause and effect. So, for example, very young children love playing peek-a-boo because it is a delightful surprise each time the person playing with them removes his or her hands to reveal his presence.

During the **preoperational stage** (approximately ages 2 to 7), children begin to use language and symbols such as numbers or measures of size—but they do not yet really understand them. At this age, Piaget found that if a child poured a cup of water from a short, wide cup into a tall, thin beaker of the same volume, the child would nonetheless assume that the tall beaker held more water *simply because it was taller.*

In the **concrete operational stage** (between ages 7 and 12), children no longer make this error and have a basic understanding of cause and effect. After about age

In the **sensorimotor stage**, understanding about the world is limited to what a child can actually see, touch, feel, smell or hear. It lasts from birth to about age 2.

The **preoperational stage** lasts from about age 2 to 7 during which children begin to use language and other symbols, such as numbers.

In the **concrete operational stage**, usually occurring between the ages of 7 and 12, children develop a basic understanding of cause and effect.

During the **formal operational stage,** usually occurring about age 12, some humans develop the capacity for abstract thought.

The **interaction school of symbolic interaction** focuses on the active role of the individual in creating the self and self-concept.

The **structural school of symbolic interaction** focuses on the self as a product of social roles.

The **looking-glass self** is the process of learning to view ourselves as we think others view us.

12, some children enter the **formal operational stage,** in which they understand not only concrete concepts such as size and cost but also abstract concepts such as truth or justice. Not all children, however, ever reach this stage.

Critics of Piaget's work question whether all children go through these stages, as Piaget claimed. They suggest that in addition to individual differences among children, there may also be cultural and gender differences in cognitive development.

Symbolic Interaction Theory

Sociologists who use symbolic interaction theory share three general premises:

1. In order to understand human behaviors, we must learn what those behaviors mean to the individual actors.
2. Meanings develop within social relationships.
3. Individuals play active roles in constructing their self-concepts.

Over the years, two distinct schools have developed within this perspective: the interaction school and the structural school (Biddle 1986; Turner 1985). The **interaction school of symbolic interaction** focuses on the active role of the individual in creating the self and self-concept. The **structural school of symbolic interaction** focuses on the self as a product of social roles.

The Interaction School

The major premise of the interaction school is that people are actively involved in creating and negotiating their own roles and self-concept. Although each of us is born into an established social structure with established expectations for our behavior, nevertheless we have opportunities to create our own self-concept. The concepts of *looking-glass self* and *role taking* illustrate how this process works.

The Looking-glass Self   Charles Horton Cooley (1902) provided a classic description of how we develop our self-concept. He proposed that we learn to view ourselves as we think others view us. He called this the **looking-glass self.** According to Cooley, there are three steps in the formation of the looking-glass self:

1. We imagine how we appear to others.
2. We imagine how others judge our appearance.
3. We develop feelings about and responses to these judgments.

The Two Schools of Symbolic Interaction

	Interaction School	Structural School
The self-concept is . . .	Negotiated	Determined by roles
The individual is . . .	Active in creating self-concept; has more freedom to choose self	Less active in creating self-concept; has less freedom to choose self
The self-concept is developed through . . .	Role taking (taking the role of others)	Performing institutionalized roles
Roles are . . .	Negotiated	Allocated
Major concepts	Looking-glass self, role taking, and self-esteem	Role identity and identity salience hierarchy

Concept Summary

For example, an instructor whose students read newspapers or doze during class and who frequently finds himself talking to a half-empty room is likely to realize that his students think he is a bad teacher. He need not, however, accept this view of himself. The third stage in the formation of the looking-glass self suggests that the instructor may either accept the students' judgment and conclude that he is a bad teacher, or reject their judgment and conclude that the students are simply not smart enough to appreciate his profound remarks.

As this suggests, our self-concept is not merely a mechanical reflection of those around us; rather it rests on our interpretations of and reactions to those judgments. We are actively engaged in defining our self-concept, using past experiences as one aid in interpreting others' responses. A person who considers herself witty will assume that others are laughing *with* her; someone used to making clumsy errors, however, will assume that others are laughing *at* her.

We also actively define our self-concept by choosing among potential looking glasses; that is, we try to choose roles and associates supportive of our self-concept (Gecas & Schwalbe 1983). The looking glass is thus a way of both forming and maintaining self-concept.

As Cooley's theory indicates, symbolic interaction considers subjective interpretations to be extremely important determinants of the self-concept. It is not only others' judgments of us that matter; our subjective interpretation of those judgments is equally important. This premise of symbolic interactionism is apparent in W. I. Thomas's classic statement: "If men define situations as real, they are real in their consequences" (Thomas & Thomas 1928, 572). People interact through the medium of symbols (words and gestures) that must be subjectively interpreted. The interpretations have real consequences—even if they are *mis*interpretations.

Role Taking   The most influential contributor to symbolic interaction theory during the last century was George Herbert Mead (1934). Mead argued that we learn social norms through the process of **role taking.** This means imagining ourselves in the role of the other in order to determine the criteria others will use to judge our behavior. This information is used as a guide for our own behavior.

According to Mead, role taking begins in childhood, when we learn the rights and obligations associated with being a child in our particular family. To understand what is expected of us as children, we must also learn our mother's and father's roles. We must learn to see ourselves from our parents' perspective and to evaluate our behavior from their point of view. Only when we have learned their roles as well as our own will we really understand what our own obligations are.

Mead maintained that children develop their role knowledge by playing games. When children play house, they develop their ideas of how husbands, wives, and children relate to one another. When the little boy comes in saying "I've had a hard day; I hope it's not my turn to cook dinner," or when the little girl warns her dolls not to play in the street and to wash their hands before eating, they are testing their knowledge of family role expectations.

In the very early years, role playing and role taking are responsive to the expectations of **significant others**—role players with whom we have close personal relationships. Parents, siblings, and teachers, for example, are decisive in forming a child's self-concept. As children mature and participate beyond this close and familiar network, the process of role taking is expanded to a larger network that helps them understand what society in general expects of them. They learn what the bus driver, their neighbors, and their employers expect. Eventually, they come to judge their behavior not only from the perspective of significant others but also from what Mead

Role taking involves imagining ourselves in the role of the other in order to determine the criteria others will use to judge our behavior.

Significant others are the role players with whom we have close personal relationships.

The **generalized other** is the composite expectations of all the other role players with whom we interact; it is Mead's term for our awareness of social norms.

The **negotiated self** is the self that we fashion to present to others in order to improve our self-esteem.

Self-esteem is the evaluative component of the self-concept; it is our judgment about our worth compared with others' worth.

Role identity is the image we have of ourselves in a specific social role.

calls the **generalized other**—the composite expectations of all the other role players with whom they interact. Being aware of the expectations of the generalized other is equivalent to having learned the norms and values of the culture. One has learned how to act like an American or a Pole or a Nigerian.

Saying that everyone learns the norms and values of the culture does not mean that everyone will behave alike or that everyone will follow the same rules. Each of us has a different set of significant others. Not only will our family experiences differ depending on the culture and subculture in which we are reared, but also as we get older, we have some freedom in choosing those whose expectations will guide our behavior. Although we may know perfectly well what society in general expects of us, we may choose to march to the beat of a different drummer.

The Negotiated Self   Role taking and the looking-glass self are ways in which the individual can become an active agent in the construction of his or her own self-concept. The self that emerges is a **negotiated self,** a self that we have fashioned by selectively choosing looking glasses and significant others.

The idea of negotiation suggests that we have an end in view. What is that end? An important one is to protect and enhance our self-esteem. **Self-esteem** is the evaluative part of the self-concept; it is our judgment of our worth compared with others' worth. Because we would all like to think well of ourselves, we strive to negotiate a self-concept that reinforces that image.

The Structural School

The structural school of symbolic interaction differs from the interaction school by stressing the importance of institutionalized social structures. Unlike the interaction school, which assumes the individual has a great deal of freedom in negotiating a self-concept, the structural school assumes that individuals are constrained and shaped in important ways by society.

Scholars from the structural school focus on institutionalized social roles such as working, parenting, and going to school, and stress the profound ability of these roles to shape both our behavior and our personality. According to this perspective, differences in the social structures we participate in help to explain why two friends who go in very different directions after high school—say, one into the army and the other into a rock band—develop into such very different people despite their initial similarities. We grow into our roles, and the parts we play eventually become a part of our self-concept.

The key concept in structural theory is **role identity,** the image we have of ourselves in a specific social role (Burke 1980, 18). For example, a woman who is a professor, a mother, and an aerobics student will have a different role identity in each setting. According to structural theorists, her self-concept will be a composite of these multiple identities (Stryker 1981). The idea of role identities draws heavily on the analogy of life as a stage. As we move from scene to scene, we change costumes, get a new script, and come out as a different character. A young man may play the role of dutiful son at home, party animal at the dorm, and serious scholar in the classroom. None of these images is necessarily false; but because the roles are difficult to carry on simultaneously, the young man does not try to do so. Most of us engage in this practice without even thinking about it. We automatically adjust our vocabularies, topics of conversation, and expressed values and concerns as we talk to elderly relatives, our friends, and our coworkers. In this sense, we are like the elephant described by the six blind men; someone trying to find the "real us" might have a hard time reconciling the different views that we present.

Identity Salience Hierarchies   The concept of role identity implies that we play one role at a time and ignore the obligations associated with other roles. In practice, however, competing demands often force us to choose between different selves. If your boss tells you a racist joke, do you laugh courteously or do you tell him you are offended? Which self will take priority?

One mechanism for making such choices is the establishment of an **identity salience hierarchy,** a ranking of your various role identities in order of their importance to you (Callero 1985). Whenever two roles come into conflict, you simply follow the role that ranks highest on your list. Research shows that in ranking role identities, we give preference to roles that provide us with the most self-esteem. This reflects two things: the social status associated with a role and our skill in performing it.

Most women, for example, who are both professors and aerobics students will rank being a professor higher in their identity salience hierarchy. A woman who was a very bad professor and a very good aerobics student, however, might reverse the order. The relative ranking of mother and professor might be more problematic. Undoubtedly, some women would rank the professor role higher than the mother role; such women would follow the norms of their job in cases of role conflicts. Differences in identity salience hierarchies help explain why two people with the same sets of roles will behave differently. Some professor/mothers will go to the office on Saturday and some will stay at home; these choices in part reflect differences in their identity salience hierarchies (Serpe 1987).

> An **identity salience hierarchy** is a ranking of an individual's various role identities in order of their importance to him or her.

Socialization Through the Life Course

Socialization is a process that occurs throughout life, beginning in childhood and continuing throughout our adult lives, even into old age.

Childhood

Early childhood socialization is called **primary socialization.** It is primary in two senses. It occurs first, and it is most critical for later development. During this period, children develop personality and self-concept; acquire motor abilities, reasoning, and language skills; and are exposed to a social world consisting of roles, values, and norms.

During the period of primary socialization, children are expected to learn and embrace the norms and values of society. They learn that conforming to rules is an important key to gaining acceptance and love, first from their family and then from a larger network. Because young children are so dependent on the love and acceptance of their family, they are under especially strong pressure to conform to their family's expectations. This is a critical step in their becoming conforming members of society. If this learning does not take place in childhood, then conformity is unlikely to develop in later life.

> **Primary socialization** is personality development and role learning that occurs during early childhood.

Adolescence

Adolescence serves as a bridge between childhood and adulthood. As such, the central task of adolescence is to begin to establish one's independence from one's parents.

■ Anticipatory socialization prepares us for the roles we will take in the future. Children everywhere, whether they dress up in their parents' old clothes or enact their fantasies through dolls or trucks, play out their visions of how mommies and daddies ought to behave.

Anticipatory socialization is role learning that prepares us for roles we are likely to assume in the future.

Professional socialization is role learning that provides individuals with both the knowledge and a cultural understanding of their profession.

Role exits are the processes by which individuals leave important social roles.

During adolescence, we often engage in **anticipatory socialization**—role learning that prepares us for roles we are likely to assume in the future. Until about 1980, for instance, all American girls were required to take "home economics" courses to learn how to sew and cook. Boys, in turn, were required to take "shop" courses to learn how to fix cars or use wood-working tools. These courses are no longer regularly required and now many boys take cooking and many girls take wood-working. Nevertheless, teenagers' household chores, part-time jobs, and volunteer work still tend to divide along traditional lines. While some girls serve as soccer coaches and some boys baby-sit, girls more often are encouraged to provide child care, cooking, housecleaning, or emotional nurturance (such as working as a hospital volunteer) whereas boys more often are involved in outdoor and mechanical tasks. Even in their volunteer activities, then, girls and boys are preparing for the family and work roles they anticipate holding as adults.

Adulthood

Because of anticipatory socialization, most of us are more-or-less prepared for the responsibilities we will face as spouses, parents, and workers. Goals have been established, skills acquired, and attitudes developed that prepare us to accept and even embrace adult roles. Because anticipatory socialization is never complete, however, anyone who wants to enter a professional field must first undergo **professional socialization.** The purpose of professional socialization is to learn not only the knowledge but also the *culture* of a profession. Medical training provides an interesting example of this process.

Most commonly, people choose to become doctors out of a desire to help others. Yet one of the primary tenets of medical culture is that doctors should be emotionally detached—distancing themselves from their patients and avoiding any show of emotion. According to sociologist and medical school professor Frederic Hafferty (1991), this cultural norm is taught right from the beginning of medical education. Through his observations, Hafferty discovered that when new students come to a medical school for the first time, second-year students almost invariably take them to the school's anatomy laboratory. There the second-year students proudly display the most grotesque, partially dissected, human cadaver available. Although officially they do so to display the school's laboratory facilities, their true purpose seems to be to elicit emotional reactions from the new students. The laughter and snickers these reactions evoke in the second-year students demonstrate to the new students that such behavior is shameful while demonstrating to the second-year students how "tough" they have become. This is a particularly vivid example of professional socialization, but every job change we make as adults requires some socialization to new responsibilities and demands.

Age 65 and Beyond

More and more Americans now live for many years past age 65. During these years, most will give up their former careers, and many will lose spouses and experience declining physical abilities. As a result, **role exits,** which can exist at any life stage, are especially common in later life.

The concept of role exits was first defined and studied by Helen Rose Fuchs Ebaugh (1988), who served for several years as a nun before leaving the convent and becoming a sociologist. Through comparing her experiences and those of other ex-nuns to the experiences of ex-alcoholics, ex-spouses, and others, she identified a common process through which people leave important social roles. Individuals first go through a stage of questioning their original role. Subsequently, they begin to consider alternative roles and to imagine what their lives would be like in those roles. During this time, they begin disengaging themselves emotionally from their original role. Eventually, the pull of the imagined new role becomes stronger than the ties to the former role.

Individuals who leave important social roles never do so completely, however; an individual can cease being a nun or a convict, but there is no way to cease being an ex-nun or ex-convict. Many ex-nuns never become comfortable wearing revealing clothing, and many ex-alcoholics decide never to return to social drinking. In addition, people who know their past may continue to interact with them on the basis of their former statuses. For example, employers may refuse to hire ex-convicts and friends may refrain from using swear words around ex-nuns.

As these examples suggest, the process of role exit is followed by the process of socialization to new roles. Even during later life, many individuals who leave earlier roles develop new careers, engage in volunteer work, or return to school. Others must adapt to new roles within intimate relationships: learning to live alone or with new spouses or learning to allow their children or others to take care of them.

Resocialization

Most of the time, socialization and role change are gradual processes. Sometimes, though, changes are abrupt and extreme. The most extreme example of role change comes about when we abandon our self-concept and way of life for a radically different one. This is called **resocialization.** Changing the social behavior, values, and self-concept acquired over a lifetime of experience is difficult, and few people undertake the change voluntarily. In this section, we look at the process of resocialization and at certain locations, known as *total institutions*, where resocialization often occurs.

Resocialization occurs when we abandon our self-concept and way of life for a radically different one.

A drastic example of resocialization occurs when people become permanently disabled. Those who become paralyzed experience intense resocialization to adjust to their handicap. All of a sudden, their social roles and capacities are changed. Their old self-concept no longer covers the situation. They may be unable to control bladder and bowels, be severely limited in their ability to get around, or be unable to function sexually in the ways they had previously. If they are single, they must face the fact that they may never marry or have children; if they are older, they may have to reevaluate their adequacy as spouses or parents. These changes require a radical redefinition of self. If self-esteem is to remain high, priorities will have to be rearranged and new, less physically active roles given prominence.

Resocialization may also be deliberately imposed by society. When an individual's behavior leads to social problems—as is the case with habitual criminals, problem alcoholics, and mentally disturbed individuals—society may decree that the individual must abandon the old identity and accept a more conventional one.

Total Institutions

Generally speaking, a radical change in self-concept requires a radical change in environment. Drug counseling one night a week is not likely to drastically alter the self-concept of a teenager who spends the rest of the week among kids who are

Connections

Historical Note

Goffman developed the idea of *total institutions* after working as an orderly in a public mental hospital in the 1950s. In that hospital, as in most others during that time, all clothing and other belongings were taken from patients at admission. Patients were given shapeless hospital gowns to wear, their hair was cut short, and they were forced to live regimented lives in which virtually all everyday decisions were taken away from them. Contact with the outside world was severely limited. In these ways, patients' individuality was diminished and their susceptibility to resocialization increased.

Total institutions are facilities in which all aspects of life are strictly controlled for the purpose of radical resocialization.

constantly "wasted." Thus, the first step in the resocialization process is to isolate the individual from the past environment.

This is most efficiently done in **total institutions**—facilities in which all aspects of life are strictly controlled for the purpose of radical resocialization (Goffman 1961a). Monasteries, prisons, boot camps, and mental hospitals are good examples. Within them, past statuses are wiped away. Social roles and relationships that formed the basis of the previous self-concept are systematically eliminated. New statuses are symbolized by regulation clothing, rigidly scheduled activity, and new relationships. Inmates are encouraged to engage in self-analysis and self-criticism, a process intended to reveal the inferiority of past perspectives, attachments, and statuses.

A Case Study: Prison Boot Camps and Resocialization

Polls repeatedly show that most people and politicians in the United States want to get tough with criminals. For some, getting tough on crime means meting out harsher punishments—locking up more people for longer periods of time. But at least one study shows that almost three times as many people believe that prevention and rehabilitation are more important in controlling crime than are longer periods of incarceration (Newcomb 1994). Advocates have proposed as much as $1 billion in new financing for prison boot camps or "shock incarceration programs" for young offenders. Perhaps their popularity arises out of the fact that among all of the proposed crime control measures, they alone simultaneously satisfy the public's urge to punish criminals with its desire to reform them.

Like all resocialization programs, prison boot camps begin with the premise that inmates must radically alter their lifestyles, values, and beliefs. To become productive members of society, inmates must come to reject their old identity and often their old relationships. How are such goals accomplished with groups of young lawbreakers, who often have a long history of prior delinquent acts?

First, new prisoners are segregated or isolated from competing social environments. Shaved heads and being called names rather than by name remind prisoners that they are leaving one self-concept behind and taking on another. Second, all aspects of daily life are strictly controlled; activities and interaction with other inmates are closely regulated. Days typically begin with an hour of strenuous calisthenics in the early morning darkness and proceed through long hours of military drilling, hard physical labor, drug counseling, and study. Meals last for eight minutes and absolutely no talking is allowed. Prisoners never walk; they run. They are taught to stand at attention before approaching a prison official and never to look him or her in the eye (Anderson 1998).

To many, boot camps sound like a great idea. They promise to teach young offenders self-discipline, respect for the law, and the value of work by placing them in a rigorous military-style setting while simultaneously satisfying the public's apparent need for "seeing some civility pounded into thugs who terrorize their neighborhoods" (Katel, Liu, & Cohn 1994).

There is some evidence that boot camps work. Offenders leave prison boot camps with stronger positive attitudes about their future than they had when they entered; in contrast, those sent to regular prison typically develop even more negative attitudes during their incarceration (MacKenzie, Shaw, & Gowdy 1993). The long-term success of boot camps is, however, far more questionable. Camp graduates are just as likely as prison parolees to end up back in trouble with the law. In a study of inmates released from Louisiana prison camps, 37 percent were arrested at least once during their first year of freedom, compared with only 25.7 percent of parolees (Katel et al. 1994).

Uniforms, shaved heads, harsh discipline, and rigorous physical demands are all designed to encourage young prison boot-camp inmates to cast off their old deviant identity and adopt a new conformist self-concept.

Given the disparity between the norms espoused in the prison camps and in the peer groups to which most inmates return, it is little wonder that so many return to crime. Both proponents of boot camps and their critics agree that without educational programs, drug and vocational counseling, and a well-designed program of postincarceration supervision, the changes in self-concept and values that prison boot camps can generate are insufficient to enable offenders to overcome the difficulties they face when they return home (MacKenzie et al. 1995).

Agents of Socialization

Socialization is a continual process of learning. Each time we encounter new experiences, we are challenged to make new interpretations of who we are and where we fit into society. This challenge is most evident when we make major role transitions—when we leave home for the first time, join the military, change careers, or get divorced, for example. Each of these shifts requires us to expand our skills, adjust our attitudes, and accommodate ourselves to new social roles. Child psychologists have noted that these periods of transition tend to herald both intellectual and moral growth in the youngster. They constitute a crisis that challenges our old assumptions about ourselves and prompts a fundamental reappraisal of who we are.

Learning takes place in many contexts. We learn at home, in school and church, on the job, from our friends, and from television. These agents of socialization have a profound effect on the development of personality, self-concept, and the social roles we assume. Each of these agents of socialization is discussed more fully in later chapters. They are introduced here to illustrate the importance of social structures for learning.

Family

The most important agent of socialization is the family. As the tragic cases of child neglect and the monkey experiments so clearly demonstrate, the initial warmth and nurturance we receive at home are essential to normal cognitive, emotional, and physical development. In addition, our family members—usually our parents but sometimes our grandparents, stepparents, or others—are our first teachers. From them we learn to tie our shoes and hold a pencil and from them we also learn the goals and aspirations that will stay with us for the rest of our lives.

The activities required to meet the physical needs of a newborn provide the initial basis for social interaction. Feeding and diaper changing give opportunities for cuddling, smiling, and talking. These nurturant activities are all vital to the infant's social and physical development; without them, the child's social, emotional, and physical growth will be stunted (Gardner 1972; Lynch 1979; Provence & Lipton 1962).

In addition to these basic developmental tasks, the child has a staggering amount of learning to do before becoming a full member of society. Much of this early learning occurs in the family as a result of daily interactions: The child learns to talk and communicate, to play house, and to get along with others. As the child becomes older, teaching is more direct, and parents attempt to produce conformity and obedience, impart basic skills, and prepare the child for events outside the family.

One reason the family is the most important agent of socialization is that the self-concept formed during childhood has lasting consequences. In later stages of development, we pursue experiences and activities that integrate and build on the foundations established in the primary years. Although the personality and self-concept are not rigidly fixed in childhood, we are strongly conditioned by childhood experiences (Mortimer & Simmons 1978).

The family is also an important agent of socialization in that the parents' religion, social class, and ethnicity influence the child's social roles and self-concept. They influence the expectations that others have for the child, and they determine the groups with which the child will interact outside the family. Thus, the family's race, class, and religion shape the child's initial experiences in the neighborhood, at school, and at work.

Schools

In Western societies, schooling has become accepted as a natural part of childhood. The central function of schools in industrialized societies is to impart specific skills and abilities necessary for functioning in a highly technological society.

Schools do much more than teach basic skills and technical knowledge, however; they also transmit society's central cultural values and ideologies. Unlike the family, in which children are treated as special persons with unique needs and problems, schools expose children to situations in which the same rules, regulations, and authority patterns apply to everyone. In schools, children first learn that levels of achievement affect status in groups (Parsons 1964, 133). In this sense, schools are training grounds for roles in the workplace, the military, and other bureaucracies in which relationships are based on uniform criteria.

Peers

In past centuries, and in some parts of the world still today, children often lived on isolated farms where their families remained almost the only important agent of socialization throughout their childhood. For the last several decades, however, compulsory education together with the late age at which most youths become full-time workers have led to the emergence of a youth subculture in modern societies. In recent years, this development has been accelerated by the tendency for both parents to work outside the home, creating a vacuum that may be filled by peer interaction.

What are the consequences of peer interaction for socialization and the development of the self-concept? Because kids who hang out together tend to dress and act a lot alike, many observers have concluded that peer pressure creates conformity; they have also assumed that conformity to peer values and lifestyles is a frequent source of family conflict, as parents and friends compete for influence. It is true that much of the youth subculture, by definition and intent, operates in opposition to adult culture; it's no fun getting your tongue pierced, for instance, if your mom decides to do the same. Nevertheless, one study suggests that the effects of peer pressure have been at least somewhat overestimated. First, it appears that peer similarity precedes group membership; in other words, kids hang around together because they share attitudes, rather than vice versa. Second, adolescents are generally more concerned about their parents' opinions than their friends'. Even if they engage in behavior with their friends that their parents would disapprove of, they usually do so only if they think their parents won't find out (Dornbusch 1989).

Many teenagers change their appearance to demonstrate their membership in a specific peer group.

© Joel Gordon

There are, however, areas of a child's development on which peer group socialization has an important influence (Gecas 1981). Once children are exposed to others their own age, they quickly come to identify with their peers and to place a high value on their acceptance (Harris 1998). Because the judgments of one's peers are unclouded by love or duty, they are particularly important in helping us get an accurate picture of how we appear to others. In addition, the peer group is often a mechanism for learning social roles and values that adults don't want to teach. For example, much of what children and teenagers know, or think they know, about sex and drinking is learned in the peer group.

Mass Media

Throughout our lives we are bombarded with impersonal messages from radios, magazines, films, and billboards. The most important mass medium for socialization, however, is undoubtedly television. Nearly every home has one, and the average person in the United States spends many hours a week watching it.

The effect of television viewing on learning is vigorously debated. Many have questioned whether the media promote violence, sexism, racism, and other problematic ideas and behaviors. The evidence is somewhat contradictory (Felson 1996). The most universally accepted conclusion is that the mass media can be an important means of supporting and validating what we already know. Through a process of selective perception, we tend to give special notice to material that supports our beliefs and self-concept and to ignore material that challenges us.

Television, however, may play a more active part than this. Studies suggest that characters seen regularly on television can become role models, whose imagined opinions become important as we develop our own roles. For example, adolescents might look to the program *Friends* for ideas about how to deal with the opposite sex. Material on television may supplement the knowledge our own experience gives us about U.S. roles and norms. These findings imply that the content of television can have an important influence.

Religion

In every society, religion is an important source of individual direction. The values and moral principles in religious doctrine give guidance about appropriate roles and behaviors. Often the values we learn through religion are compatible with the ideals we learn through other agents of socialization. For example, the golden rule ("Do unto others as you would have them do unto you") that many U.S. children learn in Sunday school is likely to fit in easily with similar messages they have heard at home and at school.

The role of religion, however, cannot be reduced to a mere reinforcer of society's norms and values. As we point out in Chapter 13, religious ideals have the power to change societies and the individuals in them. Moreover, even within modern U.S. society, there are important differences in the messages delivered by, say, the Mormon, Jewish, and Baptist religions. These differences account for some significant variability in socialization experiences.

Workplace

Almost all of us will spend a significant portion of our adult lives working outside the home for wages or salaries. The environments in which we work, however, are very different. Some of us will work with machines, others with ideas; some will work with people, others on people. Work is found in cities, factories, offices, and fields.

Much of it is impersonal, monotonous, and regulated by time clocks; but some is highly personal, challenging, and flexible.

Long-term research by Kohn and his associates indicates that the nature of our work affects our self-concept and behavior. The amount of autonomy, the degree of supervision and routinization, and the amount of cognitive complexity demanded by the job have important consequences. If your work demands flexibility and self-discipline, you will probably come to value these traits elsewhere—at home, in government, and in religion. If your work instead requires subordination, discipline, and routine, you will come to find these traits natural and desirable (Kohn et al. 1983).

Because of the importance of our work-role identity, studies demonstrate that losing one's job can be a major blow to the self-concept. Although some people who lose their jobs can protect their self-esteem by blaming the economy or the government, individuals more often assume they are to blame (Newman 1999a). As a result, unemployment increases the incidence of depression and physical illness (Shortt 1996) and reduces the sense of personal competence (Gecas 1989).

Where This Leaves Us

Each of us is unique, a product of our individual biology, abilities, personality, experiences, and choices. But each of us is also a social creation. Through socialization we come to learn the roles and values expected of us, and, more often than not, to take on those roles and values as our own. Sometimes that process is obvious: a parent slapping a child's hand for grabbing a cookie without permission, a minister preaching a sermon on the wages of sin, one girl giving another girl pointers on how to flirt. Other times, we are no more aware of the socialization process than a fish is aware of water; it is simply a part of the life around us. The typical American, for example, now spends several hours each day watching television. During those hours we not only learn who murdered this week's victim on *Law & Order*, who is this season's *American Idol*, and who married Joe Millionaire, but we also learn ways of looking at the world: to fear random violence and trust the police; to value success, talent, and fame; to honor wealth; and so on.

Understanding the effects of socialization does not mean accepting that we are victims of that socialization and have no choices in our lives. But unless we understand the ways in which we have been socialized, we will be unable to see our choices clearly and to turn those choices into realities.

Summary

1. The self is a combination of unique qualities and shared norms. The two key components of the self are the I and the me.

2. Although biological capacities enter into human development, our identities are socially bestowed and socially sustained. Without human relationships, even our natural capacities would not develop.

3. Socialization, the process of learning the norms and values necessary for participation in social institutions, occurs throughout life at four basic levels: primary socialization, anticipatory socialization, professional socialization, and resocialization.

4. Freudian theory links social development to biological cues. Freud believed that to become a healthy adult, children must develop a reasonable balance between id and superego.

5. Piaget theorized that cognition develops through a series of stages. Only in the last stage do children develop the capacity to understand and think abstractly, and some children may never reach that stage.

6. Symbolic interaction is the dominant theoretical framework in sociological studies of human development. It emphasizes that learning takes place through subjectively interpreted interaction and that the development of the self and self-concept depend on the quality of our relationships with others.

7. The interaction school of symbolic interaction attributes to the individual a very active role in constructing the self. The idea of the negotiated self suggests that we use selective interpretations to construct and maintain our self-concept with a special eye to enhancing our self-esteem.

8. The structural school of symbolic interaction stresses the profound impact of conventional roles and statuses on the development of the self-concept. The concept of role identity suggests that our self and self-concept depend in important ways on the situations or roles in which we find ourselves.

9. The family is the major agent of socialization. It is responsible for the early nurturance that helps the infant develop into a human being, and it is central in laying the foundation of the self-concept. In addition, the family provides a background (social class, religion, and so on) that determines much of the child's other interactions. Other important agents of socialization include peers, schools, mass media, the workplace, and religion.

Thinking Critically

1. What role identity is most important to you now? Do you think that might change over the course of your life? Why or why not?

2. Think about the student role in your own life. To what extent did you *take* it? To what extent did you *make* it?

3. List some specific ways you think that a family's social class might influence the way a child is socialized. Can you think of any ways that living in the city versus living in the country might matter?

4. Thinking back to your childhood, what values might you have learned from the four or five television shows that you watched the most? Did you learn from them? Why or why not?

Sociology on the Net

 ### The Wadsworth Sociology Resource Center: Virtual Society

http://sociology.wadsworth.com/

The companion Web site for this book includes a range of enrichment material. Further your understanding of the chapter by accessing Online Practice Quizzes, Internet Exercises, InfoTrac College Edition Exercises, and many more compelling learning tools.

Chapter-Related Suggested Web Sites

Sigmund Freud
http://www.utm.edu/research/iep/f/freud.htm

Piaget Society
http://www.piaget.org

Media Scope
http://www.mediascope.org

 ## InfoTrac College Edition

http://www.infotrac-college.com/wadsworth

Access the latest news and research articles online—updated daily and spanning four years. InfoTrac College Edition is an easy-to-use online database of reliable, full-length articles from hundreds of top academic journals and popular sources. Conduct an electronic search using the following key search terms:

Socialization

Nature and nurture

Social isolation

Self concept

Peer pressure

Media influence

 For more information related to this chapter, visit the Opposing Viewpoints Resource Center. Be sure you check all the options on the toolbar—viewpoints, references, statistics, and so on—and search under the following subject:

Media violence

Suggested Readings

Benedict, Ruth. 1961. *Patterns of Culture*. Boston: Houghton Mifflin. Originally published in 1934. This classic book draws on several different cultures to illustrate how behavior and personality are consistent with the culture in which a person is reared. The emphasis is on the continuity of socialization.

Croteau, David, and Hoynes, William. 2000. *Media/Society: Industries, Images, and Audiences*. Thousand Oaks, Calif: Pine Forge Press. Includes excellent sections on how the media represent different social problems and social issues and how audiences interpret and respond to media portrayals.

Goffman, Erving. 1961. *Asylums*. Garden City, N.Y.: Anchor/Doubleday. A penetrating account of total institutions and the significance of social structure in producing conforming behavior. It is primarily an analysis of mental hospitals and mental patients, although the analysis is applicable to other total institutions.

Thorne, Barrie. 1993. *Gender Play: Girls and Boys at School*. New Brunswick, N.J.: Rutgers University Press. Detailed observations of young girls and boys in classrooms and schoolyards identify how not only teachers but also children themselves teach and create ideas about what it means to be male or female.

Social Structure and Social Interaction

© Robert Caputo/Stock Boston

Intertwining Forces: Social Structure and Social Interaction

Most people who become sociologists do so because they are interested in studying particular social problems, such as homelessness, mental illness, or racial inequality. Each of these problems has roots and consequences in both broad social structures and everyday social interactions. For example, racial inequality in the United States in part stems from the nature of our national economy and political institutions: There simply aren't enough good-paying jobs near nonwhite communities, and these communities rarely have enough political power to entice corporations to bring in good jobs. But racial inequality is also reinforced on a day-to-day basis whenever teachers spend less time with nonwhite than with white students or police officers assume that nonwhites are more likely than whites to be criminals. As this example suggests, to fully understand society and social problems, sociologists must look at both social structure and social interaction. This chapter describes these two basic features of society. As we will see, research on social structures typically draws on structural-functionalist or conflict theories, whereas research on social interaction typically draws on symbolic interaction theory.

Social Structures

Many of our daily encounters occur in patterns. Every day we interact with the same people (our family or coworkers) or with the same kinds of people (salesclerks or teachers). These patterned relationships, these common dramas of everyday life, are called social structures. Each of these dramas has a set of actors (mother/child or buyer/seller) and a set of norms that defines appropriate behavior for each actor.

Formally, a **social structure** is a recurrent pattern of relationships. Social structures can be found at all levels in society. Baseball games, friendship networks, families, and large corporations all have patterns of relationships that are repeated day after day. Some of these patterns are reinforced by formal rules or laws, but many more are maintained by force of custom.

The patterns in our lives are both constraining and enabling (Giddens 1984). If you would like to be free to set your own schedule, you will find the nine-to-five, Monday-to-Friday work pattern a constraint. On the other hand, preset patterns provide convenient and comfortable ways of handling many aspects of life. They help us to get through crosstown traffic, find dating partners and spouses, and raise our children.

Whether we are talking about a Saturday afternoon ball game, families, or the workplace, social structures can be analyzed in terms of three concepts: status, role, and institution.

Status

The basic building block of society is the **status**—a specialized position in a group. Sociologists who want to study the status structure of a society include two types of statuses: achieved and ascribed. An **achieved status** is a position that a person can

A **social structure** is a recurrent pattern of relationships.

A **status** is a specialized position within a group.

An **achieved status** is optional, one that a person can obtain in a lifetime.

An **ascribed status** is fixed by birth and inheritance and is unalterable in a person's lifetime.

attain in a lifetime. Being a father or president of the PTA is an achieved status. An **ascribed status** is a position generally assumed to be fixed by birth or inheritance and unalterable in a person's lifetime. For example, although some people have "gender reassignment" surgery and some people "pass" as members of a different race, we assume that sex and race are unchangeable. Here, sex and race are considered to be ascribed statuses.

Sociologists who analyze the status structure of a society are typically concerned with four related issues (Blau 1987; Blau & Schwartz 1984): (1) identifying the number and types of statuses that are available in a society; (2) assessing the distribution of people among these statuses; (3) determining how the consequences—the rewards, resources, and opportunities—are different for people who occupy one status rather than another; and (4) ascertaining what combinations of statuses are likely or even possible.

A Case Study: Race as a Status

To illustrate how our lives are structured by status membership, we apply this approach to one particular ascribed status and ask how being African American affects relationships and experiences in the United States.

To begin: How many racial statuses are there in the United States? The 2000 census asked us to identify ourselves as belonging to one of six racial categories: white, African American, Native American, Asian, mixed race, or other. Although the labels may change, the same question appears on many of the other forms you will fill out as well as on almost every social survey. The apparent consensus on which racial statuses are significant and the nearly universal concern about them should alert us to the possible consequences that racial status has in our daily lives.

It is not just the number of statuses that has consequences. The numerical distribution of the population among racial statuses also encourages or discourages certain patterns of behavior. In 2000, for example, the U.S. census identified 2.1 million African American residents in New York City, but only 3 African American residents of Worland, Wyoming. In New York City, there are 1.6 whites for every African American; in Worland, the ratio is 1,566 to 1. This means that half of all whites in New York could marry or be best friends with an African American; in Worland, it is impossible for more than a few whites to be this closely linked to an African American.

Of course, numbers alone do not tell the whole story. By nearly every measure that one might choose, there is substantial inequality in the rewards, resources, and opportunities available to African American and white people in the United States. African American unemployment is twice that of whites; the infant mortality rate is more than twice as high; the likelihood of being murdered is six times higher (U.S. Bureau of the Census 2002). Obviously, racial status has enormous consequences on the structures of daily experiences.

Although racial inequality persists, racial status does not correspond as directly with occupational and educational statuses as it once did, and different combinations of statuses are possible. Forty years ago, being African American meant having much less education and very different kinds of jobs than whites. Today, knowing a person's ascribed status (race) is not such an accurate guide to his or her achieved statuses (education or occupation). Nevertheless, 34 percent of all nurse's aides and orderlies but only 6 percent of all physicians in the United States are African American (U.S. Department of Labor 2002). The processes through which these overlapping racial, political, and economic statuses are maintained are discussed further in Chapter 9.

Roles

The status structure of a society provides the broad outlines for interaction. These broad outlines are filled in by **roles**, sets of norms that specify the rights and obligations of each status. To use a theatrical metaphor, the status structure is equivalent to the cast of characters, whereas roles are equivalent to the scripts that define how the characters ought to act, feel, and relate to one another. This language of the theater helps to make a vital point about the relationship between status and role: People occupy statuses, but they play roles. This distinction is helpful when we analyze how structures work in practice—and why they sometimes don't work. A man may occupy the status of father, but he may play the role associated with it very poorly.

Sometimes people fail to fulfill role requirements despite their best intentions. This failure is particularly likely when people are faced with incompatible demands owing to multiple or complex roles. Sociologists distinguish between two types of incompatible role demands: When incompatible role demands develop within a single status, we refer to **role strain**; when they develop because of multiple statuses, we refer to **role conflict.** Some of the clearest examples of role strain occur within the family. Many parents, for instance, face role strains when their role of providing for their children's physical needs interferes with their role of being there when their children need them. When an employer's expectations that a good employee will not miss work and will put in long hours of overtime clashes with a parent's need to take time off to care for a sick child, role conflict occurs. When people simultaneously lack the resources and opportunities they need to meet basic role requirements—it is hard to be a good provider, for instance, when there are simply no jobs available—they may come to see role expectations as irrelevant or impossible.

As this suggests, social roles are always changing and flexible. We do not simply play the parts we are assigned with machinelike conformity. Instead, each individual plays a given role differently, depending on their other social statuses and roles, their resources, and the social rewards or punishments they will face depending on how they play their role.

Institutions

Social structures vary in scope and importance. Some, such as those that pattern the Friday night poker game, have limited application. The players could change the game to Saturday night or up the ante, and it would not have a major effect on the lives of anyone other than members of the group. If a major corporation changed seniority or family leave policies, it would have somewhat broader consequences, affecting not only employees of that firm but also setting a precedent for the way other complex organizations might come to do business. Changes in other structures, however, have the power to shape the basic fabric of our lives. We call these structures social institutions.

An **institution** is an enduring and complex social structure that meets basic human needs. Its primary features are that it endures for generations; includes a complex set of values, norms, statuses, and roles; and addresses a basic human need. Embedded in the statuses and roles of the family institution, for example, are enduring patterns for dating and courtship, child rearing, and care of the elderly. Because the institution of family is composed of many separate family groups, however, the exact rules and behaviors surrounding dating or elder care will vary.

Nevertheless, institutions provide routine patterns for dealing with predictable problems of social life. Because these problems tend to be similar across societies, we find that every society tends to have the same types of institutions.

Roles are sets of norms specifying the rights and obligations associated with status.

Role strain is when incompatible role demands develop within a single status.

Role conflict is when incompatible role demands develop because of multiple statuses.

An **institution** is an enduring social structure that meets basic human needs.

Connections

Personal Application

Many college students experience role conflict due to the multiple roles they play. Your teachers expect you to read assigned books and your boss expects you to show up on time and work cheerily and efficiently. If you are on a team, your coach expects you to get enough sleep and come to practices prepared to work hard. Your girlfriend or boyfriend expects you to go out with him or her at least once a week. If you sometimes feel there aren't enough hours in the day, you are probably experiencing role conflict.

Religion is one of the five basic institutions. Although doctrines and rituals vary enormously, all cultures and societies include a structured pattern of behavior and belief that provides individuals with explanations for events and experiences that are beyond their own personal control.

Basic Institutions

There are five basic social institutions:

- The family, to care for dependents and rear children.
- The economy, to produce and distribute goods.
- Government, to provide community coordination and defense.
- Education, to train new generations.
- Religion, to supply answers about the unknown or unknowable.

These institutions are basic in the sense that every society provides some set of enduring social arrangements designed to meet these important social needs. These arrangements may vary from one society to the next, sometimes dramatically. Government institutions may be monarchies or democracies or dictatorships. But a stable social structure that is responsible for meeting these needs is common to all societies.

In simple societies, all of these important social needs—political, economic, education, and religious—are met through one major social institution, the family or kinship group (Adams 1971). Social relationships based on kinship obligations serve as a basis for organizing production, reproduction, education, and defense.

As societies grow larger and more complex, the kinship structure is less able to furnish solutions to all the recurrent problems. As a result, some activities are gradually transferred to more specialized social structures outside the family. The economy, education, religion, and government become fully developed institutionalized structures that exist separately from the family. (The institutions of the contemporary United States are the subjects of Chapters 12 to 14.)

As the social and physical environments of a society change and the technology for dealing with that environment expands, the problems that individuals have to face change. Thus, institutional structures are not static; new structures emerge to cope with new problems.

Among the more recent social structures to be institutionalized in Western society are medicine, science, sport, the military, law, and the criminal justice system. Each of these areas can be viewed as an enduring social structure, complete with interrelated statuses and a unique set of norms and values.

Institutional Interdependence

Each institution of society can be analyzed as an independent social structure, but none really stands alone. Instead, institutions are interdependent; each affects the others and is affected by them.

In a stable society, the norms and values embodied in the roles of one institution will usually be compatible with those in other institutions. For example, a society that stresses male dominance and rule by seniority in the family will also stress the same norms in its religious, economic, and political systems. In this case, interdependence reinforces norms and values and adds to social stability.

Sometimes, however, interdependence is an important mechanism for social change. Because each institution affects and is affected by the others, a change in one tends to lead to change in the others. Changes in the economy lead to changes in the family; changes in religion lead to changes in government. For example, when years of schooling become more important than hereditary position in determining occupation, hereditary position will also be endangered in government, the family, and religion.

Institutions as Agents of Stability or Oppression

Sociologists use two major theoretical frameworks to approach the study of social structures: structural functionalism and conflict theory. The first focuses on the part that institutions play in creating social and personal stability; the second focuses on the role of institutions in legitimizing inequality (Eisenstadt 1985). Because each framework places a different value judgment on stability and order, each prompts us to ask different questions about social structures. Structural functionalism prompts us to ask how an institution contributes to order, to stability, and to meeting the needs of society and the individual. Conflict theory prompts us to ask which groups are benefitting the most from an institution and how those groups seek to maintain their advantage.

A Structural-Functional View of Institutions   Institutions provide ready-made patterns to meet most recurrent social problems. These ready-made patterns regulate human behavior and are the basis for social order. Because we share the same patterns, social life tends to be stable and predictable.

Structural-functional theorists point out that stability and predictability are important both for societies and individuals (Berger & Luckmann 1966). These theorists argue that institutions create a "liberating dependence." By furnishing patterned solutions to our most pressing everyday problems, they free us for more creative efforts. They save us from having to reinvent the social equivalent of the wheel each generation and thus they facilitate our daily lives. Moreover, because these patterns have been sanctified by tradition, we tend to experience them as morally right. As a result, we find satisfaction and security in social institutions.

A Conflict View of Institutions   By the very fact that they regulate human behavior and direct choices, institutions also constrain behavior. By producing predictability, they reduce innovation; by providing security, they reduce freedom. Some regard this as a necessary evil—the price we pay for stability.

This benign view of constraint is challenged by conflict theorists. They accept that institutions meet basic human needs, but they wonder, Why this social pattern rather than another? Why this family system instead of another? To explain why one social arrangement is chosen over another, conflict theorists ask, Who benefits?

From the viewpoint of conflict theory, institutions represent a camouflaged form of inequality—one that is supported by norms and tradition. Because institutions have existed for a long time, we tend to think of our familial, religious, and political systems as not merely one way of fulfilling a particular need but as the only acceptable way. Just as a tenth-century Mayan may have thought, "Of course, virgins should be sacrificed if the crops are bad," so we tend to think, "Of course, women earn less than men." In both cases, the cloak of tradition obscures our vision of inequality or oppression, making inequality seem normal and even desirable. As a result, conflict theorists argue that institutions stifle social change and help maintain inequality.

Types of Societies

Institutions give a society a distinctive character. In some societies, the church is the dominant institution; in others, it is the family or the economy. Whatever the circumstance, recognizing the institutional framework of a society is critical to an understanding of how it works.

The history of human societies is the story of ever-growing institutional complexity. In simple societies, we often find only one major social institution—the family or kinship group. Many modern societies, however, have as many as a dozen institutions. What causes this expansion of institutions? The triggering event appears to be economic change. When changes in technology, physical environment, or social arrangements increase the level of economic surplus, the possibilities of institutional expansion arise (Lenski 1966). In this section we sketch a broad outline of the institutional evolution that accompanied three revolutions in production.

Hunting, Fishing, and Gathering Societies

The chief characteristic of hunting, fishing, and gathering societies is that they have subsistence economies. This means that in a good year they produce barely enough to get by; that is, they produce no surplus. In some years, of course, game and fruit are plentiful, but in many years scarcity is a constant companion.

The basic units of social organization are the household and the local clan, both of which are based primarily on family bonds and kinship ties. Most of the activities of hunting and gathering are organized around these units. A band rarely exceeds 50 people in size and tends to be nomadic or seminomadic. Because of their frequent wanderings, members of these societies accumulate few personal possessions.

The division of labor is simple—based on age and sex. The common pattern is for older boys and men (other than the elderly) to participate in hunting and deep-sea fishing and for older girls and women to participate in gathering, shore fishing, and preserving. Aside from inequalities of status by age and sex, few structured inequalities exist in subsistence economies. Members possess little wealth; they have few, if any, hereditary privileges; and the societies are almost always too small to develop class distinctions. In fact, a major characteristic of subsistence societies is that individuals are homogeneous, or alike. Apart from differences occasioned by age and sex, members generally have the same everyday experiences.

Horticultural Societies

The first major breakthrough from subsistence economy to economic surplus was the development of agriculture. When people began to plant and cultivate crops, rather than just harvesting whatever nature provided, stable horticultural societies developed. The technology was often primitive—a digging stick, occasionally a rudimentary hoe—but it produced a surplus.

The regular production of more than the bare necessities revolutionized society. It meant that some people could take time off from basic production and turn to other pursuits: art, religion, writing, and frequently warfare. Of course, the people who participate in these alternate activities are not picked at random; instead, a class hierarchy develops between the peasants, who must devote themselves full time to food production, and those who live off the surplus produced by the peasants.

Because of relative abundance and a settled way of life, horticultural societies tend to develop complex and stable institutions outside the family. Some economic activity is carried on outside the family, a religious structure with full-time priests may develop, and a stable system of government—complete with bureaucrats, tax collectors, and a hereditary ruler—often develops. Such societies are sometimes very large. The Inca Empire, for example, included an estimated population of more than 4 million.

Agricultural Societies

Approximately 5,000 to 6,000 years ago, a second agricultural revolution occurred, and the efficiency of food production was doubled and redoubled through better technology. The advances included the harnessing of animals, the development, in time, of metal tools, the use of the wheel, and improved knowledge of irrigation and fertilization. These changes dramatically altered social institutions.

The major advances in technology meant that even more people could be freed from direct production. The people not tied directly to the land congregated in large urban centers and developed a complex division of labor. Technology, trade, reading and writing, science, and art grew rapidly as larger and larger numbers of people were able to devote full time to these pursuits. Along with greater specialization and occupational diversity came greater inequality. In the place of the rather simple class structure of horticultural societies, a complex class system developed, with merchants, soldiers, scholars, officials, and kings—and, of course, the poor peasants on whose labor they all ultimately depended. This last group still contained the vast bulk of the society, probably at least 90 percent of the population (Sjoberg 1960).

One of the common uses to which societies put their new leisure time and other new technology was warfare. With the domestication of the horse (cavalry) and the invention of the wheel (chariot warfare), military technology became more advanced and efficient. Military might was used as a means to gain even greater surplus through conquering other peoples. The Romans were so successful at this that they managed to turn the peoples of the entire Mediterranean basin into a peasant class that supported a ruling elite in Italy.

Industrial Societies

The third major revolution in production was the advent of industrialization 200 years ago in Western Europe. The substitution of mechanical, electrical, and fossil-fuel energy for human and animal labor caused an explosive growth in productivity, not only of goods but also of knowledge and technology. In the space of a few decades, agricultural societies were transformed. The enormous increases in energy, technology, and knowledge freed the bulk of the workforce from agricultural production and increasingly also from industrial production. The overall effect on society has been to transform its political, social, and economic character. Old institutions such as education have expanded dramatically, and new institutions such as science, medicine, and law have emerged.

A Case Study: When Institutions Die

Throughout most of history, changes in production, reproduction, education, and social control occurred slowly. When these changes occurred gradually and harmoniously, institutions could continue to support one another and to provide stable patterns that met ongoing human needs. On other occasions, however, old institutions—along with old roles and statuses—are destroyed before new ones can evolve. When this happens, societies and the individuals within them are traumatized, and societies and people fall apart.

In 1985, Anastasia Shkilnyk chronicled just such a human tragedy when she described the plight of the Ojibwa Indians of Northwestern Ontario in the book

© Frank Bigbear, Jr., Bockley Gallery

■ The three scenes in this painting by Native American artist Frank Bigbear, Jr. illustrate the vibrant culture that existed prior to Western conquest, the destruction of that culture, and the consequences of its destruction.

A Poison Stronger Than Love. Although the details are specific to the Ojibwa, her story is helpful in understanding what happened to Native Americans in general and to other traditional societies when rapid social change altered social institutions.

A Broken Society

In 1976, Anastasia Shkilnyk was sent by the Canadian Department of Indian Affairs to Grassy Narrows, an Ojibwa community of 520 people, to advise the band on how to alleviate economic disruption caused by mercury poisoning in nearby lakes and rivers. Grassy Narrows was a destroyed community. Drunken 6-year-olds roamed winter streets when the temperatures were 40 degrees below zero. The death rate for both children and adults was very high compared to that for the rest of Canada. Nearly three-quarters of all deaths were linked directly to alcohol and drug abuse. A quote from Shkilnyk's journal evokes the tragedy of life in Grassy Narrows:

> *Friday.* My neighbor comes over to tell me that last night, just before midnight, she found 4-year-old Dolores wandering alone around the reserve, about 2 miles from her home. She called the police and they went to the house to investigate. They found Dolores's

3-year-old sister, Diane, huddled in a corner crying. The house was empty, bare of food, and all the windows were broken. The police discovered that the parents had gone to Kenora the day before and were drinking in town. Both of them were sober when they deserted their children (Shkilnyk 1985, 41).

Like Dolores and Diane's parents, most of the adults in Grassy Narrows were binge drinkers. When wages were paid or the welfare checks came, many drank until they were unconscious and the money was gone. Often children waited until their parents had drunk themselves unconscious and then drank the liquor that was left. If they could not get liquor, they sniffed glue or gasoline.

Yet 20 years before, the Ojibwa had been a thriving people. How was a society so thoroughly destroyed?

Ojibwa Society Before 1963

The Ojibwa have been in contact with whites for two centuries. In 1873, they signed the treaty that defines their relationship with the government of Canada and that established the borders of their reservation.

In the decades that followed, the Ojibwa continued their traditional lives as hunters and gatherers. The family was their primary social institution. A family group consisted of a man and his wife, their sons and their wives and children, or of several brothers and their wives and children. The houses or tents of this family group would all be clustered together, perhaps as far as a half mile from the next family group.

Economic activities were all carried out by family groups. These activities varied with the season. In the late summer and fall, there was blueberry picking and harvesting of wild rice. In the winter, there was hunting and trapping. In all of these endeavors, the entire family participated, with everybody packing up and going to where the work was. The men would trap and hunt, the women would skin and prepare the meat, and the old people would come along to take care of the children and teach them. They used their reserve only as a summer encampment. From late summer until late spring, the family was on the move.

Besides being the chief economic and educational unit, the family was also the major agent of social control. Family elders enforced the rules and punished those who violated them. In addition, most religious ceremonials were performed by family elders. Although a loose band of families formed the Ojibwa society, each family group was largely self-sufficient, interacting with other family groups only to exchange marriage partners and for other ceremonial activities.

The earliest changes brought by white culture did not disrupt this way of life particularly. The major change was the development of boarding schools, which removed many Indian children from their homes for the winter months. When they returned, however, they would be accepted back into the group and educated into Indian ways by their grandparents. The boarding schools took the children away, but did not disrupt the major social institutions of the society they left behind.

The Change

In 1963, the government decided that the Ojibwa should be brought into modern society and given the benefits thereof: modern plumbing, better health care, roads, and the like. To this end, they moved the entire Ojibwa community from the old reserve to a government-built new community about 4 miles from their traditional encampment. The new community had houses, roads, schools, and easy access to "civilization." The differences between the new and the old were sufficient to destroy the fragile interdependence of Ojibwa institutions.

First, all the houses were close together in neat rows, assigned randomly without regard for family group. As a result, the kinship group ceased to exist as a physical unit. Second, the replacement of boarding schools with a community school meant that a parent (the mother) had to stay home with the children instead of going out on the trap line. As a result, overnight adult women became consumers rather than producers, shattering their traditional relationships with their husbands and community. As a consequence of the women's and children's immobility, men had to go out alone on the trap line. Because they were by themselves rather than with their family, the trapping trips were reduced from several weeks to a few days, and trapping ceased to be a way of life for the whole family. The productivity of the Ojibwa reached bottom with the government order in May 1970 to halt all fishing because of severe mercury contamination of the water caused by a white-owned paper mill. Then the economic contributions of men as well as women were sharply curtailed; the people became heavily dependent on the government rather than on themselves or on each other.

What happened was the total destruction of old patterns of doing things—that is, of social roles, statuses, and institutions. The relationships between husbands and wives were no longer clear. What were their rights and obligations to each other now that their joint economic productivity was at an end? What were their rights and obligations to their children when no one cared about tomorrow?

The Future of the Ojibwa

In 1985, the Ojibwa finally reached an out-of-court settlement with the federal and provincial governments and the mercury-polluting paper mill. The $8.7 million settlement was to compensate for damages to their way of life arising from government policies and mercury pollution. The band is using some of this money to develop local industries that it hopes will provide an ongoing basis for a productive and thriving society. Today, Ojibwa society has begun the process of healing and recovery. This process, however, will be a difficult one, for in losing their social institutions, the Ojibwa lost some of the basic building blocks of social life.

Connections

Social Policy

Like the Ojibwa, many Native Americans in the United States have lost contact with their culture and suffered greatly as a result. One response to this has been the development of the Native American Church, which combines Christian beliefs with the cultural and religious traditions of various tribes. The Church has been particularly active in U.S. prisons, where it has given many young Native Americans both pride in their culture and hope for the future, and helped individuals to exchange abuse of alcohol for controlled, ritual use of peyote.

A Sociological Response

Unfortunately, the Ojibwa are not an exceptional case. Their tragedy has been played out in tribe after tribe, band after band, all over North America. Although some Native American groups are almost entirely alcohol-free, in others, alcoholism touches nearly every family. Compared to the national average, Native Americans are 5.6 times more likely to die from alcohol-related deaths, 3.9 times more likely to have chronic liver disease and cirrhosis, and 3 times more likely to die in alcohol-related motor vehicle accidents (Beauvais 1998).

High levels of alcohol use are a health problem, an economic problem, and a social problem. Among the related issues are fetal alcohol syndrome, child and spouse abuse, unemployment, teenage pregnancy, nonmarital births, and divorce. How can these interrelated problems be addressed?

To paraphrase C. Wright Mills (see Chapter 1), when one or two individuals abuse alcohol, this is an individual problem, and for its relief we rightfully look to clinicians and counselors. When large segments of a population have alcohol problems, this is a public issue and must be addressed at the level of social structure.

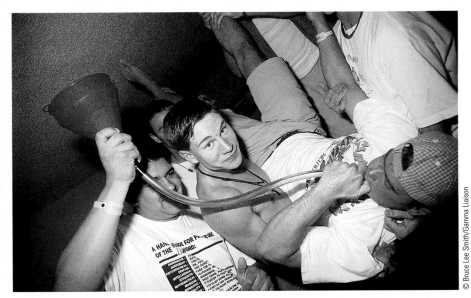

■ Native Americans are hardly the only subculture in American society to have problems with alcohol abuse. Heavy drinking and drinking games are characteristic of the high school and college years, and the expectation of drinking is built into many collegiate activities, such as parties and spring break.

© Bruce Lee Smith/Gamma Liaison

A sociological response to reducing alcoholism among Native Americans begins by asking what social structures encourage alcohol use? Conversely, why don't social structures reward sobriety?

The answer depends on one's theoretical framework. Structural functionalists focus on the destruction of Native American culture and the absence of harmony between their institutions and those of white society. Conflict theorists see the current situation as the result of a violent conflict over scarce resources, a conflict in which victorious Europeans systematically stripped Native Americans of their means of economic production and hence destroyed their society (Fisher 1987).

Regardless of theoretical position, it is obvious that Native Americans are severely economically disadvantaged. Unemployment is often a way of life; on some reservations, as few as 15 percent of the adults have jobs. Lack of employment is a critical factor in alcoholism in all populations. Having a steady, rewarding job is an incentive to stay sober; it also reduces the time available for drinking, which is essentially a leisure-time activity. From this perspective, the solution to high levels of alcoholism among Native Americans must include changing economic institutions to provide full employment and bolstering Native American culture and pride, as well as hiring more doctors, counselors, and others to help individuals fight addiction.

In many ways, fighting alcoholism is like fighting measles. We cannot eradicate the disease by treating people after they have it; we have to *prevent* it in the first place. When alcoholism is epidemic in a community, it requires community-wide efforts for prevention. Statuses, roles, and institutions must be re-formed so that people have a reason to stay sober.

Social Interaction and Everyday Life

Why do people do what they do? The answer depends in part on which social roles they are playing, but it also depends on the individual's social status and resources, on the situation, and on the individual role player. Two people playing the role of

The **sociology of everyday life** focuses on the social processes that structure our experience in ordinary face-to-face situations.

physician will do it differently, and the same individual will play the role differently depending on the patient's attitude and the circumstances. Social structure explains the broad outlines of why we do what we do, but it doesn't deal with specific concrete situations. This is where the sociology of everyday life comes in. Researchers who study the **sociology of everyday life** focus on the social processes that structure our experience in ordinary face-to-face situations.

Assumptions about Everyday Life

The sociology of everyday life is closely identified with the interaction school of symbolic interaction theory. Like that school, it stresses the importance of subjective meanings assigned to symbolic communications and of individuals' active involvement in negotiating their own roles and identities. Scholars who use the everyday life perspective, however, emphasize three additional assumptions: the problematic nature of culture, the I/me dialectic, and the importance of biography.

The Problematic Nature of Culture

In Chapter 2, we likened culture to a design for living and to a tool kit. If you consider these two images, you will see that they offer subtly different meanings of culture. The "design for living" image suggests a set of blueprints that need only to be correctly followed; the "tool kit" image, on the other hand, suggests a more dynamic approach. You do not follow a tool kit, you *use* it.

The sociology of everyday life is based on the tool kit image. It assumes that day-to-day behavior is not a matter of following clear cultural scripts but of improvising, negotiating, and adjusting to the general outline. From this point of view, culture is problematic.

Culture does furnish a great many rules and rituals, but it is not always clear which rules apply when. Being honest and standing by your friends are both norms; what do we do when telling the truth would get a friend in trouble? This predicament emphasizes that conformity to cultural norms is also problematic: It requires a continual stream of choices (Oberschall & Leifer 1986).

The I/Me Dialectic

The rules that govern our behavior may contradict each other; they may also contradict our own wishes.

Scholars from the everyday life perspective take seriously the I/me split of the self proposed by Mead (see Chapter 3). They recognize that, on the sidelines of every social encounter, the I stands ready to assert itself over the me—to barge in and do something impulsive and perhaps selfish. Any concrete situation can be envisioned as a negotiation between these two parts of ourselves.

This negotiation can be viewed as a dialectic, a process of conflict between individual freedom and social constraint (Bensman & Lilienfeld 1979). Here is a commonplace example: You are hurrying down the sidewalk, rushing to catch a movie, and you pass an acquaintance with a quick "Hi there, Lori, how are you?" Instead of replying as expected, Lori stops and proceeds to tell you that she is really depressed because she has just heard that her mother has cancer. So now what do you do? The socialized aspect of your self (the me) tells you that you must pause and show interest and concern. Your I, the spontaneous, impulsive part of your self, wants to keep hurrying to the movie. After all, you hardly know Lori, much less her mother, and you really want to see the film. In this case and in many others, your behavior represents a dialectic, a conflict between social convention and individual impulses.

A dialectic is not a simple contradiction but a process in which opposing forces engage, meet, and produce change. Viewing interaction episodes as a dialectic leads to the proposition that social interactions are never completely programmed by social structures. The outcome of a conflict is never fully predictable.

The Importance of Biography

Each of us possesses a personal history, a biography, that makes us unique. Although we share many things, none of our experiences exactly duplicate the experiences of another. Although we may all face similar social constraints as in the awkward episode with Lori, we will each bring a different self to the dialectic. Because of this, no two encounters are identical.

The uniqueness of each encounter does not mean that there is no patterned regularity in everyday life. Almost all of our daily encounters follow recognizable routines that allow us to interact without awkwardness, even with strangers. These routines explain the similarities in our encounters; biography explains why each one is just a little different from the rest.

A **frame** is an answer to the question, what is going on here? It is roughly identical to a definition of the situation.

Managing Everyday Life

Much of our daily life is covered by routines. The most important routines we use for interaction with others are carried out through talk. We all learn dozens of these verbal routines and can usually pull out an appropriate one to suit each occasion. Small rituals such as

"Hello. How are you?"
"Fine. How are you?"
"Fine."

will carry us through dozens of encounters every day. If we supplement this ritual with half a dozen others, such as "thanks/you're welcome" and "excuse me/okay," we will be equipped to meet most of the repetitive situations of everyday life. Nevertheless, each encounter is potentially problematic, and successful interaction requires selecting the appropriate ritual plus the skill and motivation to carry it out.

At the beginning of any encounter, individuals must resolve two issues: (1) What is going on here—what is the nature of the action? and (2) What identities will be granted—who are the actors? All action depends on our answers to these questions. Even the decision to ignore a stranger in the hallway presupposes that we have asked and answered these questions to our satisfaction. How do we do this?

Frames

The first step in any encounter is to develop an answer to the question, What is going on here? The answer forms a frame, or framework, for the encounter. A **frame** is roughly identical to a definition of the situation—a set of expectations about the nature of the interaction episode that is taking place.

All face-to-face encounters are preceded by a framework of expectations—how people will act, what they will mean by their actions, and so on. Even the most simple encounter, say, approaching a salesclerk to buy a package of gum, is covered by dozens of expectations: We expect that the salesclerk will speak English, will wait on the person who got to the counter first, will charge exactly the sum on the package and will not try

■ What is going on here? In order to plan an effective course of action, one of the first issues to resolve is defining just what kind of a situation it is. Our strategy for responding to the disaster would certainly be different if we believed that this Los Angeles scene had resulted from an earthquate rather than from the riots.

© David Young-Wolff/PhotoEdit

to barter with us, and will not comment on the fact that we are overweight or need a haircut. These expectations—the frame—give us guidance on how we should act and allow us to evaluate the encounter as normal or as deviant.

Our frames will be shared with other actors in most of our routine encounters, but this is not always the case. We may simply be wrong in our assessment of what is going on, or other actors in the encounter may have an entirely different frame. The final frame that we use to define the situation will be the result of a negotiation between the actors.

Identity Negotiation

After we have put a frame on an encounter, we negotiate an answer to the second question: What identities will be acknowledged? This question is far more complex than simply attaching names to the actors. Because each of us has a repertoire of roles and identities from which to choose, we are frequently uncertain about which identity an actor is presenting *in this specific situation*.

To some extent, identities will be determined by the frame being used. If a student's visit to a professor's office is framed as an academic tutorial, then the professor's academic identity is the relevant one. If the visit is framed as a social visit, then other aspects of the professor's identity (hobbies, family life, and so on) become relevant.

In many routine encounters, identities are not problematic. Although confusion about identities is a frequent device in comedy (for example, many of Shakespeare's comedies), in real life, a few verbal exchanges are usually sufficient to resolve confusion about actors' identities. In some cases, however, identity definitions are a matter of serious conflict. For example, Jennifer may want Mike to regard her as an equal, but Mike may prefer to treat her as an inferior.

Resolving the identity issue involves negotiations about both your own and the other's identity. How do we negotiate another's identity? We do so by trying to manipulate others into playing the roles we have assigned them. Mostly we handle this through talk. For example, "Let me introduce Mary, the computer whiz," sets up a different encounter than "Let me introduce Mary, the best party giver I know." Of course, others may reject your casting decisions. Mary may prefer to present a different identity than you have suggested. In that case, she will begin to try to renegotiate her identity.

Identity issues can become a major hidden agenda in interactions. Imagine an incompetent man talking to a competent woman. If the man finds this situation uncomfortable, he may try to define it as a man/woman encounter rather than a competence/incompetence encounter. He may start with techniques such as "How do you, as a woman, feel about this?" To reinforce this simple device, he might follow up with remarks such as "You're so small, it makes me feel like a giant." He may interrupt her by remarking on her perfume. He may also use a variety of nonverbal strategies such as stretching his arm across the back of her chair to assert dominance. Through such strategies, actors try to negotiate both their own and the other's identity.

Dramaturgy

The management of everyday life is the focus of a sociological perspective called dramaturgy. **Dramaturgy** is a version of symbolic interaction that views social situations as scenes manipulated by the actors to convey the desired impression to the audience.

Dramaturgy is a version of symbolic interaction that views social situations as scenes manipulated by the actors to convey the desired impression to the audience.

The chief architect of the dramaturgical perspective is Erving Goffman (1959, 1963). To Goffman, all the world is a theater. Like actors, each of us uses our appearance to establish our character—something we do each morning as we choose which clothes to wear, how to style our hair, and whether this would be a good day to show off our tattoos or piercings (if we have any). And like actors, we can use facial expressions, eye contact, posture, and other body language to enhance, reinforce, or even contradict the things we say. For example, placating a worried friend by saying "Yes, you look pretty in that dress" doesn't mean as much if you say it without looking up from your newspaper.

Sociologists who use dramaturgy also point out that life, like the theater, has both a front region, the stage, where the performance is given, and a back region, where rehearsals take place and behavior inappropriate for the stage may appear. For example, waiters at expensive restaurants are acutely aware of being on stage and act in a dignified and formal manner. Once in the kitchen, however, they may be transformed back into rowdy college kids.

The ultimate back region for most of us, the place where we can be our real selves, is at home. Nevertheless, even here front-region behavior is called for when company comes. ("Oh yes, we always keep our house this clean.") On such occasions, a married couple functions as a team in a performance designed to manage their guests' impressions. People who were screaming at each other before the doorbell rang suddenly start calling each other "dear" and "honey." The guests are the audience, and they too play a role. By seeming to believe the team's act (Goffman calls this "giving deference"), they contribute to a successful visit.

© Jose Luis Pelaez, Inc./Corbis

This boy's body language radiates his dissatisfaction.

Whose Definition? The Ex-Wife at the Funeral

How we act in any encounter depends in large part on the frame we have developed and the identities we and the other participants have worked out. One of the many processes of negotiation that goes on in social encounters is determining whose definition of the situation will be accepted. The person who enters the situation with the most power and resources is most likely to win, but either way, whoever wins this negotiation gains a powerful advantage. A paper by Riedmann (1987) on the role of the ex-wife at the funeral is an insightful illustration of this process.

In this case, a 44-year-old man dies suddenly and accidentally at the home he shares with his new wife of six months. Immediately after the accident, his 20-year-old daughter, who is staying with them, calls her mother to break the news. The ex-wife is terribly upset: She went steady with Bill from age 16 to 22 and was married to him for 20 years; even after the divorce and his remarriage, they remained friends and had lunch together frequently. Although they no longer lived together, they shared a concern for their two children and half a lifetime of memories. His parents were almost as close to her as her own; she had spent many summer afternoons with his brothers and their children.

When informed of his death, therefore, she felt bereaved. Her first response was "What? Our dad is dead?" Her definition of the situation was that she had had a death in her family. Acting on this definition, she went over to the home of her parents-in-law. Here she came up against a different definition. From the in-laws' position, it was a death in their family but not in her family. Although they did not say so, they wanted her to leave quickly before the current wife came over.

Over the next few days, the ex-wife again and again ran into the dilemmas posed by contrasting definitions of the situation. The obituary did not mention her among the relatives left behind. When she went to view the body at the mortuary, she was denied admittance on the grounds that only members of the immediate family were eligible. When the minister asked everybody but the immediate family to leave the graveyard, she was supposed to leave. Although the ex-wife defined Bill's death as a death in her family, her definition was not commonly shared.

The ex-wife and the new wife each formed teams, with each trying to pursue her own definition of the situation. The "old family" team (made up of the ex-wife, one of her children, her relatives, and one brother-in-law) and the "new family" team competed to be recognized as the "real family." At the funeral service itself, each team held down one corner of the lobby. When family and friends entered the lobby, they were faced with the predicament of which team to console first. Because many of Bill's old friends knew the ex-wife well and the new wife not at all, the "old family" team may be considered to have won this round.

In the end, though, the ex-wife was unable to impose her definition of the situation. The mortician and the priest were paid agents of the "new family" team. The ex-wife was ultimately forced to reframe the situation and to recognize her powerlessness to impose her definition of the situation on others.

Identity Work

If we view day-to-day encounters as a series of negotiations, we may be satisfied with simple descriptions of what people usually do in such encounters. Most social scientists, however, wish to go beyond this and ask *why* people do what they do. The answer most often supplied by scholars studying everyday behavior is that people are trying to enhance their self-esteem. One commentator noted wryly that this research assumes an "approval-starved person hot on the trail of a compliment" (Schneider 1981). More generally, we assume that social approval is one of the most important rewards that human interaction has to offer and we try to manage our identities so that this approval is maximized.

> **Identity work** is the process of managing identities to support and sustain our self-esteem.

Managing identities to support and sustain our self-esteem is called **identity work** (Snow & Anderson 1987). It consists of two general strategies: avoiding blame and gaining credit (Tedeschi & Riess 1981).

Avoiding Blame

There are many potential sources of damage to our social identity and self-esteem. We may have lost our job or flunked a class; on a more mundane level, we may have said the most embarrassing and stupidest thing imaginable, been unintentionally rude to an older relative, or otherwise made a fool of ourselves. When we behave in ways that make us look bad, or when we fear we are on the verge of saying or doing something embarrassing, then it is important to try to protect our identities.

Most of this work is done through talk. C. Wright Mills (1940, 909) noted that we learn the vocabulary for justifying our norm violations at the same time we learn the norms themselves. We learn what the rules are, and we simultaneously learn what kinds of accounts will justify our violations of those rules. If we can successfully explain away our rule-breaking, we can present ourselves as a person who normally obeys norms and who deserves to be thought well of by ourselves and others. The two basic strategies we use to avoid blame are accounts and disclaimers.

Accounts

Many of the rule-breakings that go on in everyday life are of a minor sort that can be explained away. We do this by giving **accounts**, explanations of unexpected or untoward behavior. Accounts are of two sorts—excuses and justifications (Scott & Lyman 1968). **Excuses** are accounts in which an individual admits that the act in question is bad, wrong, or inappropriate but claims he or she couldn't help it. **Justifications** are accounts that explain the good reasons the violator had for choosing to break the rule; often these are appeals to some alternate rule (Scott & Lyman, 47).

One of the most fertile fields for both excuses and justifications occurs in the student role. Students are expected to study and turn in papers on time, but many times they don't. Kathleen Kalab, a sociology professor at Western Kentucky University, asked her students to explain in writing why they missed class. The answers she received show how students try to explain away their norm violations so that they can preserve a positive identity. For example, one student offered the following *excuse*:

> I am so sorry I missed your class, among others, Tuesday the [date]. The reason I missed is a simple yet probably unacceptable reason, I slept until 1:45 p.m. Not only did I sleep through Sociology, I also slept through these things: Geography, trash pick-up at Poland Hall, maintenance workers sawing down a tree outside my window, my alarm clock (I believe), and many other loud and interesting things. Please see your way clear of forgiving me for this horrible and strange event. (Kalab 1987, 79)

Other students *justified* sacrificing student performance in order to be good sons or employees. For example,

> Sorry I wasn't here but Friday my parents called and our entire herd of cattle broke out and were all over the county and then I got to castrate two 500-lb bull calves. So needless to say I was extremely busy. (Kalab 1987, 75)

Or

> I am really sorry for missing your class Friday. The reason being I was part of my Reserve unit's advance party to Fort Knox. I didn't know this until late Thursday night. So please excuse me for not being there. (Kalab 1987, 76)

Accounts such as these are verbal efforts to resolve the discrepancy between what has happened and what can legitimately be expected. If they are accepted, self-identity and social status are preserved, and interaction can proceed normally.

Disclaimers

A person who recognizes that he or she is likely to violate expectations may preface that action with a **disclaimer**, a verbal device used, in advance, to defeat any doubts and negative reaction that might result from conduct (Hewitt & Stokes 1975, 3). Students commonly begin a query with "I know this is a stupid question, but . . ." The disclaimer lets the hearer know that the speaker knows the rules, even though he or she doesn't know the answer.

Disclaimers occur before the act; accounts occur after the act. Nevertheless, both are verbal devices we use to try to maintain a good image of ourselves, both in our own eyes and in the eyes of others. They help us to avoid self-blame for rule-breaking, and they try to reduce the blame that others might attribute to us. If we are successful in this identity work, we can retain fairly good reputations and our social status despite a few failures in meeting our social responsibilities.

Accounts are explanations of unexpected or untoward behavior. They are of two sorts: excuses and justifications.

Excuses are accounts in which one admits that the act in question is wrong or inappropriate but claims one couldn't help it.

Justifications are accounts that explain the good reasons the violator had for choosing to break the rule; often they are appeals to some alternate rule.

A **disclaimer** is a verbal device employed in advance to ward off doubts and negative reactions that might result from one's conduct.

A Global Perspective

focus on

Identity Work Among Australian Aborigines

"**I** want to owe you five dollars," one Australian Aborigine says to another, requesting a loan of that amount. According to anthropologist Nicolas Peterson (1993), "bumming" like this is an expected part of every-day life among Aborigines living in northeastern Australia. Unlike Native Americans, Australian Aborigines retain sufficient control over enough land that some of them can continue to be nomadic hunters and gatherers. But like most hunters and gatherers, they have virtually no surpluses of food or other needed goods. In a society where food is found (or not found) on a day-to-day basis, sharing makes good sense. Consequently, sharing is an important part of Australian Aboriginal culture. Very little spontaneous sharing occurs outside of the household, how-ever. Instead, food and other things are typically shared because they are demanded.

Unspoken demands for food and other items are common among Abo-rigines. Simply presenting yourself when food is being prepared and eaten means you have to be included. (We in the United States have this norm, too.)

Adults usually present themselves for something to eat only when large quantities of food have been brought into camp by one household. Children, on the other hand, commonly "bum" food. When children hear someone talking about food, they often go to have a look. This behavior has given rise to the mild rebuke, "Your ears have fingers." Another interesting practice is known as *warmarrkane*. At a public event, A may give B a compliment, such as "You are a fine dancer." A then has the right to make a substantial demand of B.

Demand sharing puts individuals in difficult positions when they don't have enough for themselves or want to accumulate resources for hard times. Since a demand should not be directly refused, what do Australian Aborigines do, then, when they don't want to share? Aboriginal culture provides a number of acceptable strategies for avoiding excessive demands. Men can hide their resources to avoid requests. By custom, they also can stop a de-mand simply by saying the name of the demander's mother-in-law. Men also can officially give their spears and guns to their mothers or other elderly women so that they can legitimately refuse requests from others who want to use their weapons.

Demanding, relinquishing, and concealing are all routine rituals of everyday Australian Aboriginal life. An individual's prestige depends largely on being able and willing to share. Demanding, in turn, is one way of reaffirming and making others recog-nize the demander's rights. When thought of in this way, demand shar-ing comprises identity work. By de-manding, receiving, sharing, and skill-fully avoiding the demands of others, Australian Aborigines reinforce both their status and their self-esteem.

For more information, look up the following subjects in InfoTrac College Edition:
Australian Aborigines

Or visit the following Web sites:
Wikipedia
http://www.wikipedia.org/wiki/Australian_Aborigine
Aboriginal Culture
http://library.trinity.wa.edu.au/aborigines/culture.htm

Gaining Credit

Our compliment-starved selves will not be content just to be thought of as good; we want all the credit we can get. This means that we will employ a variety of ver-bal devices to associate ourselves with positive outcomes. Just as there are a variety of ways to avoid blame, there are many ways we can claim credit. One way is to link ourselves spatially or verbally to situations or individuals with high status. This ranges from making a $1,000 donation to a political fund-raiser so that we can get a photograph of ourselves with the president to aligning ourselves verbally with "our" team (when it is winning).

Claiming credit is a strategy that requires considerable tact. Bragging is gener-ally considered inappropriate, and if you pat yourself too hard on the back, you are likely to find that others will refuse to do so. The trick is to find the delicate balance where others are subtly reminded of your admirable qualities without your actually having to ask for or demand praise.

Again the classroom and the negotiation of student identity provide good examples of how we go about doing positive identity work. Daniel and Cheryl Albas (1988) studied the ways their students at the University of Manitoba managed identity when their papers were returned. "Aces" are students who have done very well on examinations, and their task is to claim all the credit they can without bragging or being condescending to classmates who have done poorly, the "bombers." The Albases note that aces differ in sophistication: while all leave their examinations and grades prominently displayed on their desks, some, in addition, announce their score to others. In the most successful strategy, the ace begins by complaining about how hard the test was and then asks how other students did. Only when asked does the ace give his or her own score. Another strategy that enhances the ace's credit while protecting the bomber's identity is to blame both success and failure on luck. When talking to a bomber, an ace may say that "it was only luck that I studied the stuff that was on the exam." This strategy gives the ace double credit: for being a good student and for being a nice person.

A Case Study: Identity Work Among the Homeless

A particularly useful illustration of identity work can be seen in individuals who have what Goffman (1961b) calls *spoiled identities*—identities that are not merely low in status but are actively rejected by society. Examples include those who are severely handicapped, traitors, and, in some communities, people with AIDS. How do people with spoiled identities sustain their self-esteem?

In 1961, Goffman argued that two general strategies for protecting one's self-esteem in the face of a spoiled identity are (1) physically withdrawing from interaction with higher-status others and (2) trying to pass as a member of some higher-status group. Homeless people can't use either of these strategies: They can't withdraw because they have nowhere to go, and they can't pass because their clothing and circumstances declare their lowered circumstances. Lacking these two options, what strategies do homeless people use to negotiate a positive identity?

A study among homeless people in Austin, Texas, investigated just this question. David Snow and Leon Anderson (1993) ate at the Salvation Army and hung out under bridges and at the plasma center. They listened to homeless people talk about themselves, trying to discover the processes that these people used to negotiate their identities and protect their self-esteem.

As they analyzed the kinds of stories they heard from the street people in Austin, Snow and Anderson found that the stories fell into three general categories: role distancing, role embracement, and storytelling. All are verbal strategies that help homeless persons maintain their self-esteem and develop a positive identity.

Role distancing refers to the process of explaining to anyone who will listen that our current status is just temporary and not a reflection of who we really are. For example, a 24-year-old man who had been on the street for only a few weeks said, "I'm not like the other guys who hang out down at the 'Sally' [Salvation Army]. If you want to know about street people, I can tell you about them; but you can't really learn about street people from studying me, because I'm different" (Snow & Anderson 1993, 215).

© Patrick Ward/Stock, Boston

■ Even those who have very high self-esteem seek out ways to enhance and maintain their credit with others. Unfortunately, few of us receive as many ribbons as this mouse breeder, and even he probably finds few opportunities to show off all of his awards. Most of us, then, have to seek somewhat more subtle ways to demonstrate what wonderful people we are.

This photograph was taken outside the Salvation Army in Austin, Texas, where David Snow and Leon Anderson did their participant observation study of the homeless. Their research shows that many of the people who hang out at the "Sally" manage to negotiate a positive identity despite being homeless. Getting by without a job, car, money, or house requires a lot of ingenuity, and some long-term homeless persons take pride in their ability to fend for themselves.

Role embracement is the reverse of role distancing. People who embrace their roles accept that their role is permanent and take pride in it. Some of the homeless people Snow and Anderson studied developed romantic notions of "brethren of the road," cultivated nicknames such as Boxcar Billy and Panama Red, and proudly introduced themselves as bums and tramps. They took pride in their independence and in their skill at panhandling and "dumpster diving" (going through trash cans for food, clothing, recyclables, and so on). These people looked down their noses at those homeless people who depended on the Salvation Army to take care of them.

A final form of identity work that homeless people engage in is *storytelling*. Individuals who use storytelling attempt to establish a positive, nonstreet identity by telling others how important or rich he or she was or would be. One homeless man who was hanging around a transient bar bumming cigarettes and beers announced often that he had been offered a job as a mechanic with Harley-Davidson at $18.50 per hour. Others told stories about the future—how they were going to set up their own business, inherit a lot of money, or buy extravagant presents for their families. Some of these stories were within the realm of reason, but most were not. Still, most homeless people were willing to courteously accept others' stories, even if they knew they were false. In this way, they worked together to support each other's identity claims.

The homeless people that Snow and Anderson studied had all the ingredients for the formation of a negative identity: They were hungry, poor, ragged, homeless, frequently drunk and sick, and unemployed. It is not too surprising that some of them had negative self-images. More surprisingly, many nevertheless managed to feel good about themselves and construct positive identities. Their success in doing so reaffirms the assumption made by the interaction school: Even in the face of a spoiled identity, we can negotiate a positive self-concept.

Where This Leaves Us

In the 1950s, structural-functional theory dominated sociology, and a great deal of emphasis was placed on the power of institutionalized norms to *determine* behavior. Durkheim, with his views on positivism and constraint, was a favorite classic theorist. Similarly, during this period, the structural school of symbolic interaction theory clearly dominated the interaction school, and theorists stressed the power of institutionalized roles to determine behavior and personality.

Beginning in the 1960s, sociologists grew increasingly concerned that this view of human behavior reflected an "oversocialized view of man" (Wrong 1961). In 1967, Garfinkel signaled rebellion against this perspective when he argued that the deterministic model presented people as "judgmental dopes" who couldn't do their own thinking.

In the last two decades, sociological thinkers have increasingly tended to view social behavior as more negotiable and less rule bound and have increasingly focused on how people resist rather than accommodate social pressures (Perrow 1986). This change is most obvious in the sociology of everyday life, but it is also evident in most other areas of sociology. Studies of mental hospitals, businesses, and complex organizations now suggest that the behavior of actors may be best understood as a game in which each player chooses a strategy to maximize her or his self-interest (Crozier & Friedberg 1980). Even bureaucracies are not seen as wholly deterministic. Employees stress some rules, ignore others, and reinterpret the rest (Fine 1984, 1996).

This does not mean that the rules don't make a difference. Indeed, they make a great deal of difference, and there are obvious limits to the extent to which we can negotiate given situations. As W. I. Thomas noted in 1923, "The child is always born into a group of people among whom all the general types of situations which may arise have already been defined and corresponding rules of conduct developed, and where he has not the slightest chance of making his definitions and following his wishes without interference" (cited in Shalin 1986, 17).

The perspective of life as problematic and negotiable is a useful balance to the role of social structure in determining behavior. Our behavior is neither entirely negotiable nor entirely determined.

Summary

1. The analysis of social structure—recurrent patterns of relationships—revolves around three concepts: status, role, and institution. Statuses are specialized positions within a group and may be of two types: achieved or ascribed. Roles define how status occupants ought to act and feel.

2. Because societies share common human needs, they also share common institutions. The common institutions are family, economy, government, education, and religion. Each society has some enduring social structure to perform these functions for the group.

3. Institutions are interdependent; none stands alone, and a change in one results in changes in others. Institutions regulate behavior and maintain the stability of social life across generations. Conflict theorists point out that these patterns often benefit one group more than others.

4. An important determinant of institutional development is the ability of a society to produce an economic surplus. Each major improvement in production has led to an expansion in social institutions.

5. The sociology of everyday life is a perspective that analyzes the patterns of human social behavior in concrete encounters in daily life. Three assumptions guide theory and research in this approach to studying human interaction: culture is problematic; individuals experience a dialectic between freedom and social constraint; and each individual is unique, possessing a biography unlike any other individual.

6. Deciding how to act in a given encounter requires answering two questions: What is going on here? What identities will be granted? These issues of framing and identity resolution may involve competition and negotiation between actors or teams of actors.

7. Dramaturgy is a perspective pioneered by Erving Goffman. It views the self as a strategist who is choosing roles and setting scenes to maximize self-interest.

8. The desire for approval is an important factor guiding human behavior. To maximize this approval, people engage in active identity work to sustain and support their self-esteem. This work takes two forms: avoiding blame and gaining credit.

Thinking Critically

1. Is social class an achieved or ascribed status? What would a structural functionalist say? A conflict theorist? Symbolic interactionist?

2. Consider religion as an institution. How would a conflict theorist view it? What might a structural-functionalist say? Which position is closest to your own and why?

3. Pick a social problem that affects you personally; for example, alcoholism, unemployment, racism, sexism, illegal immigration. Describe a social structural solution—one that focuses on changing the underlying social structural causes of the problem, rather than simply improving individuals' situations one by one.

4. Describe a time when you disagreed with someone about his or her identity. What kind of situation was it, and why was identity problematic? In the end, whose definition of identity was accepted? Why?

Sociology on the Net

The Wadsworth Sociology Resource Center: Virtual Society

http://www.sociology.wadsworth.com

The companion Web site for this book includes a range of enrichment material. Further your understanding of the chapter by accessing Online Practice Quizzes, Internet Exercises, InfoTrac College Edition Exercises, and many more compelling learning tools.

Chapter-Related Suggested Web Sites

The Psychology of Cyberspace
http://www.rider.edu/users/suler/psycyber/psycyber.html

Native Web
http://www.nativeweb.org

InfoTrac College Edition

http://www.infotrac-college.com/wadsworth/

Access the latest news and research articles online—updated daily and spanning four years. InfoTrac College Edition is an easy-to-use online database of reliable, full-length articles from hundreds of top academic journals and popular sources. Conduct an electronic search using the following key search terms:

Social interaction

Social structure

Social status

Social institutions

Social roles

For more information related to this chapter, visit the Opposing Viewpoints Resource Center. Be sure you check all the options on the toolbar—viewpoints, references, statistics, and so on—and search under the following subjects:

Native Americans

Homeless persons

Suggested Readings

Barnes, J. A. 1994. *A Pack of Lies: Toward a Sociology of Lying.* New York: Cambridge University Press. An interesting book on lying in everyday life. Barnes argues that almost all of us lie once in a while, especially in socially ambiguous situations. Barnes also looks at lying among politicians, bureaucrats, and medical practitioners.

Goffman, Erving. 1959. *The Presentation of Self in Everyday Life.* New York: Doubleday. The book in which Goffman lays out the basic ideas behind dramaturgy. Each of Goffman's books is enjoyable reading and easily accessible to the average undergraduate.

Lenski, Gerhard. 1966. *Power and Privilege: A Theory of Social Stratification.* New York: McGraw-Hill. A major work distinguishing the fundamental characteristics of different types of societies, particularly in terms of socially structured inequality.

Shkilnyk, Anastasia. 1985. *A Poison Stronger Than Love: The Destruction of an Ojibwa Community.* New Haven, Conn: Yale University Press. An ethnographic community study that focuses on the social structures of a Native American community in Canada. This book provides a powerful illustration of the extent to which individual well-being depends on stable institutions.

Snow, David A., and Anderson, Leon. 1993. *Down on Their Luck: A Case Study of Homeless Street People.* Berkeley: University of California Press. A comprehensive look at the structural causes and personal costs of homelessness and an excellent example of a study that uses both quantitative and qualitative methods.

Intersections

Culture, Social Structure, and Everyday Life

A major theoretical debate in sociology has revolved around the question of which comes first, culture or social structure? As noted in Chapter 1, Karl Marx believed that the structure of economic institutions determines every other facet of social life, including culture. In contrast, Max Weber believed that values were important causes of social structure. Most sociologists agree with both points of view. On Marx's side, by regulating access to social rewards, social structures clearly reinforce certain values and norms and discourage and inhibit others. But, social institutions exist only to the extent that they are expressed in the roles that individuals play. Thus, as Weber would have it, changes in cultural values and norms eventually lead to changes in individual behavior and therefore to changes in the statuses and roles that make up the larger social structure.

A Case Study: Values, Institutions, and the Environment

According to the United Nations, 24 percent of mammals and 12 percent of bird species are currently threatened with extinction (United Nations Environment Programme 2002a). In addition, 1 percent of tropical forests are vanishing each year. Of the world's population, 40 percent now suffer from serious water shortages, and that percentage is expected to escalate dramatically in the next two decades. On Native American reservations and in many minority and poor communities in the United States, exposure to toxic waste has led to a surge in the number of children born with major birth defects. The question is, of course, how did things get this bad and what can we do about it?

Writing in 1967, history professor Lynn White, Jr., argued that the roots of the modern ecological crisis lie in Western traditions of technology and science. White believed that a technology capable of efficiently exploiting natural resources (and other human beings, for that matter) began its unparalleled growth in medieval Europe at least partly because of the Judeo-Christian philosophy that man is the master of nature rather than simply a part of it. White argues that the religious values of Hindus, Native Americans, and traditional African cultures are quite different on this key matter, and that, at least partly for this reason, their social structures have been less ecologically destructive. The traditional Hindu and African belief has been that human beings are connected to all elements of nature by insoluble spiritual and psychological bonds (Moshoeshoe II 1993; Singh 1993); Native American philosophy maintains that no action should be taken without assessing its consequences for the next seven generations. These beliefs call into question the wisdom of the wasteful lifestyles that have accompanied technological and institutional developments in the West.

In Western culture, belief in individualism, materialism, "progress and growth," and the idea that government should have a very limited role in controlling business complements a religious ideology that tends to emphasize humankind's dominion over all other creatures. As Weber noted, these beliefs are important cultural correlates of capitalist economic structures; as ecofeminists point out, they also tend to be associated with male-dominated family structures. Even though scholars debate which came first—technology and social institutions or values—and theologians question just how instrumental Western religious values have been in the destruction of the environment, research clearly shows that commitment to these dominant political and economic values is associated with lower levels of environmental concern among U.S. citizens (Dunlap & Van Liere 1984).

Despite their different histories, cultural traditions, and social structures, many people living in both the industrialized West and the developing nations of Africa and South America are now starting to show a high level of concern for environmental deterioration and widespread support for environmental protection. People in poorer countries recognize that overpopulation creates ecological problems; residents of wealthier nations realize that technology and wasteful lifestyles have devastating consequences for the environment. Furthermore, results from an international survey indicate that a very high percentage of respondents throughout the world also say they are willing to pay higher prices for goods and services if doing so will protect the environment (Dunlap, Gallup, & Gallup 1993).

If cultural values toward the environment have changed, why hasn't there been a corresponding change in social structures and social institutions? Part of the answer is that current, environmentally harmful ways of doing business are profitable. Another part of the answer is that the attitudes of private citizens have changed faster than business and governmental leaders realize. A final part of the answer is that even when individual values change, it is hard to change behavior if supporting institutional structures are not in place. It is difficult, for instance, to conserve energy and reduce pollution if there is no system of mass transportation in place that would allow people to get to work or school on time without relying on a private automobile. No matter how committed one is to the principle of recycling, it is difficult to do so if there are no groups or organizations that routinely pick up or process recyclables. Change in cultural values or beliefs does lead to institutional change, but the process is slow, often depending on changes in one set of statuses and roles, one cluster of groups and group memberships, at a time.

MicroCase Online Exercise

For an interactive exercise using MicroCase data sets, go to the text companion website at http://sociology.wadsworth.com/brinkerhoff/essentials6e. Choose Chapter 4 and then select "Intersection Exercise" from the left navigation bar.

CHAPTER 5

Groups, Networks, and Organizations

© Tom Carter/PhotoEdit

Human Relationships

Sociology is the study of relationships. We are concerned with how relationships are formed and the consequences these relationships have for the individual and the community. In this chapter we review the basic types of human relationships, from small and intimate groups to large and formal organizations, and discuss some of the consequences of these relationships.

Social Processes

Some relationships are characterized by harmony and stability; others are made stressful by conflict and competition. We use the term **social processes** to describe the types of interaction that go on in relationships. This section looks closely at four social processes that regularly occur in human relationships: exchange, cooperation, competition, and conflict.

Exchange

Exchange is voluntary interaction in which the parties trade tangible or intangible benefits with the expectation that all parties will benefit (Stolte, Fine, and Cook 2001). A wide variety of social relationships include elements of exchange. In friendships and marriages, exchanges usually include intangibles such as companionship, moral support, and a willingness to listen to the other's problems. In business or politics, an exchange may be more direct; politicians, for example, openly acknowledge exchanging votes on legislative bills—I'll vote for yours if you'll vote for mine.

Exchange relationships are based on the expectation that people will return favors and strive to maintain a balance of obligation in social relationships (Molm & Cook 1995). This expectation is called the **norm of reciprocity** (Gouldner 1960; Uehara 1995). If you help your sister-in-law move, then she is obligated to you. Somehow she must pay you back. If she fails to do so, the social relationship is likely to be strained. A corollary of the norm of reciprocity is that you avoid accepting favors from people with whom you do not wish to enter into a relationship. For example, if someone you do not know very well volunteers to type your term paper, you will probably be suspicious. Your first thought is likely to be, "What does this guy want from me?" If you do not want to owe this person a favor, you will say that you prefer to type your own paper. Nonsociologists might sum up the norm of reciprocity by concluding that there's no such thing as a free lunch.

Exchange is one of the most basic processes of social interaction. Almost all voluntary relationships are entered into with the expectation of exchange. In traditional U.S. families, these exchanges were clearly spelled out in social norms. Although not always followed in practice, the husband was expected to support the family, which obligated the wife to keep house and look after the children; or, conversely, she bore the children and was expected to keep house, which obligated him to support her.

An exchange relationship persists only if each party to the interaction is getting something out of it. This does not mean that the rewards must be equal; in fact,

---

Social processes are the forms of interaction through which people relate to one another; they are the dynamic aspects of society.

Exchange is voluntary interaction from which all parties expect some reward.

The **norm of reciprocity** is the expectation that people will return favors and strive to maintain a balance of obligation in social relationships.

Connections

Personal Application

The norm of reciprocity also applies in dating relationships. If you are a man and buy your date dinner or a movie, you may feel that she now owes you something in return—gratitude, a good night kiss, or more. If you are the woman, you also may believe that you are now indebted to your date, and so may do things you really don't want to do in exchange. When couples disagree on who owes what to whom, situations like these can escalate to sexual assault or even date rape.

rewards are frequently very unequal. Nor does this mean that each party to the exchange relationship has equal power; rather, the actor with greater control over a more valuable resource always has more power. You have probably seen children's play groups, for example, in which one child is treated badly by the other children and is permitted to play with them only if he agrees to give them his lunch or allows them to use his bicycle. If this boy has no one else to play with, he may find this relationship more rewarding than the alternative of playing alone. The continuation of very unequal exchange relationships usually rests on a lack of desirable alternatives (Emerson 1962; Molm 2003; Stolte, Fine, and Cook 2001).

Cooperation

Cooperation occurs when people work together to achieve shared goals. Exchange is a trade: I give you something and you give me something else in return. Cooperation is teamwork. It is characteristic of relationships in which people work together to achieve goals that they cannot achieve alone. Consider, for example, a four-way stop sign. Although it may entail some waiting in line, in the long run we will all get through more safely and more quickly if we cooperate and take turns. Most continuing relationships have some element of cooperation. Spouses cooperate in raising their children; children cooperate in tricking their substitute teachers.

Cooperation is interaction that occurs when people work together to achieve shared goals.

Competition is a struggle over scarce resources that is regulated by shared rules.

Competition

It is not always possible for people to reach their goals by exchange or cooperation. If your goal and my goal are mutually exclusive (for example, I want to sleep and you want to play your stereo), we cannot both achieve our goals. Similarly, in situations of scarcity, there may not be enough of a desired good to go around. In these situations, social processes are likely to take the form of either competition or conflict.

A struggle over scarce resources that is regulated by shared rules is **competition** (Friedsam 1965). The rules usually specify the conditions under which winning will be considered fair and losing will be considered tolerable. When the norms are violated and rule-breaking is uncovered, competition may erupt into conflict.

Many of the good things in life are in short supply: They are scarce resources. When the struggle for scarce resources is not regulated by norms that specify the rules of fair play, conflict often results.

© M. Greenlar/The Image Works

Competition is a common form of interaction in individualistic societies like the United States. Jobs, grades, athletic honors, sexual attention, marriage partners, and parental affection are only a few of the scarce resources for which individuals or groups compete. In fact, it is difficult to identify many social situations that do not entail competition. One positive consequence of competition is that it stimulates achievement and heightens people's aspirations. It also, however, often results in personal stress, reduced cooperation, and social inequalities (elaborated on in Chapters 7 through 11).

Because competition often results in change, groups that seek to maximize stability often devise elaborate rules to avoid the appearance of competition. Competition is particularly problematic in informal groups such as friendships and marriages. Friends who want to stay friends will not compete for anything of high value; they might compete over computer game scores, but they won't compete for each other's spouses. Similarly, couples who value their marriage will not compete for their children's affection or loyalty. To do so might destroy the marriage.

Conflict

Conflict is a struggle over scarce resources that is not regulated by shared rules; it may include attempts to destroy, injure, or neutralize one's rivals.

When a struggle over scarce resources is not regulated by shared rules, **conflict** occurs (Coser 1956). Because no tactics are forbidden and anything goes, conflict may include attempts to neutralize, injure, or destroy one's rivals. Conflict creates divisiveness rather than solidarity.

Conflict with outsiders, however, may enhance the solidarity of the group. Whether the conflict is between warring superpowers or warring street gangs, the us-against-them feeling that emerges from conflict with outsiders causes group members to put aside their jealousies and differences to work together. Groups from nations to schools have found that starting conflicts with outsiders is a useful device for restraining the more quarrelsome members of their own group. Critics of U.S. foreign policy suggest that our foreign adventures (such as invading Iraq in 2003) often seem to be timed to improve politicians' approval ratings by diverting the public's attention away from economic recessions and other divisive domestic issues.

A **group** is two or more people who interact on the basis of shared social structure and recognize mutual dependency.

Social Processes in Everyday Life

Exchange, cooperation, competition, and even conflict are important aspects of our relationships with others. Few of our relationships involve just one type of group process. Even friendships usually involve some competition as well as cooperation and exchange; similarly, relationships among competitors often involve cooperation.

An **aggregate** is people who are temporarily clustered together in the same location.

We interact with people in a variety of relationships. Some of these relationships are temporary and others permanent; some are formal and others informal. In the rest of this chapter, we discuss the wide variety of relationships we have under three general categories: groups, social networks, and organizations.

A **category** is a collection of people who share a common characteristic.

Groups

A **group** is two or more people who interact on the basis of a shared social structure and who recognize mutual dependency. Groups may be large or small, formal or informal; they range from a pair of lovers to IBM. In all of them, members share a social structure specifying statuses, roles, and norms, and they share a feeling of mutual dependency.

The distinctive characteristics of a group stand out when we compare the group to two collections of people that do not have these characteristics: aggregates and categories. An **aggregate** is people who are temporarily clustered together in the same location (for example, all the people on a city bus, those attending a movie, or shoppers in a mall). Although these people may share some norms (such as walking on the right when passing others), they are not mutually dependent. In fact, most of their shared norms have to do with procedures to maintain their independence despite their close physical proximity. The other nongroup is a **category**—a collection of people who share a common characteristic. Dorm residents, Greeks, bald-headed men, and Samoans are categories of people. Most of the people who share category membership will never meet, much less interact.

The distinguishing characteristics of groups hint at the rewards of group life. Groups are the people we take into account and the people who take us into account. They are the people with whom we share many norms and values. Thus, groups can be a major source of solidarity and cohesion, reinforcing and strengthening our integration into society. When groups function well, they offer benefits ranging all the way from sharing basic survival and problem-solving techniques to satisfying personal and emotional needs. Conversely, when groups function poorly, they can create anxiety, conflict, and social stress.

How Groups Affect Individuals

When a man opens a door for a woman, do you see traditional courtesy or sexist condescension? When you listen to Nelly or to the Dave Matthews Band, do you hear good music or mindless noise? Like taste in music, many of the things we deal with and believe in are not true or correct in any absolute sense; they are simply what our groups have agreed to accept as right.

The tremendous impact of group definitions on our own attitudes and perceptions was cleverly documented in a classic experiment by Asch (1955). In this experiment, the group consists of nine college students, all supposedly unknown to each other. The experimenter explains that the task at hand is an experiment in visual judgment. The subjects are shown two cards similar to those in Figure 5.1 and are asked to judge which line on Card B is most similar to the line on Card A. This is not a difficult task; unless you have a bad squint or have forgotten your glasses, you can tell that Line 2 most closely matches the line on the first card.

FIGURE 5.1
The Cards Used in Asch's Experiment
In Asch's experiment, subjects were instructed to select the line on Card B that was equal in length to the line on Card A. The results showed that many people will give an obviously wrong answer in order to conform to the group.
SOURCE: From "Opinions and Social Pressure," by Solomon E. Asch. Copyright © November 1955 by Scientific American, Inc. All rights reserved.

Card A

Card B

© Joel Gordon

In this sequence of pictures a subject (in the middle) shows the strain that comes from disagreeing with the judgments of other members of the group. Some subjects in the Asch experiment disagreed on all 12 trials of the experiment. However, 75 percent of the experimental subjects agreed with the majority on at least one trial. Subjects who initially yield to the majority find it increasingly difficult to make independent judgments as the experiment progresses.

The experimental part of this research consists of changing the conditions of group consensus under which the subjects make their judgments. Each group must make 15 decisions and, in the first few trials, all of the students agree. In subsequent trials, however, the first eight students all give an obviously wrong answer. They are not subjects at all but paid stooges of the experimenter. The real test comes in seeing what the last student—the real subject of the experiment—will do. Will he go along with everybody else, or will he publicly set himself apart? Photographs of the experiment show that the real subjects wrinkled their brows, squirmed in their seats, and gaped at their neighbors; in 37 percent of the trials the naive subject publicly agreed with the wrong answer, and 75 percent gave the wrong answer on at least one trial.

In the case of this experiment, it is clear what the right answer should be. Many of the students who agreed with the wrong answer probably were not persuaded by group opinion that their own judgment was wrong, but they decided not to make waves. When the object being judged is less objective, however—for example, whether Eminem is better than Britney Spears—the group is likely to influence not only public responses but also private views. Whether we go along because we are really convinced or because we are avoiding the hassles of being different, we all have a strong tendency to conform to the norms and expectations of our groups.

Yet small groups rarely have access to legal or formal sanctions—they usually can't throw those who disagree with them in jail or the like—so why do individuals so often go along with groups' opinions? The answer is fear—fear of not being accepted by the group (Douglas 1983). The major weapons that groups use to punish nonconformity are ridicule and contempt, but their ultimate sanction is exclusion from the group. From "you can't sit at our lunch table anymore" to "you're fired," exclusion is one of the most powerful threats we can make against others. This form of social control is most effective in cohesive groups, but Asch's experiment shows that fear of rejection can induce conformity to group norms even in artificial laboratory settings (Asch 1955).

Interaction in Small Groups

We spend much of our lives in groups. We have work groups, family groups, and peer groups. In class we have discussion groups, and everywhere we have committees. Regardless of the type of group, its operation depends on the quality of interaction among members. This section reviews some of the more important factors that affect the kind of interaction we experience in small groups. As we will see, interaction is facilitated by small size, physical proximity, and open communication patterns.

Size

The smallest possible group is two people. As the group grows to three, four, and more, its characteristics change. With each increase in size, each member has fewer opportunities to share opinions and contribute to decision making or problem solving. In many instances, the larger group will be better equipped for solving problems and finding answers, but this practical utility may be gained at the expense of individual satisfaction. Although there will be more ideas to consider, the likelihood that our own ideas will be influential diminishes. As the group gets larger, interaction becomes more impersonal, more structured, and less personally satisfying.

Physical Proximity

Dozens of laboratory studies demonstrate that interaction is more likely to occur among group members who are physically close to one another. This effect is not limited to the laboratory.

In a classic demonstration of the role of proximity in group formation and interaction, Festinger, Schacter, and Back (1950) studied a married-student housing project. All of the residents had been strangers to one another before being arbitrarily assigned to the housing unit. The researchers wanted to know what factors influenced friendship choices within the project. The answer: physical proximity. Festinger et al. found that people were twice as likely to choose their next-door neighbors as friends as they were to choose people who lived only two units away. In general, the greater the physical distance, the less likely friendships were to be formed. An interesting exception to this generalization is that people who lived next to the garbage cans were disproportionately likely to be chosen as friends. Why? Because many of their neighbors passed by their units daily and therefore had many chances to interact and form friendships.

Communication Patterns

Interaction of group members can be either facilitated or hindered by patterns of communication. Figure 5.2 shows some common communication patterns for five-person groups. The communication pattern allowing the greatest equality of participation is the *all-channel network*. In this pattern, each person can interact with every other person with approximately the same ease. Each participant has equal access to the others and an equal ability to become the focus of attention.

The other two common communication patterns allow for less interaction. In the *circle pattern*, people can speak only to their neighbors on either side. This pattern reduces interaction, but it does not give one person more power than others have. In the *wheel pattern*, on the other hand, not only is interaction reduced but a single, pivotal individual gains greater power in the group. The wheel pattern is characteristic of the traditional classroom. Students do not interact with one another; instead they interact directly only with the teacher, thereby giving that person the power to direct the flow of interaction.

Communication patterns are often created, either accidentally or purposefully, by the physical distribution of group members. The seating of committee members at a round table tends to facilitate either an all-channel or a circle pattern, depending on the size of the table. A rectangular table gives people at the ends and in the middle of the long sides an advantage. They find it easier to attract attention and are apt to be more active in interactions and more influential in group discussions. Consider the way communication is structured in the classes and groups you participate in. How do seating structures encourage or discourage communication?

Cohesion

One of the important dimensions along which groups vary is their degree of **cohesion** or solidarity. A cohesive group is characterized by higher levels of interaction and by strong feelings of attachment and dependency. Because its members feel that their happiness or welfare depends on the group, the group may make extensive claims on the individual members (Hechter 1987). Cohesive adolescent friendship groups can enforce dress codes on their members; cohesive youth gangs can convince new male members to commit random murders and can convince new female members to submit to gang rapes.

Marriage, church, and friendship groups differ in their cohesiveness. What makes one marriage or church more cohesive than another? Among the factors that contribute to cohesion are small size, similarity, frequent interaction, long duration, a clear distinction between insiders and outsiders, and few ties to outsiders (Hechter 1987; Homans 1950; McPherson, Miller, Popielarz, and Drobnic

FIGURE 5.2
Patterns of Communication
Patterns of communication can affect individual participation and influence. In each figure the circles represent individuals and the lines are flows of communication. The all-channel network provides the greatest opportunity for participation and is more often found in groups in which status differences are not present or are minimal. The wheel, by contrast, is associated with important status differences within the group.

All-channel

Circle

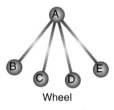
Wheel

Cohesion in a group is characterized by high levels of interaction and by strong feelings of attachment and dependency.

1992). Although all marriages in our society are the same size (two members), a marriage where the partners are more similar, spend more time together, and so on will generally be more cohesive than one where the partners are dissimilar and see each other for only a short time each day.

Decision Making in Small Groups

One of the primary research interests in the sociology of small groups is how group characteristics (size, cohesion, and so on) affect group decision making. This research has focused on a wide variety of actual groups: flight crews, submarine crews, business meetings, and juries, to name a few (Davis & Stasson 1988; Robert & Carnevale 1997).

Generally, groups strive to reach consensus; they would like all their decisions to be agreeable to every member. As the size of the group grows, consensus requires lengthy and time-consuming interaction so that everybody's objections can be clearly understood and incorporated. Thus, as groups grow in size, they often adopt the more expedient policy of majority rule. This policy results in quicker decisions but often at the expense of individual satisfaction. It therefore reduces the cohesiveness of the group.

Choice Shifts

One of the most consistent findings of research is that it is seldom necessary to resort to majority rule in small groups. Both in the laboratory and in the real world, there is a strong tendency for opinions to converge. One of the classic experiments on convergence was done by Sherif in 1936. In this experiment, strangers were put into a totally dark room. A dot of light was flashed onto the wall, and each participant was asked to estimate how far the light moved during the experimental period. After the first session, the participants recorded their own answers and then shared them with the other participants. There was quite a bit of variation in the estimates. Then they did the experiment again. This time there was less difference. After four trials, all participants agreed on an estimate that was close to the average of the initial estimates. (The dot of light was, in fact, stationary.)

This tendency toward convergence has been demonstrated in dozens of studies since. Convergence, however, is not always to a middle position. Sometimes, the group reaches consensus on an extreme position. This is called the *risky shift* when the group converges on an adventurous option and the *tame shift* when the choice is extremely conservative. Sometimes these choice shifts depend on persuasive arguments put forward by one or more members, but often they result from general norms in the group that favor either conservatism or risk (Davis & Stasson 1988). For example, one might expect a church steering committee to choose the safest opinion and a terrorist group to choose the riskiest option.

A special case of choice shift is *groupthink* (Janis 1982). Groupthink occurs when pressures to agree are so strong that they stifle critical thinking. For example, Diane Vaughan (1996) showed how groupthink contributed to the tragic 1986 explosion of the space shuttle *Challenger*. The engineers working on the *Challenger* all knew before the launch that the shuttle's O-rings probably would suffer some damage. But political pressures to launch the shuttle, coupled with a culture within NASA that rewarded risk taking, helped to create a situation in which the engineers essentially convinced each other and themselves that the risk of O-ring failure was within acceptable limits. As this example illustrates, groupthink often results in bad decisions.

One of the most powerful mechanisms of social control is the threat of exclusion from valued groups. To avoid the threat of exclusion, we conform: we dress as other members do, think as they think, and do as they do.

Types of Groups

Some groups have a more important impact than others on our lives. Almost all of you, for example, belong to a family group as well as to the student body of your college or university. But except for an occasional student activist, membership in the family is far more important than membership in the student body and will have a more lasting effect. Sociologists call small, intimate, and lasting groups *primary groups;* they call larger, impersonal groups *secondary groups.*

Primary Groups

Primary groups are characterized by intimate, face-to-face interaction (Cooley [1909] 1967). These groups represent our most complete experiences in group life. The closest approximation to an ideal primary group is probably the family, followed by adolescent peer groups and adult friendships. The relationships formed in these groups are relatively permanent and constitute a basic source of identity and attachment.

The ideal primary group tends to have the following characteristics: (1) personal and intimate relationships, (2) face-to-face communication, (3) permanence, (4) a strong sense of loyalty or we-feeling, (5) small size, (6) informality, and (7) traditional or nonrational decision making. In addition to the family, friendship networks, coworkers, and gangs may be primary groups. Groups such as these are major sources of companionship, intimacy, and belongingness, conditions that strengthen our sense of social integration into society.

Secondary Groups

By contrast, **secondary groups** are formal, large, and impersonal. Whereas the major purpose of many primary groups is simply to provide companionship, secondary groups usually form to accomplish some specific task. The perfect secondary group is entirely rational and contractual in nature; the participants interact solely to accomplish some purpose (earn credit hours, buy a pair of shoes, get a paycheck). Their

Primary groups are groups characterized by intimate, face-to-face interaction.

Secondary groups are groups that are formal, large, and impersonal.

Differences between Primary and Secondary Groups

Concept Summary

	Primary Groups	Secondary Groups
Size	Small	Large
Relationships	Personal, intimate	Impersonal, aloof
Communication	Face to face	Indirect—memos, telephone, etc.
Duration	Permanent	Temporary
Cohesion	Strong sense of loyalty, we-feeling	Weak, based on self-interest
Decisions	Based on tradition and personal feelings	Based on rationality and rules
Social structure	Informal	Formal—titles, officers, charters, regular meeting times, etc.

interest in each other does not extend past this contract. If you have ever been in a lecture class of 300 students, you have firsthand experience of a classic secondary group. The interaction is temporary, anonymous, and formal. Rewards are based on universal criteria, such as scores on multiple-choice examinations, not on such particularistic grounds as your effort or need. The Concept Summary shows the important differences between primary and secondary groups.

Comparing Primary and Secondary Groups

Primary and secondary groups serve very different functions for individuals and societies. From the individual's point of view, the major purpose of primary groups is **expressive** activity, giving individuals social integration and emotional support. Your family, for example, usually provides an informal support group that is bound to help you, come rain or shine. You should be able to call on your family and friends to bring you some soup when you have the flu, to pick you up in the dead of night when your car breaks down, and to listen to your troubles when you are blue.

Because we need primary groups so much, they have tremendous power to bring us into line. From the society's point of view, this is the major function of primary groups: They are the major agents of social control. The reason most of us don't shoplift is because we would be mortified if our parents, friends, or coworkers found out. The reason most soldiers go into combat is because their buddies are going. We tend to dress, act, vote, and believe in ways that will keep the support of our primary groups. In short, we conform. The law would be relatively helpless at keeping all the millions of us in line if we weren't already restrained by the desire to stay in the good graces of our primary groups. One corollary of this, however, which Chapter 6 addresses, is that if our primary groups consider shoplifting or tax evasion acceptable, then our primary-group associations may lead us into deviance rather than conformity.

The major functions of secondary groups are **instrumental** activities, the accomplishment of specific tasks. If you want to build an airplane, raise money for a community project, or teach introductory sociology to 2,000 students a year, then

Expressive describes activities or roles that provide integration and emotional support.

Instrumental describes activities or roles that are task oriented.

■ Many of the groups we participate in combine characteristics of primary and secondary groups. The elementary school classroom is a secondary group, yet many of the friendships developed there will last for 6, 12, or even 40 years.

© Bob Daemmrich / Stock Boston

secondary groups are your best bet. They are responsible for building our houses, growing and shipping our vegetables, educating our children, and curing our ills. In short, we could not do without them.

The Shift to Secondary Groups

In preindustrial society, there were few secondary groups. Vegetables and houses were produced by families, not by Del Monte or Del Webb. Parents taught their own children, and neighbors nursed one another's ills. Under these conditions, primary groups served both expressive and instrumental functions. As society has become more industrialized, more and more of our instrumental needs are the obligation of some secondary group rather than of a primary group.

In addition to losing their instrumental functions to secondary groups, primary groups have suffered other threats in industrialized societies. In the United States, for example, approximately 16 percent of all households move each year (Schachter 2001). This fact alone means that our ties to friends, neighborhoods, and coworkers are seldom really permanent. People change jobs, spouses, and neighborhoods. One consequence of this breakdown of traditional primary groups is that many people rely on secondary contacts even for expressive needs; if they have marriage problems, they may hire a counselor rather than talk to a parent, for example.

Many scholars have suggested that these inroads on the primary group represent a weakening of social control; that is, the weaker ties to neighbors and kin mean that people feel less pressure to conform. They don't have to worry about what the neighbors will say because they haven't met them; they don't have to worry about what mother will say because she lives 2,000 miles away and what she doesn't know won't hurt her. There is some truth in this suggestion, and it may be one of the reasons that small towns with stable populations are more conventional and have lower crime rates than do big cities with more fluid populations—an issue addressed more fully in Chapter 15.

Social Networks

A **social network** is an individual's total set of relationships.

Each of us belongs to a variety of primary and secondary groups. Through these group ties we develop a **social network.** This social network is the total set of relationships we have. It includes our family, our insurance agent, our neighbors, our classmates and coworkers, and the people who belong to our clubs. Through our social network, we are linked to hundreds of people in our communities and perhaps across the country.

Strong and Weak Ties

Strong ties are relationships characterized by intimacy, emotional intensity, and sharing.

Although our insurance agent and our mother are both part of our social network, there is a qualitative difference between them. We can divide our social networks into two general categories of intimacy: strong ties and weak ties. **Strong ties** are relationships characterized by intimacy, emotional intensity, and sharing. We have strong ties with the people we would confide in, for whom we would make sacrifices, and whom we expect to make sacrifices for us. **Weak ties** are relationships that are characterized by low intensity and intimacy (Granovetter 1973). Coworkers, neighbors, fellow club members, distant cousins, and in-laws generally fall in this category.

Weak ties are relationships with coworkers, neighbors, and distant relatives that are characterized by low intensity and intimacy.

Your social network does not include everybody with whom you have ever interacted. Many interactions, such as those with some classmates and neighbors, are so superficial that they cannot truly be said to be part of a relationship at all. Unless contacts develop into personal relationships that extend beyond the simple exchange of services or a passing nod, they would not be included in your social network.

Research suggests that social networks are vital for integration into society; they encourage conformity and build a firm sense of self-identity (Wellman 1999; Wellman & Berkowitz, 1988). Because of their importance for the individual and society, documenting the trends in social networks has been a major focus of sociological study.

■ A critical part of our social network is our strong ties—the handful of people to whom we feel intense loyalty and intimacy. Like these two sisters, women are somewhat more apt than men to choose their strong ties from among family members.

David Brinkerhoff

The Relationship between Ties and Groups

The distinction between strong and weak ties obviously parallels the distinction between primary and secondary groups. The difference between these two sets of concepts is that strong and weak apply to one-to-one relationships whereas primary and secondary apply to the group as a whole. We can have both strong and weak ties within the primary as well as the secondary group.

For example, the family is obviously a primary group; it is relatively permanent, with strong feelings of loyalty and attachment. We are not equally intimate with every family member, however. We may be very close to our mother but estranged from our brother. Similarly, although the school as a whole is classified as a secondary group, we may have developed an intimate relationship, a strong tie, with one of our schoolmates. *Strong* and *weak* are terms used to describe the relationship between two individuals; *primary* and *secondary* are characteristics of the group as a whole.

Strong Ties

Although we have a large research literature on strong ties, there is little consensus on how to define them operationally. As a result, different studies yield different pictures of the extent of strong ties. For example, one study asked a national sample to name those individuals they could "discuss important matters with." Using this definition, the average individual had only three strong ties (Marsden 1987). Another major study used a definition that included every adult you lived with, engaged in social activities with, or would borrow from or confide in. This study found that the average person had 15 to 19 ties (Fischer 1982).

Despite these differences, all studies agree on the factors that affect the number and composition of strong ties. The most important of these factors is education. People with more education have more strong ties, have a greater diversity of strong ties, and are less reliant on kin. The number of ties also varies by residence and age. Urban residents have more strong ties than rural residents, perhaps in part because they have a greater variety from which to choose. Older respondents consistently report the fewest

Connections

Example

In 2003, sociologists asked 24,613 volunteers to send an e-mail message to one of 18 people in 13 countries. Each volunteer was asked to forward the message to someone he or she knew, with the goal of eventually delivering the message to the target person. Only 384 e-mail chains reached their target, with most chains breaking down quickly because people didn't forward the message. Successful chains were more likely than unsuccessful chains to rely on weak ties: Individuals forwarded the message to anyone they knew in the same occupation or country as the target, rather than sending it to individuals with whom they had strong ties.

Strong communities are built by dense networks of weak ties, such as those in small, stable towns where everybody knows everybody else.

Ed Kashi

A Global Perspective

Finding Jobs, Changing Jobs in Singapore and China

How will you go about looking for a job after you graduate? After you have read over all of the classified ads and gone to the Career Planning and Placement Office, what will you do next? For most of us, the answer seems obvious: ask friends and family if they are aware of any challenging, well-paying, secure job opportunities for persons with just our credentials. Research conducted in the United States suggests that this might be the wrong strategy.

Mark Granovetter (1973, 1974) and others have shown that while individuals are likely to use their personal network of strong ties to search for work, they are more likely to actually find work through their weak ties. How does this work? This is how Granovetter explains "the strength of weak ties": The people to whom we are most strongly tied are a lot like us; our friends and family tend to be interested in the same things we are,

share similar educational and economic backgrounds, and know the same people we do. Consequently, they are unlikely to offer much information about job possibilities that we didn't already know or couldn't find out. In contrast, our weak ties— people with whom we interact less frequently, whom we know less well— are more likely to have information about jobs that weren't listed in the newspaper and that our friends didn't know about. Furthermore, because our weak ties are most likely to bridge us to people with higher status than our own, weak ties give us access to information about more prestigious jobs and to people who may have some influence in hiring decisions (Lin 1990). But are weak ties equally important in finding work in other countries with different economic and social structures?

In the late 1980s and early 1990s, Yanjie Bian and Soon Ang (1997) went to Tianjin, China and to Singapore to study this question. Tianjin is the third largest city in China (Map 5.1) and in the late 1980s had still not begun the transition to a more Westernized

market economy. Jobs in Tianjin and elsewhere in China were assigned by the government, and although new workers could express their job preferences, these preferences did not necessarily affect their final assignments. In Singapore, the situation was somewhat different. Because of labor shortages, Singaporean workers had a very broad range of jobs from which to choose. And because advertising for new employees was costly, Singaporean employers had a strong stake in trying to identify potential employees who would be loyal to the company and stay with them for a long time.

In both Tianjin and Singapore, individuals rely much more directly and explicitly on their social networks for favors of all kinds than is often true in the United States. The guiding principle for individual social networks is *guanxi* (pronounced gwan-she). *Guanxi* literally means "relationship" or "relation" and refers to a set of interpersonal connections that are based on the reciprocal exchange of favors. *Guanxi* relations are strong ties, and those who fail to reciprocate

---

strong ties. Gender does not appear to make much difference in the number of strong ties that people have, but it does affect the source: Women's ties are more likely than men's to be drawn from the kin group (Marsden 1987; Moore 1990).

Weak Ties

Weak ties are also important to social life. For example, research indicates that many people first hear about jobs and career opportunities through weak ties (Granovetter 1974; McPherson & Smith-Lovin 1982; Newman 1999b). In this and other instances, the more people you know, the better off you are.

Voluntary associations are nonprofit organizations designed to allow individuals an opportunity to pursue their shared interests collectively.

Voluntary Associations

In addition to relationships formed with individuals, many of us voluntarily choose to join groups and associations. We may join the PTA, a bowling team, the Elks, or the Sierra Club. These groups, called **voluntary associations,** are nonprofit organizations designed to allow individuals an opportunity to pursue their shared interests col-

exchanges lose face (Bian 1997; Bian & Soon 1997).

So what did Bian and Ang find? In contrast to the United States, finding work and changing jobs in Singapore and China depends more on strong ties than on weak ones. In China, the trust embedded in *guanxi* allows people to "bend government rules" in helping job seekers find more desirable employment; in Singapore, employers have more faith in the future loyalty of potential employees who are personally recommended to them through *guanxi* networks. Nevertheless, in both China and Singapore, the best strategy of all appears to be one that involves both weak ties and *guanxi*. Getting help in the job search process depends first and foremost on ties of trust and reciprocity. The highest level job placements occur, however, when a job seeker uses strong ties to approach a higher-ranked person with whom the job seeker has only a weak tie. What is true in China and Singapore may also be true in the United States. Looking for work is not nearly as hard when friends have well-placed friends.

MAP 5.1
Tianjin, China and Singapore.

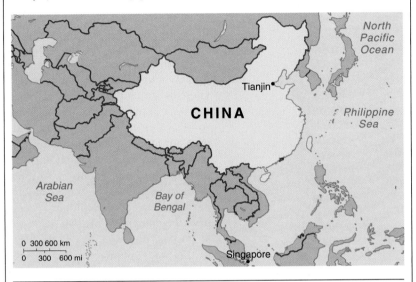

 For more information, look up the following subjects in InfoTrac College Edition:
Strong ties
Job seeking

 Or visit the following Web site:
Guanxi
http://www.projectjanel.org/china/guanxi.html

lectively. They vary considerably in size and formality. Some—for example, the Elks and the PTA—are very large and have national headquarters, elected officers, formal titles, charters, membership dues, regular meeting times, and national conventions. Others—for example, bowling teams and quilting groups—are small, informal groups that draw their membership from a local community or neighborhood.

Voluntary associations are an important mechanism for enlarging our social networks. Most of the relationships we form in voluntary associations will be weak ties. But voluntary associations also can introduce us to people with whom we will develop strong ties as close friends and intimates.

Voluntary associations perform an important function for individuals. Studies document that people who participate in them generally report greater satisfaction and personal happiness, longer life, greater self-esteem, more political effectiveness, and a greater sense of community (Burman 1988; Ellison 1991; Moen, Dempster-McLain, & Williams 1989; Prestby et al. 1990; Rietschlin 1998). The correlation between high participation and greater satisfaction does not necessarily mean that joining a voluntary association is the road to happiness. At least part of the relationship is undoubtedly due to the fact that happy persons who feel politically effective and attached to their communities are more likely than others to join voluntary associations. It also appears to be true, however, that greater participation can be an avenue for achievement and lead to feelings of integration and satisfaction.

The Mediation Hypothesis

An important characteristic of voluntary associations is that they combine some of the features of primary and secondary groups—for example, the companionship of a small group and the rational efficiency of a secondary group. Some scholars have therefore suggested that voluntary associations mediate (provide a bridge) between primary and secondary groups (Pollock 1982). They allow us to pursue instrumental goals without completely sacrificing the satisfactions that come from participation in a primary group. Through participation in voluntary associations, we meet some of our needs for intimacy and association while we achieve greater control over our immediate environment. Take, for example, the sportsman who wishes to protect both wildlife habitat and the right to have guns. This individual can write letters to his representatives in Congress, but he will believe, rightly, that as an individual he is unlikely to have much clout. If this same individual joins with others in, say, the National Wildlife Federation or the National Rifle Association, he will have the enjoyment of associating with other like-minded individuals and the satisfaction of knowing that a paid lobbyist is representing his opinions in Washington. In this way, voluntary associations provide a bridge between the individual and large secondary associations.

Participation in Voluntary Associations

Although some social critics have argued that membership in U.S. voluntary associations has declined—a thesis popularized in the book *Bowling Alone*, by Robert Putnam (2000)—most observers believe that, if anything, participation has increased (Rich 1999). It is true that some large voluntary associations, such as the Elks and bowling leagues, have seen declines in membership. Other groups, however, are burgeoning, especially small local associations, groups focused on ethnicity or gender issues; alternative religious organizations, and Internet-based groups (Rich).

Most people in the United States belong to at least one voluntary association, and approximately one-fourth participate in three or more (Curtis, Grabb, & Baer 1992). Among those who report membership, a large proportion are passive participants—they belong in name only. They buy a membership in the PTA when pressured to do so, but they don't go to meetings. Similarly, anyone who subscribes to *Audubon* magazine is automatically enrolled in the local Audubon Club, but few subscribers become active members. Because so many of our memberships are superficial, they are also temporary. Nevertheless, most people in the United States maintain continuous membership in at least one association.

Active participation in voluntary associations exhibits much the same pattern we noted earlier for strong ties. Urban residents, the middle-aged, and the well-educated are the groups most frequently involved (Curtis et al. 1992). Although women belong to about as many associations as men do, the types of organizations they join are different; women tend to join smaller, less formal, and more expressive organizations. The most important reason for these patterns of belonging appears to be the different opportunities that men and women, higher and lower social classes, the young and old, and urban and rural residents have to participate meaningfully in the full range of voluntary associations (McPherson & Smith-Lovin 1982; Williams & Ortega 1986).

Community

In everyday life, we hear a lot of talk about the benefits of having a "sense of community." In mourning the contemporary loss of community, commentators seem to regard community as a good thing, but they usually don't give a very specific definition about what community is. More importantly, they don't give a very specific meaning for what community *was*.

According to sociologists, a strong **community** is characterized by dense, cross-cutting social networks (Wellman & Berkowitz 1988). A community is strongest when all members are linked to one another through complex overlapping ties. Many scholars believe that communities of the past were characterized by such networks, but one important study challenges this assumption. Using network data collected in 1939 from one neighborhood in Bloomington, Indiana, Karen E. Campbell (1990) concludes that ". . . this neighborhood, and perhaps others like it, was not significantly more sociable than contemporary settings" (151).

Furthermore, studies show that network ties need not be strong to have important consequences for individuals and the community. For example, research shows that neighbors help each other in many ways, even though they share few confidences (Wellman & Wortley 1990). Studies also show that as the proportion of people who simply know one another increases, deviance and fear of crime are reduced, better control is exercised over children, and the weak and handicapped are more likely to be cared for (Freudenberg 1986). Finally, research suggests that voluntary associations promote community cohesion and solidarity by creating a network of weak ties among members (McPherson 1983; Sampson 1988).

> A **community** is characterized by dense, cross-cutting social networks.

Computer Networks and Communities

With the exponential rise in use of the Internet, many individuals now seek and find on-line networks and communities (DiMiaggio et al. 2001; Wellman 1999; Wellman et al. 1996). Through on-line discussion groups and chat rooms, anyone can quickly send out a comment or request to a large and diverse audience. Although the quality

■ Participating in computer games can increase social networks by helping each individual make new friends and by cementing existing friendships.

© Mark Peterson/CORBIS

of information and relationships obtained via the Internet can vary widely, the Internet does provide a wide network of weak ties to many people who might otherwise be isolated. Furthermore, because individuals often forward the comments or requests they receive to others, this network of weak ties can grow both broadly and quickly.

On-line networks also can provide strong ties and a true sense of community, although this is less common. Even when individuals initially enter on-line groups simply to obtain information, those who stay typically do so because they enjoy the social support, companionship, and sense of community the group offers. Relatively strong on-line communities can form over everything from organizing political efforts to writing and sharing personal journals, each group fulfilling a different combination of instrumental and expressive functions. Many of the most popular on-line groups link people who share a health problem. Within these groups, individuals share not only suggestions regarding medical treatment but also their fears, sorrow, and triumphs as they grapple with their injuries or illnesses.

Interestingly, even participating in computer games—seemingly a highly individual activity—can *increase* individuals' social networks. A national survey (Jones 2003) conducted in 2002 found that 65 percent of college students regularly or occasionally play computer games (including Internet games and computerized video games). Surprisingly, two-thirds of computer gamers believe that playing computer games does not reduce the time they spend with family or friends, and 20 percent believe that gaming helped them make new friends and cement existing friendships. In observations on college campuses, the researchers found that students often sit together in computer labs and trade tips on new games or new strategies for old popular games. Other students participate in chat options on interactive, multi-player games. Even computer games, it seems, can increase both strong and weak ties.

Complex Organizations

Few people in our society escape involvement in large-scale organizations. Unless we are willing to retreat from society altogether, a major part of our lives is organization-bound. Even in birth and death, large, complex organizations (such as hospitals and vital statistics bureaus) make demands on us. Throughout the in-between years, we are constantly adjusting to organizational demands (Kanter 1981).

Sociologists use the term **complex organizations** to refer to large, formal organizations with complex status networks (Brinkerhoff & Kunz 1972). Examples include universities, governments, business corporations, and voluntary associations such as the Masons and the Catholic Church.

These complex organizations make a major contribution to the overall quality of life within society. Because of their size and complexity, however, they don't supply the cohesion and personal satisfaction that smaller groups do. They may make their members feel as if they are simply cogs in the machine rather than important people in their own right. This is nowhere more true than in a bureaucracy.

Complex organizations are large formal organizations with complex status networks.

Bureaucracy is a special type of complex organization characterized by explicit rules and hierarchical authority structure, all designed to maximize efficiency.

Classic Bureaucracies

Bureaucracy is a special type of complex organization characterized by explicit rules and a hierarchical authority structure, all designed to maximize efficiency. In popular usage, bureaucracy often has a negative connotation: red tape, silly rules, and unyielding rigidity. In social science, however, it is simply an organization in which the roles of each actor have been carefully planned to maximize efficiency.

The Classic View

Most large, complex organizations are also bureaucratic: IBM, General Motors, U.S. Steel, the Catholic Church, colleges, and hospitals. The major characteristics of bureaucracies were outlined almost 100 years ago by Max Weber ([1910] 1970a):

1. *Division of labor and specialization.* Bureaucratic organizations employ specialists in each position and make them responsible for specific duties. Job titles and job descriptions specify who is to do what and who is responsible for each activity.

2. *Hierarchy of authority.* Positions are arranged in a hierarchy so that each position is under the control and supervision of a higher position. Frequently referred to as chains of command, these lines of authority and responsibility are easily drawn on an organization chart, often in the shape of a pyramid.

3. *Rules and regulations.* All activities and operations of a bureaucracy are governed by abstract rules or procedures. These rules are designed to cover almost every possible situation that might arise: hiring, firing, and the everyday operations of the office. The object is to standardize all activities.

4. *Impersonal relationships.* Interactions in a bureaucracy are supposed to be guided by the rules rather than by personal feelings. Consistent application of impersonal rules is intended to eliminate favoritism and particularism.

5. *Careers, tenure, and technical qualifications.* Candidates for bureaucratic positions are supposed to be selected on the basis of technical qualifications such as high scores on civil service examinations, education, or experience. Once selected for a position, persons should advance in the hierarchy by means of achievement and seniority.

6. *Efficiency.* Bureaucratic organizations coordinate the activities of a large number of people in the pursuit of organizational goals. All have been designed to maximize efficiency. From the practice of hiring on the basis of credentials rather than personal contacts to the rigid specification of duties and authority, the whole system is constructed to keep individuality, whim, and favoritism out of the operation of the organization.

Organizational Culture

Weber's classic theory of bureaucracy almost demands robots rather than individuals. A list of rules that covered every possible situation would be unwieldy and impossibly long. Not surprisingly, therefore, we find that few organizations try to be totally bureaucratic. Instead, organizations strive to create an atmosphere of goodwill and common purpose among their members so that they all will apply their ingenuity and best efforts to meeting organizational goals (DiTomaso 1987). This goodwill is as essential to efficiency as are the rules. In most organizations, in fact, working exactly according to the rules is considered a form of sabotage. For example, unions of public employees (such as the police) that cannot legally strike engage in "working to the rule" as a form of protest: They follow every nit-picking rule and fill out every form carefully. The result is usually a sharp slowdown in work and general chaos.

Sociologists use the term *organizational culture* to refer to the pattern of norms and values that affects how business is actually carried out in an organization. The key to a successful organizational culture is cohesion, and most organizations strive to build cohesion among their members. They do this by encouraging interaction among employees (providing lunchrooms, sponsoring after-hours sports leagues, having company picnics and newsletters, and developing unifying symbols, such as mascots, company colors, or uniforms). This is most clearly apparent when you think about the large bureaucratic organization represented by a university, but it is also characteristic of transnational corporations, especially in Japan.

In many organizations, the formal rules have very little to do with the day-to-day activity of the members. In situations as varied as classrooms and shop floors, people evolve their own way of doing business and may have little use for the formal rules (Ouichi & Wilkins 1985). They may not even know what the rules are. What determines whether an organization works by the rules or not?

The degree of bureaucratization in an organization is related to the degree of uncertainty in the organization's activities. When activities tend to be routine and predictable, the organization is likely to emphasize rules, central planning, and hierarchical chains of command. When activities change rapidly in unpredictable ways, there is more emphasis on flexibility and informal decision making (Simpson 1985). This explains why, for example, classrooms tend to be less bureaucratic and factories more bureaucratic.

Criticisms of Bureaucracies

Bureaucracy is the standard organizational form in the modern world. Organizations from churches to governments are run along bureaucratic lines. Despite the widespread adoption of this organizational form, it has several major drawbacks. Three of the most widely acknowledged are as follows:

1. *Ritualism.* Rigid adherence to rules may mean that a rule is followed regardless of whether it helps accomplish the purpose for which it was designed. The rule becomes an end in itself rather than a means to an end. For example, individuals may struggle to arrive at 9 A.M. and leave at 5 P.M. when they could work more effectively from 10 to 6. Although the existence of a bureaucracy per se doesn't always breed rote rule adherence (Foster 1990), an overemphasis on bureaucratic rules can stifle initiative and prevent the development of more efficient procedures (Blau & Meyer 1971).

2. *Alienation.* The emphasis on rules, hierarchies, and impersonal relationships can sharply reduce the cohesion of the organization. This has several drawbacks: It reduces social control, reduces member satisfaction and commitment, and increases staff turnover. All of these may interfere with the organization's ability to reach its goals.

3. *Structured inequality.* Critics charge that the modern bureaucracy with its multiple layers of authority is a profoundly antidemocratic organization. The purpose of the bureaucratic form is to concentrate power in one or two decision makers whose decisions are then passed down as orders to subordinates. Some observers believe that the amassing of concentrated power in the name of efficiency and rationalism is incompatible with a democratic society (Perrow 1986).

McDonaldization is the process by which the principles of the fast-food restaurant—efficiency, calculability, predictability, and control—are coming to dominate more sectors of American society.

In his highly influential writings, George Ritzer (1996) claimed that many modern bureaucratic organizations are now experiencing McDonaldization. **McDonaldization** refers to "the process by which the principles of the fast-food restaurant are coming to dominate more and more sectors of American society as well as of the rest of the world" (1). The emphasis in McDonaldization is on four elements: efficiency, calculability, predictability, and control. For example, McDonald's streamlined its procedures to serve customers extremely rapidly, advertised how many ounces are in a Big Mac rather than how it tastes (which is harder to calculate), and guaranteed that a Big Mac in New York tastes exactly like a Big Mac in Des Moines. It also has controlled not only its employees (through strict guidelines on exactly how to do various tasks) but also its customers (through limited menus and uncomfortable seats that encourage customers to order and eat quickly). These principles have now been adopted by all kinds of organizations around the world, from drop-off laundries to "dial-a-porn" sex businesses.

The attempt to rationalize bureaucratic structures through McDonaldization often produces *irrational* consequences. The disadvantages of McDonaldization stem directly from each supposed advantage. For instance, although it is more efficient for businesses to use voice-mail systems instead of operators, it is less efficient for the customers who must listen to a series of menus, hoping that they will reach the department they seek before getting disconnected. Similarly, businesses like McDonald's make decisions based on calculations of how they can best generate a profit, but they do not calculate the impact of their business decisions on the environment or on the quality of life of their customers or workers. The predictability that chain stores and restaurants offer makes the world a less interesting place, as large national businesses drive out unique local businesses. And the control that McDonaldized organizations offer is frequently dehumanizing—something you have probably experienced every time your name and identity are replaced by an institutional identification number.

Japanese Bureaucracies

Over the last few decades, Japan has come to dominate dozens of areas of international trade, from automobiles to video recorders. The astounding success of Japanese manufacturers has led to intense scrutiny of Japanese bureaucracies to discover their secret (Dore 1973; Lincoln and McBride 1987; Marsh & Mannari 1976; Masatsugu 1982).

Like their U.S. counterparts, Japanese firms are large, complex bureaucracies. They have hierarchies of authority, lots of rules, hiring on the basis of technical qualifications, and many of the other features of classic bureaucracies. In addition, they have several distinct characteristics:

Bureaucracies are characterized by alienating, rigid job conditions, such as those experienced by these customer service representatives.

1. *Permanent employment.* Japanese firms have a strong norm of permanent employment. On leaving school, employees typically enter a firm and remain there until retirement. The company is expected to make a lifetime commitment to the employee, and the employee is expected to make a lifetime commitment to the firm. Recent economic problems in Japan have threatened the guarantee of lifetime employment, but it is still far more common in Japan than in the United States.

2. *Internal labor markets.* Higher-level positions are filled from below from within the same firm. Thus, people can make a lifetime commitment to the same firm and still have the opportunity for promotion and upward mobility. This system differs from the common U.S. practice of pursuing careers by changing employers. Because of lifetime commitment and internal labor markets, seniority in the firm and age are highly correlated—and generally highly rewarded.

3. *Relative absence of white-collar/blue-collar distinction.* In Japan, all workers are expected to be committed to the firm; thus all workers work on salary rather than for an hourly wage. There are no separate cafeterias and parking lots for the supervisors and for the line workers. This also means that the internal labor market crosses the white-collar/blue-collar line: Although one might start as a blue-collar worker, seniority is likely to move one into management ranks.

■ The bottom-up style of management characteristic of Japanese organizations gives workers a greater investment in their product. It may also be the key to higher-quality products.

© Eiji Miyazawa/Black Star

4. *Participatory decision making.* Unlike the classic bureaucratic model in which decisions flow in only one direction (down), a more circular pattern of decision making is the norm in Japanese organizations. Suggestions may originate among workers' circles and then rise to the top (called bottom-up management), or they may originate at the top but be implemented only after there is substantial consensus among the workers. This participatory decision-making process is yet another mechanism that reduces the distinction between management and line workers and increases the cohesion and commitment of the employees.

Unlike the classic bureaucratic model, which pretty much ignores the human component of the workplace, the Japanese bureaucratic model recognizes that complex organizations are *human groups* rather than machines. Thus, it takes into consideration issues such as cohesion and group dynamics. Ironically, the realization that even corporations such as IBM are groups and are affected by group processes is moving many U.S. companies toward the Japanese model at the exact same time that economic pressures are leading Japanese firms toward the U.S. model.

Where This Leaves Us

Humans are social beings. Our lives are lived within relationships, groups, networks, and—whether we like it or not—complex organizations. Without these human connections we cannot survive, let alone thrive. Groups, networks, and organizations help us obtain the very basics of life—food, clothing, work, shelter, companionship, love. They also enable us to make our mark on the world, as we raise our children within families, create better communities through voluntary associations, strive for success within complex organizations from schools to corporations, and so on. Yet working with others also carries risks, for exchange and cooperation can turn to competition and conflict, and groups can affect our ideas and behaviors in ways we may not even recognize. A sociological understanding of groups, networks, and organizations can help us understand, prevent, and, where necessary, counteract these effects.

Summary

1. Relationships are characterized by four basic social processes: exchange, cooperation, competition, and conflict.

2. Groups are distinguished from aggregates and categories in that members take one another into account, and their interactions are shaped by shared expectations and interdependency.

3. Group interaction is affected by group size and the proximity and communication patterns of group members. The amount of interaction in turn affects group cohesion, the amount of social control the group can exercise over members, and the quality of group decisions.

4. Primary groups are characterized by intimate, face-to-face interaction. They are essential to individual satisfaction and integration, and they are also primary agents of social control in society. Secondary groups are large, formal, and impersonal. They are generally task-oriented and perform instrumental functions for societies and individuals.

5. Each person has a social network that consists of both strong and weak ties. The number of strong ties is generally greater for individuals who are urban, middle-aged, and highly educated.

6. In the United States, weak ties are important in learning about new jobs; in China and Singapore, strong ties appear to provide better help to job seekers.

7. Voluntary associations are nonprofit groups that bring together people with shared interests. They combine some of the expressive functions of primary groups with the instrumental functions of secondary groups.

8. When individuals are linked by dense, cross-cutting networks, they form a community.

9. Bureaucracy is a rationally designed organization whose goal is to maximize efficiency. The chief characteristics of a bureaucracy are division of labor and specialization, a hierarchy of authority, a system of rules and regulations, impersonality in social relations, and emphasis on careers, tenure, and technical qualifications.

10. Although most contemporary organizations are built on a bureaucratic model, many are far less rational than the classic model suggests. Critics of McDonaldization suggest that bureaucratic emphasis on rationality can have irrational consequences. All effective bureaucracies rely on an organizational culture to inspire employees to give their best efforts and to help meet organizational goals.

11. Japanese organizations fit somewhere between the extreme rationalism of the classic bureaucracy and the people-centered collectivist organizations. Japanese organizations try to maximize employees' commitment to the firm through four mechanisms: lifetime employment, internal labor markets, little distinction between blue- and white-collar workers, and participatory decision making.

Thinking Critically

1. As people around the world have been brought together by computer networks, something we might call "electronic proximity" has developed. Do you think that electronic proximity will have the same effects on friendship as physical proximity does?

2. Can you think of a situation in your life in which the primary source of social control was a secondary rather than a primary group?

3. Suppose you were trying to get help for a family member's substance-abuse problem. Who would you turn to, your strong or your weak ties?

4. From your experience, what are some of the functions of bureaucracy? What are some of the dysfunctions? Would you classify the dysfunctions as manifest or latent? Why?

5. Why do you suppose both union leaders and managers often oppose participatory decision making? What other obstacles do you think there will be for U.S. firms implementing Japanese management techniques?

Sociology on the Net

The Wadsworth Sociology Resource Center: Virtual Society

http://sociology.wadsworth.com/

The companion Web site for this book includes a range of enrichment material. Further your understanding of the chapter by accessing Online Practice Quizzes, Internet Exercises, InfoTrac College Edition Exercises, and many more compelling learning tools.

Chapter-Related Suggested Web Sites

Center for the Study of Group Processes
http://www.uiowa.edu/~grpproc/

E-Commerce Now: Japanese Management
http://www.ecommerce-now.com/images/ecommerce-now/Japanesemanagement.htm

Historical Background of Organizational Behavior
http://web.cba.neu.edu/~ewertheim/introd/history.htm

InfoTrac College Edition

http://www.infotrac-college.com/wadsworth

Access the latest news and research articles online—updated daily and spanning four years. InfoTrac College Edition is an easy-to-use online database of reliable, full-length articles from hundreds of top academic journals and popular sources. Conduct an electronic search using the following key search terms:

Bureaucracy

Organizational behavior

Social network

Interpersonal communication

Small groups

Suggested Readings

Fischer, Claude S. 1992. *America Calling: A Social History of the Telephone to 1940.* Berkeley: University of California Press. Fischer's research on social networking has motivated him to investigate the social history of the telephone in the United States, a cultural artifact that facilitates networking in modern society.

Menjivar, Cecilia. 2000. *Fragmented Ties.* Berkeley, Calif.: University of California Press. Extensive interviews and observations of Salvadoran immigrants to California show how social networks can both help and hinder recent immigrants.

Rheingold, Howard. 2002. *Smart Mobs: The Next Social Revolution.* New York: Perseus Publishing. "Smart mobs" ("mobs" rhymes with "robes") are individuals who use mobile phones and Internet devices to work and play together. Smart mobs include everyone from teenagers looking for others to "hang out" with to prostitutes seeking customers and terrorists coordinating attacks.

Ritzer, George. 1996. *The McDonaldization of Society,* revised edition. Thousand Oaks, Calif.: Pine Forge Press. A fascinating account by a scholar and social activist on how McDonaldization is changing everything from how we are born to how we die.

Putnam, Robert. 2000. *Bowling Alone: The Collapse and Revival of American Community.* New York: Simon & Schuster. An influential and controversial argument about the decline of voluntary associations, neighborhoods, and other community ties in American culture today.

CHAPTER 6

Deviance, Crime, and Social Control

AP/Wide World Photos

Outline

Conformity and Deviance

In providing a blueprint for living, our culture supplies norms and values that structure our behavior. These norms and values tell us what we ought to believe in and what we ought to do. Because we are brought up to accept them, for the most part we do what we are expected to do and think as we are expected to think. Only "for the most part," however, because none of us follows all the rules all the time.

Previous chapters concentrated on how norms and values structure our lives and how we learn them through socialization. This chapter considers some of the ways individuals break out of these patterns—from relatively unimportant eccentric behaviors to serious violations of others' rights.

Understanding Conformity

Most people, most of the time, conform to social norms. To understand why people *break* social norms, we first must understand why people usually conform. The forces and processes that encourage conformity are known as **social control.** Social control takes place at three levels:

1. Through self-control, we police ourselves.
2. Through informal controls, our friends and intimates reward us for conformity and punish us for nonconformity.
3. Through formal controls, the state or other authorities discourage nonconformity.

Self-control occurs because individuals **internalize** the norms and values of their group. They make conformity to these norms part of their self-concept. Thus, most of us do not murder, rape, or rob, not simply because we are afraid the police would catch us but because it never occurs to us to do these things; they would violate our sense of self-identity. A powerful support of self-control is **informal social control,** self-restraint exercised because of fear of what others will think. Thus, even if your own values did not prevent you from cheating on a test, you might be deterred by the thought of how embarrassing it would be to be caught. Your friends might sneer at you or drop you altogether; your family would be disappointed in you; your professor might publicly embarrass you by denouncing you to the class. If none of these considerations is a deterrent, you might be scared into conformity by the thought of **formal social controls,** i.e., administrative sanctions such as fines, expulsion, or imprisonment. Cheaters, for example, face formal sanctions such as automatic failing grades and dismissal from school.

Whether we are talking about cheating on examinations or murder, social control rests largely on self-control and informal social controls. Few formal agencies have the ability to force compliance to rules that are not supported by individual or group values. Sex is a good example. In many states, sex between unmarried persons is illegal, and you can be fined or imprisoned for it. Even if the police devoted a substantial part of their energies to stamping out illegal sex, however, they would probably not succeed. In contemporary United States, a substantial proportion of unmarried people are not embarrassed about having sexual relations; they do not care if their friends know about it. In such conditions, formal sanctions cannot enforce conformity. Prostitution, marijuana use, seat-belt laws—all are examples of situations for which laws unsupported by public consensus have not produced conformity.

Social control consists of the forces and processes that encourage conformity, including self-control, informal control, and formal control.

Internalization occurs when individuals accept the norms and values of their group and make conformity to these norms part of their self-concept.

Informal social control is self-restraint exercised because of fear of what others will think.

Formal social controls are administrative sanctions such as fines, expulsion, or imprisonment.

Defining Deviance

People may break out of cultural patterns for a variety of reasons and in a variety of ways. Whether your nonconformity is regarded as deviant or merely eccentric depends on the seriousness of the rule you violate. If you wear bib overalls to church or carry a potted palm with you everywhere, you will be challenging the rules of conventional behavior. Probably nobody will care too much, however; these are minor kinds of nonconformity. We speak of **deviance** when norm violations exceed the tolerance level of the community and result in negative sanctions. Deviance is behavior of which others disapprove to the extent that they believe something significant ought to be done about it.

Defining deviance as behavior of which others disapprove has an interesting implication: It is not the *act* that is important but the *audience*. The same act may be deviant in front of one audience but not another, deviant in one place but not another.

Few acts are intrinsically deviant. Even taking another's life may be acceptable in war, police work, or self-defense. Whether an act is regarded as deviant often depends on the time, the place, the individual, and the audience. For this reason, sociologists stress that *deviance is relative*. For example, alcohol use is deviant for adolescents but not for adults, having two wives is deviant for the United States but not for Nigeria, carrying a gun to town is deviant in the late twentieth century but was not in the nineteenth century, and wearing a skirt is deviant for a U.S. male but not for a U.S. female.

The sociology of deviance has two concerns: why people break the rules of their time and place and the processes through which the rules become established. In the following sections, we review several major theories of deviance before looking at crime—a type of deviance—in the United States.

■ Although your parents are not likely to be pleased if you pierce your nose, doing so does not violate any major norms or arouse too much public disapproval. It is an example of nonconformity but not of deviance.

Sociological Theories About Deviance

Biological and psychological explanations for deviant behavior typically focus on how processes going on within the individual lead to deviance. Such theories often look for the causes of deviance in genetics, neurochemical imbalances, or childhood failures to internalize appropriate behavior or attitudes. Most sociologists agree that biology and psychology play a role in causing deviance, but consider social forces even more important. Sociological theories, therefore, search for the causes of deviance within the social structure rather than within the individual (see the Concept Summary "Theories of Deviance").

Deviance refers to norm violations that exceed the tolerance level of the community and result in negative sanctions.

Structural-Functional Theories

In Chapter 1, we said that the basic premise of structural-functional theory is that the parts of society work together like the parts of an organism. From this point of view, deviance is alien to society, an indication that the parts are not working right.

This perspective was first applied to the explanation of deviance by Emile Durkheim in his classic study of suicide ([1897] 1951). Durkheim was trying to explain why people in industrialized societies are more likely to commit suicide than

Theories of Deviance

Concept Summary

	Major Question	Major Assumption	Cause of Deviance	Most Useful for Explaining Deviance of
Structural-Functional Theory				
Strain theory	Why do people break rules?	Deviance is an abnormal characteristic of the social structure.	A dislocation between the goals of society and the means to achieve them	The working and lower classes who cannot achieve desired goals by prescribed means
Conflict Theory	How does unequal access to scarce resources lead to deviance?	Deviance is a normal response to competition and conflict over scarce resources.	Inequality and competition	All classes: Lower class is driven to deviance to meet basic needs and to act out frustration; upper class uses deviant means to maintain their privilege
Symbolic Interaction Theories				
Differential association theory	Why is deviance more characteristic of some groups than others?	Deviance is learned like other social behavior.	Subcultural values differ in complex societies; some subcultures hold values that favor deviance; these are learned through socialization	Delinquent gangs and those integrated into deviant subcultures and neighborhoods
Deterrence theories	When is conformity not the best choice?	Deviance is a choice based on cost/benefit assessments.	Failure of sanctioning system (benefits of deviance exceed the costs)	All groups, but especially those lacking a "stake in conformity"
Labeling theory	How do acts and people become labeled *deviant*?	Deviance is relative and depends on how others label acts and actors.	People whose acts are labeled *deviant* and who accept that label become career deviants	The powerless who are labeled *deviant* by more powerful individuals

are people in other societies. He suggested that in traditional societies the rules tend to be well known and widely supported. As a society grows larger, becomes more heterogeneous, and experiences rapid social change, the norms of society may be unclear or no longer applicable to current conditions. Durkheim called this situation **anomie**; he believed that it was a major cause of suicide in industrializing nations.

Strain Theory

Anomie is a situation in which the norms of society are unclear or no longer applicable to current conditions.

Strain theory suggests that deviance occurs when culturally approved goals cannot be reached by culturally approved means.

The anomie idea was broadened to apply to all sorts of deviant behavior in Robert Merton's (1957) **strain theory.** Strain theory suggests that deviance results when culturally approved goals cannot be reached by culturally approved means. Strain theory is most useful for understanding the consequences of our strong cultural emphasis on economic success. Although this goal is widely shared by Americans, the means to obtain this goal are not. In particular, Merton argued that people from the lower social classes have less opportunity to achieve success through culturally approved means,

Merton's Types of Deviance

Merton's strain theory of deviance suggests that deviance results whenever there is a disparity between goals and the institutionalized means available to reach them. Individuals caught in this dilemma may reject the goals or the means or both. In doing so, they become deviants.

Modes of Adaptation	Cultural Goals	Institutional Means
Innovation	Accepted	Rejected
Ritualism	Rejected	Accepted
Retreatism	Rejected	Rejected
Rebellion	Rejected/replaced	Rejected/replaced

Concept Summary

such as attending school to become a lawyer or computer programmer. According to Merton, lower-class persons turn to crime not because they *reject* American values but because they *accept* them: They believe that only through crime can they achieve our shared cultural goal of economic success.

Of course, most people who find society's norms inapplicable to their situation nevertheless do not turn to a life of crime. Merton identifies four ways in which people adapt to situations of anomie (see the Concept Summary "Merton's Types of Deviance"): innovation, ritualism, retreatism, and rebellion. The mode of adaptation depends on whether an individual accepts or rejects society's cultural goals and accepts or rejects appropriate ways of achieving them.

People who accept both society's goals and its norms about how to reach them are conformists. Most of us conform most of the time. When people cannot successfully reach society's goals using society's rules, however, deviance is a likely result. One form deviance may take is innovation; people accept society's goals but develop alternative means of reaching them. Innovators include not only lower-class criminals but also students who pursue academic achievement through cheating or athletic achievement through steroids. In these instances, deviance rests on using illegitimate means to accomplish socially desirable goals.

Other people who are blocked from achieving socially desired goals respond by rejecting those goals. Ritualists slavishly go through the motions prescribed by society, but their goal is security, not success. Their major hope is that they will not be noticed. Thus, they do their work carefully, even compulsively. Although ritualists may appear to be overconformers, Merton says they are deviant because they have rejected our society's values on achievement and upward mobility. They have turned their back on normative goals but are clinging desperately to procedure. Retreatists, by contrast, adapt by rejecting both procedures and goals. They are society's dropouts: the vagabonds, drifters, and street people. The final mode of adaptation—rebellion—involves the rejection of society's goals and means and the adoption of alternatives that challenge society's usual patterns. Rebels are the people who start communes or revolutions to create an alternative society. Unlike retreatists, they are committed to working toward a different society.

Strain theory explicitly defines deviance as a social problem rather than a personal trouble; it is a property of the social structure, not of the individual. As a consequence, the solution to deviance lies not in reforming the individual deviant but in reducing the mismatch between cultural goals and culturally approved means. This theory underscores the sociological view that society, not the individual, is an important cause of deviant behavior.

Conflict Theory

Strain theory suggests that deviance results from a lack of integration among the parts of a social structure (norms, goals, and resources); it is viewed as an abnormal state produced by extraordinary circumstances. Conflict theorists, however, see deviance as a natural and inevitable product of competition in a society in which groups have different access to scarce resources. They suggest that the ongoing processes of competition should be the real focus of deviance studies (Lemert 1981).

Conflict theory proposes that deviance results from competition and class conflict. Class conflict affects deviance in two ways (Reiman 1998): (1) Class interests determine how the criminal justice system responds to crime and (2) economic pressures can lead to crime, particularly property crimes, among the poor.

Responding to Crime

Conflict theorists argue that the law is a weapon used by the ruling class to maintain the status quo (Arrigo 1998; Liska, Chamlin, & Reed 1985; Reiman 1998). This interpretation fits in with the general Marxist notion that all social institutions, including law, have been created to rationalize and support the current distribution of power and economic resources.

Supporters of this position note that the criminal justice system spends more money deterring muggers than embezzlers and more money arresting prostitutes than their clients. Courts impose much more severe sentences for street crimes than for corporate crimes and impose much heavier sentences against those who use drugs favored by the poor (such as "crack" cocaine) than those who use drugs favored by the more affluent (such as other forms of cocaine). Police are more likely to arrest those who assault members of the ruling class (well-off whites) than those who assault the powerless (nonwhites and the poor) (Reiman 1998). Finally, even when people from the upper and lower classes commit similar crimes, those from the lower class are more likely to be arrested, prosecuted, and sentenced (Reiman). The system clearly seems to benefit the upper classes.

Similarly, most conflict theorists reject strain theory's assumption that an inability to meet socially accepted goals through legitimate means makes poor people unusually likely to commit crimes. Instead, research suggests, most poorer people adjust their goals downward sufficiently so that they can meet their goals through respectable means (Simons & Gray 1989). At the same time, many highly successful individuals adjust their goals so far upward that they cannot reach them by legitimate means. Recent court cases that reveal Microsoft's illegal attempts to gain a monopoly over Internet services and tobacco manufacturers' attempts to make cigarettes more addictive provide clear evidence that the means-versus-goals discrepancy is not limited to the lower class. Conflict theorists argue that it only appears that rich people commit few crimes because rich people control the state, schools, and courts, and so are often able to avoid criminal labels (Reiman 1998).

Lower-Class Crime

Although the preceding view of the way crime is defined would be accepted by all conflict theorists, some believe that individuals in the lower class really are more likely to commit criminal acts. One critical criminologist has declared that crime is a rational response for the lower class (Quinney 1980). These criminologists generally seem to agree with Merton's strain theory that a means/ends discrepancy is particularly acute among the poor and that it may lead to crime (Greenberg 1985). They believe, however, that this is a natural condition of an unequal society rather than an unnatural condition.

Symbolic Interaction Theories

Symbolic interaction theories of deviance suggest that deviance is learned through interaction with others and involves the development of a deviant self-concept. Deviance is not believed to be a direct product of the social structure but of specific face-to-face interactions. This argument takes three forms: differential association theory, deterrence theories, and labeling theory.

Differential Association Theory

To explain the common observation that children who grow up in neighborhoods where there are many delinquents are more likely to be delinquent themselves, Edwin Sutherland (1961) developed differential association theory. **Differential association theory** argues that people learn to be deviant when more of their associates favor deviance than favor conformity.

How does differential association encourage deviance? There are two primary mechanisms. First, if our interactions are mostly with deviants, we may develop a biased image of the generalized other. We may learn that, of course, everybody steals or, of course, the ability to beat other people up is the most important criterion for judging a person. The norms that we internalize may be very different from those of conventional society. The second mechanism has to do with reinforcements. Even if we learned conventional norms, a deviant subculture will not reward us for following them. In fact, a deviant subculture may reward us for violating the norms. Through these mechanisms, we can learn that deviance is acceptable and rewarded.

Differential association theory stems largely from the structural school of symbolic interaction. People develop a deviant identity because they are thrust into a deviant subculture. One's situation determines one's identity.

Deterrence Theories

Differential association theory can only explain deviance that occurs in settings and groups that encourage it. Deterrence theories provide a broader explanation of deviance. These theories suggest that deviance results when social sanctions, formal and informal, provide insufficient rewards for conformity. **Deterrence theories** (also known as "rational choice theories") combine elements of structural-functional and symbolic interaction theories. Although they place the primary blame for deviance on an inadequate (dysfunctional) sanctioning system, they also assign the individual an active role in choosing whether to deviate or conform. This theory assumes that the actor assesses the relative balance of positive and negative sanctions and makes a cost/benefit decision about whether to conform or be deviant (McCarthy 2002; Paternoster 1989; Piliavin et al. 1986). When social structures do not provide adequate rewards for conformity, a larger portion of the population will choose deviance.

Differential association theory points out that people who grow up in crime-ridden neighborhoods are more likely to grow up to be criminals themselves. It is easy to see how differential association theory might apply to gang members like these, but can you think of ways it might also help to explain white-collar crime?

Differential association theory argues that people learn to be deviant when more of their associates favor deviance than favor conformity.

Deterrence theories suggest that deviance results when social sanctions, formal and informal, provide insufficient rewards for conformity.

■ Young women like these, who have received significant rewards for conventional behavior, are highly unlikely to be tempted by opportunities for deviance.

Empirical studies show that three kinds of rewards are especially important in deterring deviance: instrumental rewards, family ties, and self-esteem.

Instrumental Rewards   Unemployment and low wages are among the very best predictors of crime rates at any age (Crutchfield 1989; Devine, Shaley, & Smith 1988; McCarthy 2002). People with no jobs or with dead-end jobs have little to lose and perhaps much to gain from deviance. People who have or can look forward to good jobs, on the other hand, are likely to conclude that they have too much to lose by being deviant. Nevertheless, research suggests that most people with low-wage, dead-end jobs are unwilling to risk what little they have by engaging in deviant activities (Newman 1999b).

Family Ties   Consistent evidence shows that young people with strong bonds to their parents are more likely to conform (Crosnoe, Erikson, & Dornbusch 2002; Messner & Krohn 1990). Parents are in a very strong position to exert informal sanctions that encourage conformity, and young people who are close to their parents are vulnerable to these informal sanctions. There is an important corollary to this rule: deviant parents often produce delinquent children, in large part because they use harsh and inconsistent punishment and form only weak family attachments. Interestingly, family ties have even stronger effects on delinquency in China than in the West probably because in China the family remains the most important institution of social control (Zhang & Messner 1995).

Self-esteem   On a more symbolic level, deterrence theory suggests that people choose deviance or conformity depending on which will do the most to enhance their self-esteem (Kaplan, Martin, & Johnson 1986). For most of us, self-esteem is enhanced by conformity; we are rewarded for following the rules. People whose efforts are not rewarded, however, may find deviance an attractive alternative in their search for positive feedback. Among lower-class boys, delinquency has been found to improve self-esteem (Rosenberg, Schooler, & Schoenbach 1989).

According to deterrence theorists, positive sanctions give individuals a "stake in conformity"—something to lose, whether it's a job, parental approval, or self-esteem. When social structures fail to reward conformity, individuals have less to lose by choosing deviance.

Labeling Theory

A third theory of deviance, which combines symbolic interaction and conflict theories, is labeling theory. **Labeling theory** is concerned with the processes by which the label *deviant* comes to be attached to specific people and specific behaviors. This theory takes to heart the maxim that deviance is relative. As the chief proponent of labeling theory puts it, "Deviant behavior is behavior that people so label" (Becker 1963, 90).

The process through which a person becomes labeled as deviant depends on the reactions of others toward nonconforming behavior. The first time a child acts up in class, it may be owing to high spirits or a bad mood. This impulsive act is *primary deviance*. What happens in the future depends on how others interpret the act. If teachers, counselors, and other children label the child a troublemaker and if she accepts this definition as part of her self-concept, then she may take on the role of a troublemaker. Continued rule violation because of a deviant self-concept is called *secondary deviance*.

The major limitation of labeling theory is that (1) it doesn't explain why primary deviance occurs and (2) it cannot explain repeated deviance by those who haven't been caught, that is, labeled.

Power and Labeling   A crucial question for labeling theorists is how an individual comes to be labeled deviant or criminal. Many labeling theorists take a conflict perspective when answering this question. They assume that one of the strategies groups use in competing with one another is to get the other groups' behavior labeled as deviant rather than its own behavior. Naturally, the more power a group has, the more likely it is to be able to brand its competitors as deviant. This, labeling theorists allege, explains why lower-class deviance is more likely to be subject to criminal sanctions than is upper-class deviance.

In a classic study, Becker (1963) describes how this competition between interest groups caused marijuana users in the United States to be labeled criminal. Before 1937, marijuana use was not illegal in the United States. In 1937, however, a powerful vested-interest group, the Federal Bureau of Narcotics (FBN), campaigned to have it declared illegal. (Since Prohibition had ended, the bureau either had to find a new enemy or go out of business.) In conjunction with another vested-interest group, the Consolidated Brewers, who wanted to shift public attention away from the dangers of alcohol, the FBN launched a major media campaign to stigmatize marijuana use by associating marijuana with violence and other criminal behaviors. Because at the time most marijuana users were disenfranchised poor, Hispanic, or African American, the FBN faced little opposition to its campaign. As a result of its success, it created a new group of criminals. Becker refers to those who are in a position to create and enforce new definitions of morality as **moral entrepreneurs.**

From Sin to Sickness   Labeling theory's emphasis on subjective meanings gives us a framework for understanding the changing definitions of deviance. In recent years, there has been an increasing tendency for behaviors that used to be labeled deviant to be labeled illnesses instead. For example, many now consider alcoholism to be a disease. When a form of deviance comes to be viewed as illness, social reaction changes. It is no longer appropriate to put people in jail for being public drunks; instead they are put in hospitals. Physicians and counselors, rather than judges and sheriffs, treat them. Other

---

Labeling theory is concerned with the processes by which labels such as *deviant* come to be attached to specific people and specific behaviors.

Moral entrepreneurs are people who are in a position to create and enforce new definitions of morality.

Connections

Example

In the last decade, a growing anti-smoking movement has fought to outlaw tobacco smoking in public buildings, private businesses such as restaurants and bars, and even on the street. Anti-smoking activists have also battled to stigmatize smokers by, for example, developing advertising campaigns that accuse pregnant women who smoke of endangering their fetuses and parents who smoke of endangering their children. Sociologists would describe these activists as moral entrepreneurs, who are working to create and enforce new definitions of morality and deviance.

MARIHUANA
THE ASSASSIN OF YOUTH

© David Brinkerhoff

Posters such as this one were widely distributed in the 1930s in the successful campaign—sponsored by breweries—to criminalize marijuana use.

Crime is behavior that is subject to legal or civil penalties.

forms of deviance, such as child abuse, gambling, murder, and rape, also may be regarded as forms of mental illness, better treated by physicians than sheriffs (Conrad & Schneider 1992). At present we have reached few firm decisions on these issues. The public seems to believe that although some murderers, rapists, and so on are mentally ill and should be treated by physicians, others are just bad and should be put in jail.

Individuals who acquire the *sick* label rather than the *bad* label are entitled to treatment rather than punishment and are excused from blame for their behavior (Conrad & Schneider 1992). As you might expect, people in positions of power are more apt to be successful in claiming the sick label. For example, the upper-class woman who shoplifts is likely to be labeled *neurotic*, whereas the lower-class woman who steals the same items is likely to be labeled *shoplifter*. The middle-class boy who acts up in school may be called *hyperactive* and given medication, whereas the lower-class boy is called a *juvenile delinquent* and arrested.

Crime

Most deviant behavior is subject only to informal social controls. When deviance becomes labeled crime, it becomes subject to legal penalties. This is, in fact, the definition of **crime**: behavior that is considered so unacceptable that it is subject to legal penalties. Most, though not all, crimes violate social norms and are subject to informal as well as legal sanctions. In this section, we briefly define the different types of crimes, look at crime rates in the United States, and examine the findings about who is most likely to commit these crimes.

Major Crimes Involving Violence or Property

Each year the federal government publishes the Uniform Crime Report (UCR), which summarizes the number of crimes known to the police for eight major index crimes (U.S. Department of Justice 2001a):

- *Murder and nonnegligent manslaughter.* Overall, murder is a rare crime; yet some segments of society are touched by it much more than others. More than 46 percent of all murder victims in 2001 were African American and 76 percent were male.
- *Rape.* Rape accounted for 6.3 percent of all reported violent crimes in 2001. Even though most rapes still go unreported, more than 90,000 women reported being raped that year. A large national sample surveyed in 1995/96 found that 15 percent of all American women and 2 percent of all men had been raped at some point in their lives (Tjaden & Thoennes 1998).
- *Robbery.* Robbery is defined as taking or attempting to take anything of value from another person by force, by threat of force, by violence, or by putting the victim in fear. Unlike simple theft or larceny, robbery involves a personal confrontation between the victim and the robber. Thus, robbery is considered a crime of violence.
- *Assault.* Aggravated assault is an unlawful attack for the purpose of inflicting severe bodily injury. Kicking and hitting are included in assault, but in 72 percent of cases during 2001, assault involved a weapon.
- *Burglary, larceny-theft, motor-vehicle theft,* and *arson* are the four property crimes included in the UCR. Property crimes are much more common than crimes of violence and account for 88 percent of the crimes covered in the UCR.

Figures 6.1 and 6.2 depict trends in seven crimes since 1980. They show that crime rates have been declining more or less steadily for the last decade or so; they

FIGURE 6.1

Changes in Violent Crime Rates, 1980–2001

Violent crime rose during the late 1980s and early 1990s, but has been declining sharply since then. Murder, rape, and robbery are now less common than they were in 1980.
SOURCE: U.S. Department of Justice (1995, 2001a).

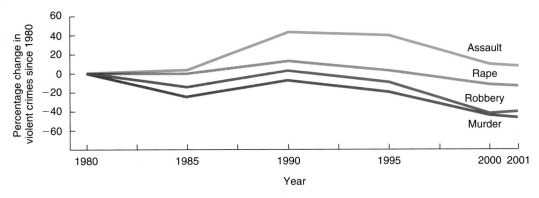

FIGURE 6.2

Changes in Property Crime Rates, 1980–2001

Burglary, larceny, and motor vehicle theft have all declined steadily since 1990, and are now less common than they were in 1980.
SOURCE: U.S. Department of Justice (1995, 2001a).

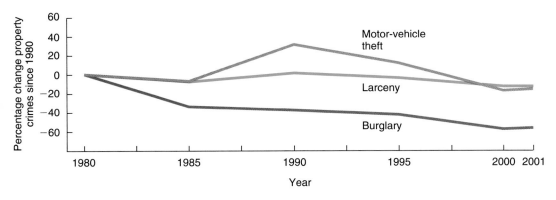

are now lower than they were 30 years ago. While the cause of this decline is hotly debated, most observers agree that a major reason is that young people commit most crimes, and there are now fewer young people than in earlier generations.

Victimless Crimes

The so-called **victimless crimes**—such as drug use, prostitution, gambling, and pornography—are voluntary exchanges between persons who desire goods or services from one another (Schur 1979). They are called victimless crimes because participants in the exchange typically do not see themselves as being victimized or as suffering from the transaction: There are no complaining victims.

There is substantial debate about whether these crimes are truly victimless. Some believe that such activities are legitimate areas of free enterprise and free

Victimless crimes such as drug use, prostitution, gambling, and pornography are voluntary exchanges between persons who desire goods or services from each other.

The Environment

focus on

Crimes Against the Environment

When most of us think of crime, the first thing that comes to mind is an image of gang members, drug traffickers, or thieves. Yet white-collar crime also takes an enormous toll on society. One particular category—environmental crime—has particularly devastating, long-term consequences. Here we look at two examples of the extent of this damage and the criminal justice system's response.

In March 1989, the Exxon *Valdez* struck a reef in Prince William Sound, spilling more than 11 million gallons of crude oil over a 1,600-square-mile area and damaging more than 800 miles of shoreline in one of the world's richest wildlife areas (Clinard 1990, 44–53). Two abundant fisheries were devastated; more than half a million birds in 90 species died, as did approximately 30 percent of the local sea otter population (Kay 1994).

Exxon was clearly criminally negligent in this case. The captain of the *Valdez* had a known history of alcoholism, including several arrests for drunken driving; 9 hours after the accident, testing revealed that his blood alcohol content was still unacceptably high. Furthermore, the tanker itself had not been equipped with the double-bottom safety hull that even then was required by law.

In a 1991 settlement, Exxon agreed to pay $900 million for environmental restoration, the largest payment that had ever been made to settle an environmental damage case. Three years later, a federal jury in Anchorage awarded another $286.8 million to fishers in compensation for the damage the spill had caused to their livelihood ("Awarded" 1994).

Critics wonder, however, whether Exxon executives were truly punished for their crime. For instance, although companies are legally required to clean up their spills, Exxon was allowed to take back $40 million of the $900 million settlement in payment for its cleanup efforts. In fact, Exxon received this money before most of the restoration projects were even started (Kay 1994).

On December 2, 1984, over 40 tons of lethal gases escaped from the Union Carbide pesticide plant in Bhopal, India. At least 8,000 were killed, and 120,000 suffered lasting lung, brain, kidney, muscular, immunological, reproductive, and genetic damage (Vosters 2003).

Like Exxon, Union Carbide was criminally negligent. In the 4 years before the disaster, maintenance and safety crews in the unit had been more than halved, despite the fact that an internal safety audit had shown a total of 61 hazards, 30 of them major. On the night of the deadly leak, the refrigeration unit, which should have controlled temperatures, had been shut off by company officials to save on electricity bills, the safety systems were known to be broken, and the factory siren had been switched off so that the neighborhood community would not be "unduly alarmed" ("International Campaign for Justice" n. d.).

In 1989, the India Supreme Court announced a settlement with Union Carbide. In exchange for a payment of US $475 million, the court absolved Union Carbide of all future criminal liabilities. Because of widespread victim dissatisfaction with this settlement, the court revised the settlement in 1991, reinstating criminal cases against the company and its officials and ordering Union Car-

choice and that the only reason these acts are considered illegal is because of some self-righteous busybodies (Jenness 1990). Others argue that prostitutes, drug abusers, and pornography models *are* victims (Chapman & Gates 1978; Dworkin 1981): Even though there is an element of choice in the decision to engage in these behaviors, individuals are usually forced or manipulated into it by their disadvantaged class position. Although it is easy to see how laws against pimping or selling drugs might protect prostitutes or drug users from victimization, it is more difficult to see how prostitutes and drug users benefit from laws that make their own activities illegal, as well.

Because there are no complaining victims, these crimes are difficult to control. The drug user is generally not going to complain about the drug pusher, and the illegal gambler is unlikely to bring charges against a bookie. In the absence of a complaining victim, the police must find not only the criminal but also the crime. Efforts to do so are costly and divert attention from other criminal acts. As a result,

bide to build a 500-bed hospital to care for victims of the disaster. Through August 1999, compensation has averaged approximately $3,300 for loss of life and $800 for those who suffered permanent disability (www. corpwatch.org/trac/bhopal/merger. html). The nominal loss to Union Carbide shareholders (prior to its merger with Dow Chemical) was approximately 50 cents per share (www. corpwatch.org/trac/bhopal/factsheet. html).

The Exxon *Valdez* and the Bhopal disaster are not isolated cases. The criminal justice system continues to prosecute environmental criminals with far less vigor than it prosecutes drug dealers and convenience store bandits.

© AFP/Corbis

■ Carrying posters with images of the victims, Bhopal protesters demand that Union Carbide provide more adequate compensation for the enormous human and environmental consequences of corporate acts of criminal irresponsibility.

For more information, look up the following subjects in InfoTrac College Edition:
Environmental justice
Exxon *Valdez* oil spill disaster

Or visit the following Web sites:
Center for Health, Environment, and Justice
www.chej.org
Corp Watch
www.corpwatch.org

laws relating to victimless crimes are irregularly and inconsistently enforced, most often in the form of periodic crackdowns and routine harassment.

White-Collar Crimes

Crime committed by respectable people of high social status in the course of their occupation is called **white-collar crime** (Sutherland 1961). White-collar crime occurs at several levels. It is committed by employees against companies, by companies against employees, by companies against customers, and by companies against the public (for example, by dumping toxic wastes into the air, land, or water).

When companies are the perpetrators, white-collar crime is often referred to as *corporate crime*. Sometimes, corporate crime more closely parallels organized crime than it does anything else. For example, accountants, auditors, and executives working for Enron Corporation worked together to hide the company's debts, exaggerate

White-collar crime is crime committed by respectable people of high status in the course of their occupation.

its profits, and pull in money from investors whom they tricked into buying their stock for much more than it was worth. Meanwhile, corporate executives took home multi-million-dollar salaries. When its false bookkeeping became known and the company was forced into bankruptcy, Enron employees lost their jobs, Enron retirees lost their pensions, and investors lost millions. By definition, the only thing distinguishing this type of corporate crime from organized crime is that the thrift embezzlers wore Brooks Brothers suits and expensive white shirts while "racketeers," at least in popular mythology, often wear black shirts and white ties.

Because most white-collar crime goes unreported, its total economic cost is difficult to assess. However, most scholars and law enforcement officials believe that the dollar loss due to corporate crime dwarfs that lost through street crime (Hagan 2002). In addition to the economic cost, there are social costs as well. Exposure to repeated tales of corruption tends to breed distrust and cynicism and, ultimately, to undermine the integrity of social institutions. If you think that all members of Congress are crooks, then you quit voting. If you think that every police officer can be bought, then you cease to respect the law. Thus, the costs of such crime go beyond the actual dollars involved in the crime itself.

The reasons for white-collar crime are similar to those for street crimes: People want more than they can legitimately get and think the benefits of a crime outrun its potential costs (Coleman 1988). Differential association also plays a role. In some corporations, organizational culture winks at or actively encourages illegal behavior. Speaking of the insider trading scandals that rocked Wall Street during the 1980s, one observer notes:

> You gotta do it. . . . Everybody else is. [It] is part of the business. . . . You work at a deli, you take home pastrami every night for free. It's the same thing as information on Wall Street. . . . I know you want to help your mother and provide for your family. This is the way to do it. Don't be a schmuck. Nobody gets hurt. (As quoted in Reichman 1989, 198)

The magnitude of white-collar crime in our society makes a mockery of the idea that crime is predominantly a lower-class phenomenon. Instead, it appears that people of different statuses simply have different opportunities to commit crime. Those in lower statuses are hardly in the position to engage in price fixing, stock manipulation, or tax evasion. They are in a position, however, to engage in high-risk, low-yield crimes such as robbery and larceny. In contrast, higher-status individuals are in the position to engage in low-risk, high-yield crimes (Reiman 1998; Schur 1979). Because of the complexity of the transactions involved, white-collar crime is difficult to detect. Even if detected, white-collar offenders usually receive more lenient treatment than street criminals. For example, one study found that health-care professionals guilty of Medicaid fraud were much less likely to be incarcerated than persons charged with grand theft, despite the fact that the dollar losses from the Medicaid crimes, on the average, were much greater (Tillman & Pontell 1992).

Conflict theorists argue that the absence of white-collar crime statistics from the UCR and the relative absence of white-collar criminals from our prisons reflect the fundamental class bias in our criminal justice system (Braithwaite 1985).

Correlates of Crime: Age, Sex, Class, and Race

In 2001, only 20 percent of the crimes reported in the UCR were cleared by an arrest. Murder was the crime most likely to be cleared and burglary was least likely. This means that the people arrested for criminal acts represent only a sample of those who commit reported crimes; they are undoubtedly not a random sample. The low level of

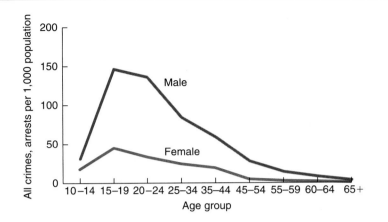

FIGURE 6.3
Arrest Rates by Age and Sex, 2001
Arrest rates in the United States and most other nations show strong and consistent age and sex patterns. Arrest rates peak sharply for young people ages 15 to 24; at all ages, men are considerably more likely than women to be arrested.
SOURCE: U.S. Department of Justice (2001a).

arrests coupled with the low levels of crime reporting warn us to be cautious in applying generalizations about arrestees to the larger population of criminals. With this caution in mind, we note that the persons arrested for criminal acts are disproportionately male, young, and from minority groups. Figure 6.3 shows the pattern of arrest rates in 2001 by sex and age. As you can see, crime rates, especially for men, peak sharply during ages 15 to 24; during these peak crime years, young men are about three to four times more likely to be arrested than women of the same age. Minority data are not available by age and sex, but the overall rates show that African Americans and Hispanics are over three times more likely than whites to be arrested.

What accounts for these differentials? Can the theories reviewed earlier help explain these patterns?

Age Differences

The age differences in arrest rates noted in Figure 6.3 are both longstanding and characteristic of nearly every nation in the world that gathers crime statistics (Cook & Laub 1998; Hirschi & Gottfredson 1983). Researchers disagree over the reasons for the high arrest rates of young adults, but deterrence theories have the most promise for explaining this age pattern.

In many ways, adolescents and young adults have less to lose than other people. They don't have a "stake in conformity"—a career, a mortgage, or a credit rating (Steffensmeier et al. 1989). When young people do have jobs and especially when they have good jobs, their chances of getting into trouble are much less (Allan & Steffensmeier 1989).

Delinquency is basically a leisure-time activity. It is strongly associated with spending large blocks of unsupervised time with peers (Agnew & Petersen 1989; Osgood & Wilson 1989). When there is "nothing better to do," a substantial portion of young people will get in trouble. On the other hand, deviance is deterred by spending a lot of time with one's parents or with conforming peers (Gardner & Shoemaker 1989).

Sex Differences

The sex differential in arrest rates has both social and biological roots. Women's smaller size and lesser strength make them less likely to engage in violence or personal confrontation; they have learned that, for them, these are ineffective strategies. Evidence linking male hormones to aggressiveness indicates that biology also may be a factor in women's lower inclination to engage in violent behavior.

Among social theories of deviance, deterrence theory seems to be the most effective in explaining these differences. Generally, girls are supervised more closely than boys, and they are subject to more social control; this is especially true in the lower class (Chesney-Lind & Shelden 2004; Hagan, Gillis, & Simpson 1985; Heidensohn 1985; Thompson 1989). Whereas parents may let their boys wander about at night unsupervised, they are much more likely to insist on knowing where their daughters are and with whom they will be associating. The greater supervision that girls receive increases their bonds to parents and other conventional institutions; it also reduces their opportunity to join gangs or other deviant groups.

These explanations raise questions about whether changing roles for women will affect women's participation in crime. Will increased equality in education and labor-force participation and increased smoking and drinking also carry over to greater equality of criminal behavior? So far, the answer appears to be no (Chesney-Lind & Shelden 2004; Steffensmeier & Allan 1996). It is true that the crime rate for women has increased faster than the crime rate for men with regard to minor property crimes and drug-related crimes. But violent crimes increased considerably faster among men than among women, and most of the female drug-related arrests were for possession, not drug dealing (Chesney-Lind & Shelden; Maher & Daly 1996). Furthermore, female arrest rates overall may actually still be lower now than they were in the mid-1800s (Boritch & Hagan 1990). Finally, it is important to note that during this same time period, the gender gap in rates of violent and major property crime has actually increased.

This pattern of change lends support to some recent feminist theories of crime. Whereas deterrence theory has focused on the importance of women's strong bonds to conventional authority in reducing their involvement in crime relative to that of men, feminist theories have shifted the focus. They argue that it is men's strong bonds to conventional gender roles that *increase* their involvement in crime (Bourgois 1995; Katz 1988; Messerschmidt 1993). According to these theories, to be considered "masculine," boys and men must challenge authority and must act aggressively or even violently, at least in certain times and places. This theory may be particularly useful for explaining crimes against women by groups of men, such as gang rapes (Lefkowitz 1997; Sanday 1990).

Feminist criminologists also have noted that *victimization* of females by males explains a significant proportion of crime among females (Chesney-Lind & Shelden 2004). Girls and women who have been sexually or physically abused by men (including male relatives) are more likely to run away from home, turn to drugs, enter prostitution, and respond violently to their abusers and others.

Social-Class Differences

The effect of social class on crime rates is complex. Although sociologists have historically held that social class is an important correlate of criminality (Braithwaite 1981; Elliott & Ageton 1980; Thornberry & Farnworth 1982), some studies have found that the relationship is not very strong and in some cases is nonexistent (Johnson 1980; Krohn et al. 1980; Tittle & Meier 1990). Much of the inconsistency appears to center on difficulties in measuring both social class and crime.

Braithwaite's (1981) review of more than 100 studies leads to the conclusion that lower-class people commit more of the direct interpersonal types of crimes normally handled by the police than do people from the middle class. These are the types of crimes reported in the UCR. Middle-class people commit more of the crimes that involve the use of power, particularly in the context of their occupational roles: fraud, embezzlement, price fixing, and other forms of white-collar crime. There is also evidence that the social-class differential may be greater for adult crime than for juvenile delinquency (Thornberry & Farnworth 1982).

Nearly all the deviance theories we have examined offer some explanation of the social-class differential. Strain theorists and some conflict theorists suggest that the lower class is more likely to engage in crime because of blocked avenues to achievement, which explains why crime rises along with unemployment (Grant & Martinez-Ramiro 1997). Deterrence theorists attribute greater crime among the lower class to these people receiving fewer rewards from conventional institutions such as school and the labor market. All of these theories accept and seek to explain the social-class pattern found in the UCR, where, indeed, the lower class is overrepresented.

Labeling and conflict theories, on the other hand, argue that this overrepresentation is not a reflection of underlying social-class patterns of deviance but of bias in the law and within social control agencies (Williams & Drake 1980). Evidence suggests, for instance, that the disproportionately high lower-class homicide rates found in most modern societies result from governmental failure to provide the least privileged with the same legal means of conflict resolution as is typically provided to the social elite (Cooney 1997). Overrepresentation of the lower class also reflects the particular mix of crimes included in the UCR; if embezzlement, price fixing, and stock manipulations were included in the UCR, we might see a very different social-class distribution of criminals.

Race Differences

Although African Americans compose only 13 percent of the population, they make up 53 percent of those arrested for murder, 38 percent of those arrested for rape, and 36 percent of those arrested for assault (U.S. Department of Justice 2001a). Hispanics, who compose about 12 percent of the total population, represent about 28 percent of those imprisoned for violent crimes. These strong differences in arrest and imprisonment rates are explained in part by social-class differences between minority and white populations. Even after this effect is taken into account, however, African Americans and Hispanics are still much more likely to be arrested for committing crimes.

The explanation for this is complex. As we will document in Chapter 9, race continues to represent a fundamental cleavage in U.S. society. The continued and even growing correlation of race with poverty, unemployment, inner-city residence, and female-headed households reinforces the barriers between African Americans and whites in U.S. society. An international study confirms that the larger the number of overlapping dimensions of inequality, the higher the "pent-up aggression which manifests itself in diffuse hostility and violence" (Messner 1989). The root cause of higher minority crime rates, from this perspective, is the low quality of minority employment—which leads directly to unstable families and neighborhoods (Newman 1999b; Sampson 1987; Sampson & Groves 1989).

Poverty and segregation combine to put African American children in the worst neighborhoods in the country, where getting into trouble is a way of life and where lack of resources makes conventional achievement almost impossible (Matsueda & Heimer 1987; Newman 1999b). Differential association theory thus explains a great deal of the racial difference in arrest rates. Deterrence theory is also important. African American children are much more likely to live in a fatherless home and thus lack an important social bond that might deter deviant behavior.

In addition to these factors that may increase the propensity to deviance among minorities, there is also evidence that whether we are talking about troublemaking in school, stealing cars, or petty theft, minority-group members are more likely than whites to be *labeled* deviant and, if apprehended by the police, more apt to be cited, prosecuted, and convicted (Austin & Allen 2000; Cureton 2000).

Mental Illness

Think about your first response to hearing of an individual whose idea of a good time is strolling naked through the park or whose idea of a good meal is something retrieved from a dumpster and washed down with a bottle of cheap wine. We might characterize each of them as "one sick individual." Indeed, the fact that their behavior is also illegal may well escape us.

Changes in the language we use to describe deviance are paralleled by changes in the ways we end up treating those who have violated the norms. Not only is public drunkenness no longer considered a crime in most municipalities, but offenders are also likely to end up being treated by mental health professionals in alcohol detoxification units rather than by police officers in the city jail. This is the process we described earlier as the medicalization of deviance.

Defining Mental Illness: Labeling and the Medical Model

One important point about medicalizing deviance is that behaviors ranging from true psychosis to alcohol abuse or eccentricity all may end up being considered mental illness. For this reason, some sociologists have suggested that mental illness should be defined as *residual deviance* (Scheff 1966). According to this perspective, mental illness is simply the name we give to all those nonconforming behaviors that do not fall neatly into any other category. This definition is most often used by sociologists who use the labeling perspective.

Other sociologists adopt what has been called the **medical model of mental illness.** Even though they clearly believe that social factors such as stress and a lack of social support contribute to the development of psychological impairments, they also believe that mental illness, like physical illness, can be objectively defined, has an identifiable cause (even if it has not been discovered yet), and will get worse if left untreated. Sociologists who are interested in understanding how mental illness is distributed across different social groups typically use some version of the medical model.

The **medical model of mental illness** holds that mental illness can be objectively defined, has identifiable causes, and if left untreated will become worse.

Correlates of Mental Illness

The relationship between social class and mental illness is very similar to that for crime. Although the relationships are complex in both instances, mental disorder and crime appear to be somewhat more common among the lower class. In contrast, the relationship between sex and mental disorder appears to be the mirror image of that for crime: Although there are differences by type of crime and type of mental disorder, men, overall, are more likely to engage in crime and women, overall, are more likely to experience mental illness.

Social Class Differences

Since the 1920s, when the very earliest sociological studies of mental disorder were conducted, research has been clear: Rates of mental disorder are highest among the lowest social class. Researchers are divided, however, about why this is. Some argue that the social stress associated with lower-class life causes mental disorder. Others believe that the onset of mental illness causes people to lose their jobs and drift downwards in social class. Still others believe that the lower classes are no more likely to experience psychiatric difficulties than any other group, they are simply more likely to be labeled as mentally ill. Probably all of these processes are at work.

Connections

Historical Note

Sociologists who use the labeling perspective point out that definitions of mental illness, like definitions of deviance, vary across time and culture. In the nineteenth century, men who masturbated and women who sought higher education were labeled mentally ill by psychiatrists. These days, such individuals are considered perfectly normal. On the other hand, psychiatrists now assign the diagnosis of antisocial personality disorder to children who lie, skip school, fight, and steal. In the past, such children would simply have been labeled bad.

Research clearly shows that the lower class does, in fact, experience more of the types of stress that can cause some forms of mental disorder (Turner, Wheaton, & Lloyd 1995). For instance, mental health problems increase after people lose their jobs (Ali and Avison 1997), and unpleasant work environments can precipitate mental illness among vulnerable individuals (Link, Dohrenwend, & Skodol 1986). The stresses of poverty and economic insecurity appear to be particularly important in understanding the causes of disorders such as major depression (Ortega & Corzine 1990).

At the same time, research shows that the onset of disorders such as schizophrenia makes it difficult for people to keep a job. Not only may individuals lose their initial job, once potential employers discover that an individual has a history of mental disorder, employers may be reluctant to hire them for anything other than a minimum wage job (Link et al. 1987, 1997). In these cases, mental disorder clearly causes people to drift into a lower social class (Ortega & Corzine 1990).

Finally, although it is not clear that lower-class individuals are any more likely to be unfairly labeled as mentally ill, it does appear that, once labeled, lower-class individuals receive harsher treatment. They are hospitalized for longer periods of time and receive less effective treatment (Ortega & Corzine 1990).

Sex Differences

Depression is the most common form of mental illness, affecting about 10 percent of all adults living in the United States. Depression (and anxiety disorders) is much more likely to be diagnosed in women; as a result, overall rates of mental disorder are higher for women than for men. A number of factors appear to be involved. First, women may simply be more willing than men to admit to feeling anxious or depressed (Mirowsky & Ross 1995). Second, women, especially working women with young children, may have more stress in their lives than men. Alternatively, women may be more emotionally vulnerable to both their own stresses and to the stresses of their loved ones (Conger et al. 1993). On the other hand, some scholars point out that men show consistently higher rates of substance abuse and dependence and personality disorders (Kessler et al. 1994). Like feminist theorists who study crime, these scholars argue that differences in gender roles explain differences in mental disorders. Because the traditional male role makes it more appropriate for men to respond to stress with aggression or substance abuse, when their behavior does not end up being labeled criminal, it may be labeled as paranoid schizophrenic or antisocial personality instead. In contrast, the traditional female role encourages women to respond to problems passively, directing stress inward—resulting in anxiety and depression—rather than outward, where it might result in anger or violence (Horowitz & White 1987).

The Sociology of Law

We have reviewed theories about deviance and examined current findings about crime and mental disorder. In the last sections of this chapter, we will look at one particular formal mechanism of social control—the criminal justice system. Among the questions of interest are: Why punish? What is a just punishment? How does the criminal justice system work? Can we reduce crime? We begin by taking a broad theoretical overview of the sociology of law.

Theories of Law

The cornerstone of the formal control system is the law. Generally, law is seen as serving three major functions: It provides formal sanctions to encourage conformity and discourage deviance, it helps settle disputes, and it may be an instrument for social change (Vago 1989). Beyond this simple summary, there is substantial discussion among scholars about how the law operates.

Most citizens, and probably even most sociologists, take a general structural-functional approach to law (Rich 1977). By clearly spelling out expected behaviors and punishing violators, the law helps maintain society. Although some may benefit more than others from particular laws and laws may be unequally enforced, law itself is a good thing, a benefit to society.

Conflict theorists, of course, take a somewhat different position: They suggest that the legal apparatus was designed to maintain and reproduce the system of inequality. Law, in this view, is a tool used by elites to dominate and control others (Chambliss 1978).

Both perspectives have obvious merit. Although law does serve the general interest by maintaining order, it is not surprising to find that it serves some interests better than others. The relationship between law and inequality is a central concern of sociologists of law.

What Is Justice?

Justice is an enormously difficult concept to define. Some argue that justice is served when everyone is treated equally, for example, when everyone who commits first-degree murder gets 30 years with no exceptions and no parole. Others believe that justice should be more flexible, taking individual differences and specific circumstances into account.

In this as in many other fields, Max Weber's contributions are insightful. Weber ([1914] 1954) distinguished between two types of legal procedures: rational and substantive. Rational law is based on strict application of the rules, regardless of fairness in specific cases. Substantive law, on the other hand, takes into account the unique circumstances of the individual case. For example, although the penalty for motor-vehicle homicide might be 3 years, substantive law might levy a lower penalty on a grief-stricken father who has accidentally killed his child than on someone who has killed a stranger for kicks.

Studies of actual sentencing outcomes suggest that law tends to be much more substantive than rational. If law were rational, one would expect that sentences would be highly correlated with the nature of the crime; they are not. If law were a tool of the elite, one would expect sentences to be affected by the race and class of the offender and the victim. Research suggests that this is indeed true. Convicted murderers, for example, are less likely to get the death penalty if their victims were African American (Reiman 1998; Schaefer, Hennessy, & Ponterotto 1999; U.S General Accounting Office 1996; Williams & Holcomb 2001). Racial disparities are also apparent in data on individuals wrongfully convicted of crimes and later acquitted based on DNA evidence. Although only 15 percent of all sexual assaults and murders involve African American assailants and white victims, 40 percent of those falsely convicted of these crimes are African American men accused of attacking white women (Scheck, Neufeld, & Dwyer 2000, 204–205). In general, however, sentencing has only a rough association with the crime or the characteristics of the victim or offender. Decisions seem to depend more on the individual judge than on any characteristic of a particular case.

Why Punish?

Any assessment of American justice must come to grips with the question: Why punish? Traditionally, there have been five major rationalizations for punishment (Conrad 1983):

1. *Retribution.* Society punishes offenders to avenge the victim and society as a whole; this is a form of revenge and retaliation.
2. *Reformation.* Offenders are not punished but rather are corrected and reformed so that they will become conforming members of the community.
3. *Specific deterrence.* Punishment is intended to scare offenders so that they will think twice about violating the law again.
4. *General deterrence.* By making an example of offenders, society scares the rest of us into following the rules.
5. *Prevention.* By incapacitating offenders, society keeps them from committing further crimes.

Today, social control agencies in the United States represent a mixture of these different philosophies and practices. In the 1980s and 1990s, however, as a result of crowded court dockets and prisons, some scholars argued that the goal of prisons and community corrections shifted from the punishment or rehabilitation of individual criminals to the identification and management of unruly "high risk" groups (Feeley & Simon 1992).

The Criminal Justice System

The formal mechanisms of social control mentioned at the beginning of the chapter are administered through the criminal justice system. In the United States, this system consists of a vast network of agencies set up to deal with persons who deviate from the law: police departments, probation and parole agencies, rehabilitation agencies, criminal courts, jails, and prisons.

The Police

Police officers occupy a unique and powerful position in the criminal justice system because they are empowered to make arrests in a context of low visibility: Often there are no witnesses to police encounters with suspected offenders. Although they are supposed to enforce the law fully and uniformly, everyone realizes that this is neither practical nor possible. In 1999, there were 3 full-time police officers for every 1,000 persons in the nation. This means that the police ordinarily must give greater attention to more serious crimes. Minor offenses and ambiguous situations are likely to be ignored.

Police officers have a considerable amount of discretionary power in determining the extent

In most situations, the police officer is out on her or his own, away from supervision and direction, and must make snap decisions about whether to pursue violations or disregard them.

focus on Technology

Does It Take a Rocket Scientist to Do Police Work?

When a young Manhattan Beach police officer was killed during what appeared to be a routine traffic stop, videotapes from the surveillance cameras at an ATM near the crime scene could provide investigators with only blurry clues. At a nearby aerospace laboratory, some of America's brightest scientists and engineers were hard at work developing systems that could produce clear images of objects hundreds and thousands of miles away. When one of the scientists heard about the case, he offered his company's image-enhancement services to police investigators. Ultimately, aerospace engineers, using technology developed for the space program, helped clarify the images taken at the ATM and, in turn, were able to identify the car used by the killer, giving police a real break in the case (Preimsberger 1996).

Although such specific partnerships between cops and rocket scientists remain relatively rare, technology is increasingly changing the way that law enforcement work is done. In 1964, only one U.S. city, St. Louis, Missouri, had a police computer system. Today, police departments in nearly every city with a population over 50,000 have computer support services (Siegel 1995). The Dick Tracy or *Dragnet* image of the detective—notepad and pencil in hand, carefully sketching where the body lay and where the bullets entered the wall—is increasingly being replaced by one of a police technician, using a palmtop computer and a computer-aided drafting system to create a three-dimensional reconstruction of the crime scene (Breuninger 1995).

Identification imaging now permits many police officers to obtain photo identification and fingerprint images without transporting a suspect to the station. Prints and mug shots can be immediately cross-checked via computers to find out whether the suspect has a previous record or an outstanding warrant in almost any jurisdiction. With integrated ballistics identification systems, police officers can determine whether a spent bullet originated from a weapon used in an earlier crime, a technique particularly useful in police efforts to control gang activity (Boyle & O'Connor 1996). Ironically, the proliferation of one type of technology—the videocam—led to the apprehension of a rather unexpected group of offenders: the police officers involved in the Rodney King beating.

Still, despite the many benefits technology offers, it cannot replace other aspects of law enforcement. We need to be careful that the emphasis on hardware and software does not lead us to ignore the importance of police-community relations (Siegel 1995).

 For more information, look up the following keywords in InfoTrac College Edition:
Law enforcement technology

 Or visit the following Web sites:
Justice Technology Information Network
http://www.nlectc.org/justnet.html

---

to which the policy of full enforcement is carried out. Should a drunk and disorderly person be charged or sent home? Should a juvenile offender be charged or reported to parents? Should a strong odor of marijuana in an otherwise orderly group be overlooked or investigated? Unlike decisions meted out in courts of law, decisions made by police officers on the street are relatively invisible and thus hard to evaluate.

The Courts

Once arrested, an individual starts a complex journey through the criminal justice system. This trip can best be thought of as a series of decision stages. A significant proportion of those who are arrested are never prosecuted. Of those who are prosecuted, however, almost 90 percent are eventually convicted, with almost all convictions resulting from pretrial negotiations, rather than public trials (U.S. Department of Justice 2001b). Thus, the pretrial phases of prosecution are far more crucial to arriving at judicial decisions of guilt or innocence than are court trials themselves. Like the police, prosecutors have considerable discretion in deciding whom to prosecute and on what charges.

Throughout the entire process, the prosecution, the defense, and the judges participate in negotiated plea bargaining. The accused is encouraged to plead guilty in the interest of getting a lighter sentence, a reduced charge, or, in the case of multiple offenses, the dropping of some charges. In return, the prosecution is saved the trouble of assembling evidence sufficient for a jury trial.

Prisons

For most people, getting tough on crime means locking criminals up and throwing away the key. Presidential politics, the increasing strength of the Republican Party, the rise of conservative religious denominations, and overall public opinion have contributed to rapid expansion in the number of law enforcement officers, paramilitary police units, and incarcerations (Curry 1996; Jacobs & Helms 1997; Kraska & Kappeler 1997). Almost three-quarters (74 percent) of the U.S. public, for instance, believes courts do not deal harshly enough with criminals (U.S. Department of Justice 2001b). In response to political pressures and public demand, the rate of imprisonment—especially for low-level drug crimes—has risen sharply in the last two decades. As of 2002, there were more than 2 million people in U.S. prisons and jails—more than three times the number in 1980, and almost twice as many as in 1990 (Harrison & Karberg 2003). Rates of imprisonment are now higher in the United States than anywhere else in the world, primarily due to harsher sentencing policies, especially for drug-related crimes, such as "mandatory minimums" and "three strikes and you're out" laws (Figure 6.4).

Residents of U.S. federal and state prisons are disproportionately young men who are uneducated, poor, and African American. As of 2002, about 30 percent of

■ Prisons are total institutions where inmates are assigned numbers, wear identical uniforms, live in identical cells, and follow the same routines. They are also environments full of anger, hatred violence, boredom, and insecurity, which breed further deviance.

© A. Ramey/Woodfin Camp & Associates

FIGURE 6.4

Number of People in Prison During 2002 (per 100,000 Population)

The United States leads the world in imprisoning its own population. Not only do we imprison more people than do similar countries like Canada, we even imprison more people than do dictatorships like China and Cuba.

SOURCE: International Centre for Prison Studies 2002.

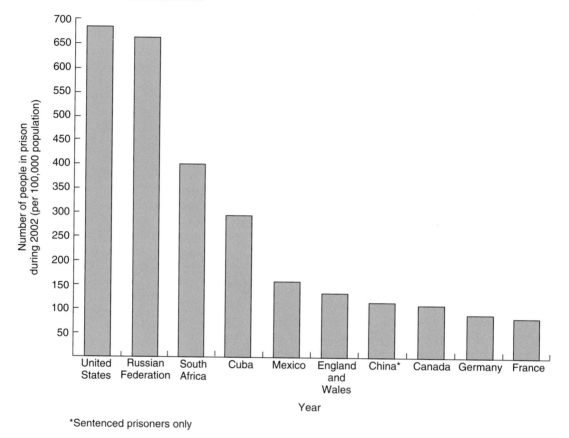

*Sentenced prisoners only

all prisoners are African American males in their 20s (Harrison & Karberg 2003). Even more shockingly, 12.9 percent of all African American males ages 25 to 29 are in prison, compared to 4.3 percent of Hispanics and 1.6 percent of whites.

The sharp increase in the use of prison to control crime has resulted in a crisis in prison (and jail) conditions. Many facilities are housing twice as many inmates as they were designed to hold, in inhumane conditions; in Arizona, jail inmates—most of whom have not even been tried yet, let alone convicted—are housed in tents in the desert with temperatures rising up to 125 degrees. These conditions have been shown to be the chief cause of violence in prisons (Gaes & McGuire 1985). As a result, prisons in more than 30 states are under court order to reduce crowding and improve conditions. This is an enormously expensive undertaking. Do we really want to spend billions and billions of dollars to build more prisons to warehouse a growing proportion of those accused or convicted of crime? Do we need to?

A growing number of empirical studies demonstrate that the certainty of getting caught has more deterrent effect on crime than the length of the sentence (McCarthy

American Diversity

Is Capital Punishment Racist?

In 1972, in the case of *Furman* v. *Georgia*, three African American defendants appealed their death sentences to the U.S. Supreme Court on the grounds that capital punishment constituted cruel and unusual punishment. Their argument was that other defendants, many of whom were white, committed equally or more serious crimes and were not sentenced to death.

There was, in fact, good statistical support for their claim that capital punishment is racist. Between 1930 and 1967, nonwhites accounted for 54 percent of executed prisoners and 89 percent of those executed for rape (Radelet 1981). These figures are much higher than the percentage of nonwhites in the general population and also much higher than the percentage of nonwhites among convicted rapists. In a 5 to 4 decision, the Supreme Court agreed with the defendants, holding that the uncontrolled discretion of judges and juries in capital cases denied defendants constitutionally guaranteed rights to due process (Bell 1992).

The *Furman* decision put a temporary stop to capital punishment, but states attempted to solve the problem of disparity in sentencing by giving judges and juries less discretion in these cases. Have the new laws reducing discretion eliminated earlier racial discrepancies in capital punishment?

Studies conducted in the post-*Furman* era continue to show that race is a strong determinant of who is sentenced to death. Between 1977 and 1990, 39 percent of executed convicts were African American—about three times the proportion of African Americans in the population (Culver 1992). New research shows, however, that race of the victim has more of an effect on sentencing than does race of the defendant: those convicted of killing whites are significantly more likely to receive the death penalty than those convicted of killing African Americans (Reiman 1998; Schaefer et al. 1999; U.S. General Accounting Office 1996; Williams & Holcomb 2001).

On the basis of evidence such as this, Warren McCleskey, an African American who had been convicted and sentenced to death for the 1978 murder of a white police officer, appealed to the U.S. Supreme Court, arguing that his sentence was a product of racial discrimination. In its decision, the Court admitted that there is "some risk of racial prejudice influencing a jury's decision in a criminal case," but rejected McCleskey's appeal on the basis that the criminal justice system would be immobilized if it were forced to make itself fully even-handed (Bell 1992, 332).

For more information, look up the following subjects in InfoTrac College Edition:
Capital punishment

Or visit the following Web sites:
The Sentencing Project
http://www.sentencingproject.org
American Civil Liberties Union
http://www.aclu.org/DeathPenalty

2002). These findings suggest that we are pursuing the wrong strategy. Rather than building more prisons to warehouse criminals for longer periods of time, we need to put more money into law enforcement. Today, most experts agree that increasing the certainty that criminals will be caught will reduce crime more than will clobbering the few that we do catch.

Community-Based Corrections

Another approach to solving the prison crisis is to change the way we punish convicted criminals. Only one-quarter of the convicted offenders under the jurisdiction of social control agencies are actually in jail or prison. The other three-quarters are on probation or parole. The public has been generally negative about probation and parole, believing—often rightly—that probation has meant giving criminals a "slap on the wrist" and parole has meant letting criminals out without effective supervision.

As the cost of imprisoning larger numbers of people balloons to crisis proportions, there has been increased interest in effective community-based corrections. New intensive supervision probation programs are being used across the country. They include curfews, mandatory drug testing, supervised halfway houses, mandatory community service, frequent reporting and unannounced home visits, restitution, electronic surveillance, and split sentences (incarceration followed by supervised probation) (Lurigio 1990). A recent review conducted for the U.S. National Institute of Justice found that when these programs included treatment for drug addiction and other supportive services, they increased the chances of rehabilitation and reduced overall costs to the system (Petersilia 1999).

Where This Leaves Us

The conservative approach to confronting deviance and crime has generally been to pass laws making deviance into crime and to increase penalties for convicted criminals. This approach has dominated since the 1970s, which is why prison populations have soared. An alternative approach is, first, to develop greater tolerance for victimless crimes and other forms of deviance that are relatively inconsequential. For more serious forms of deviance and crime, we can address the social problems that give rise to these activities. A leading criminologist (Currie 1998) advocates five major strategies for doing so:

1. Reduce inequality and social impoverishment.
2. Replace unstable, low-wage, dead-end jobs with decent jobs.
3. Prevent child abuse and neglect.
4. Increase the economic and social stability of communities.
5. Improve the quality of education in all communities.

These strategies would require a massive commitment of energy and money. They are not only expensive but also politically risky. Whereas law-and-order advocates want to get tough on crime by sending more criminals to jail, a policy incorporating Carrie's five strategies would channel dollars and beneficial programs into high-crime neighborhoods. Such a policy calls for teachers, not police officers, and good jobs rather than more prisons.

Observers from all sociological perspectives and all political parties recognize that social control is necessary. They recognize that rape, assault, and drug-related crimes are serious problems that must be addressed. The issue is how. The sociological perspective suggests that crime can be addressed most effectively by examining social institutions rather than individual criminals.

Summary

1. Most of us conform most of the time. We are constrained to conform through three types of social control: (1) self-restraint through the internalization of norms and values, (2) informal social controls, and (3) formal social controls.

2. Nonconformity occurs when people violate expected norms of behavior. Acts that go beyond eccentricity and challenge important norms are called deviance. Crimes are a specific kind of deviance for which there are formal sanctions.

3. Deviance is relative. It depends on society's definitions, the circumstances surrounding an act, and the particular groups or subcultures one belongs to.

4. Structural functionalists use strain theory to blame deviance on social disorganization; symbolic interactionists propose differential association, deterrence, and labeling theories, which lay the blame on interaction patterns that encourage a deviant self-concept; conflict theorists find the cause of deviance in inequality and class conflict.

5. Most crimes are property crimes rather than crimes of violence. Rates of index crimes generally rose from 1960 to 1990 but have fallen in the last decade.

6. Many arrests are for victimless crimes—acts for which there is no complainant. Laws relating to such crimes are the most difficult and costly to enforce.

7. The high incidence of white-collar crimes, those committed in the course of one's occupation, indicates that crime is not merely a lower-class behavior.

8. Males, minority-group members, lower-class people, and young people are disproportionately likely to be arrested for crimes. Some of this disparity is due to their greater likelihood of committing a crime, but it is also explained partly by their differential treatment within the criminal justice system.

9. Women and lower-class people have higher rates of most types of mental disorder. Although the reasons are complex, differences in exposure to stress appears to be the primary reason. Men have higher rates of substance abuse and personality disorders than women.

10. The sociology of law is concerned with how law is established and how it works in practice. Law reflects economic and political institutions and operates differently depending on social class. In practice, legal decisions are highly variable rather than determined by formal rules.

11. The criminal justice system includes the police, the courts, and the correctional system. Considerable discretion in the execution of justice is available to authorities at each of these levels.

12. The United States faces a "crisis of penalty," as our "get-tough" approach to crime is populating prisons far beyond capacity. Evidence suggests that longer sentences may not be necessary. Alternatives to imprisonment include community-based corrections and social change to reduce the causes of crime.

Thinking Critically

1. Can differential association theory explain why some boys/girls who grow up in bad neighborhoods do not become delinquent? Can you?

2. Why do you think most Americans view street crime as more serious than corporate crime? What would a conflict theorist say? A structural functionalist?

3. Describe a deviant that you have known well—someone who got in trouble with the law or should have. Evaluate the theories of deviance in light of this one person. Which theory best explains why your acquaintance deviated rather than conformed?

4. Devise a strategy for deterring white-collar or corporate crime.

5. Why would the race of the victim be as important as the race of the defendant in predicting whether a convicted killer will be sentenced to death? What does this research finding tell us about how our society values racial and ethnic diversity? If racial discrimination exists in the death sentencing, is that a good reason to stop capital punishment altogether? Why or why not?

Sociology on the Net

The Wadsworth Sociology Resource Center: Virtual Society

http://sociology.wadsworth.com/

The companion Web site for this book includes a range of enrichment material. Further your understanding of the chapter by accessing Online Practice Quizzes, Internet Exercises, InfoTrac College Edition Exercises, and many more compelling learning tools.

Chapter-Related Suggested Web Sites

U.S. Department of Justice
http://www.usdoj.gov

Federal Bureau of Investigation
http://www.fbi.gov

National Criminal Justice Reference Service
http://www.ncjrs.org

Federal Bureau of Prisons Quick Facts
http://www.bop.gov/fact0598.html

InfoTrac College Edition

http://www.infotrac-college.com/wadsworth/

Access the latest news and research articles online—updated daily and spanning four years. InfoTrac College Edition is an easy-to-use online database of reliable, full-length articles from hundreds of top academic journals and popular sources. Conduct an electronic search using the following key search terms:

White collar crime

Deviance/deviant behavior

Delinquency

Mental illness

Criminal justice

For more information related to this chapter, visit the Opposing Viewpoints Resource Center. Be sure you check all the options on the toolbar—viewpoints, references, statistics, and so on—and search under the following subjects:

Capital punishment

Crime

Prisons

Suggested Readings

Bourgois, Philippe. 1995. *In Search of Respect: Selling Crack in El Barrio*. New York: Cambridge University Press. A book that demonstrates how young men's search for dignity and meaning in the poverty-stricken minority ghettoes of New York can lead them to adopt a street culture based on drugs and violence. It is based on in-depth interviews and observations.

Butterfield, Fox. 1995. *All God's Children: The Bosket Family and the American Tradition of Violence*. New York: Alfred Knopf. A tracing of one family's history of violence and crime back to its origins in the slave and plantation culture of the Old South. This book offers a fascinating account of the ways that poverty and violence affect not only children but children's children.

Conover, Ted. *Newjack: Guarding Sing Sing*. New York: Alfred Knopf. Investigative journalist Ted Conover spent a year working as a prison guard at Sing Sing and, more briefly, elsewhere. His book details the impact of prison life on guards as well as prisoners, and the impact of "prison culture" on all of us.

Prejean, Helen. 1993. *Dead Man Walking: An Eye Witness Account of the Death Penalty in the United States*. New York: Vintage. A deeply moving, thoughtful book written by a Roman Catholic nun who has worked for many years both with convicted murderers on death row and with the families of their victims.

Reiman, Jeffrey. 1998. *The Rich Get Rich and the Poor Get Prison: Ideology, Class, and Criminal Justice* (5th ed.). Boston: Allyn & Bacon. The classic analysis of how the police and legal system ignore the crimes of the wealthy while searching out and cracking down on both nonconventional and criminal behavior among the poor.

CHAPTER 7

Stratification

AP/Wide World Photos

Structures of Inequality

Inequality exists all around us. Maybe your mother loves your sister more than you or your brother received a larger allowance than you did. This kind of inequality is personal. Sociologists study a particular kind of inequality called stratification. **Stratification** is an institutionalized pattern of inequality in which social statuses are ranked on the basis of their access to scarce resources.

If you are female and your parents gave your brother more money for college tuition because they decided he was nicer than you, this inequality was not stratification. But if they gave him more money because they consider it more important for sons than for daughters to go to college, that was stratification. Inequality becomes stratification when two conditions exist:

1. The inequality is *institutionalized*, backed up both by social structures and by long-standing social norms.
2. The inequality is based on membership in a status (such as oldest son or blue-collar worker) rather than on personal attributes.

The scarce resources that we focus on when we talk about inequality are generally of three types: material wealth, prestige, and power. When inequality in one of these dimensions is supported by social structures and long-standing social norms, and when it is based on status membership, then we speak of stratification.

Types of Stratification Structures

Stratification is present in every society. All societies have norms specifying that some categories of people ought to receive more wealth, power, or prestige than others. There is, however, wide variety in the ways in which inequality is structured.

A key difference among structures of inequality is whether the categories used to distribute unequal rewards are based on ascribed or achieved statuses. As noted in Chapter 4, *ascribed statuses* are those that are fixed by birth and inheritance and are unalterable during a person's lifetime. *Achieved statuses* are optional ones that a person can obtain in a lifetime. Being African American or female, for example, is an ascribed status; being an ex-convict or a physician is an achieved status.

Every society uses some ascribed and some achieved statuses in distributing scarce resources, but the balance between them varies greatly. Stratification structures that rely largely on ascribed statuses as the basis for distributing scarce resources are called **caste systems**; structures that rely largely on achieved statuses are called **class systems.**

Caste Systems

In a caste system, whether you are rich or poor, powerful or powerless depends entirely on who your parents are (Smaje 2000). Whether you are lazy and stupid or hardworking and clever makes little difference. Your parents' position determines your own.

This system of structured inequality reached its extreme form in nineteenth-century India. The level of inequality in India was not very different from that in many European nations at the time, but the system for distributing rewards was

---

Stratification is an institutionalized pattern of inequality in which social statuses are ranked on the basis of their access to scarce resources.

Caste systems rely largely on ascribed statuses as the basis for distributing scarce resources.

Class systems rely largely on achieved statuses as the basis for distributing scarce resources.

Connections

Personal Application

Everyone has both ascribed and achieved statuses. You now have the achieved statuses of high school graduate and college student, and hope to have the achieved status of college graduate. If your parents graduated college, you also have the ascribed status of coming from an educated family, which you will keep whether or not you graduate college. How others view you will depend on both your achieved and your ascribed statuses.

markedly different. The Indian population was divided into castes, roughly comparable to occupation groups, that differed substantially in the amount of prestige, power, and wealth they received. The distinctive feature of this caste system was that caste membership was unalterable; it marked one's children and one's children's children. The inheritance of position was ensured by rules specifying that all persons should (1) follow the same occupation as their parents, (2) marry within their own caste, and (3) have no social relationships with members of other castes (Weber [1910] 1970b).

Class Systems

In a class system, achieved statuses are the major basis of unequal resource distribution. Occupation remains the major determinant of rewards, but it is not fixed at birth. Instead, you can achieve an occupation far better or far worse than those of your parents. The amount of rewards you receive is influenced by your own talent, ambition, and work—or lack thereof.

The primary difference between caste and class systems is not the level of inequality but the opportunity for achievement. The distinctive characteristic of a class system is that it permits **social mobility**—a change in social class. Technically, mobility that occurs from one generation to the next is **intergenerational mobility.** Change in occupation and social class during an individual's own career is **intragenerational mobility.** Both kinds of mobility may be downward as well as upward.

Even in a class system, ascribed characteristics have an influence. Whether you are male or female, Hispanic or non-Hispanic, Jewish or Protestant is likely to influence which doors are thrown open and which barriers have to be surmounted. Nevertheless, these factors are much less important in a class society than in a caste society. Because class systems predominate in the modern world, the rest of this chapter is devoted to them.

Classes—How Many?

A class system is an ordered set of statuses. Which statuses are included? And how are they divided? Two theoretical answers to these questions are presented here.

Marx: The Bourgeoisie and the Proletariat

Karl Marx (1818–1883) believed that there were only two classes. We could call them the haves and the have-nots; Marx called them the bourgeoisie (boor-zhwah-zee) and the proletariat. The **bourgeoisie** are those who own the tools and materials necessary for their work—the means of production. The **proletariat** are those who do not. The latter must therefore support themselves by selling their labor to those who own the means of production. In Marx's view, **class** is determined entirely by one's relationship to the means of production.

Relationship to the means of production obviously has something to do with occupation, but it is not the same thing. According to Marx, your college instructor, the manager of the Sears store, and the janitor are all proletarians because they work for someone else. Your garbage collector is probably also a proletarian who sells his labor to an employer; if he owns his own truck, however, your garbage collector is a member of the bourgeoisie. The key factor is not income or occupation but whether individuals control their own tools and their own work.

Marx, of course, was not blind to the fact that in the eyes of the world, store managers are regarded as more successful than truck-owning garbage collectors. Probably

Social mobility is the process of changing one's social class.

Intergenerational mobility is the change in social class from one generation to the next.

Intragenerational mobility is the change in social class within an individual's own career.

The **bourgeoisie** is the class that owns the tools and materials for their work—the means of production.

The **proletariat** is the class that does not own the means of production. They must support themselves by selling their labor to those who own the means of production.

In Marxist theory, **class** refers to a person's relationship to the means of production.

■ Removing garbage is both unpleasant and absolutely essential to modern life, yet most garbage collectors are paid low wages. Structural-functional theory attributes their low wages to their lack of skill whereas conflict theory attributes it to their lack of power.

False consciousness is a lack of awareness of one's real position in the class structure.

Class consciousness occurs when people are aware of their relationship to the means of production and recognize their true class identity.

managers think of themselves as being superior to garbage collectors. In Marx's eyes, this is **false consciousness**—a lack of awareness of one's real position in the class structure. Marx, a social activist as well as a social theorist, hoped that managers and janitors could learn to see themselves as part of the same oppressed class. If they developed **class consciousness**—an awareness of their true class identity—he believed a revolutionary movement to eliminate class differences would be likely to occur.

Weber: Class, Status, and Power

Several decades after Marx wrote, Max Weber developed a more complex system for analyzing classes. Instead of Marx's one-dimensional ranking system, which provided only two classes, Weber proposed three independent dimensions on which people are ranked in a stratification system (Figure 7.1). One of them, as Marx suggested, is

FIGURE 7.1
Weber's Model of Social Class
Weber identifies three independent dimensions of stratification. This multidimensional concept is sometimes called social class.

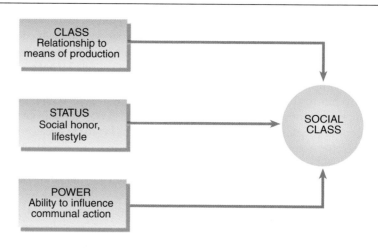

class. The second is **status,** or what Weber called social honor, expressed in lifestyle. Unlike people united by a common class, people united by a common lifestyle form a community. They invite one another to dinner, marry one another, engage in the same kinds of recreation, and generally do the same things in the same places. The third dimension is **power,** the ability to overcome resistance and to influence or force others to do what you want them to do, regardless of their own wishes.

Weber argued that although status and power often follow economic position, they may also stand on their own and have an independent effect on social inequality. In particular, Weber noted that status often stands in opposition to economic power, depressing the pretensions of those who "just" have money. Thus, for example, a member of the Mafia may have a lot of money. He may in fact own the means of production (a brothel, a heroin manufacturing plant, or a casino). But he will not have social honor.

Most sociologists use some version of Weber's framework to guide their examination of stratification systems. Rather than speaking of class (the Marxian dichotomy), we speak of *social* class. **Social class** is a category of people who share roughly the same class, status, and power and who have a sense of identification with one another. When we speak of the upper class or of the working class, we are speaking of social class in this sense.

Social class differs from *class* in two ways. First, it recognizes the importance of status and power as well as that of class. Second, it includes the element of self-awareness. Although people may be ignorant of their class situation, they are usually well aware of their social-class position, often using it as an important means to map the social world and their own place in it. People recognize that they are similar to others of their own social class but different in important ways from those in other social classes. The manager of the Sears store and the garbage collector, for example, are likely to be aware that they are members of the middle and working class, respectively.

Status is social honor, expressed in lifestyle.

Power is the ability to direct others' behavior even against their wishes.

Social class is a category of people who share roughly the same class, status, and power and who have a sense of identification with each other.

Inequality in the United States

Stratification exists in all societies. In Britain, India, and China alike, social structures ensure that some social classes routinely receive more rewards than do others. This section considers how the system works in the United States.

Measuring Social Class

If you had to rank all the people in your classroom by social class, how would you do it? There are many different strategies you could use: their incomes, their parents' incomes, the size of their savings accounts, or the way they dress and the cars they drive. Some students would score highly no matter how you ranked them, but others' scores might differ depending on your measurement procedure. The same thing is true when we try to rank people in the United States; the picture of inequality we get depends on our measurement procedure.

Self-Identification

A direct way of measuring social class is simply to ask people what social class they belong to. Given our definition of *social class* as a self-aware group, people should be able to tell you which social class they are in. Sure enough, when we ask people

FIGURE 7.2
Social-Class Identification in the United States

Social class is a very real concept to most Americans. They are aware of their own social-class membership: They feel that, in a variety of important respects, they are similar to others in their own social class and different from those in other social classes. The great majority of Americans place themselves in either the working or the middle class.

SOURCE: General Social Survey, 2000 (http://sda.berkeley.edu:7502/archive.htm).

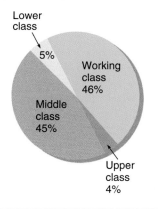

Socioeconomic status (SES) is a measure of social class that ranks individuals on income, education, occupation, or some combination of these.

"Which of the following social classes would you say you belong to?" fewer than 1 percent say they don't know. Hardly anybody says, "What do you mean, 'social class'?" and even fewer tell you that we don't have social classes in the United States. The concept of social class is meaningful to most people in the United States, and they have an opinion about where they fit in the hierarchy.

The results of a 2000 survey are presented in Figure 7.2. As you can see, only tiny minorities see themselves as belonging to the upper and lower classes, and the bulk of the population is split nearly evenly between working- and middle-class identification. Studies show that the difference between working- and middle-class identification has important consequences, affecting what church you go to, how you vote, and how you raise your children. Some of these differences are discussed in later parts of this book.

Socioeconomic Status

An alternative way to measure social class is by **socioeconomic status (SES)**, which ranks individuals on income, education, occupation, or some combination of these. SES measures do not produce self-aware social-class groupings, but result in a ranking of the population from high to low on criteria such as years of school completed, family income, or occupation.

Many scholars use occupation alone as their indicator of social-class position. The device most often used to rank occupations is the Occupational Prestige Scale. The scale is based on survey research in which large random samples are given lists of occupations and asked to choose the statement that best reflects their personal opinion of the general standing of each job: excellent, good, average, somewhat below average, or poor. The prestige of an occupation rests on the overall evaluation that sample respondents give to it. Repeated tests have shown that this procedure yields consistent results; almost the same ordering of occupations has been demonstrated on U.S. samples since 1927 as well as in other Westernized societies, from urban Nigeria to Great Britain (Hodge, Siegel, & Rossi 1964; Hodge, Treiman, & Rossi 1966). Although the question specifically refers to *men* who hold these occupations, occupations are ranked the same way for women (Bose & Rossi 1983). Thus, we can be confident that the scale produces a reliable ordering of occupations (see Table 7.1 for a partial list of ranked occupations).

Economic Inequality

Income inequality is very high in all class systems, but is especially high in the United States. Of the 29 industrialized nations that participate in the long-term Luxembourg Income Study (2000), only two, Mexico and Russia, have more income inequality than the United States.

Income inequality in the United States has increased steadily across the board since 1970, even for white men with full-time, year-round jobs (DeNavas-Walt & Cleveland 2002). The income gap, however, is most pronounced at the two ends of the income spectrum: The poorest 10 percent of the population has become significantly poorer, while the wealthiest 10 percent has become significantly wealthier. When we divide the U.S. population into five equal-sized groups, we find that the poorest 20 percent of American households now receive only 3.5 percent of all personal income, whereas the richest 20 percent receive 50 percent—more than 14 times as much (Figure 7.3). In contrast, in Sweden, for example, doctors and lawyers earn on average only about twice what waitresses and gas station mechanics earn.

TABLE 7.1
Occupational Prestige Ratings

Occupation	Score	Occupation	Score	Occupation	Score
Physician	86	Mathematician	63	Railroad conductor	42
Lawyer	75	Veterinarian	62	Postal clerk	42
College professor	74	Computer programmer	61	Tailor	42
Systems analyst	74	Sociologist	61	Mechanic	40
Architect	73	Legislator	61	Typist	40
Chemist	73	Editor or reporter	60	Advertising agent	39
Physicist	73	Police officer	60	Dock worker	37
Biologist	73	Statistician	56	Brickmason	36
Dentist	72	Librarian	54	Barber	36
Aeronautical engineer	72	Fire fighter	53	Dressmaker	36
Judge	71	Social worker	52	Shoe repair	36
Chief executive officer	70	Office manager	51	Meat cutter or butcher	35
Geologist	70	Electrician	51	Baker	35
Civil engineer	69	Funeral director	49	Salesclerk	34
Minister	69	Real estate agent	49	Housekeeper or butler	34
Psychologist	69	Soldier	49	Bulldozer operator	34
Pharmacist	68	Mail carrier	47	Bus driver	32
Registered nurse	66	Machinist	47	Truck driver	30
Secondary school teacher	66	Musician	47	Cashier	29
Athletes	65	Secretary	46	Cab driver	28
Accountant	65	Bank teller	43	Bartender	25
Elementary school teacher	64	Nursing aide	42	Gas station attendant	21

SOURCE: Davis, James A., and Smith, Tom W. National Data Program for the Social Sciences: General Social Survey, Cumulative File, 1972–1991. Ann Arbor, Mich.: Inter-University Consortium for Political and Social Research, 1992. Appendix F. Reprinted with the permission of the Inter-University Consortium for Political and Social Research.

The rise in income inequality is primarily due to changes in the U.S. economic structure (Morris & Western 1999). As we will discuss in more detail in Chapter 14, 80 percent of all Americans now work in service or retail jobs. These jobs typically pay far less than the manufacturing jobs that once dominated the U.S. economy. Meanwhile, across all economic sectors, employers are laying off permanent employees and replacing them with lower-paid temporary or part-time workers. Other employers are firing well-paid American workers and, in search of lower wages, moving jobs either to southern states or, increasingly, to foreign countries. Combined with declining union membership and influence and a stagnant minimum wage, these changes have kept the incomes of poor and working-class Americans from rising.

As bad as income inequality is, looking only at that measure actually understates the levels of economic inequality in the United States. If we measure inequality by the distribution of *wealth*—all that a given household accumulates over the years (savings, investments, homes, land, cars, and other possessions)—we find that the richest

FIGURE 7.3

Income Inequality in the United States, 2001

Distributions of income in the United States show an increase in income inequality since World War II. When we divide the U.S. population into five equal-sized groups (quintiles), we find that half of the total U.S. income goes to the richest 20 percent of the population, whereas the poorest 20 percent of the population receive less than 4 percent.

SOURCE: DeNavas-Walt & Cleveland, 2002.

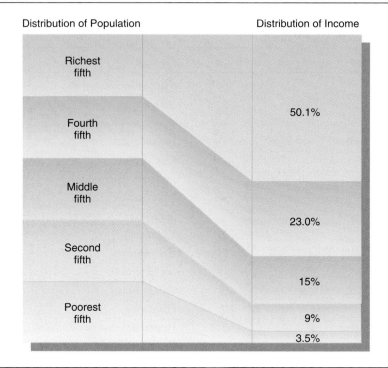

Distribution of Population | Distribution of Income

Richest fifth — 50.1%
Fourth fifth — 23.0%
Middle fifth — 15%
Second fifth — 9%
Poorest fifth — 3.5%

20 percent of households own 84 percent of all wealth. Historical research suggests that this unequal distribution of wealth is a longstanding pattern in the United States, dating back at least to 1810. However, wealth inequality has increased over the last two decades, and is now higher in the United States than in any other industrialized nation (Keister & Moller 2000).

The Consequences of Social Class

The following chapters point out the influence of social class in a number of areas—among them religious participation, divorce, prejudice, and work satisfaction. Here it suffices to say that almost every behavior and attitude we have is related to our social class. Do you prefer bowling to tennis? foreign films or American? beer or sherry? These choices and nearly all the others you make are influenced by your social class. Knowledge of a person's social class will often tell us more about an individual than any other single piece of information. This is why "What do you do for a living?" almost always follows "Glad to meet you."

Quality of Life

Some social-class differences are merely subcultural differences in tastes and lifestyles. If you prefer football or even professional wrestling to the symphony, there is no objective way to say that your tastes are worse than those who subscribe to elite culture; they are just different. On many dimensions, however, social-class differences are more meaningful. Consider the following:

- People with incomes below $10,000 a year report twice as many days on which they can't work because of disability as do those with incomes of $35,000 or more (U.S. Bureau of the Census 1999a).

- People with incomes of less than $7,500 a year are two and a half times as likely to have been the victim of a violent crime as those with incomes over $75,000 (Rennison 2002).
- Infants whose mothers fail to graduate from high school are 60 percent more likely to die before their first birthday as infants whose mothers attend college (U.S. Department of Health and Human Services 2002).

On these and many other indicators, individuals who have more money enjoy a higher quality of life.

Change and Continuity in the Consequences of Social Class

Over the last few decades, there have been some striking changes in the consequences of social class. We can see the clearest evidence in three studies of "Middletown" (Muncie, Indiana), conducted in 1924, 1977, and 1999. After the 1924 investigation, the researchers concluded that social class

> is the most significant single cultural factor tending to influence what one does all day long throughout one's life: whom one marries; when one gets up in the morning; whether one belongs to the Holy Roller or Presbyterian church; or drives a Ford or a Buick; whether or not one's daughter makes the desirable high school Violet Club; or one's wife meets with the Sew We Do Club or with the Art Students' League; whether one belongs to the Odd Fellows or to the Masonic Shrine; whether one sits about evenings with one's necktie off; and so on indefinitely throughout the daily comings and goings of a Middletown man, woman, or child. (As quoted in Caplow & Chadwick 1979)

In contrast, the 1977 and 1999 studies found dramatic declines in social-class differences (Table 7.2). By 1999, working-class women were only marginally more

TABLE 7.2
Changes in Lifestyles by Social Class in Middletown Between 1924 and 1999
Differences in social class declined sharply in Middletown between 1924 and 1999. The lifestyles of working-class families and business-class families were much more similar in 1999 than they were in 1924.

	1924	1977	1999
Percentage of mothers rising before 6 AM on workdays			
Business class	15%	13%	50%
Working class	93	27	53
Percentage of mothers who work			
Business class	3	42	77
Working class	45	58	87
Percentage of mothers who attend religious services once a week			
Business class	40	57	46
Working class	20	45	39
Percentage of mothers stressing independence in children			
Business class	15	28	24
Working class	6	24	21

SOURCE: 1924 and 1977 data from Caplow & Chadwick, 1979, Inequality and Life-Styles in Middletown, 1970–1978, *Social Science Quarterly* 60(3) © 1979 by University of Texas Press. Reprinted by permission of the authors and publisher. 1999 data by permission of Louis Hicks.

likely than middle-class women to get up early and only marginally less likely to stress independence in their children. Overall, there was marked convergence on all measures.

To some extent, this decline in social class differences is a result of significant income increases among the working and middle classes since 1924. The increases have been particularly important for those who were barely keeping their heads above water. Although the cars, televisions, and homes of the working class are not of the same quality as those of the middle class, the working class now does have them. An additional factor in reducing some of the major differences in life chances is the extension of public services. Public schools, the GI Bill, and veterans' benefits have helped reduce some of the more severe consequences of lower social class. As we will demonstrate later in this chapter, however, there are still big differences between rich and poor.

Explanations of Inequality

According to *Forbes* magazine, Steven Spielberg is now worth $2.2 *billion* dollars, and earns many millions each year. Meanwhile, the average police officer or teacher earns about $39,000, and 20 percent of American families have annual incomes below $17,970 (DeNavas-Walt & Cleveland 2002). How can we account for such vast differences in income? Why isn't somebody doing anything about it? We begin our answers to these questions by examining the social structure of stratification—that is, instead of asking about Steven Spielberg or Officer Malloy, we ask why some *statuses* routinely get more scarce resources than others. After we review these general theories of stratification, we will turn to explanations about how individuals are sorted into these various statuses.

Structural-Functional Theory

The structural-functional theory of stratification begins (as do all structural-functional theories) with the question, Does this social structure contribute to the maintenance of society? This theoretical position is represented by the work of Davis and Moore (1945), who conclude that stratification is necessary and justifiable because it contributes to the maintenance of society. Their argument begins with the premise that each society has essential tasks (functional prerequisites) that must be performed if it is to survive. The tasks associated with shelter, food, and reproduction are some of the most obvious examples. Davis and Moore argue that we need to offer high rewards as an incentive to make sure that people are willing to perform these tasks. The size of the rewards must be proportional to three factors:

1. *The importance of the task.* When a task is very important, very high rewards may be necessary to guarantee that it is done.
2. *The pleasantness of the task.* When the task is relatively enjoyable, there will be no shortage of volunteers and high rewards need not be offered.
3. *The scarcity of the talent and ability necessary to perform the task.* When relatively few have the ability to perform an important task, high rewards are necessary to motivate this small minority to perform the necessary task.

Let us apply this reasoning to two tasks, health care and reproduction. The tasks of the physician require quite a bit of skill, intelligence beyond the average, long years of training, and long hours of work in sometimes unpleasant and stressful circumstances. To motivate people who have this relatively scarce talent to undertake such a demanding and important task, Davis and Moore would argue that we must hold out the incentive of very high rewards in prestige and income. Society is likely to decide, however, that little reward is necessary to motivate women to fill the even more vital task of having and raising children. Although the function is essential, the potential to fill the position is widespread (most women between 15 and 40 can do it), and the job has sufficient noncash attractions that no shortage of volunteers has arisen.

In many ways, this is a supply-and-demand argument that views inequality as a rational response to a social problem. This theoretical position is sometimes called *consensus theory* because it suggests that inequality is the result of societal agreement about the importance of social positions and the need to pay to have them filled.

Criticisms

This theory has generated a great deal of controversy. Among the major criticisms are these: (1) High demand (scarcity) can be artificially created by limiting access to good jobs. For example, keeping medical schools small and making admissions criteria unnecessarily stiff reduce supply and increase demand for physicians. (2) Social-class background, sex, and race or ethnicity probably have more to do with who gets highly rewarded statuses than do scarce talents and ability. (3) Many highly rewarded statuses (rock stars and professional athletes, but also plastic surgeons and speechwriters) are hardly necessary to the maintenance of society.

The Conflict Perspective

A clear alternative to the Davis and Moore theory is given by scholars who adhere to conflict theory. They explain inequality as the result of class conflict rather than as a result of consensus about how to meet social needs. We review traditional Marxist thought first and then describe more recent applications of conflict theory to the study of inequality.

Marxist Theory

Marx argued that inequality was rooted in private ownership of the means of production. Those who own the means of production seek to maximize their own profit by minimizing the amount of return they must give to the proletarians, who have no choice but to sell their labor to the highest bidder. In this view, inequality is an outcome of private property, where the goods of society are owned by some and not by others. In Marxist theory, stratification is neither necessary nor justifiable. Inequality does not benefit society; it benefits only the rich.

Although Marx did not see inequality as either necessary or justifiable, he did see that it might be nearly inevitable. The reason lies in the division of labor. Almost any complex task, from teaching school to building automobiles, requires some task specialization: Some people build fuel pumps and others install them, some teach algebra and others teach poetry. To make such a division of labor function effectively, somebody has to coordinate the efforts of all the specialists. Individuals who do this coordination are in a unique position to pursue their own self-interest—to hire their

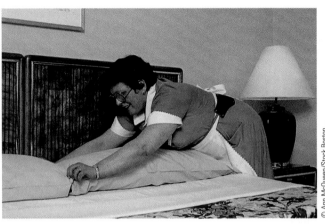

■ Ironically, people who work the hardest often earn the least income and the least honor. Conflict theorists point out that social-class background and race are often more important than ability in determining who gets the good jobs and who gets the hard, low-paying jobs.

own children in preference to others', to give themselves more rewards than they give others, and generally to increase the gap between themselves and those they coordinate. Marx's coauthor, Friedrich Engels, explained it this way:

> It is therefore the law of the division of labor that lies at the basis of the division into classes. But this does not prevent the division into classes from being carried out by means of violence, and robbery, trickery and fraud. It does not prevent the ruling class, once having the upper hand, from consolidating its power at the expense of the working class, from turning its social leadership into an intensified exploitation of the masses. (Engels [1880] 1965, 79)

Modern Conflict Theory

During the early days of industrialization when Marx was writing, those who owned the factories also managed them. In the modern economy, however, control and ownership may be independent; ownership may be divided among many stockholders, while management is controlled by a handful of hired experts (Wright 1985).

Like the earlier Marxist theory, modern conflict theory recognizes that the powerful can oppress those who work for them by claiming the profits from their labor (Wright 1985). It goes beyond Marx's focus on ownership, however, by considering how control also may affect the struggle over scarce resources (Grimes 1989). In addition, modern conflict theory looks at noneconomic sources of power, especially gender and race. These theorists argue, for example, that in the same way that capitalists gain benefits from the productive labor of workers, men can gain benefits from the "reproductive" labor of women. The term **reproductive labor** is usually used to describe traditionally female tasks such as cooking, cleaning, and nurturing—those tasks that often make it possible for others to work and play. Modern conflict theorists point out that in most families, those with the least power do the most reproductive labor; as a result, these individuals end up having fewer opportunities to earn the good incomes that might otherwise increase their power within the family (Cancian & Oliker 2000).

Reproductive labor refers to traditionally female tasks such as cooking, cleaning, and nurturing that make it possible for a society to continue and for others to work and play.

A Comparison of Two Models of Stratification

Basis of Comparison	Structural-Functional Theory	Conflict Theory
1. Society can best be understood as…	Groups *cooperating* to meet common needs	Groups *competing* for scarce resources
2. Social structures…	Solve problems and help society adapt	Maintain current patterns of inequality
3. Causes of stratification are…	Importance of vital tasks, unequal ability, pleasantness of tasks	Unequal control of means of production maintained by force, fraud, and trickery
4. Conclusion about stratification…	Necessary and desirable	Unnecessary and undesirable, but difficult to eliminate
5. Strengths…	Consideration of unequal skills and talents and necessity of motivating people to work	Consideration of conflict of interests and how those with control use the system to their advantage
6. Weaknesses…	Ignores importance of power and inheritance in allocated rewards; functional importance overstated	Ignores the functions of inequality and importance of individual differences

Criticisms

There seems to be little doubt that people who have control (through ownership or management) systematically use their power to extend and enhance their own advantage. Critics, however, question the conclusion that this means that inequality is necessarily undesirable and unfair. This is certainly a debatable assumption. First, people *are* unequal. Some people are harder working, smarter, and more talented than others. Unless forcibly held back, these people will pull ahead of the others—even without force, fraud, and trickery. Second, coordination and authority *are* functional. Organizations work better when those trying to do the coordinating have the power or authority to do so.

Symbolic Interaction Theory

Unlike structural-functional theory and conflict theory, symbolic interaction theory does not attempt to explain why some statuses are so much better rewarded than others. Instead, it asks *how* these inequalities are perpetuated in everyday life.

One of the major contributions of symbolic interaction theory is its identification of the importance of **self-fulfilling prophecies.** Self-fulfilling prophecies occur when something is *defined* as real and therefore *becomes* real in its consequences. This social dynamic is one of the ways that social class statuses are reinforced. For example, when teachers assume that lower-class students are less intelligent and less capable of doing intellectual work, the teachers are less likely to spend time helping them learn. Instead, teachers may shuffle lower-class students off to vocational classes that emphasize discipline and mechanical skills rather than intellectual skills. After several years of such "schooling," lower-class students may, in fact, have fewer intellectual skills than others.

Symbolic interaction theory also helps us understand how everyday interactions reinforce inequality by constantly reminding us of our place in the social order. For example, in most restaurants, waiters and waitresses must enter through the back doors (Paules 1991). They often must use separate bathrooms that are far less pleasant than

Self-fulfilling prophecies occur when something is *defined* as real and therefore *becomes* real in its consequences.

those used by customers, and must take their breaks and get their orders ready in windowless rooms that lack air conditioning. In some restaurants, waiters and waitresses must wear clothes that make them look like maids and butlers; in bars, waitresses often must wear clothes that make them look like prostitutes. Customers often speak rudely (or crudely) to serving staff, who are expected to smile in response. And, at the end of the evening, the customer decides whether the waiter or waitress deserves a tip. In all these ways, normal restaurant interactions reinforce customers' sense of social superiority and servers' sense of social inferiority.

The Determinants of Social-Class Position

With each generation, the social statuses in a given society must be allocated anew. Some people will get the good positions and some will get the bad ones; some will receive many scarce resources and some will not. In a class system, this allocation process depends on two things: the characteristics of the individuals (their education, aspirations, skills, parents' income, and so on) and the characteristics of the labor market. We refer to these, respectively, as micro- and macro-level factors that affect achievement.

Microstructure: Status Attainment

If *Sports Illustrated* gave you the job of predicting the top 20 college football teams in the country next year, you could go to the trouble of finding out the average height, weight, and experience level of each team's members, the dollars allocated to the athletic department, the years of coaching experience, and the attendance at games. From this information you could devise some complex system of predicting the winners. You would probably do a better job for a lot less trouble, however, if you predicted that last year's winners will be this year's winners. The same thing is true in predicting winners and losers in the race for class, status, and power. The simplest and most accurate guess is based on social continuity.

U.S. culture values achievement more than inherited wealth and status, and occupations are not directly inherited. Yet people tend to have occupations of a status similar to that of their parents. How does this come about? The best way to describe the system is as an **indirect inheritance model.** Parents' occupations do not directly cause children's occupations, but the family's status and income help determine children's aspirations and opportunities.

The **indirect inheritance model** argues that children have occupations of a status similar to that of their parents because the family's status and income determine children's aspirations and opportunities.

Inherited Characteristics: Help and Aspirations

The best predictor of your eventual social class is your parents' income (Corcoran 1995). Your parents' income affects your life chances in many ways (Corcoran; Harris 1996). If your parents are middle or upper class, you are more likely to be born healthy and more likely to get good nutrition and health care during childhood. As a result, you are less likely to have mental or physical disabilities that might reduce your potential income. Your parents will have the time and money to give you a stimulating environment in which your intellectual capacities can thrive, and will most likely live in neighborhoods with good schools, where teachers will assume you are "college material."

Class differences in home environment and in parents' support for school also have important effects on children's success. Bright and ambitious lower-class children often find it hard to do well in school when they have to study at a noisy kitchen table amid others who think that studying for a future career is less important than helping the family put food on the table now. In contrast, middle-class children who grow up in supportive environments often find it hard to fail even if their ambitions and talents are modest.

In addition, if your parents went to college or have middle-class jobs, then you have probably always assumed that you, too, would go to college, and automatically signed up for algebra and chemistry in high school. If your parents didn't graduate from high school and tend to think that education is a necessary evil, then you probably bypassed algebra for a shop or sewing class. If you later decided you wanted to go to college, you first had to make up the classes you missed in high school.

If you finish high school, the chance that you will attend college is substantially greater if your parents attended college. If they did, they are likely to have both the income and the contacts that will help you get into a good school. After you graduate, they are likely to know people who can help you get good jobs. They may also help you buy clothes for your job interviews, purchase your first home, or pay for family vacations, allowing you to invest your earnings in a new business. They might even invest in the business themselves. All these factors make parents' income a powerful predictor of their children's eventual income (Corcoran 1995).

The Wild Cards: Achievement Motivation and Intelligence

The social-class environment in which a child grows up is the major determinant of educational attainment. There are, however, two wild cards that keep education from being directly inherited: achievement motivation and intelligence. Neither of these factors is strongly related to parents' social class, and both act as filters that allow people to rise above or fall below their parents' social class (Duncan, Featherman, & Duncan 1972).

Achievement motivation is the continual drive to match oneself against standards of excellence. Students who have this motivation are always striving for A's, are never satisfied with taking easy courses, and have a real need to compete. Not surprisingly, students with high achievement motivation do better than others in school. Because achievement motivation is not strongly related to parents' social class, it is one way for people to move out of their inherited position.

Intelligence is another important factor in determining educational and occupational success (Duncan et al. 1972). Because intelligent people are born into all social classes, intelligence is a factor that allows for both upward and downward intergenerational mobility. (Chapter 13 looks at the issue of social class and intelligence in greater detail.)

Achievement motivation is the continual drive to match oneself against standards of excellence.

Macrostructure: The Labor Market

The indirect inheritance model summarizes the processes of individual status attainment. It shows how some people come to be well prepared to step into good jobs, whereas others lack the necessary skills or credentials. By themselves, however, skills and credentials do not necessarily lead to class, status, or power. The other variable in the equation is the labor market: If there is a major economic depression, you will not be able to get a good job no matter what your education, achievement motivation, or aspirations. The character of the labor market and the structure of occupations it provides have a significant effect on individual achievements.

FIGURE 7.4
The Changing Occupational Structure, 1900–2000
Since the turn of the century, the occupational structure of the United States has shifted away from farming, fishing, and forestry. Today there are many more white-collar, professional, and managerial jobs.
SOURCE: U.S. Bureau of the Census, 1999a, 424 and U.S. Census Bureau, Census 2000 summary file 3, matrices P49, P50 and P51.

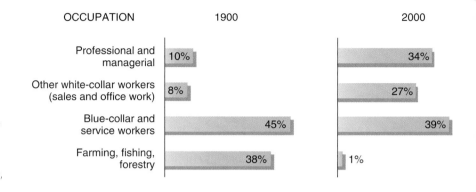

OCCUPATION	1900	2000
Professional and managerial	10%	34%
Other white-collar workers (sales and office work)	8%	27%
Blue-collar and service workers	45%	39%
Farming, fishing, forestry	38%	1%

As Figure 7.4 shows, the proportion of positions at the top of the U.S. occupational structure has increased dramatically over the last century. Not everyone, however, has benefitted equally from these new opportunities for upward mobility. Although women and minorities now have an easier time entering high-earning occupations, they tend to find themselves in the lower-earning positions within those occupations. For example, compared to men and whites, women and non-white lawyers are more likely to work as public defenders and less likely to work as corporate lawyers.

Labor market theorists suggest that the United States has a *segmented labor market*: one labor market for good jobs (usually in the big companies) and one labor market for bad jobs (usually in small companies). Women and minorities are disproportionately directed into companies with low wages, low benefits, low security, and short career ladders. (Segmented labor markets and the contemporary economic structure are discussed in more detail in Chapter 14.)

■ When the number of jobs available in a geographic area or occupation shrink, finding employment becomes difficult regardless of one's education, achievement, motivation, or aspirations. This is the situation facing these men, who were all seeking jobs at Microsoft.

AP/Wide World Photos

The American Dream: Ideology and Reality

A system of stratification is an organized way of ensuring that some categories of individuals get more social rewards than others. As we have seen, sometimes this means a great disparity not only in income but also in health, honor, and happiness. Yet in most highly stratified systems, the poor don't sit around plotting class revolutions. For the most part, inequality is accepted as fair and natural.

This consensus about inequality indicates the role of the normative structure in reinforcing and justifying a system of stratification. Each system furnishes an **ideology**— a set of norms and values that rationalizes the existing social structure (Mannheim 1929). The ideology is built into the dominant cultural values of the society—often into its religious values. For example, the Hindu religion maintains that a low caste in this life is a punishment for poor performance in a previous life. If you live well in this life, however, you can expect to be born into a higher caste in the next life. Thus, the Hindu religion offers mobility (extragenerational mobility, we might call it) and also an incentive to accept one's lot in life. To attack the caste system would be equivalent to saying that the gods are unfair or the Hindu religion is stupid.

In the United States, the major ideology that justifies inequality is the *American Dream*. This ideology proposes that equality of opportunity exists in the United States and that your position in the class structure is a fair reflection of what you deserve: That is, if you are worthy and if you work hard, you can succeed. Conversely, because your position comes entirely from your own efforts, no one but you can be blamed for your failures.

The American Dream is, in part, a reality: most Americans who were poor as children do not remain poor as adults. On the other hand, poor children are several times more likely than others to remain poor as adults, especially if they are African American or if their family was poor for many years. One large, long-term study found that 46 percent of African Americans who grew up in persistently poor families and 24 percent of whites from persistently poor families also experienced poverty as adults (Corcoran 1995).

The upper class is the most likely to believe that the United States is a land of opportunity and that everybody receives a fair shake, but most others believe this, too. Nevertheless, there are some who question this ideology, such as the disenchanted person who stated, "The rich stole, beat, and took. The poor didn't start stealing in time, and what they stole, it didn't value nothing, and they were caught with that!" (Huber & Form 1973).

These dissenters, however, are few, and most of them are less interested in changing the rules of the game than in being dealt into it. Instead, most people in the United States believe in "fair shares" rather than in "equal shares" (Ryan 1981). They believe that people who work harder or are smarter deserve to get ahead. Nevertheless, 44 percent of U.S. adults interviewed in 2000 for the General Social Survey (a well-regarded national survey) believed at least to some extent that it is the government's responsibility to reduce income differences between the rich and poor.

An **ideology** is a set of norms and values that rationalizes the existing social structure.

Social Class and Social Life

The United States is a middle-class nation. If given only three categories for self-identification, more than two-thirds of the population consider themselves middle class. U.S. norms and values are the norms and values of the middle class.

Everybody else becomes a subculture. This section briefly reviews the special conditions of each of the classes in the United States.

The Upper Class

In 2001, a family living in the United States required an income of $150,499 to be in the richest 5 percent of Americans (DeNavas-Walt & Cleveland 2002). Thus, a variety of more-or-less ordinary salespersons, doctors, lawyers, and managers in towns and cities across the nation qualify as very rich compared to the majority. Although their incomes are nothing to sneeze at, most of this upper 5 percent is still middle class. Like members of the working class, they would have a hard time making their mortgage payments if they—or their spouses—lost their jobs and were out of work for a few months. This is because although their current income is quite high, their wealth—their investments, savings, and assets they could easily sell—may not add up to much more than their debts.

The true upper class, on the other hand, is made up of two overlapping groups: those whose families have had high incomes and statuses for more than a generation and those who themselves earn incomes in the millions of dollars. The central institution that cements the first group, whose upper-class status is inherited from their parents, is the private preparatory school, especially New England boarding schools such as Philips Academy, Groton, St. Paul, and Choate (Higley 1995). Many graduates of these schools attend Ivy League colleges, such as Harvard, Yale, and Princeton. Once there, many join secret undergraduate societies such as Skull and Bones and Scroll and Key. (Both George H. and George W. Bush attended Philips Academy and Yale and belonged to Skull and Bones.) After graduation, these individuals are likely to join selective country clubs and urban dining clubs, such as New York's University Club. They join Episcopalian or perhaps Presbyterian churches and do volunteer fund-raising for high-culture organizations such as art museums, symphonies, opera companies, and the like.

■ Only a small elite ever have the opportunity to drink champagne and eat hors d'ouevres in a skybox at a football game.

TABLE 7.3
The Ten Richest People in the United States, 2002
Each year *Forbes* magazine publishes a list of the 400 richest people in the United States. About 40 percent of these fabulously wealthy individuals inherited their fortunes; two-thirds are largely individually responsible for generating their vast wealth. Their major avenues to riches were the stock market, manufacturing, and the media.

Name, Age	College	Estimated Net Worth	Major Wealth Source
William H Gates III, 46	Harvard (dropout)	$43 billion	Microsoft Corporation
Warren Edward Buffett, 72	Columbia	36 billion	Berkshire Hathaway
Paul Gardner Allen, 49	Harvard	21 billion	Microsoft Corporation
Alice L. Walton, 53	Trinity	18.8 billion	Wal-Mart inheritance
Helen R. Walton, 83	Oklahoma	18.8 billion	Wal-Mart inheritance
Jim C. Walton, 54	None given	18.8 billion	Wal-Mart inheritance
John T. Walton, 56	None given	18.8 billion	Wal-Mart inheritance
S. Robson Walton, 58	Columbia	18.8 billion	Wal-Mart inheritance
Lawrence Joseph Ellison, 58	Illinois (dropout)	15.2 billion	Oracle
Steven Anthony Ballmer, 46	Harvard	11.9 billion	Microsoft Corporation

SOURCE: The Forbes Four Hundred 2003.

Unlike those who inherit their millions, other members of the true upper class earned their wealth—most often in computers, the stock market, or mass media (Table 7.3). There are about a half million millionaires in the United States. Few went from rags to riches, however. Most had middle- or upper-class parents who sent them to excellent schools and helped them financially in many ways.

Although the true upper class is only a small portion of the population, it has a great impact on society. By controlling media resources, contributing money to political campaigns, building museums, or deciding to open or close manufacturing plants, the rich have the power to affect the lives of millions. The power they wield makes us very interested in their politics and attitudes. Yet, we know relatively little about the lives of the rich. Although their pictures may occasionally appear in the news, they rarely allow social scientists into their homes and clubs to study them.

The Middle Class

The middle class is a large and diverse group. Ranging from professionals with graduate degrees to salespersons and secretaries, middle-class workers have widely varying incomes, with some earning less than the typical working-class individual. Compared to those in the working class, however, middle-class workers tend to have more job security and more opportunities for promotions and advancement. Middle-class workers also can expect to have important benefits such as health insurance and sick leave and to have incomes that will continue to increase over most of their working lives. The middle class is also united by having at least a high school education and, in most cases, at least some college.

Middle-class culture differs from both elite upper-class culture and working-class culture. Compared to working-class individuals, middle-class individuals are less likely to decorate their homes with religious icons or to belong to bowling leagues, and are more likely to value education and equality between the sexes; compared to upper-class individuals, members of the middle class are less likely to decorate their homes with modern art or to belong to golf leagues (Halle 1993).

The Working Class

Who are the members of the working class? The answer is determined partly by income, but mostly by occupation, education, self-definition, and lifestyle. Generally, the working class includes those who work in blue-collar industries and their families. They are the men and women who work in chemical, automobile, and other manufacturing plants; they load warehouses, drive trucks, work as secretaries, and build houses. Although they sometimes receive excellent wages and benefits, it is the working class that has 10 to 15 percent unemployment during economic recessions and slumps. And although a majority are high school graduates, an eleventh-grade education is more common than a year of college.

Economic Prospects

Quite a few members of the working class have incomes as good as or better than the lower-middle class. Truck drivers, for example, often make more than nurses and public school teachers. As a result, working-class families may live in the same neighborhoods as middle-class families. Their economic prospects differ, however, in three ways. First, working-class people have little or no chance of promotion. The barrier between manual labor and management is virtually impassable. The height of working class individuals' earning power may be reached at age 25. Second, layoffs and plant closings expose them to more economic uncertainty. With American corporations moving more and more production sites to countries where labor is less expensive, and with the American economy increasingly shifting away from manufacturing to service industries, in recent years job loss and unemployment have become even more common in the lives of the working-class (Newman 1999a). Third, and at least partially as a result of these changes, working-class individuals are much less likely than members of the middle class to receive pensions, health insurance, and other benefits. They are also less likely to have much savings or other assets. As a result, layoffs or illnesses can quickly drive working class families into poverty (Newman).

As a result of low prospects and economic uncertainty, members of the working class tend to place a higher value on security than others. One aspect of this is home ownership. In the 2000 General Social Survey, 56 percent of those Americans who considered themselves working class owned their own homes, compared to 69 percent of middle-class Americans—less of a difference than we might expect. For working-class homeowners, their homes are their major financial asset and their only effective savings plan. Their resulting interest in property taxes, property values, and neighborhood maintenance drives much of their political activity.

Working-class culture differs significantly from middle-class culture. Compared to middle-class men, working-class men are more likely to enjoy spending their free time in sex-segregated, traditionally male activities, such as watching football and drinking beer with their buddies.

The Poor

Each year, the U.S. government sets an official poverty level: the minimum amount of money a family needs to have a decent standard of living. The poverty level adjusts for family size, and in 2001 the poverty level for a family of four was $17,960 (Proctor & Dalaker 2002). Under this definition, 32.9 million people, almost 12 percent of all Americans, were classified as poor in 2001. Almost all observers agree that the official poverty level is set much too low and so underestimates the percentage of Americans who are poor.

Who Are the Poor?

Poverty cuts across several dimensions of society. It is found among white Americans as well as among nonwhites, in small towns and big cities, and in traditional nuclear families as well as in female-headed households. But poverty does not affect all groups equally. As Table 7.4 indicates, African Americans and Hispanics are far more likely to be poor than are whites or Asians; children are more likely to be poor than are middle-aged or elderly persons; and inner-city residents are more likely to be poor than are suburbanites or rural dwellers. In addition, households run by single mothers are significantly more likely to be poor: 26.4 percent of female-headed families are poor, compared to 4.9 percent of families that include a married couple (Proctor & Dalaker 2002). Finally, having a job is no guarantee against poverty: 38 percent of poor persons over age 16 hold a job, and many hold more than one job.

How Poor Are the Poor?

An important issue that arises in discussing U.S. poverty is how poor the poor actually are. Two concepts are important here: absolute poverty and relative poverty. **Absolute poverty** means the inability to provide the minimum requirements of life. A growing segment of the absolute poor are those who are homeless.

Absolute poverty is the inability to provide the minimum requirements of life.

TABLE 7.4
Americans Living Below the Poverty Level, 2001

	Millions of People	Percentage of poor who are:	Percentage of Group in Poverty
Total	32.9	100.0	11.7
Ethnicity			
White non-Hispanic	15.3	46.4	9.9
African American	8.1	24.7	22.7
Hispanic	8.0	24.3	21.4
Asian/Pacific	1.3	3.8	10.2
Residence			
Central cities	13.4	40.7	16.5
Suburbs	12.1	36.6	8.2
Rural areas	7.5	22.7	14.2
Age			
Under 18	11.73	35.7	16.3
18–64	17.8	54.0	10.1
65 and over	3.4	10.4	10.1

SOURCE: Proctor & Dalaker, 2002.

■ Although we often think of poverty as a problem of racial minorities or of the inner cities, the simple facts are that most poor people in the United States are white and almost equally likely to live in nonmetropolitan areas as in the central cities.

© Tim Carlson/Stock Boston

Connections

Historical Note

Homelessness has been a major problem for the United States since the early 1980s, when the federal government slashed funds for low-income housing while increasing subsidies for "gentrifying" good-quality older buildings in inner-city neighborhoods. Although the latter policy was intended to improve quality of life in these neighborhoods, its unintended consequence was to raise rents. Meanwhile, the value of the minimum wage (adjusted for inflation) declined, and public assistance became harder to get and lower in value. As a result, Americans now must earn *twice* the mandated minimum wage to afford a modest, two-bedroom apartment. The result has been an epidemic of homelessness.

In any given year, about 3.5 million Americans—almost one-third of whom are children—experience homelessness ("How Many People" 2002). Not only do they lack a roof over their heads, but they are often hungry and ill. In Los Angeles, 35 percent of homeless people have active tuberculosis; in New York City's shelters for the homeless, 38 percent of children have asthma, a condition that can be life threatening for them (Cousineau 1997; Herbert 1999; Kleinman et al., 1996). By any standards, they are absolutely poor.

Relative poverty means the inability to maintain what your society regards as a decent standard of living. The poor in the United States come in both forms. Those who live at or close to the poverty level have a roof over their heads and food to eat most of the time. On the other hand, their car is broken down and so is the television, the landlord is threatening to evict them, they are eating too much macaroni, and they cannot afford to take their children to the doctor or the dentist. Although they may not be absolutely poor, they are deprived in terms of what is regarded as a decent standard of living.

MAP 7.1
Percentage of State Residents Living Below Poverty Line, 2000–01
SOURCE: Proctor & Dalaker, 2002.

District of
Columbia

Less than 10%
10% to 14.9%
15% to 20%

The Concentration of Poverty

Beginning in the 1970s, poor persons have increasingly become geographically concentrated in neighborhoods where at least 40 percent of residents are poor. In a very influential book entitled *The Truly Disadvantaged*, William J. Wilson (1987) argued that the concentration of poverty makes it more difficult for poor people to improve their circumstances because it isolates them from middle-class mentors, role models, job networks, and other resources. More recent research suggests that living in a poor neighborhood during early childhood and late adolescence is especially harmful (Small & Newman 2002).

Causes of Poverty

Earlier in this chapter, we said that both micro- and macro-level processes are at work in determining social-class position. The causes of poverty are simply a special case of these larger processes. At the micro level, some believe poverty can be explained by various "cultures of poverty"; at the macro level, some believe poverty is better explained by the lack of an adequate structure of opportunity.

Cultures of Poverty   The idea that poverty is caused (or perpetuated) by a **culture of poverty** was first promoted by anthropologist Oscar Lewis (1969). Lewis argued that poor people hold a set of values that protects their self-esteem and maximizes

Relative poverty is the inability to maintain what your society regards as a decent standard of living.

The **culture of poverty** is a set of values that emphasizes living for the moment rather than thrift, investment in the future, or hard work.

their ability to extract enjoyment from dismal circumstances. This set of values—the culture of poverty—emphasizes living for the moment rather than thrift, investment in the future, or hard work. Recognizing that success is not within their reach and that no matter how hard they work or how thrifty they are, they will not make it, the poor come to value living for the moment.

The culture of poverty hypothesis fits neatly with the American Dream ideology, and a substantial majority of the people of the United States agree that the poor are poor because of their values. In one survey, 77 percent of Americans agreed that most unemployed people could find work if they really tried; 75 percent agreed that most welfare recipients would find paid work if they weren't on welfare; and 45 percent agreed that the poor are mainly to blame for their problems (Harris Poll 2000). When the question is phrased more broadly, 65 percent of Americans agree that those who fail in life have only themselves to blame—a far higher percentage than in almost any other country (Pew Research Center 2003a).

Similarly, and more recently, other theorists have argued that families remain in poverty over generations because of "welfare culture" or "oppositional culture." The "welfare culture" argument asserts that children who grow up on welfare never learn to value holding a job. Instead, they learn that there is no shame in receiving welfare and that it makes sense to have more babies and get welfare instead of seeking employment (Mead 1986, 1992; Murray 1984). The "oppositional culture" argument asserts that poor youths grow up to be poor adults because they actively reject work, education, and marriage as symbols of a middle-class white culture that they despise.

Comprehensive reviews of 30 years of research on poverty provide very limited support for any of the culture of poverty theories (Corcoran 1995; Small & Newman 2001). Researchers have found that poor people overwhelmingly share the same attitudes toward welfare, work, education, and marriage as do middle-class people. In fact, far from devaluing work, most of America's poor work as many hours as they can find employment, many working far more than 40 hours a week (Newman 1999b).

The Changing Labor Market   The culture of poverty theories implicitly blame the poor for perpetuating their condition. Critics of these theories suggest that we cannot explain poverty by looking at micro-level processes. To understand poverty, they argue, we need to look at the changing labor market. If there are no good-paying jobs available, then we don't need to psychoanalyze people in order to figure out why they are poor.

In fact, the changing labor market is particularly critical for understanding contemporary poverty. As we documented in Figure 7.4, the shift from an agricultural to an industrial society produced major structural pressure for upward mobility earlier in this century. Now, at the end of the century, the deindustrialization of the United States is squeezing the lower middle of the U.S. occupational structure and creating structural pressure for downward mobility among the traditional working class (Newman 1999a, 1999b). Good jobs have virtually disappeared for the high school graduate who has no advanced training. Instead of the good union jobs that their parents held, today's high school graduates often find themselves working at dead-end jobs, with no benefits, for the minimum wage. A little arithmetic shows that the minimum wage means poverty.

This macrostructural approach to poverty suggests that a major cause of poverty is the absence of good jobs. This is a critical issue, and we will look at the changing occupational structure in more detail in Chapter 14.

■ When children come to school from very unequal backgrounds, their achievements in school are also likely to be unequal, even if they attend schools of equal quality.

New Theories of Poverty   As we have seen, research provides little support for cultural theories of poverty. Theories that stress the changing labor market seem much more useful for explaining why many people in a society are likely to be poor.

It is less clear, however, how important labor market changes are in explaining why it is so difficult for the children of poor parents to escape from poverty when they grow up. There is some evidence, for instance, that labor market forces are less important than parents' income in explaining continued poverty (Corcoran 1995), but we are still not certain exactly why. Only part of the relationship between the poverty of parents and of their adult children is explained by the direct and indirect inheritance factors we discussed earlier—parents' education, children's education, neighborhood and school characteristics, and access to welfare benefits (Corcoran). Among the other possibilities that still need to be studied are that poor children are handicapped by worse physical and mental health, by more stress, by the lack of both strong and weak ties to persons who could help them obtain good jobs, and by the lack of the "cultural capital"—middle-class fashion sense or speech patterns, for example—needed to get good jobs.

Social Class and Public Policy: Fighting Poverty

If the competition is fair, inequality is acceptable to most people in the United States. The question is how to ensure that no one has an unfair advantage. Politicians, activists, and social scientists have promoted three different approaches to fostering equality: income redistribution, outlawing discrimination, and increasing educational opportunities.

American Diversity

focus on

Subsidies, Incentives, and Welfare: Who Benefits?

Each year, politicians get elected to office by promising to reduce taxes and "unnecessary" federal spending. Their campaigns succeed because many Americans believe a disproportionate share of tax dollars go to services for the poor, such as Medicaid, Aid to Families with Dependent Children, and Food Stamps. But just exactly who gets what benefits?

Can you answer the following questions about "welfare" benefits (Waldman 1992, 56)?

1. Does the government spend more on (a) job training for the poor, (b) Head Start for low-income children, (c) Women, Infants, and Children nutrition subsidies, or (d) health care for the richest 10 percent of elderly Medicare beneficiaries?
2. Do government housing subsidies give (a) 10 times as much to the poor as to the middle class, (b) 5 times as much to the poor, (c) about equally to both groups, or (d) more to the middle class than to the poor?
3. Considering spending programs and tax subsidies together, on the average, which of the following

individuals gets the most money in federal benefits? Someone with income (a) less than $10,000, (b) from $10,000 to $40,000, (c) from $40,000 to $100,000, or (d) more than $100,000?

The correct answer to all three questions is "d". Each year, the government spends more for medical care for well-off seniors than it does for all of the other programs combined. Taking tax credits for mortgage interest and other home ownership tax breaks into account, the federal government spends four times as much to support middle- and upper-income families' housing than it does to house the poor. Finally, if the entire range of subsidies, incentives, and welfare programs is considered, an average upper-income individual will get considerably more than a typical poor person, even though it is precisely those middle- and upper-middle class individuals who are the most likely to complain that the government is ignoring them.

Even studies by conservative taxpayers' groups show that the average person with an income over $100,000 receives direct cash benefits—such as social security—that are slightly higher than those received by persons with income of less than $10,000. If such tax breaks as mortgage interest deductions, health care reimbursement accounts, and the benefits of IRAs are also considered, the poor receive substantially less than the middle class, and the wealthy receive more than any other group.

For at least a decade, angry taxpayers across the nation have insisted that welfare mothers be required to work for their assistance. As we grapple with questions related to taxes, the national debt, and appropriate federal expenditures, we might be wise to consider again the question of what distinguishes a subsidy from a welfare benefit and ask, "Just who are those welfare 'kings and queens'?"

For more information, look up the following subjects in InfoTrac College Edition:
Welfare
Tax incidence

Or visit the following Web site:
Welfare Information Network
http://www.financeprogjectinfo.org/win

Income Redistribution

Income redistribution can foster equality by adding income to those on the lower end of the social scale through welfare programs or by subtracting it from those on the upper end of the social scale through estate taxes.

In the United States, various welfare programs attempt to alleviate the economic burdens of the poor, working, and middle classes. These programs include direct income payments, such as Aid to Families with Dependent Children (AFDC), and indirect income, such as food stamps and Medicaid. In addition, although not normally thought of as welfare, Social Security pensions and tax deductions for home mortgage payments are also designed to shift wealth from some groups to others (see Focus on American Diversity box). Americans staunchly support tax deductions, Social Security, and other *entitlement* programs (i.e., programs to which all Americans are automatically entitled). Other welfare programs, however, have come under strong attack over the past quarter century.

If welfare programs are designed to supplement the income of those at the bottom or in the middle of the class system, estate taxes (taxes on inheritances) are designed to prevent those at the top from having an unfair advantage. Even though Americans generally agree that it is acceptable for ambitious, lucky, or clever persons to amass large fortunes, historically they have not believed that it is fair that their children should start their race with such a large advantage. Thus, since 1931, the United States has had a progressive estate tax. The maximum tax has varied over the years from 50 to 90 percent of the estate.

Although estate taxes certainly help to redistribute wealth, they have neither seriously harmed the economic well-being of affluent families nor significantly reduced unequal advantage. If wealthy individuals die at 70, their children are already middle aged. The $130,000 or so that the parents spent to send their children to the best private schools and universities that money could buy, the new homes they bought for the children when they married, the businesses they set them up in, the gifts of cash, and the trusts they set up for their grandchildren—none of these are part of the estate. By the time the parents die, the children have already been established as rich themselves (Lebergott 1975). Unless we ban private schools, transfers of money to one's children, and giving one's children good jobs, this inheritance is outside the scope of public policy.

Outlawing Discrimination

Because people start out with unequal backgrounds, they do not start the "success race" with equal chances of winning. Antidiscrimination and affirmative action laws do not attempt to reduce those inequalities. Instead, they attempt to ensure that no unfair obstacles are thrown in anyone's way during the race. These laws have had considerable effect; newspapers, for instance, no longer can advertise that some jobs are open only to men or to whites. Although some very able people are still held back unfairly, antidiscrimination and affirmative action laws help at least some of those who work very hard and who are very able to overcome the handicaps they began the race with.

Increasing Educational Opportunities

Education is widely believed to be the key to reducing unfair disadvantages associated with poverty. Pre-kindergarten classes designed to provide intellectual stimulation for children from deprived backgrounds, special education courses for those who don't speak standard English, and loan and grant programs to enable the poor to go as far in school as their ability permits—all these are designed to increase the chances of students from lower-class backgrounds getting an education.

These programs have had some success: Colleges and universities have many more students from disadvantaged backgrounds than they used to. Because students spend only 35 hours a week at school, however, and another 130 hours a week with their families and neighbors, the school cannot overcome the entire deficit that hinders disadvantaged children. For example, researchers have found that poor children and better-off children perform at almost the same level in first- and second-grade mathematics while school is in session. For children in poverty, however, every summer means a loss in learning, whereas every summer means a gain for those out of poverty (Entwisle & Alexander 1992). The home environments of less-advantaged children do not often include trips to the library and other activities that encourage them to use and remember their schoolwork. Consequently, for every step they take at school, they slide back half a step at home during the summer.

Where This Leaves Us

Research on stratification leads to one basic conclusion: As long as some people are born in tenements or shacks, as long as their parents are uneducated and have bad grammar and small vocabularies, and as long as they have no encyclopedias or intellectual stimulation—while others are born to wealthy, educated parents with excellent connections and "cultural capital"—there can never be true equality of opportunity. Thus, the only way to create equal opportunity is to attack these underlying problems. The questions for Americans is, do we want equal opportunity badly enough to pay the costs?

Summary

1. Stratification is distinguished from simple inequality in that (a) it is based on social roles or membership in social categories rather than on personal characteristics, and (b) it is supported by norms and values that justify unequal rewards.

2. There are two types of stratification systems. In a caste system, your social position depends entirely on your parents' position. In a class system, social position is based on educational and occupational attainment so individuals can have higher or lower positions than their parents.

3. Marx believed that there was only one important dimension of stratification: class. Weber added two further dimensions, and most sociologists now rely on his three-dimensional view of stratification: class, status, and power.

4. Inequality in income and wealth is substantial in the United States and has increased during the last quarter century. This inequality has widespread consequences and affects every aspect of our lives. Although the negative consequences of being in the working class or lower class are less than they were 50 years ago, lower-class and working-class individuals continue to be disadvantaged in terms of health, happiness, and lifestyle.

5. Structural-functional theorists use a supply-and-demand argument to suggest that inequality is a functional way of sorting people into positions; inequality is necessary and justifiable. Conflict theorists believe that inequality arises from conflict over scarce resources, in which those with the most power manipulate the system to enhance and maintain their advantage. Symbolic interaction theory focuses on how social status is reinforced through self-fulfilling prophecies and everyday interaction.

6. Allocation of people into statuses includes macro and micro processes. At the macro level, the labor market sets the stage by creating demands for certain statuses. At the micro level, the status attainment process is largely governed by indirect inheritance.

7. Despite high levels of inequality, most people in any society accept the structure of inequality as natural or just. This shared ideology is essential for stability. In the United States, this ideology is the *American Dream*, which suggests that success or failure is the individual's choice.

8. Approximately 13 percent of the U.S. population fall below the poverty level. Many of the poor are children. Although some part of poverty may be due to micro-level processes, the structure of opportunity determines how extensive poverty is in a society. How poverty is transmitted across generations seems best explained by a combination of structural opportunities and the many health and social disadvantages indirectly inherited from poor parents.

9. Because families pass their social class on to their children, any attempt to reduce inequality must reduce the ability of parents to pass on their economic status and values.

Thinking Critically

1. Can you think of any ways in which the U.S. system of stratification resembles a caste system?

2. To what social class do you belong? How do you know?

3. Can you think of any aspect of individual behavior or of social life that is likely to be unaffected by social class?

4. What macro-level processes do you think are most responsible for the increasing numbers of homeless in the United States? What micro-level processes?

5. People who own their own homes or apartment complexes can deduct the amount of their mortgage interest and the property taxes they pay from their federal income tax liability. Renters pay enough rent each month to cover the costs of their landlords' tax bill and mortgage, but they cannot deduct that amount from their own taxes. Is this a form of welfare? Do you think it's fair?

Sociology on the Net

 The Wadsworth Sociology Resource Center: Virtual Society

http://sociology.wadsworth.com/

The companion Web site for this book includes a range of enrichment material. Further your understanding of the chapter by accessing Online Practice Quizzes, Internet Exercises, InfoTrac College Edition Exercises, and many more compelling learning tools.

Chapter-Related Suggested Web Sites

U.S. Bureau of the Census
http://www.census.gov

National Coalition for the Homeless
http://www.nationalhomeless.org/inequality.org
http://www.inequality.org

The Urban Institute
http://www.urban.org

 InfoTrac College Edition

http://www.infotrac-college.com/wadsworth/

Access the latest news and research articles online—updated daily and spanning four years. InfoTrac College Edition is an easy-to-use online database of reliable, full-length articles from hundreds of top academic journals and popular sources. Conduct an electronic search using the following key search terms:

Welfare reform

Poverty

Social class

Inequality

Homelessness

 For more information related to this chapter, visit the Opposing Viewpoints Resource Center. Be sure you check all the options on the toolbar—viewpoints, references, statistics, and so on—and search under the following subject:

Poverty

Suggested Readings

Ehrenreich, Barbara. 2001. *Nickel and Dimed: On (Not) Getting By in America*. New York: Metropolitan. Investigative journalist Barbara Ehrenreich spent a year trying to support herself at minimum wage jobs: waitress, maid, nursing home aide, and Wal-Mart salesclerk. Her book details the extraordinary physical and emotional toll exacted by working- and lower-class jobs. A gripping read.

Hochschild, Jennifer L. 1995. *Facing Up to the American Dream: Race, Class, and the Soul of the Nation*. Princeton, N.J.:

Princeton University Press. A book that examines the extent to which the American dream is inaccessible to many people in the United States and addresses the issue of why there is so little support for redistribution among the poor of the United States.

LeBlanc, Adrian Nicole. 2003. *Random Family: Love, Drugs, Trouble, and Coming of Age in the Bronx.* New York: Scribner. Journalist LeBlanc spent ten years following an extended family that lives in one of the country's worst slums. Her richly detailed book paints a vivid portrait of life at the bottom, where 13-year-old girls choose to have babies to keep their boyfriends and their brothers enjoy the quiet they can find only in a prison's solitary confinement.

Patterson, James T. 1994. *America's Struggle against Poverty, 1900–1994.* Cambridge, Mass.: Harvard University Press. A history of poverty, poor people, and poverty legislation in the United States during this century.

Wilson, William Julius. 1987. *The Truly Disadvantaged: The Inner City, the Underclass, and Public Policy.* Chicago: University of Chicago Press. A strong statement about the growing U.S. underclass by one of the nation's most prominent experts. Wilson argues convincingly for new government programs to put a floor under all citizens.

CHAPTER 8

Global Inequality and Globalization

© AFP/Corbis

Outline

Inequality and Development

The central fact in the international political economy is the vast inequality that exists in today's world. In 2001, gross domestic product per capita was $34,320 in the United States but only $470 in Sierra Leone (United Nations Development Programme 2003). Average life expectancy in the United States is 77 years; the average in Sierra Leone is 35. The massive disparities not only in wealth and health but also in security and justice are the driving mechanism of current international relationships

Because massive inequality leads to political instability and to unjustifiable disparities in health and happiness, nearly every nation—whether more or less developed—thinks that reductions in international inequality are desirable. The most accepted way to do this is through development—that is, by raising the standard of living of the less-developed nations.

What is development? First, development is *not* the same as Westernization. It does not necessarily entail monogamy, three-piece suits, or any other cultural practices associated with the Western world. **Development** refers to the process of increasing the productivity and the standard of living of a society, leading to longer life expectancies, better diets, more education, better housing, and more consumer goods.

Importantly, development is not a predictable, unidirectional process. Some countries, such as South Korea, have developed at faster rates than others. Other countries, such as the nations of the former Soviet Union, for example, have become *less* developed over time.

Three Worlds: Most- to Least-Developed Countries

Almost all societies in the world have development as a major goal: They want more education, higher standards of living, better health, and more productivity. Just as social scientists often think of three social classes in the U.S. stratification system—upper, middle, and lower—nations of the world can also be stratified into roughly three levels.

The **most-developed countries** are those rich nations that have relatively high degrees of economic and political autonomy. Examples include the United States, the Western European nations, Japan, Canada, Australia, and New Zealand. Taken together, these nations make up roughly 20 percent of the world's population, produce 86 percent of the gross world product, and own 74 percent of the world's telephone lines (United Nations Development Programme 2003). Politically, economically, scientifically, and technologically, they dominate the international political economy.

Less-developed countries include the former Soviet Union and the former Communist bloc nations of Eastern Europe, plus several nations in Southeast Asia and Central and South America that have seen substantial improvements in standard of living during the past few years. These countries hold an intermediate position in the world political economy. They have far lower living standards than the most-developed nations but are substantially better off than the least-developed nations.

The remaining 75 percent or so of the world's population lives in the **least-developed countries.** These countries are characterized by poverty and political weakness. Although they vary in population, political ideologies, and resources, they are considerably behind on every measure of development.

Development refers to the process of increasing the productivity and standard of living of a society—longer life expectancies, more adequate diets, better education, better housing, and more consumer goods.

Most-developed countries are those rich nations that have relatively high degrees of economic and political autonomy.

Less-developed countries are those nations that have lower living standards than the most-developed countries but are substantially better off than the least-developed nations.

Least-developed countries are those nations that are characterized by poverty and political weakness and that are considerably behind on every measure of development.

MAP 8.1
Human Development Around the World
The human development index measures the average achievements in a country in three basic dimensions of human development—
life expectancy, educational attainment, and a decent standard of living.
SOURCE: United Nations Human Development Report 2003.

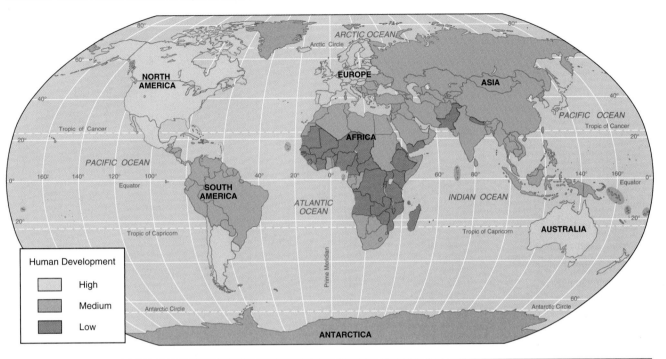

The Human Development Index

The differences among the world's nations are obvious: In the most-developed countries, people are healthier, more educated, and richer. But how important are these differences in the day-to-day quality of the average person's life?

One approach to answering this question is to develop an index that measures the average achievements of a country along the basic dimensions of human experience: life expectancy, educational attainment, and a decent standard of living. Another approach focuses not only on these three aspects of development but also takes into account the unequal opportunities of men and women. Map 8.1 shows the location of the most- to least-developed countries worldwide. Table 8.1 compares several basic quality-of-life indicators for 15 nations, representing most-developed, less-developed, and least-developed countries. In addition to information about longevity and economic productivity, Table 8.1 also includes each country's overall ranking on the composite Human Development Index and the Gender-Related Development Index. These indexes are based on information about adult literacy rates and educational attainment, life expectancy, and per capita gross domestic product; the greater the disparity between men's and women's quality of life, the lower a country's Gender-Related Development ranking will be compared to its overall Human Development ranking.

In general, the more productive a nation is, the better its quality of life. Norway, with a per capita gross national product of $29,620, also has one of the lowest infant mortality rates in the world: Each year, for every 1,000 live births in Norway, 4 children die before their first birthday. In contrast, in the world's poorest nation—Sierra Leone—

TABLE 8.1
The Extent of International Inequality: Comparing 15 Countries

Type of Country	GDP Per Capita (U.S. dollars)	Life Expectancy at Birth	Infant Mortality Rate/1,000 Live Births	Human Development Ranking*	Gender-Related Development Ranking*
Most-developed countries					
Canada	$27,130	79	5	8	6
Japan	25,130	81	3	9	13
Norway	29,620	79	4	1	1
Singapore	22,680	78	3	28	28
United States	34,320	77	7	7	5
Less-developed countries					
Brazil	$7,360	68	31	65	58
China	4,020	71	31	104	83
Ecuador	3,280	71	24	97	84
Russian Federation	7,100	67	18	63	56
Saudi Arabia	13,330	72	23	73	68
Least-developed countries					
Congo, Dem. Rep.	$680	41	129	167	136
Haiti	1,860	49	79	150	122
Nepal	1,310	59	66	143	119
Pakistan	1,890	60	84	144	120
Sierra Leone	470	35	182	175	NA

*Out of 175 countries
**Out of 144 countries

SOURCE: Human Development Report (2003).

per capita gross national product is only $470, and 182 out of every 1,000 children die before their first birthday. Note, however, that the relationship between economic productivity and quality of life is not perfect. For example, per capita income is 15 percent higher in the United States than in Norway, but, according to the United Nations Development Programme (2003), Norwegians enjoy a higher quality of life than do Americans. In large part, Norway's rankings on human development reflect the fact that access to health care, education, and adequate nutrition is more universally available there than in the more affluent United States.

The gap between quality of life for the world's wealthiest and poorest is enormous. The world's richest 1 percent of people earn as much income each year as do the world's poorest 57 percent. Put another way, the ratio of the income of the top 20 percent of the world's population to that of the poorest 20 percent is 71 to 1. Women and children are particularly at risk in poor nations. A half-million women in the developing nations die each year during childbirth, at rates 10 to 100 times those found in the most-developed nations. Worldwide, more than 160 million children are moderately or severely malnourished (United Nations Development Programme 2003). International inequality is indeed dramatic.

A Global Perspective

Women and Development

Around the world, women's work matters. In many parts of the developing world, women are the main producers of food crops. In several parts of sub-Saharan Africa, for instance, women are responsible for at least 70 percent of the total production, processing, and marketing of food, as well as responsible for almost all cooking. In addition to providing food and caring for children, women also have the added burden of collecting firewood and water, tasks that can take up to 2 hours daily (United Nations Population Fund 1991).

Although women's labor accounts for at least two-thirds of the total work hours expended on any given day, they receive only one-tenth of the world's income in return and own only about one-hundredth of the world's property (AID Horizons 1983). Yet at least until the mid-1970s, development agencies routinely directed educational programs about the advantages of using new types of seed, fertilizer, and cropping systems to male audiences. No wonder these programs failed to increase agricultural productivity when the people doing the actual farming were female!

Most specialists now recognize that sustainable development will only become possible when women are full partners in the process. Research clearly shows that overall economic growth has been fastest in those areas of the world where women have higher status and slowest where they face the greatest disadvantage (United Nations Population Fund 1992). Where women attain higher levels of education, their productivity in both the paid and unpaid labor market is enhanced. Furthermore, because women appear to be more likely than men to invest new resources in the nutrition, medical care, and education of their children (Schultz 1993), investments in women's education may be particularly crucial to development efforts.

As a result of the growing awareness of the critical role that women play in societal as well as domestic economies, new development strategies are more often aimed at women. Indeed, the advancement of women is one of the main goals the United Nations Development Programme now pursues in all of its efforts to help countries build sustainable development. Among the promising programs are those designed to improve women's literacy, provide training and small loans to women wishing to start their own microbusinesses, promote legal changes to allow women to own and manage property, and give women access to contraception so they can control their reproduction. Such programs cannot take the place of broader programs designed to reduce massive international and national disparities, but including women in development programs will help to ensure that benefits reach down to the household level.

Ed Kashi

■ Although women and girls constitute only about half of the world's population, they do about two-thirds of the work. In return for their labor, women worldwide receive approximately one-tenth of the income earned and control about one-hundredth of the property.

For more information, look up the following subjects in InfoTrac College Edition:
United Nations Development Fund for Women
Microloans

Or visit the following Web sites:
United Nations Development Programme
http//www.undp.org/gender/
The UN Internet Gateway on the Advancement and Empowerment of Women
http//www.un.org/womenwatch/index.html

Although many of the differences between nations are matters of culture, there is general consensus that rich is better than poor, security is better than insecurity, and health is better than sickness. No nation wants to be poor and underdeveloped. What are the causes of current inequalities and what are the prospects for their relief? We examine two general theories of development—modernization and world-systems theory—and discuss their implications in the resolution of global inequalities.

Structural-Functional Analysis: Modernization Theory

Modernization theory sees development as the natural unfolding of an evolutionary process in which societies go from simple to complex economies and institutional structures. This is a structural-functional theory based on the premise that adaptation is the chief determinant of social structures. According to this perspective, developed nations are merely ahead of the developing nations in a natural evolutionary process. Given time, the developing nations will catch up.

Modernization theory emerged in the 1950s and 1960s, when many believed that developing nations would follow pretty much the same path as the developed nations.

Modernization theory sees development as the natural unfolding of an evolutionary process in which societies go from simple to complex economies and institutional structures.

■ Despite the rapidity with which new ideas and technologies can be diffused, in Vietnam as in many other countries around the world, rickshaws, bicycles, mules, and feet remain more common forms of transportation than automobiles.

Greater productivity through industrialization would lead to greater surpluses, which could be used to improve health and education and technology. Initial expansion of industrialization would lead to a spiral of ever-increasing productivity and a higher standard of living. These theorists believed this process would occur more rapidly in the least-developed nations than it had in Europe because of the direct introduction of Western-style education, health care, and technology (Chodak 1973).

Events have shown, however, that development is far from a certain process. Some "developed" nations such as the former Soviet Union have regressed over time. Thailand and Singapore have modernized quickly, Haiti has modernized hardly at all, and Mexico has gone through wild economic upswings and downturns. The least-developed countries have not caught up with the developed world and, in many cases, the poor have simply become poorer, while the rich have become richer.

Why haven't the less-developed nations followed in the footsteps of developed nations? The primary reason is that they encounter many obstacles not faced by nations that developed earlier: population pressures of much greater magnitude, environments ravaged by the developed nations since they were colonial powers, and the disadvantage of being latecomers to a world market that is already carved up. These formidable obstacles have given rise to an alternative view of world modernization—world-systems theory.

Conflict Analysis: World-Systems Theory

The entire world may be viewed as a single economic system that has been dominated by capitalism for the past 200 years. Nation-states and large **transnational corporations** (i.e., corporations that produce and distribute goods in more than one country) are the chief actors in a free-market system in which goods, services, and labor are organized to maximize profits (Chirot 1986; Turner & Musick 1985). This system includes an international division of labor in which some nations extract raw materials and others fabricate raw materials into finished products.

Nation-states can pursue a variety of strategies to maximize their profits on the world market. They can capture markets forcibly through invasion, they can try to manipulate markets through treaties or other special arrangements (such as NAFTA), or they can simply do the international equivalent of building a better mousetrap. The Japanese auto industry (indeed, all of Japanese industry) is a successful example of the last strategy.

On a global basis, capitalism operates with less restraint than it does within any single nation. There is no organized equivalent of welfare or Medicaid to take care of indigent nations. In the absence of a world political structure, economic activity is regulated only by market forces such as supply and demand (Turner & Musick 1985).

World-systems theory is a conflict analysis of the economic relationships between developed and developing countries. It looks at this economic system with a distinctly Marxist eye. Developed countries are the bourgeoisie of the world capitalist system, and underdeveloped and developing countries are the proletariat. The division of labor between them is supported by a prevailing ideology (capitalism) and kept in place by an exploitive ruling class (rich countries and transnational corporations) that seeks to maximize its benefits at the expense of the working class (underdeveloped and developing countries).

World-systems theory distinguishes two classes of nations: core societies and peripheral societies. **Core societies** are rich, powerful nations that are economically diversified and relatively free from outside control. They arrive at their position of dominance, in part, through exploiting the peripheral societies. On an international

Transnational corporations are large corporations that produce and distribute goods internationally.

World-systems theory is a conflict perspective of the economic relationships between developed and developing countries, the core and peripheral societies.

Core societies are rich, powerful nations that are economically diversified and relatively free from outside control.

Peripheral societies are poor and weak, with highly specialized economies over which they have relatively little control.

level, they are very similar to the industrial core of the United States' dual economy (Chapter 14).

Peripheral societies, by contrast, are poor and weak, with highly specialized economies over which they have relatively little control (Chirot 1977). Like the small-business sector in the U.S. dual economy, peripheral societies are vulnerable to change and have little control over their environment. Some of the poorest countries rely heavily on a single cash crop for their export revenue. For example, 70 percent of Guinea-Bissau's export earnings come from cashews and more than half of Uganda's earnings come from coffee (Central Intelligence Agency 2002). The economies of these and many other developing nations are vulnerable to conditions beyond their control: world demand, crop damage from infestation, flooding, drought, and so on.

A key element of world-systems theory is the connectedness between core society prosperity and peripheral society poverty. According to this theory, our prosperity is their poverty. In other words, our inexpensive shoes, transistors, bananas, and the rest depend on someone in a least-developed nation receiving low wages, often while working for a company based in one of the developed nations. Were their wages to rise, our prices would rise and our standard of living would drop. Furthermore, if their wages were to rise they also would want more consumer goods; the ensuing demand on resources would simply be more than the most-developed economies or the planet could sustain.

Economic Development and the Environment

Development refers to the process of raising the standard of living and improving the quality of life of a society. In theory, development programs should be as concerned with improving access to education and health care and with ensuring the rights of all citizens—regardless of age, gender, race, or ethnicity—as they are with increasing labor productivity. In practice, however, aid to least-developed nations has focused almost exclusively on fostering economic development, and its success is most often measured simply in terms of increases in gross national product or access to consumer goods. Under this model of development, affluent lifestyles become the goal to which poor nations aspire.

The Threat to the Environment

What happens when everyone around the world wants a refrigerator, a car, and a single-family home, and many more people have the funds needed to purchase them? Can the planet afford this level of consumption? The United Nations estimates that if current trends in population growth and consumption continue, the world will start running out of aluminum, copper, oil, and other natural resources during this century (United Nations Environmental Programme 2002).

Paradoxically, threats to the so-called "renewable" resources may be even more severe. Water and trees, for example, are renewable resources. With wise use, they are self-replenishing. Overuse, however, can destroy such resources within a generation. It is estimated that by the year 2025, 48 countries (most in Africa) will face severe water shortages, affecting millions of people (United Nations Environment

Industrial development in the absence of environmental protection creates grave health risks. This Indonesian child at play in a trash-filled waterway near his home is at risk for numerous life-threatening illnesses.

© AFP/Corbis

Programme 2002b); currently, 43 countries are deemed by the United Nations to have insufficient water for their people (Joynt and Poe 2003).

Agriculture accounts for more water use by people—43 percent worldwide—than any other single factor. Industry accounts for only 15 percent of worldwide water use, and households use only about 2 percent. Although agriculture and industry have improved nutrition and access to goods in the short run, in the long term both may have serious detrimental consequences. Agriculture is a singularly inefficient user of water. Sixty-two percent of the water used for irrigation in some developing countries is lost before it ever reaches the fields (Joynt and Poe 2003); this runoff leaches important nutrients from the soil and contributes to increased soil salinity. In the long run, then, irrigation may reduce agricultural productivity.

Industry is the second largest user of water; moreover, the kinds of industry often held up as models for development—food processing, petroleum refining, and pulp and paper refineries—are particularly water intensive. Thus, in many of the least-developed countries, economic development is likely to place extraordinary demands on already fragile water supplies. Water is used by factories not only for cooling, generating steam to run equipment, and as a method of transport but also as a dumping ground for industrial waste. Industrial contamination of rivers and lakes has reached such epidemic proportions in Eastern Europe and the former Soviet Union that it is estimated that 75 percent of the drinking water in these nations is polluted (Feshbach 1999).

The Need for Sustainable Development

With a population of more than 6 billion and growing, it is clear that planet Earth simply cannot sustain the kind of energy-hungry, wasteful, pollution-riddled development that has driven the economies of most less-developed nations. In 1987, the U.N.–established Brundtland Commission defined **sustainable development** as "development that meets the needs of the present without compromising the ability of future generations to meet their own needs" ("Brundtland Report" 2000). There is

Sustainable development is development that meets the needs of the present without compromising the ability of future generations to meet their own needs.

Technology

Information Technology and Global Culture

Even the most creative society discoverers or invents only a small portion of all its innovations. Much of the content of U.S. culture, for instance, was acquired from successive waves of immigrants, each bringing their own language, values, and customs to American shores. In the early twenty-first century, one of the primary vehicles for cultural diffusion is the new information technology. The information revolution, which began in the United States, is rapidly spreading over the world. Although huge disparities persist, virtually every nation on every continent now has some access to the Internet and to e-mail technology (World Bank 2003). Worldwide annual sales of personal computers exceed 90 million units (Samuelson 1999); to gain perspective on that number, consider that car sales, globally, are probably about half that number. In the space of one year, the number of Internet users increased by nearly 80 percent, from 171 million users in 1999 to 304 million users in March 2000 (U.S. Internet Council and ITTA 2000).

Not only has the use of information technology spread throughout the world, but the fact that it has done so means that other aspects of cultural diffusion also take place much more rapidly. Ideas can sweep the world within days and be introduced into the most remote villages within weeks and months. A fervor for democracy, for example, swept the world in 1989. The year began with prodemocracy student protests in Tiananmen Square in Beijing, China and ended with the fall of the Berlin Wall and the toppling of communist governments in Eastern Europe. Many of those seeking freedom and democracy relied on the ideals and symbols of the American Revolution; the Statue of Liberty lent symbolic support to the demonstration at Tiananmen Square. By the same token, Islamic fundamentalism is spreading globally in part through Internet chat rooms, Web sites, and the like.

The spread of information technology not only means more rapid change in all areas of life, it also means growing international similarity. In Moscow, Beijing, Nairobi, and Boston, business leaders are wearing the same kinds of suits. In the western United States and in Africa, plans to develop virtual universities are well under way. Throughout the world, young people are listening to the same kinds of music. White Americans now enjoy salsa dancing, and Russian youth listen to *khard-roka* and *khevimetallu* (i.e., hard rock and heavy metal). On a threatening level, the speed of diffusion has meant that nuclear weapons and terrorist technologies are also widespread.

Peter Turnley/Black Star

■ The "Goddess of Liberty," which symbolized the prodemocracy demonstrations in Tiananmen Square in Beijing in 1989, bears a marked resemblance to the U.S. Statue of Liberty. The ideals of democracy, its symbols, and even its constitutional forms have diffused throughout the world. Nevertheless, economic and political realities may repulse ideas that do not fit established patterns.

 For more information, look up the following subjects in InfoTrac College Edition:
Global culture (as keywords)
Internet, demographic aspects

 Or visit the following Web sites:
NUA Internet Surveys
http://www.nua.ie/surveys
Taking It Global
http://www.takingitglobal.org

broad agreement within the international community that two major changes will be necessary to achieve this type of development.

First and foremost, sustainable development will require the elimination of absolute poverty so that the 1.2 billion poorest of the world's people can produce or buy the food, clothing, and housing necessary to ensure health and self-respect. Ending absolute poverty will depend, then, on improving access to education, health care, clean water, and sanitation. It will depend, too, upon increasing the status of women and reducing the buildup of military arms that already costs least-developed nations an enormous share of their scarce resources (Sen & Grown 1987). Ending poverty is a goal in itself, but because poverty and rapid population are often found together, it will also enhance development by limiting population size.

In addition to eliminating destitution, sustainable development will require meeting the legitimate aspirations of the 3.6 billion people who are neither very poor nor rich. In today's world, the most-affluent 1.2 billion satisfy their aspirations with little regard for sustainability, but the world will not be able to satisfy the aspirations of billions more people in this way. Consequently, in order to meet the needs of the middle third of the world's population, the economic inequalities that exist within and between countries will have to be reduced and the benefits of development more fairly distributed.

Second, sustainable development will require slower population growth. Smaller increases in world population, coupled with an evolution to lower-consumption lifestyles and more efficient production, will reduce the environmental impact of development and help to ensure equal access to "the good life" across generations (United Nations Population Fund 1992).

A Case Study: Economic Development and the Environment in Eastern Europe

The current situation in Eastern Europe and the former Soviet Union offers good examples of what can happen to the environment when economic development is pursued at any cost.

- Chemical pollution, especially dioxin, has reduced life expectancy for men and women in the Russian town of Dzerzhinsk to about 50 years—lower than the life expectancy in India or Bolivia and about equivalent to that in Haiti (Feshbach 1999).
- The Vistula River, which runs through Poland, is so laden with poisons and corrosive chemicals that it is considered unusable, even for factory coolant systems (Jensen 1990).
- Industrial accidents and releases of radioactive waste from the Mayak Nuclear Plant have exposed more than 400,000 Siberians to more or less continuous radiation over the last half century. Rates of leukemia and other cancers are very high throughout the region (Little 1998).

These and other ecological woes stem from the "growth-at-any-cost" mentality that drove Soviet economic development at least since the late 1920s (Feshbach 1999). Coupled with concerns about national security and the rapid rise of a huge military-industrial complex, Soviet bloc emphasis on production over efficiency created both a badly outdated industrial sector and environmental hazards of unparalleled magnitude (Stanglin 1992). Of course, this creates a real quandary: How can the countries of Eastern Europe and the former Soviet Union revitalize their economies *and* clean up the environment when repairing the ecological damage alone would cost billions of dollars?

Forms and Consequences of Economic Dependence

Most contemporary scholars use some form of world-systems theory to understand development. Although many would reject the Marxist implications of the theory, there is general agreement that international inequality is the result of competition for scarce natural resources within an international capitalistic economic framework. Here we discuss the forms and consequences of economic dependence that occur within this framework.

The Forms of Economic Dependence

The primary characteristic of the peripheral nations is that they have relatively little control over their economies. They are dependent. In many cases, these dependent relationships represent a triple alliance among three actors: transnational corporations (generally based in the most-developed nations), the governments of the peripheral societies, and local elites. Scholars who study global dependence identify three types of dependence (Bradshaw 1988): the classic "banana republic," industrial dependence, and foreign-capital dependence.

The Banana Republic

The classic case of dependence occurs in a least-developed nation whose economy is dependent on the export of raw materials—bananas, fruit, or minerals. African nations, the Indian subcontinent, and Latin American countries have all fallen into this situation at different points in their histories; many of the world's peripheral nations are still in it. Because the cost of manufactured imports is much higher than the prices paid for exported raw materials, these countries accumulate huge debts. In order to pay the debts, they must produce and export more "luxury" crops such as coffee, sugar, or even broccoli; this emphasis on export crops reduces the nation's ability to feed itself and retards the development of economic diversity.

Industrial Dependence

Increasingly, transnational corporations are making use of one of the other major assets of less- and least-developed countries: cheap labor. Assembly plants located in developing countries have reduced labor costs dramatically for these companies (say, from $10 an hour to $3 a day) and have also saved them money on the transportation costs of goods that will be sold in these countries. Increasingly, too, transnational corporations are moving *skilled* jobs, such as data processing and computer programming, to nations like India, Mexico, and the Philippines.

Some observers have hoped that investment in industrial plants in developing countries would improve local health and transportation facilities, provide jobs, spur development of indigenous subsidiary industries, and generally galvanize local economies. Empirical studies demonstrate that few of these good effects have been realized (Farmer 1999; Lappé et al. 1998).

Foreign-Capital Dependence

When there is a relatively strong indigenous elite, dependence may take the form of dependence on loans and investment from foreign banks, the International Monetary Fund (IMF), and the World Bank (Wimberly 1990). In this case, the firms will be

Connections

Personal Application

If you have ever traveled to a less-developed country, you have seen the consequences of economic dependence. You probably were warned not to drink water from the faucets, because these nations lack the economic resources to provide safe drinking water. Because wages are so low in these nations, you could buy meals, clothes, and souvenirs very cheaply and could afford to purchase services like taxis that you could never use at home. And because even at these low wages, many can find no jobs at all, you might have seen beggars or prostitutes and been warned to watch out for thieves.

■ These workers are assembling Nikes in a factory outside Ho Chi Minh City. By locating plants in the least-developed nations, transnational firms realize huge savings in labor costs and U.S. consumers get cheaper products. Among the undesirable side effects, however, are loss of jobs for U.S. workers and the growth of global inequality.

owned and operated by local residents, rather than by foreign corporations. In theory, this should encourage development by providing additional capital. Instead, the extension of loans to less- and least-developed nations has usually resulted in a dramatic debt crisis. High inflation and worldwide recessions have decreased the demand for products produced in these countries. At the same time, the IMF and the World Bank have increased the interest rates debtor nations have to pay on existing loans and forced these nations to cut back on social programs such as food subsidies and health care for the poor in exchange for further economic aid (Kolko 1999; Peabody 1996). To meet debt payments, countries have had to reduce investment in their own economies and squeeze their own people even harder (Wimberly 1990).

The Consequences of Economic Dependence

By definition, economic dependence is a bad thing. It means that you do not have control over your economy and, perhaps, over your government. Although ties to richer, stronger economies might be expected to benefit less-developed nations, research does not bear out this expectation. Scholars have identified at least three unfortunate consequences of foreign-capital penetration.

First, efforts to attract foreign capital usually result in excessive investment in urban areas and almost no investment in the rural areas, where most of the nation's population is likely to live. One long-term consequence of this investment pattern is a substantial decrease in the overall standard of living: per capita food production actually decreases as people migrate to huge urban areas that are ill-equipped to receive them (Stokes & Anderson 1990).

Second, only the urban elites and the small urban middle class actually benefit from foreign capital (Bradshaw 1988). After all, low wages are what attract foreign investment in the first place. Thus, transnational corporate penetration and foreign aid dependence increase income inequality. Partially as a result of that inequality, infant mortality rates rise (Wimberly 1990; Lappé et al. 1998).

A final result of dependence on foreign capital is that governments often become more authoritarian as they ally themselves with local elites in an attempt to ensure continued investment. Because this alliance occurs at the expense of the majority of the people, anger and conflict increase; transnational penetration has been shown to increase the risk of political violence (London & Robinson 1989).

Some scholars have questioned whether or not foreign-capital penetration actually has all of the negative effects outlined above. Asking the question "Does foreign capital really do more harm than good?" virtually all studies have at least agreed that domestic investment in poor nations is far more beneficial than is foreign investment (Dixon & Boswell 1996; Firebaugh 1996).

Competition, Change, and International Relationships

Serious income and wealth inequalities can cause conflict both within and between nations. In this section we look at these challenges to the status quo and at the developed nations' responses to them. We then offer a brief history of Haiti as a case study of these processes.

Challenges to the Status Quo

Awareness of vast inequalities within a nation can lead to organized nationalist movements, resulting in violent, revolutionary challenges to the status quo (Kerbo 1991, 513). During the last four decades, there have been nationalist revolutions or violent class struggles in Vietnam, Angola, Chile, Algeria, Ethiopia, Nicaragua, El Salvador, Zimbabwe, and Afghanistan, to name just some.

Inequality also can cause conflict *between* nations. Information technology, the media, and increasing international travel all contribute to a rising sense of deprivation and frustration as citizens of peripheral nations are exposed to the lifestyles of more affluent countries. They also can lead citizens of the less-developed nations to feel that their culture is under attack by western culture. Now that the Cold War is over and the Soviet Union no longer poses a military threat, many observers believe that the greatest threats to the United States will come from the less-developed nations—as the events of September 11 demonstrated. Nearly a half-dozen peripheral and semi-peripheral nations—Argentina, Brazil, India, Pakistan, and South Africa—now have nuclear weapons capacities. Even more important than the threat of nuclear attack is the threat of terrorism (including bioterrorism), since many individuals in developing nations, like the terrorists who destroyed the World Trade Center, can easily obtain the resources needed to attack the United States and other developed countries.

The Developed World's Response

Responses to the threats and opportunities posed by the least-developed nations can be placed in three broad categories: military, humanitarian, and market-based.

Military Responses

One response to possibly revolutionary challenges from peripheral nations has been military action to stop unrest. Such action can be either *overt* or *covert*. The U.S. invasions of Iraq in 1991 and 2003, for example, illustrate overt intervention. Covert actions include propaganda campaigns, rigged elections, and help in staging coups. CIA involvement in the Chilean coup that ousted President Allende and its consideration of a plan to damage Fidel Castro's image by giving him LSD before a major public address (Kerbo 1991, 517) are examples of covert action.

Humanitarian Responses

In recent years, the United States has begun to use its military for humanitarian interventions as well as for "regime change." The food relief efforts in Iraq and Afghanistan following U.S. military interventions in those countries are examples. Meanwhile, citizens in core nations provide both tax dollars and private donations to aid people in poor nations, and many people in the United States volunteer in agencies such as the Peace Corps to improve living conditions in less-fortunate parts of the world.

Market-Based Responses

Why do developed nations interfere in least-developed countries? Why have countries such as Nicaragua and Vietnam become battlegrounds for more-developed nations? Why, as in Afghanistan, do the more-developed nations find themselves at war with the

The distribution of U.S. grain in India, Haiti, and Somalia and in Rwandan refugee camps serves many purposes: helping those in need, supporting U.S. farm prices, and furthering U.S. foreign policy objectives.

© Betty Press/Woodfin Camp & Associates

least-developed nations they have armed? Although ideology plays a part, the importance of the least-developed nations as a market provides a more compelling answer.

Nearly one-third of all world trade is with the least-developed world. These nations are both an important market for manufactured goods and an important source of cheap labor and raw materials. Any closure of trade with the least-developed world would strike a major blow to international trade and world capitalism. Although the avowed aim of U.S. foreign policy has been to ensure each nation's right to self-determination, in practice U.S. policy has been more concerned with keeping nations open to capitalist interests than with protecting political liberty. In an effort to keep markets in the developing world open to U.S.-owned transnational corporations, the United States has provided economic and military support to repressive, right-wing, totalitarian regimes in nations such as South Korea, Vietnam, Chile, Nicaragua, and Iran (Sluka 2000). Despite evidence of continuing human rights violations, the Clinton administration renewed China's most-favored-nation trade status, presumably because a market for U.S. goods that consists of 1.26 billion people is simply too lucrative to penalize or ignore. Likewise, it is hard to assess the extent to which the presence of European and U.S. peacekeeping forces in Somalia was a result of humanitarian concerns and how much was a result of pressure from the European and U.S. petroleum firms that hold large oil exploration contracts in the country.

Nowhere are the political and economic interdependencies of most-to-least developed nations more clear than in the 1991 and 2003 U.S.-Iraqi conflicts. Ever since oil prices quadrupled in 1973, most- and less-developed nations have exchanged arms for oil, with little attention to either their own diplomatic objectives or those of their clients' states. The United States, one of the world's largest arms exporters, enamored with cheap energy and unwilling to pay a higher price for oil and gasoline, faced the grand irony of waging two wars against an Iraqi army equipped with weapons provided by U.S., British, French, German, and Italian arms suppliers. As diplomats continue to struggle with the political consequences of the most recent conflicts in the Middle East and possible new ones, defense industries in the developed world continue to arm and rearm the region.

Haiti: A Case Study

The history of Haiti, a Caribbean least-developed nation, illustrates the complex relationships between developed and less-developed countries and the tensions in those relationships.

Haiti has a predominantly agricultural economy. Approximately 70 percent of the labor force is involved in agriculture, most on small family plots worn away by severe soil erosion. Coffee is the major cash crop, but light manufacturing—cosmetics, textiles, toys, and baseballs—provides the major source of export revenues (Central Intelligence Agency 2002). With a population of about 6.4 million people and an annual growth rate of 1.7 percent, Haiti is one of the fastest growing and poorest countries in the Western Hemisphere. Per capita gross domestic product is $1,860 (United Nations Development Programme 2003). Yet even this figure does not accurately reflect Haitian poverty, given the wide earnings gap between local elites and the remaining population who make almost nothing. About 51 percent of Haitians 15 years of age and older are illiterate. Life expectancy is only 49 years, and the infant mortality rate is 10 to 20 times higher than that found in many of the most-developed nations (United Nations Development Programme 2003).

Despite the restoration of the Aristide presidency, most Haitians remain very poor.

Background

From the beginning Haiti's role in the world system has been unique. Originally a French colony, the Haitian Republic gained its independence in 1804 as the result of a slave uprising. As the first black republic in a world system still dominated by slavery, Haiti was diplomatically isolated for more than three decades from both the United States and European powers, who feared that the Haitian revolution would set a dangerous example for African Americans in their own countries and colonies (Plummer 1985; Farmer 2003). This concern did not, however, stop foreign companies from trading with Haiti, and Haitian migrant labor played a major role in the Cuban and Dominican Republican sugar economies.

Because political unrest threatened U.S. economic interests and because the United States wanted to use Haiti as a base to protect its routes to the Panama Canal, the U.S. Marines occupied Haiti from 1915 to 1934. Working with local elites, the Marines helped found the modern Haitian military and ran the customs and banking business of the country (Wilentz 1993). Overt racism also played a role in the U.S. decision to occupy Haiti. U.S. Assistant Secretary of State William Phillips explained at the time that the occupation was necessary because of ". . . the failure of an inferior people to maintain the degree of civilization left them by the French, or to develop any capacity of self-government entitling them to international respect and confidence" (Schmidt 1971, 63).

The Duvalier Dictatorship

Between 1957 and 1986, Haiti was governed by the Duvalier family. In cooperation with foreign investors, the Duvaliers and a small economic and military elite amassed fortunes. The Duvaliers robbed the country of somewhere between $20 and $800 million before a popular uprising led to their ouster. Although the United States issued strong verbal protests about human rights violations during the Duvaliers' reign, U.S. foreign policy provided tacit support for the dictators and even helped François Duvalier (Papa Doc) stay in power in exchange for supporting sanctions against Cuba and its "communist threat."

Aristide and Modern Haiti

In December 1990, the Haitian people elected the socialist Jean-Bertrand Aristide as president by a 67 percent majority, in the first free and fair election since the Duvaliers took power. In February 1991, Aristide was sworn in, but in September he was overthrown by the military in a coup led by two U.S.-trained Haitian generals. Although they had previously expressed some concern about Aristide's socialist leanings, U.S. leaders played the key role in negotiating and militarily enforcing his return to power, perhaps in hope that his return would stem the tide of Haitians seeking refuge in the United States. Ironically, the international embargo spearheaded by the United States in the period preceding the reinstatement of Aristide had little impact on the Haitian elite; it did, however, substantially increase the suffering of Haiti's poor (Post 1993). Constitutionally prohibited from seeking a second term in office, Aristide handed power to the opposition party of Rene Prevalin in 1996. Amid charges of serious election irregularities, Aristide was sworn in for a second five-year presidential term in 2000, only to be ousted again in 2004. In the aftermath of violent political rivalries, new leaders are now in charge of a country where poverty seems endless.

Haiti in the Twenty-First Century

Haiti remains very much a least-developed country, one that may, in fact, be more vulnerable and more dependent than most. Haiti has no valuable natural resources; the land has been all but stripped bare of its once rich forests, and its soil is damaged almost beyond repair. Although Haiti can offer extremely cheap labor, better-educated labor in Mexico and in the newly developing nations of Southeast Asia make Haitian labor less attractive. Furthermore, the near impossibility of earning a living has forced many young men and women to support themselves through prostitution, contributing to skyrocketing rates of AIDS (Farmer 1999). Poverty also makes it difficult for individuals to fight off illnesses such as tuberculosis, and each illness drives families even further into debt (Farmer 1999). Thus, illness becomes yet another barrier to employment opportunities and economic development. Finally, with the end of the Cold War, the United States no longer needs Haiti as a strategic foothold against communism. In short, Haiti has virtually nothing to trade on the world market. The primary reason for U.S. intervention seems to be its desire to stem the flood of Haitian refugees to the United States.

If it is to survive as a nation, let alone develop, Haiti will need resources from abroad. Research suggests, however, that those resources must take the form of foreign aid (gifts) if they are to help Haiti on the road to development (Wimberly 1990). Foreign investment and loans are only likely to help a very determined elite to maintain its power.

Globalization

Chapter 2 briefly introduced the concept of the globalization of culture—the increasing spread of cultural elements (including fashion trends, musical styles, and cultural values) around the globe. The globalization of culture is part of a broader process. More generally, **globalization** refers to the process through which ideas, resources, practices, and people increasingly operate in a worldwide rather than local framework. In this section, we briefly discuss the sources of globalization and the cultural, economic, and political effects of globalization.

The Sources of Globalization

Globalization stems from a combination of political and technological forces. The collapse of the Soviet Union made it possible for the nations that emerged in its wake (like the Ukraine and Estonia) as well as the nations that had been restrained by its political power (like Poland and Czechoslovakia) to move toward a more capitalistic economic system. To do so, they needed to seek out economic, political, and cultural ties to other nations that could either serve as sources of raw goods and labor or markets for their products.

The collapse of the Soviet Union also reduced political tensions that had pressed nations to adopt international trade barriers. Now that the nations of Europe are no longer fearful of the Soviet might, they have united into what is virtually a continental government, in the form of the European Union. Within this union, goods, individuals, and services can flow more freely than ever before. Polish doctors can now seek higher paying jobs in Finland, Finnish doctors seek work in Sweden, and Swedish doctors seek work in England, with little concern about visas or immigration laws. German factories can transport and sell their products in Spain, and Spanish factories can send their products to Greece with minimal paperwork or tariffs to pay. Similarly, the North American Free Trade Agreement (NAFTA) was adopted in 1994 to reduce trade barriers between Canada, the United States, and Mexico.

All these changes were made more feasible by new technological developments. The Internet, e-mail, cellular telephone service, fax machines, and the like all made it easier, cheaper, and faster for corporations and individuals to invest and work internationally.

Cultural Impact

In an African urban nightclub, young people listen to American rock and roll and drink Pepsi. In New York City, young people go to Jamaican reggae concerts and watch *Queer as Folk*, a British television show. In India, Hollywood films are as popular as "Bollywood" (Bombay-produced) films and French wines are preferred by those who can afford them. All of these are examples of the global spread of culture. Increasingly, movies, television shows, music, literature, and other arts are distributed and enjoyed around the world.

These elements of popular culture carry with them not only entertainment but also cultural values. As Indian adolescents watch American films, they not only learn about the latest fashions and music, but also learn to question traditional Indian practices and beliefs like arranged marriages, the subservience of women, obedience to parents, and the idea that the family is more important than the individual.

Globalization refers to the process through which ideas, resources, practices, and people increasingly operate in a worldwide rather than local framework.

Connections
Social Policy

France is an example of a country that has fought hard to protect itself against the cultural impact of globalization. The Académie Français is officially in charge of policing the French language against the encroachment of non-French words (like *computer* and *hamburger*) and creating new French words to fit new technologies (such as *téléphone cellulaire* for cell phone). Similarly, to combat the impact of mass media from other cultures, the government subsidizes the French film and television industries, and requires that 40 percent of films shown on French television must be in French.

■ Globalization is spreading both American products and American cultural values around the world, as evidenced by this FedEx billboard in a Vietnamese rice paddy.

© Les Stone/The Image Works

Economic Impact

Globalization has also had a striking economic impact. Increasingly, economic activity takes place between people who live in different nations as goods and services are *sold* internationally. These days, Russians and Chinese buy Coca-Cola, Americans buy Volvos and Toyotas. Globalization also exists when goods are *produced* internationally. Transnational corporations, such as General Motors, may buy raw goods in one country, own a factory that puts smaller pieces together in a second country, assemble the full product in a third country, contract out its data processing to a firm in a fourth country, and then sell its product worldwide.

There is considerable debate about the possible effects, good and ill, of such international economic enterprises. Some observers hope that ties of international finance will create a more interdependent (and peaceful) world, while stimulating economic growth and improving everyone's standard of living (Stiglitz 2003). Others are concerned that transnationals are exercising a thinly veiled imperialism, in which cheap labor as well as raw materials are extracted from poorer countries for the benefit of wealthier countries. In addition, these critics allege that the practice of having labor-intensive work done in the less- and least-developed nations exposes workers in those countries to dangers that are banned by law in most Western nations (Michalowski & Kramer 1987).

Critics have also raised questions about the impact of economic globalization even within the developed nations. Here in the United States, hundreds of thousands of workers have lost their jobs when corporations found it cheaper to move those jobs overseas ("NAFTA" 2003). Other workers have been forced to accept cuts in benefits or pay in order to keep their jobs. The question is whether this global movement of jobs raises the standard of living overall by shifting work from wealthier to poorer countries or instead depresses wages to the level of the cheapest bidders.

Political Impact

How has globalization affected the balance of political power within nations and across nations? Some observers have noted that transnational corporations now dwarf many national governments in size and wealth. Their ability to move capital, jobs, and prosperity from one nation to another gives them power that transcends the law of any particular country (Michalowski & Kramer 1987). When a nation's economy depends on a transnational corporation, that nation can't afford to alienate the corporation. For example, Guatemala has limited ability to constrain the labor practices of United Fruit Company because the corporation could cripple the country's economy if it wanted to.

Another aspect of globalization is the sharp rise in international nongovernmental organizations. These organizations include the World Bank, the World Trade Organization, the International Monetary Fund, and the United Nation's International Criminal Court. The underlying premise of these organizations is that they will diminish the independent power of national governments and press nations to conform to international goals (such as ending torture of political prisoners, prosecuting war criminals, or reducing trade barriers). Although sometimes this benefits the individual nation, in other cases less-developed nations have suffered as a result of this loss of autonomy (Stiglitz 2003).

World Opinion

Popular support for globalization is surprisingly strong, according to a survey of 38,000 people in 44 developed and less-developed countries conducted by the Pew research Center (2003a). In almost every country surveyed, from the United States to Uzbekistan, Venezuela, and Mali, three-quarters or more of respondents believed that growing international trade and business ties were good for their country. Support for economic globalization is particularly strong in the poor countries of Africa. Similarly, in most countries, strong majorities of respondents consider increased access to foreign television, movies, and music a good thing. Support for cultural globalization is weakest among older people and in a handful of Muslim countries.

Around the world, people generally also link globalization to improvements in their standard of living. In most of the countries surveyed, majorities believe that the overall impact of globalization on their country has been good and that it has led to greater availability of food and access to medical care. Even multinational corporations and the major nongovernmental organizations get favorable ratings by majorities in most of the surveyed nations. The major exceptions occur in the conflict-ridden countries of the Middle East and Asia (such as Jordan and Pakistan) and the nations that are experiencing severe and sudden economic downturns (such as Argentina and Russia).

A Case Study: Globalization, Global Inequality, and 9/11

The terrorist attacks of September 11, 2001, which destroyed the World Trade Center and damaged the Pentagon, partly resulted from globalization, global inequality, and opposition to these forces (Amanat 2001; Barber 2001; Jacquard 2002; Stern 2003).

One underlying cause of the 9/11 attacks is the deepening belief among many Muslims that their nations and religion are under political attack. This belief has roots in the Russian invasion of Afghanistan and the civil war in Bosnia that pitted Muslims against Christians. Actions taken by the United States have also played a

■ Pro-Taliban supporters in Pakistan one week after the 9/11 attacks on the United States. Such demonstrations reflected fear not only of American cultural and economic power but also of American political and military power.

large role in creating this sense of victimhood among many Muslims. The United States consistently has supported Israel against the Palestinians. It has also used an economic blockade to try to starve out the Iranian and Iraqi governments, invaded Iraq, and based military troops in Saudia Arabia, where the most holy sites of Islam are located. These actions on the part of the United States and other non-Muslim nations have left many Muslims feeling that not only are individual Muslim governments under attack, but their religion is as well. It also has contributed to a sense of wounded pride among Muslims who feel that they no longer control their own national destinies.

To these external problems are added internal economic and political problems. The Middle East and the Muslim countries of Asia have been wracked by war for the last 50 years. Poverty is very high, inequality is extreme, governments by and large are corrupt, and access to education is often limited. As a result, millions of individuals are suffering, and thousands are willing to become the foot-soldiers of Al Qaeda and other similar groups. Even members of the middle and upper classes, who are protected from the worst impacts of these forces, still live in a culture of alienation, despair, and wounded pride. These are the individuals, like Osama bin Laden and the 19 terrorists who attacked the United States on 9/11, who become the leaders and lieutenants in global terrorism. It is probably no coincidence that the terrorists of 9/11 attacked the World Trade Center and the Pentagon—icons of American economic and military power.

Finally, the globalization of culture also underlay the sentiments that led to 9/11. Because of the mass media and information technology, people throughout the Muslim world are inundated with American culture. In countries where women are expected to cover themselves from head to toe, American television shows display nearly naked women. Rap music boasts of sexual conquests, and Hollywood romances feature independent women and men whose lifestyles are the antithesis of traditional Muslim values. Around the world, America has become the symbol of the good life, but also a symbol of materialism, violence, promiscuity, and the attack on traditionalism. The terrorist actions of 9/11 were designed as blows against all these forces.

Where This Leaves Us

As we have seen throughout this textbook, power is one of the most important determinants of the distribution of scarce resources, internationally as well as locally. Nations that lack power have lower life expectancies and educational attainments and higher infant mortality rates. In these nations, women have lower status, housing is poor quality, access to consumer goods is limited, and malnutrition and even starvation are common.

Importantly, quality of life and levels of economic and political power in the developing and developed nations are directly related to each other. We enjoy comfortable lifestyles in the United States and in other developed nations in part because of raw goods and products we obtain cheaply from countries where people work for pennies an hour. We are secure in our political and military power because other nations cannot counter it. Similarly, American culture is spreading around the world in part because of our economic and political power and because other cultures lack the power to oppose it. The events of 9/11 demonstrate this interconnectedness, and demonstrate what can happen when resentment of our power rises internationally.

Summary

1. Inequality is the key fact in the international political economy. Reducing this disparity through the development of less-developed and least-developed countries is a common international goal. Development is not the same as Westernization; it means increasing productivity and raising the standard of living.

2. The world's nations can be divided into the rich, diversified, independent, most-developed nations; the less-developed nations, including not only the former Soviet Union and former Communist-bloc countries of Eastern Europe but also some countries in Southeast Asia and South America that are experiencing economic growth; and the least-developed nations of the periphery. Least-developed nations can be further subdivided into banana republic, industrial dependence, and foreign-capital dependence.

3. The Human Development Index and the Gender-Related Development Index use literacy and educational attainment, life expectancy, and economic productivity to assess quality of life overall, as well as quality of life adjusted for the effects of gender inequality.

4. In many least-developed countries, women are the principal food producers, but at least until the mid-1970s they were largely ignored by development programs. New more successful development strategies are more often aimed at women.

5. Modernization theory, a functionalist perspective of social change, rests on the assumption that less-developed countries will evolve toward industrialization by adopting the technologies and social institutions used by the developed countries.

6. World-systems theory, a conflict perspective, views the world as a single economic system in which the already industrialized countries, known as core societies, control world resources and wealth at the expense of less-developed, peripheral societies. Empirical studies document that foreign-capital penetration has adverse consequences for developing nations.

7. Where economic development has been pursued without attention to resource management, pollution, or wasteful consumption, the environmental consequences have been devastating. In order to improve the quality of life in least-developed nations, more attention will need to be paid to these issues.

8. Sustainable development is the ability to meet the needs of the present without compromising the ability of future generations to meet their own needs. Sustainable development will require (a) the elimination of absolute poverty and the reduction of inequality within and between nations, and (b) a reduction in population growth, coupled with more efficient production and less wasteful lifestyles.

9. As the international political economy undergoes competition, conflict, and change, the most-developed nations' responses can be classified as military, humanitarian, and market-based.

10. The decline in trade barriers and the rise of new information technologies have fostered the international spread of cultural ideas, the political power of transnational corporations and organizations (such as the World Bank), the internationalization of the economy, and the movement of people around the globe. This process is known as globalization.

Thinking Critically

1. Consider how the major themes of U.S. stratification covered in Chapter 7 apply to international inequality. *You* are a hundred times better off than the average person in the Democratic Republic of Congo. Is this a necessary and just reflection of your greater contribution to society? Is it surprising that so many people in the least-developed world ascribe to a conflict theory of international inequality?

2. Critically evaluate the components of the Human Development Index. Which seems most important to you? Could this index be used to understand group differences in quality of life within the United States?

3. Some people speculate that the most serious drain on world resources in the next decades will not be caused by simple population growth but rather by the growing affluence of the Chinese—who can now afford more food, cars, and material goods. Consider both the advantages and disadvantages of growing Chinese affluence for people in the United States.

4. Suppose Egypt can generate more revenue by buying the rice it needs for food with profits derived from exporting strawberries than it can by growing its own rice. Do you think this would be a good economic development strategy? What problems, if any, do you think such a strategy might create?

5. One reason nations use coercion to force their will on others is that there is no authority to compel them to get along. Consider how relations among U.S. states might differ if federal power was as weak as that of the United Nations to control international affairs.

Sociology on the Net

The Wadsworth Sociology Resource Center: Virtual Society

http://sociology.wadsworth.com/

The companion Web site for this book includes a range of enrichment material. Further your understanding of the chapter by accessing Online Practice Quizzes, Internet Exercises, InfoTrac College Edition Exercises, and many more compelling learning tools.

Chapter-Related Suggested Web Sites

The United Nations
http://www.un.org

The United Nations Children's Fund
http://www.unicef.org

The World Bank
http://www.worldbank.org

InfoTrac College Edition

http://www.infotrac-college.com/wadsworth

Access the latest news and research articles online—updated daily and spanning four years. InfoTrac College Edition is an easy-to-use online database of reliable, full-length articles from hundreds of top academic journals and popular sources. Conduct an electronic search using the following key search terms:

Developing countries

Sustainable development

International poverty

International debt

 For more information related to this chapter, visit the Opposing Viewpoints Resource Center. Be sure you check all the options on the toolbar—viewpoints, references, statistics, and so on—and search under the following subjects:

Terrorism

United States foreign relations

Suggested Readings

Chirot, Daniel. 1986. *Social Change in the Modern Era*. New York: Harcourt Brace Jovanovich. A historical and comparative approach to understanding social changes in the world. Chirot gives an excellent introduction to a complex area.

Farmer, Paul. 1999. *Infections and Inequalities: The Modern Plagues*. Berkeley, Calif.: University of California Press. A brilliant analysis of the link between disease and social inequality, written by a physician/anthropologist who for many years has divided his time between a clinic in inner-city Boston and one in rural Haiti.

Lappé, Frances Moore, Collins, Joseph, and Rosset, Peter. 1998. *World Hunger: Twelve Myths*. New York: Grove Press. An excellent explanation of the critical role inequality within and between nations plays in world hunger.

Are Economic Development and Traditional Cultural Values Incompatible?

Intersections

In some ways, Karl Marx was modernization theory's greatest proponent. Marx believed that the industrializing, economically developed countries of the nineteenth century would surely lead the way to a new, universal, world economic and cultural order. Today, few people probably place much stock in Marx's predictions. A worker-led revolution, for instance, no longer seems inevitable, and the increasing political significance of fundamentalist Christian and Islamic groups casts doubt on Marx's prediction of the end of organized religion. Nevertheless, Marx did correctly predict the impact of industrialization on virtually every nation on earth, and he did accurately foresee the profound cultural changes it would bring in its wake: everything from rising educational levels to changing gender roles and attitudes toward authority.

The question that social scientists continue to debate is whether economic development makes these cultural changes inevitable, as Marx believed, or whether, as Max Weber suggested, traditional cultural values persist and in fact shape other cultural and institutional responses to modernization. Using data on basic values and attitudes from 65 societies, representing 75 percent of the world's population, measured at three different points in time, Ronald Inglehart and Wayne E. Baker (2000) explore the intersection between changing economic structures and institutions and traditional cultural values and beliefs.

Consistent with points made in this chapter, Inglehart and Baker begin their study by noting that economic development does not occur in a simple linear fashion. Some societies, such as the Russian Federation, become less developed over time. Perhaps even more importantly, economic development does not simply mean more and more industrialization. Instead, as economic development proceeds, fewer workers are employed in manufacturing jobs and more workers are employed in service and knowledge-based industries. This is what happened in the United States in 1956 and what has happened in most of the rest of the developed world since that time. This shift is often referred to as the transition from an industrial to a postindustrial society, or from an industrial to a service-based economy, and we will discuss its economic and work-related consequences in more detail in Chapter 14. For Inglehart and Baker, the shift is important because it also affects the relationship between economic development and traditional cultural values.

Scholars such as Daniel Bell (1973) noted that when societies shift from agrarian to industrial, people become less dependent on the uncontrollable forces of nature. As a result of an increased capacity to control the environment, during the early industrializing stage, economic development often undermines the role of religious interpretations of history and human destiny. As a result, traditional values may come to be replaced by more bureaucratic and secular ways of dealing with nature and one another.

Individuals living in industrial societies spend most of their working lives producing objects, and in ways consistent with the logic and rules of the assembly line. In contrast, in a postindustrial society, most people spend their working lives processing information and dealing with one another. This kind of work requires individuals who have good communication skills and who can make decisions on their own. Reflecting the changing requirements of work, members of the most developed, postindustrial societies may come to more highly value autonomy, flexibility, tolerance, and self-expression. Furthermore, because of their wealth, they may also be less concerned about economic and physical security than would be true in agrarian or even industrialized societies, and more concerned with their subjective well-being and overall quality of life (Inglehart 1997). Thus, Inglehart and Baker (2000) argue that the relationship between economic development and cultural change is different for industrializing societies and "postindustrializing" societies.

Inglehart and Baker found that the least-developed, least-industrialized societies in their sample had the most traditional values. They had the lowest levels of tolerance for abortion, divorce, and homosexuality. They tended to most heavily emphasize parental authority and male dominance in political and economic life, and they also placed the highest importance on religion. In contrast, industrialized societies were less traditional, with values emphasizing both increased rationalization and increased secularization. Although the transition from an industrial to a postindustrial, service-based economy appeared to bring little change in traditional versus secular values, members of postindustrial societies did more highly value self-expression and quality of life over economic and physical security; they also expressed increased levels of tolerance and happiness. In an interesting paradox, Inglehart and Baker also found that although postindustrialization is often accompanied by decreasing involvement in organized religion, the postindustrial concern for quality of life means that there is also an increased interest in spiritual matters!

Although economic development is generally associated with the cultural changes outlined above, Inglehart and Baker found that the broad cultural heritage of a society— Protestant, Roman Catholic, Orthodox, Confucian, or Communist—leaves a lasting imprint on culture. For instance, historically Protestant societies value self-expression much more than the historically Catholic societies, no matter what proportion of their labor forces are engaged in the service sector. Similarly, the ex-Communist societies are much more secular than others at comparable levels of economic development, and English-speaking societies are much more traditional.

Within societies, however, religious differences are much smaller than cross-national differences. Values of Nigerian Muslims, for example, are closer to those of their Christian counterparts than they are to those of Indian Muslims. This is probably because Nigerian values, including those that are linked to religion, have become part of a national culture that continues to be transmitted to all Nigerians by schools and the media.

For this reason, despite economic development and globalization of the media, Inglehart and Baker believe that national culture is likely to remain a key element of human experience. Industrializing societies may *not* generally become more like the United States. In fact, in some ways, the United States may be the deviant case, its people holding much more traditional values and beliefs than those in any other equally prosperous, economically developed, postindustrial society (Inglehart & Baker 2000). Although economic development seems to push cultural values in a common direction, in the United States as elsewhere in the world, a society's cultural heritage continues to shape the direction and rate at which values and beliefs can change.

MicroCase Online Exercise

For an interactive exercise using MicroCase data sets, go to the text companion website at http://sociology.wadsworth.com/brinkerhoff/essentials6e. Choose Chapter 8 and then select "Intersection Exercise" from the left navigation bar.

CHAPTER 9

Racial and Ethnic Inequality

Cleo Photography/Photoedit

Race and Ethnicity

Race and ethnicity are ascribed characteristics that define categories of people. Each has been used in various times and places as bases of stratification; that is, cultures have thought it right and proper that some people receive more scarce resources than others simply because they belong to one category rather than another. In the following section, we provide a basic framework for looking at racial and ethnic inequality; then we turn to a separate analysis of the patterns of inequality that exist in the United States. We conclude the chapter by looking at ethnic relations elsewhere, through a case study of Bosnia.

A Framework for Studying Racial and Ethnic Inequality

How is it possible for groups to interact on a daily basis within the same society and yet remain separate and unequal? In this section, we begin by defining race and ethnicity. We then introduce sociological concepts that help explain how societies maintain and reinforce group differences.

The Social Construction of Race and Ethnicity

A **race** is a category of people treated as distinct because of *physical* characteristics to which *social* importance has been assigned. An **ethnic group** is a category whose members are thought to share a common origin and to share important elements of a common culture—for example, a common language or religion (Marger 2003). Both race and ethnicity are handed down to us from our parents, but the first refers to the presumed genetic transmission of physical characteristics whereas the second refers to socialization into cultural characteristics. In actuality, racial differences among humans account for only a tiny fraction of all genes.

In sum, both race and ethnicity are based loosely if at all on physiological characteristics such as skin color. For this reason, sociologists talk of the **social construction of race and ethnicity:** the process through which a culture defines what constitutes a race or an ethnic group. As this suggests, this process is based more on social ideas than biological facts. Both individual self-identity and institutional forces play a role in the creation and maintenance of racial and ethnic statuses. For example, the idea of a "white" race, as opposed to separate Anglo Saxon, Mediterranean (Italian and Greek), Hebrew (Jewish), and Slavic races, only emerged around 1930 (Jacobson 1998). At around the same time, the U.S. Bureau of the Census declared that those with Mexican background should be classified as nonwhite. The Mexican government complained, and the Bureau reversed itself. Currently the census bureau defines Hispanic Americans as an ethnic group, whose members can belong to any race. Similarly, the shift from "black" to "African American" is an example of changing from a racial to an ethnic group identification.

A **race** is a category of people treated as distinct on account of *physical* characteristics to which *social* importance has been assigned.

An **ethnic group** is a category whose members are thought to share a common origin and to share important elements of a common culture.

The **social construction of race and ethnicity** is the process through which a culture (based more on social ideas than on biological facts) defines what constitutes a race or an ethnic group.

The growing number of multiracial births in the United States is beginning to blur the very concept of race (Kalish 1995; Morganthau 1995). Yet many Americans continue to be quite uncomfortable when they cannot wedge an individual into a predetermined racial slot. Golf superstar Eldrick "Tiger" Woods has had to fight constantly against journalists and others who want to describe him simply as African American, even though two of his eight great-grandparents were Native American, four were Asian, and one was European American. The 2000 census was the first U.S. census that allowed individuals to describe themselves as multiracial, rather than forcing them to choose a single racial category.

Elsewhere in the world, new ethnic identities are forged by changing national borders. Only during the twentieth century did Sicilians, Napolitanos, Milanese, and other groups begin the process of developing a common Italian language, culture, and ethnic identity. Conversely, after the break-up of the former Soviet Union, Lithuanians, Latvians, Kazakhs, and others began rebuilding ethnic, linguistic, and cultural traditions that had been suppressed or even abandoned during the Soviet years (Nagel 1994). As these examples illustrate, racial and ethnic statuses are not fixed. Over time, individuals may change their racial and ethnic identification, and society, too, may change the statuses it recognizes and uses.

The Semi-Caste Model

Most contemporary scholars use some form of conflict theory to explain how racial and ethnic inequalities are developed and maintained. This theory suggests that in the conflict over scarce resources, historical circumstances such as access to technology and slavery gave some groups advantages while holding other groups back. To maintain their power, those who have advantages work to keep others from getting access to them (Tilly 1998). These inherited advantages have left us with two stratification systems, class and a caste-like system based on race. These two systems combine together into a **semi-caste system**. A semi-caste system is a hierarchical ordering of social classes within racial categories that are also hierarchically ordered (Figure 9.1).

The American semi-caste system is further illustrated in Table 9.1. As the table shows, the races display very similar patterns of internal inequality. In all three populations, the wealthiest 20 percent of families receive almost half of all income; but the median income of white families is more than one and one-half times that of Hispanic and African American families. Also, racial differences in *wealth*—all assets owned by a household—are even greater than income differences. In 2000, the median net worth of white non-Hispanic families was $79,400, compared to a

A **semi-caste system** is a hierarchical ordering of social classes within racial categories that are also hierarchically ordered.

FIGURE 9.1
The Semi-Caste Model
Race and ethnicity are factors in the stratification system of the United States. There is a social class hierarchy within each race, but there is also a castelike barrier between races. Upper-class nonwhites are not as upper class as upper-class whites, and the lowest positions in the social-class hierarchy are reserved for nonwhites.

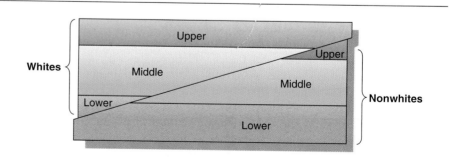

net worth of $7,500 for African American families and $9,750 for Hispanic families. These racial differences in wealth, not racial differences in income, are primarily responsible for the continuing U.S. racial divide (Oliver & Shapiro 1997).

Majority and Minority Groups

Rather than talk about white and African American or Jew and Arab, sociological theories of race and ethnic relationships usually refer to majority and minority groups. A **majority group** is one that is culturally, economically, and politically dominant. A **minority group** is a group that, because of physical differences, is regarded as inferior and is kept culturally, economically, and politically subordinate. Although minority groups are often smaller than majority groups, that is not always the case. In the Republic of South Africa, for example, whites made up only 15 percent of the population but until recently were the majority group, controlling all major political and social institutions. Similarly, some scholars regard women as a minority group because they have been economically, politically, and culturally subordinate to men.

A **majority group** is a group that is culturally, economically, and politically dominant.

A **minority group** is a group that is culturally, economically, and politically subordinate.

Patterns of Interaction

Relations between majority and minority groups may take one of four general forms: conflict, accommodation, acculturation, or assimilation.

Conflict As we discussed in Chapter 5, conflict is a struggle over scarce resources that is not regulated by shared rules; it may include attempts to neutralize, injure, or destroy one's rivals. Although some intergroup conflicts are expressed in violence (for example, genocide), conflict may also be expressed in laws forbidding social, political, or economic participation by the minority group. When African Americans or Hispanics are systematically excluded from voting or from some forms of political or social participation, this is a form of conflict. It excludes them from the competition for scarce resources.

TABLE 9.1
Income and Wealth of Families by Race and Ethnicity, 2000
United States income data support the semi-caste model. Within each racial or ethnic group, the richest 20 percent receive almost 50 percent of all income for that group, indicating real social class differences. But it is also true that whites as a group have considerably more income and wealth than do African Americans or Hispanics, indicating real race and ethnic differences.

Income Quintile	Percentage of Income Received		
	African American	Hispanic	White
Poorest fifth	3%	4%	5%
Second fifth	9	9	10
Third fifth	16	15	16
Fourth fifth	25	23	23
Richest fifth	47	49	47
Median income	$34,192	$35,054	$53,256
Median wealth	$7,500	$9,750	$79,400

SOURCE: U.S. Bureau of the Census 2002b, 2003a.

Connections

Example

In 1492, all Jews in Spain were ordered by royal decree to either convert to Christianity or leave the country. Those who did neither were hanged or burned at the stake as heretics. Over the centuries, descendants of Jews who stayed in Spain and converted came to blend more and more into Spanish culture. Today, they are fully acculturated. The only signs of their ancestors' religion are a few remnant rituals, like lighting candles on Friday nights. These remnants have become simply family traditions, separate from the religious meanings and prayers that once accompanied them.

Accommodation occurs when two groups coexist as separate cultures in the same society.

Assimilation is the full integration of the minority group into the institutions of society and the end of its identity as a subordinate group.

Social distance is the degree of intimacy and equality in relationships between two groups.

Accommodation   When two groups coexist as separate cultures in the same society, we speak of **accommodation.** They are essentially parallel cultures, each with its own institutions. The term was originally developed to apply to situations such as Canada's French and English provinces. As any French Canadian will tell you, separate is seldom truly equal. Nevertheless, accommodation gives at least outward support to the norm of equality. Sometimes referred to as pluralism, these systems are often difficult to maintain.

Acculturation   Another possible outcome of intergroup contact is for the minority group to adopt the culture of the majority group. This process is called acculturation (see Chapter 2). It includes learning the language, history, and manners of the majority group; it also involves accepting the loyalties and values of the majority group as one's own. As middle-class African Americans have discovered, however, full acculturation does not necessarily mean full acceptance or an end to discrimination (Feagin & Sikes 1994).

Assimilation   When full acceptance comes—when the minority group is fully integrated into the institutions of society and ceases to be a subordinate group—we speak of **assimilation.** Assimilation means that category membership in a race or ethnic group no longer affects social relationships. In the case of racial and ethnic groups, it includes going to the same schools, living in the same neighborhoods, belonging to the same social groups, and being willing to marry one another.

The Maintenance of Inequality: Basic Processes

To understand the persistence of racial and ethnic inequality, we need to look at the processes that promote what sociologists call **social distance.** Social distance is operationally defined by questions such as, "Would you be willing to have members of this group as good friends?" It is a measure of the degree of intimacy and equality in the relationship between two groups.

Prejudice and discrimination are processes that allow social distance to be maintained even when physical distance is absent. Most societies also use segregation, or physical distance, as an aid to maintaining social distance.

Prejudice

The foundation of prejudice is stereotyping, a belief that people who belong to the same category share common characteristics—for example, that athletes are dumb or that African Americans are naturally good dancers. Stereotyping is not always bad. If your seatmate on a bus is an elderly woman, you might start a conversation about the high prices of groceries; if your seatmate is a 30-year-old man, it might be about the prospects of next year's football team. These conversational topics are based on stereotyped notions of what these people are likely to be interested in. The woman might be able to give you details about middle guards, and the man might be completely uninterested in sports, but stereotypes often prove useful at least as a starting point in relationships.

Prejudice moves beyond stereotyping in that it is always a negative image and always irrational. It exists despite the facts rather than because of them (Pettigrew 1982). A person who believes that all Italian Americans are associated with the Mafia will ignore all instances of the law-abiding behavior of Italian Americans. If confronted with an exceptionally honest man of Italian descent, the bigot will rationalize him as the exception that proves the rule. **Racism** is a form of prejudice. It is the belief that inherited physical characteristics associated with racial groups determine individuals' abilities and are a legitimate basis for unequal treatment.

A startling example of racist prejudice was the decision by the United States to put Japanese American citizens in concentration camps during World War II (Smith 1995). This decision, which occurred in the absence of any evidence suggesting that Japanese-Americans were disloyal to the United States, demonstrates the irrationality of prejudice. That irrationality was epitomized in the words of then-General John DeWitt (1943) who claimed that "The very fact that no sabotage [by Japanese Americans] has taken place to date is a disturbing and confirming indication that such action will be taken."

Learning Prejudice

What causes prejudice? We learn to hate and fear in the same way we learn to love and admire. Prejudice is a shared meaning that we develop through our interactions with others. Most prejudiced people learn prejudice when they are very young, along with other social norms. This prejudice may then grow or diminish, depending on whether groups and institutions encountered during adulthood reinforce these early learnings (Wilson 1986).

Prejudice arises from and is reinforced by institutionalized patterns of inequality. In a stratified society, we tend to rate ourselves and others in terms of economic worth. If we observe that no one pays highly for a group's labor, we are likely to conclude that the members of the group are not worth much. Through this learning process, members of the minority as well as the majority group learn to devalue the minority group (Wilson 1992).

Prejudice is an irrational, negative attitude toward a category of people.

Racism is the belief that inherited physical characteristics associated with racial groups determine individuals' abilities and characteristics and provide a legitimate basis for unequal treatment.

Although the groups involved may differ, nations are remarkably similar in both the causes and consequences of prejudice and discrimination. Here, Turkish victims of neo-Nazi violence mourn the victims of an arson attack that left three children injured and five adults dead.

© Peter Turnley/Corbis

Prejudice is also powerfully reinforced by patterns of segregation and discrimination. A child growing up in a society where separation by race or sex is well established is very likely to learn prejudice. "They are not like us" can lead quickly to "They are not as good as us."

Competition over scarce resources (such as good jobs, nice homes, and admission to prestigious universities) also increases prejudice. For all racial and ethnic groups, prejudicial attitudes are closely associated with the belief that gains for other racial and ethnic groups will spell losses for one's own group (Bobo & Hutchings 1996).

Maintaining Prejudice: The Self-Fulfilling Prophecy

In Chapter 7, we introduced the concept of the self-fulfilling prophecy—where acting on the belief that a situation exists causes the situation to become real. The self-fulfilling prophecy is one very important mechanism for maintaining prejudice. A classic example is the situation of American women until the last few decades. Because women were considered to be inferior and capable of only a narrow range of social roles, they were given limited education and barred from participation in the institutions of the larger society. That they subsequently knew little of science, government, or economics was then taken as proof that they were indeed inferior and suited only for a role at home. In fact, many women were unsuited for any other role: Being treated as inferiors had made them ignorant and unworldly. The same process reinforces boundaries between racial and ethnic groups. For example, if we assume that Jews are clannish, then we don't invite them to our homes. When we subsequently observe that they associate only with one another, we take this as confirmation of our belief that they are clannish.

Personal Factors and Prejudice

Even though prejudice is learned, all individuals are not equally susceptible to it. Three factors that dispose people to prejudice are authoritarianism, frustration, and beliefs about stratification.

Authoritarianism is a tendency to be submissive to those in authority and aggressive and negative to those lower in status (Pettigrew 1982). Authoritarians in the United States tend to be strongly anti-African American and anti-Semitic (as well as homophobic).

Frustration is another characteristic associated with prejudice. People or groups who are blocked in their own goal attainment are likely to blame others for their problems. This practice, called **scapegoating,** has appeared time and again. For example, many members of the current American neo-Nazi movement are undereducated, underemployed, or unemployed people who blame Jews, African Americans, and affirmative action for their troubles.

Finally, prejudice is more likely to exist among individuals who believe strongly in the American Dream. People who subscribe to the view that we can all get ahead if we work hard and that poor people have only themselves to blame are substantially more likely to attribute poverty or disadvantage to personal deficiencies. In the case of disadvantaged minorities, the American Dream ideology supports the belief that the disadvantage is the fault of undesirable traits within the minority group (Kluegel & Smith 1983; Pettigrew 1982).

Discrimination

Treating people unequally because of the categories they belong to is **discrimination.** Prejudice is an attitude; discrimination is behavior. Often, discrimination follows from prejudice, but it need not. Figure 9.2 shows the possible combinations of prejudice

---

Authoritarianism is the tendency to be submissive to those in authority, coupled with an aggressive and negative attitude toward those lower in status.

Scapegoating occurs when people or groups who are blocked in their own goal attainment blame others for their failures.

Discrimination is the unequal treatment of individuals on the basis of their membership in categories.

FIGURE 9.2
The Relationships between Prejudice and Discrimination
Prejudice is an attitude; discrimination is behavior. They do not always go hand in hand. Some people act on their attitudes, whereas others suppress their own attitudes to conform to community standards. Fair-weather friends are unprejudiced people who will discriminate anyway; timid bigots are prejudiced people who are deterred from discrimination by community standards.

PREJUDICED?

	No	Yes
DISCRIMINATE? No	Friend	Timid bigot
DISCRIMINATE? Yes	Fair-weather friend	Bigot

and discrimination. Some individuals are logically consistent: They are prejudiced, so they discriminate (bigots); or they aren't prejudiced, so they don't discriminate (friends). Some people, however, are inconsistent, usually because their own values are different from those of the dominant culture. Fair-weather friends do not personally believe in racist or sexist ideologies; nevertheless, they discriminate because of what their customers, neighbors, or parents would say. They do not wish to rock the boat by acting on values not shared by others. The fourth category, the timid bigots, have the opposite characteristics: Although they themselves are prejudiced, they hide their prejudices for fear of what others would think (Merton 1949).

Public policy directed at racism is aimed mostly at reducing discrimination—allowing fair-weather friends to act on their fraternal impulses and putting some timidity into the bigot. As Martin Luther King, Jr., remarked, "The law may not make a man love me, but it can restrain him from lynching me, and I think that's pretty important" (as quoted in Rose 1981, 90).

Segregation

Prejudice and discrimination may occur between groups in close, even intimate, contact; they create social distance between groups. Differences between groups are easier to maintain, however, if social distance is accompanied by **segregation**—the physical separation of minority- and majority-group members. Thus, most societies with strong divisions between racial or ethnic groups have ghettos, barrios, Chinatowns, and Little Italies, where, by law or custom, members of the minority group live apart.

Historical studies suggest that high levels of residential segregation of Hispanic, Asian, and African Americans are not new; they have existed since at least 1940 and have changed relatively little. Such segregation is no longer established in law, but it is no historical accident. It occurs partly as a result of social class segregation of neighborhoods. African Americans, for instance, are substantially less likely than whites to leave poor neighborhoods and substantially more likely to move into them. Even the most educated African Americans remain substantially less likely than the least-educated whites to escape "distressed" neighborhoods (South & Crowder 1997).

Connections

Social Policy

Affirmative action is an example of a policy aimed at reducing the impact of past and present discrimination and prejudice. Under federal affirmative action guidelines, employers must set hiring goals (not quotas) that match the proportion of *qualified* women and minorities in a given field. The main impact of affirmative action has been to change hiring practices. Employers now must advertise job openings rather than filling them through "old boy networks" and word of mouth and must stipulate in advance the necessary qualifications for a job rather than manipulating the requirements to fit a pre-chosen individual.

Segregation refers to the physical separation of minority- and majority-group members.

■ Racial segregation remains a fact of life in the United States. Even among the middle class, African Americans are more likely than European Americans to live in a poor neighborhood.

© Ted Spiegel/Corbis

Similarly, socioeconomic status is related to living in the suburbs and to better-quality housing for all racial and ethnic groups. Yet, the suburbs in which middle-class African Americans and Hispanics, especially Puerto Ricans, live are considerably poorer and more problem ridden than those in which middle-class Asian and European Americans reside (Alba, Logan, & Stults 2000; Rosenbaum 1996). This penalty is not so much the result of economic differences between racial and ethnic groups but rather of the political and social pressures that whites have used to keep minorities "in their place" (Saltman 1991).

Race and Ethnic Inequalities in the United States

Racial and ethnic inequality is not new. In this section, we discuss the past, present, and future social positions of racial and ethnic groups in the United States.

White Ethnic Groups

The earliest immigrants to North America were English, Dutch, French, and Spanish. By 1700, however, English culture was dominant on the entire Eastern seaboard. The English became the majority group, and everybody who came after that time became a minority group, subject at times to prejudice and discrimination.

The Melting Pot

The extent of interaction and assimilation among white ethnic groups led some idealistic observers to hope that a new race would emerge in North America, where "individuals of all nations are melted into a great race of men" (Cràvecoeur [1782] 1974). In fact, careful observers suggest that the melting pot never existed. Certainly, our language is peppered with words borrowed from other languages (*frankfurter, ombudsman, hors d'oeuvre, chutzpah*), and some of us are such mixtures of nationalities that we would be hard pressed to identify our national heritage. Instead of a blending of all cultures, however, what has occurred is a specific form of acculturation—**Anglo-conformity**, the adoption of English customs and English language. To gain admission into U.S. society and to be eligible for social mobility, one has to learn "correct" English, become restrained in public behavior, work on Saturday and worship on Sunday, and, in general, act like the American version of the English prototype.

> **Anglo-conformity** is the process of acculturation in which new immigrant groups adopt the English language and English customs.

The Future of White Ethnicity

Despite generations of acculturation, many white people in the United States still identify with their ethnic heritage. They are proud to be Italians, Greeks, Norwegians, or Poles. Nevertheless, high rates of intermarriage are blurring these identities. As a result of intermarriage, a growing segment of the population cannot identify themselves with a single ethnic group; they are simply "unhyphenated whites" (Lieberson & Waters 1993). Although ethnicity for this group may be largely symbolic and a matter of choice, it still has important consequences. In many ways, its very invisibility means that the ethnicity of the unhyphenated white becomes the national "American" identity and the standard against which all other cultural and ethnic identities are judged (Doane 1997).

Although being Irish has little impact on most Irish Americans' lives these days, many still enjoy celebrating their cultural heritage, like these boys at a St. Patrick's Day parade.

Nevertheless, ethnicity has ceased to be a basis for stratification among white, non-Hispanic people in the United States. Although ethnic differences do exist, these differences are not related to structured inequality. The integration of 69 percent of the population from diverse backgrounds and conditions is a remarkable achievement. Yet it leaves out a significant portion of these people. Here we consider the situation of the other 31 percent: the nonwhites and the Hispanics.

African Americans

African Americans now comprise 13.1 percent of the U.S. population. Until very recently, they were the largest minority group in the country. But their importance goes beyond their numbers. Next to Native Americans, African Americans have offered the greatest challenge to the United States's view of itself as a moral and principled nation.

The social position of African Americans has its roots in one central fact: almost all African Americans descended from people who came here as slaves. After slavery ended, both legal barriers (such as patently unfair "literacy tests" that barred them from voting) and illegal barriers (such as lynching any African Americans who challenged white authority) prevented most African Americans from rising in the American social and economic structure. Real change did not take place until the Civil Rights Act of 1964 and the Civil Rights activism of the late 1960s.

These days, more African Americans are middle class than ever before. At the same time, however, a troubling fissure has emerged within the African American population: on the one hand, a working- and middle-class population that is increasingly integrated into U.S. society; on the other, a population of poor African Americans living in segregated neighborhoods with few employment opportunities that has not been included in the overall improvement. In fact, for this second group, the situation has deteriorated (Wilson 1996).

Current Concerns

Since World War II, important improvements have been made in many areas of African American life. Nevertheless, neighborhood segregation has declined very little; it remains so high and has such damaging consequences that one scholar has referred to it as "American apartheid" (Massey 1990). African American infants are still almost two and one-half times as likely as white infants to die before their first birthday (U.S. Bureau of the Census 1999), and African American men's life expectancy is still 5 years less than white men's.

Similarly, although African Americans are rapidly catching up with whites in their educational attainment, their rise in education has not been matched fully by any rise in income. As of 2000, 22 percent of all African American families live below the poverty line, and the median income for African American families is only 64 percent that of white families (U.S. Bureau of the Census 2002b). This striking economic disadvantage is due to two factors: African American workers earn less than white workers, and African American families are less likely to have two earners.

Low Earnings   One of the reasons individual African Americans earn less than whites is that they are twice as likely to be unemployed. In 2003, African American unemployment was 11.8 percent compared to 5.5 percent for whites. Even when they are employed full-time and year-round, however, median income for African Americans is only 72 percent that of whites.

African Americans (especially older persons) are less well educated than whites, and a relatively high proportion live in the South, where wages are low; the average African American worker is also somewhat less experienced than the average white worker. African Americans also find themselves disproportionately in very low-paying jobs. For example, African Americans continue to make up 22 percent of the nation's maids and 21 percent of its janitors (U.S. Bureau of the Census 2002b).

These differences account for only part of the earnings gap between African American and white people in the United States (Cancio, Evans, & Maume 1996). The other part is the result of a pervasive pattern of discrimination that produces a very different occupational distribution, pattern of mobility, and earnings picture for African Americans and whites in the United States. Thanks largely to government employment opportunities, there is a growing African American middle class (Hout 1986). Yet, African American professionals often are kept outside the true corporate power structure (Collins 1993, 1997): They are given less authority at work, and receive lower economic returns than are whites employed in comparable positions (Smith 1997; Wilson 1997).

Female-headed Families   About half of the gap between African American and white family incomes is due to the fact that African American families are less likely to include an adult male. Because women earn less than men and because a one-earner family is obviously disadvantaged relative to a two-earner family, these female-headed households have income far below those of husband-wife families. The fact that so many more African American than white families are headed by females—45 percent compared to 14 percent—has led some commentators to conclude that poverty is the result of bad decisions by African American men and women. This type of argument is an example of "blaming the victim," and empirical evidence suggests that it simply isn't true. Rather than *causing* poverty, research indicates that female headship *results* from poverty (Lichter, LeClere, & McLaughlin 1991; Luker 1996; Newman 1999b). African American women are less likely to marry because so many men in their community cannot support families.

Hispanic Americans

Hispanics or Latinos are an ethnic group rather than a racial category, and a Hispanic may be white, black, or some other race. This ethnic group includes immigrants and their descendants from Puerto Rico, Mexico, Cuba, and other Central or South American countries. Hispanics constitute 13.4 percent of the U. S. population, making them the largest minority group in the country. The largest group of Hispanics is of Mexican origin (66 percent), with 14 percent from Central and South America, 9 percent from Puerto Rico, 4 percent from Cuba, and the rest from other countries.

It is almost impossible to speak of Hispanics as if they were a single group. Their experiences in the United States have been and continue to be very different. Evidence suggests that their economic and political experiences may be growing more divergent rather than more similar (Portes & Truelove 1987). Although Cubans are becoming increasingly assimilated, as signaled by a very high rate of U.S. citizenship, the Mexican American, or Chicano, population is not.

Figure 9.3 compares the various Hispanic groups to one another and to the non-Hispanic white, Asian, and African American populations on three measures: education, poverty, and family structure. On two of these measures, a Hispanic group comes out at the very bottom: Mexican Americans are the most poorly educated racial or ethnic group, and Puerto Ricans are the most likely to live in poverty. In addition, Puerto Ricans are almost as likely as African Americans to live in female-headed households.

Despite the difficulties many Hispanics now face, they remain overall optimistic about their current and future prospects in the United States (Romero & Elder 2003). In a national poll conducted in 2003 by the *New York Times* and CBS, 81 percent of Hispanics believed that Hispanics had the same or better chances of getting ahead in America as do other Americans. Perhaps because 57 percent of Hispanics who participated in the survey were immigrants, 75 percent believed that they had a better chance to succeed than did their parents. (In contrast, only 56 percent of surveyed non-Hispanics believed their chances were better than were their parents' chances.) Finally, only 28 percent of surveyed Hispanics believed that, if they had some minor trouble with the law, the police would treat them more harshly than others.

Hispanic Americans are now the largest and fastest growing ethnic minority in the United States.

© Chuck Fishman/Woodfin Camp & Associates

Current Concerns

If current immigration patterns continue, Hispanics are expected to constitute 17 percent of the U.S. population by 2020 and 24 percent by 2050 (U.S. Bureau of the Census 2002b). This rapid growth raises four concerns for the status of Hispanics: First, because the new immigrants are young and poorly educated, the socioeconomic position of the Hispanic population is falling. Second, the growing Hispanic population has raised concerns among non-Hispanics about competition for jobs and so has spurred greater prejudice and discrimination. Third, as the number of Hispanics in the country has increased, some white Americans have become fearful that Hispanic culture and the Spanish language will "take over" the country. This too, has fostered ill will toward Hispanic Americans. Finally, rapid rise in the number of Hispanics has led to increasing residential segregation (Massey & Denton 1988). Segregation, in turn, may retard the rate at which new immigrants learn English and become integrated into U.S. society.

FIGURE 9.3
Education, Poverty, and Family Structure by Race and Hispanic Origin, 2000
On two out of three measures of disadvantage, a Hispanic ethnic group comes out at the bottom. A significant portion of this disadvantage, however, can be traced to recent immigration and poor English skills.
SOURCE: U.S. Bureau of the Census 2002b.

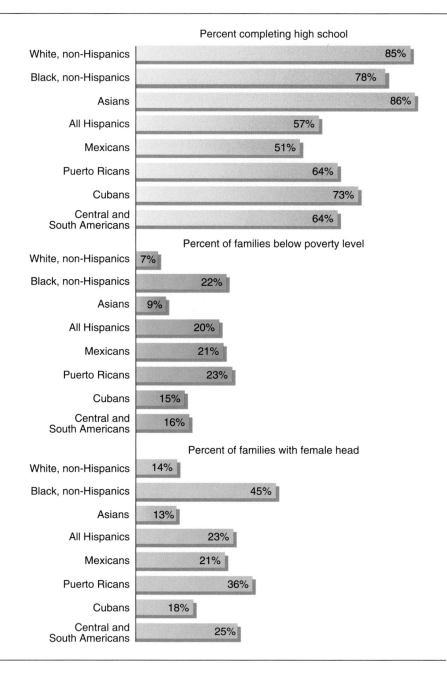

Despite these concerns, studies show that the caste-like barrier separating races operates much less dramatically for Hispanics. For white Hispanics, the problem is largely one of class. As a result, by the second and third generation, most Hispanics can translate educational attainment into well-paying jobs and leave the segregated barrios (Massey & Mullan 1984). The exception are those Hispanics (mostly Puerto Ricans and others from the Caribbean) who suffer the triple disadvantage of being Hispanic, poor, and African American (Massey & Bitterman 1985).

■ During World War II, the U.S. government imprisoned many of its own citizens in detention camps and deprived them of their property simply because they were of Japanese ancestry.

© National Archives/Meyers Photo-Art

Asian Americans

The Asian population of the United States (Japanese, Chinese, Filipinos, Koreans, Laotians, and Vietnamese) more than doubled between 1980 and the present, yet it still constitutes only 4 percent of the total population. The Asian population can be broken into three segments: descendants of nineteenth-century immigrants (Chinese and Japanese), the post-World War II immigrants (Filipinos, Asian Indians, and Koreans), and the recent refugees from Southeast Asia (Cambodians, Laotians, and Vietnamese).

A century ago, Asian immigrants were met with sharp and occasionally violent racism. Today, incidents of racial violence directed at Vietnamese and other Asians continue to make headlines. Despite these handicaps, Asian Americans have experienced high levels of social mobility. People in the United States of Japanese and Chinese descent have surpassed the educational attainment of white people there, and it appears that many of the more recent streams of Asian immigrants will follow the same path. For example, although many of the Southeast Asian refugees who came to the United States between 1975 and 1984 began their lives here on welfare, almost twice as many Vietnamese youths aged 20 to 24 are enrolled in school as are white youths of the same age.

Current Concerns

The high level of education earned by Asian Americans is a major step in opening doors to high-status occupations; Asian American families are only slightly more likely than white families to live below the poverty line (U.S. Bureau of the Census 2002b). Yet, discrimination is not all in the past. Asian American applicants are less likely to be accepted at elite colleges and universities than are whites with the same credentials (Takagi 1990). Studies have concluded that highly educated, native-born Asian males earn substantially less than similarly qualified white men (Suzuki

1989, 16) and that they are promoted more slowly (Tang 1993). Discrimination can be even more direct and extreme. In 1999, a white supremacist murdered Joseph Ileto, a Filipino American postal carrier, as part of a personal "war" against affirmative action. Earlier that day, the murderer fired 70 bullets into a Jewish community center (Finz 1999).

Native Americans

Native Americans (American Indians) are one of the smallest minority groups in the United States (about 1 percent of the entire population), and nearly half of their members live in just four states: Oklahoma, Arizona, California, and New Mexico. Native Americans are widely regarded as our most disadvantaged minority group. Compared to African Americans, for example, Native Americans have lower median incomes and are more likely to live in houses without indoor plumbing, heating, or telephones (Table 9.2). Native Americans have the highest rates of alcoholism and premature death of any U.S. racial or ethnic group. This situation exists despite hopeful new signs of economic vitality on some Indian reservations over the past 20 years—for example, development of mineral reserves on the Navajo reservation in the Southwest (Schaefer 1990, 199–200) and the advent of gambling casinos elsewhere (*New York Times* 2000).

Table 9.2 summarizes the situation on some of the larger Native American reservations. Keep in mind, though, that the status of Native Americans is highly diverse. Native Americans represent more than 200 tribal groupings, with different cultures and languages. Some have been successful: fish farmers in the Northwest, ranchers in Wyoming, and bridge builders in Maine. In urban areas, many Native Americans have entered the professions and blended into the majority culture. On isolated reservations with few economic resources (and little opportunity to draw crowds to casinos), on the other hand, socioeconomic conditions often are quite poor.

TABLE 9.2
Social and Economic Characteristics of Selected Native American Reservations, 2000

	Reservation					
	San Carlos Apache	Blackfeet	Wind River Cheyenne-Arapaho	Hopi	Navajo	Rosebud Sioux
Percent 65 years and over	5%	6%	12%	9%	7%	6%
Percent high school graduate (25 years or older)	58%	74%	83%	67%	56%	73%
Percent college graduate (25 years or older)	3%	13%	15%	10%	7%	11%
Percent families with female heads	17%	13%	8%	15%	14%	20%
Percent families below poverty level	48%	30%	16%	37%	40%	46%
Percent using wood to heat home	10%	14%	8%	37%	52%	8%
Percent lacking telephones	21%	11%	8%	32%	60%	25%
Percent with no vehicle	n/a	27%	12%	6%	24%	18%
Percent lacking indoor water or toilets	11%	2%	2%	27%	32%	3%
Median family income, 1999	$17,585	$26,832	$35,238	$22,989	$22,392	$18,673

SOURCE: U.S. Bureau of the Census 2000, American Factfinder: http://factfinder.census.gov/home/aian/aian_aff2000.html.

More than any other minority, Native Americans have remained both unacculturated and unassimilated. It is clear that Native American culture is not simply different from the dominant culture; in many ways it is a shattered culture. Those who cling to earlier values face a severe case of anomie because the means to achieve the old goals are gone forever. In addition, prejudice and discrimination stand in the way of achieving the new goals. The story of the Ojibwa presented in Chapter 2 represents an extreme case of what has happened to Native American culture.

Arab Americans

According to the 2000 U.S. census, Arab Americans comprise considerably less than 1 percent of the U.S. population. Because of recent world events, however, their status in this country is of particular political and sociological importance.

All Arab Americans are immigrants or children of immigrants from North Africa and the Middle East (including Morocco, Algeria, Saudi Arabia, and Iraq—Iran is not an Arabic country); the largest single group is from Lebanon ("Arab American Demographics" 2003). Each of these countries has its own traditions, but they share common linguistic, cultural, and historical traditions. Some Arab Americans descend from families that emigrated to the United States in the 1800s, some emigrated themselves only in the last few years. Two-thirds of Arab Americans are Christian.

Arab Americans are a highly educated population. They are as likely as other white Americans to have graduated high school and are slightly more likely to have graduated college. As a result, the majority hold professional jobs, and their median incomes are somewhat above the U.S. average.

Polls throughout the 1990s consistently have found somewhat negative attitudes toward Arabs and Arab immigrants (Jones 2001). In polls taken during the 1991 Persian Gulf Crisis, 59 percent of surveyed Americans associated Arabs with terrorists, 58 percent with violence, and 56 percent with religious fanatics. By 1993, after the first terrorist attack on the World Trade Center, only 39 percent of Americans told a Gallup Poll that they had a favorable opinion of Arabs (with

Arab Americans, like these Michigan schoolchildren, are an increasingly important minority group in the United States.

© Ed Kashi/Corbis

32 percent having an unfavorable opinion and 29 percent with no opinion). In addition, two-thirds of Americans thought there were too many Arab immigrants entering the United States. In fact, higher percentages of Americans were opposed to Arab immigrants than to immigrants from any other part of the globe.

Current Concerns

Interestingly, the terrorist attacks of 9/11 have not harmed American opinion of Arabs. Many Americans who previously had no opinion of Arabs have since gotten a "crash course" in Arabic and Muslim history and culture, and have taken pains not to discriminate against all Arabs or Muslims because of the actions of a few. A Gallup Poll taken only a month after the 9/11 attacks found that attitudes had actually improved since their 1993 poll. In 2001, 54 percent reported very or mostly favorable opinions of Arabs and 66 percent reported very or mostly favorable opinions of Muslims (Jones 2001). Similarly, a poll conducted in 2003 found that 51 percent reported favorable attitudes toward Muslim Americans—considerably lower than the approximately 70 percent who held favorable views of Protestants, Catholics, and Jews, but not much lower than the 58 percent who held favorable views of evangelical Christians (Pew Research Center 2003b)

Still, political changes since 9/11 raise questions about the current and future status of Arabs and Muslims in this country. One special concern is whether the sweeping federal USA Patriot Act, passed to help the government fight terrorism, has encouraged discrimination. During the first six months of 2003, the Justice Department received more than 1,000 reports from Arab immigrants alleging that Department employees had violated their civil rights or civil liberties, with violations ranging from verbal abuse to beatings. The department also has received more than 500 reports of "hate crimes" against Arab and Muslim Americans (Shenon 2003).

The Future of Racial and Ethnic Inequality in the United States

As we've seen, over the last few decades there has been considerable improvement in the social status of various minority groups. Yet inequality remains. What are some of the reasons for this, and what are some of the strategies that can help reduce inequality?

The Persistence of Inequality

There is no question that traditional racism in the United States has declined over time. Few Americans now believe that African Americans earn less money than whites because of their innate inferiority, and few believe that people should be discriminated against because of their race and ethnicity. So why does racial and ethnic inequality continue? Answers include the effects of subtle racism and institutionalized racism.

Subtle racism reflects widespread belief in the ideology of the American Dream. Because of this ideology, many people continue to blame the poor for their poverty; the predominant white explanation for poverty stresses the lack of motivation among African Americans. As a result of such beliefs, whites are reluctant to support governmental policies designed to promote economic equality between African Americans and whites (Bobo & Kluegel 1993; Quillian 1996).

focus on

The Environment

Environmental Racism

A growing body of research shows that members of minority communities are exposed to a disproportionately large number of health and environmental risks in their neighborhoods and workplaces (Bullard, Warren, &

© Luis Rosendo/FPG

■ Third World countries in Africa and South America, like minority communities and Indian reservations in the United States, are often on the receiving end of industrial contaminants. In the short run, nations such as Argentina may find storing toxic waste a viable way of earning the hard currency necessary to pay off large foreign debts. In the long run, however, the communities that host hazardous-waste disposal sites bear all of the health costs and receive far fewer economic benefits (jobs) than the communities that produce it.

Johnson 2001; Camacho 1998). Landfills for hazardous waste are disproportionately located in African American and Hispanic communities. Farm workers (mostly members of minority groups) and their children are exposed to the poisons of pesticides sprayed on crops. Uranium mining has poisoned thousands on Native American reservations, and many more will likely be poisoned by radioactive waste sites now being established on reservations. Researchers have found that exposure to environmental pollution is more highly correlated with race than with any other factor, including poverty (Bullard, 1993; Stretesky & Hogan, 1998).

Similarly, the United States and other Western industrial powers export hazardous waste and household garbage to the less-developed nations of Africa, Asia, and Latin America. American manufacturers currently export about a third of their total domestic production of pesticides; about one-fourth of these exports are products that cannot be sold in the United States for any use whatsoever (Bright 1990).

In December 1991, Lawrence Summers, the chief economist of the World Bank, sent a note to his colleagues that may well summarize the

view of many movers and shakers about the environment. An excerpt follows:

> Just between you and me, shouldn't the World Bank be encouraging more migration of the dirty industries to the LDCs [Less-Developed Countries]? I can think of three reasons:
> (1) The costs of health-impairing pollution depend on the forgone [lost] earnings from increased morbidity and mortality. From this point of view a given amount of health-impairing pollution should be done in the country with the lowest cost which will be the country of the lowest wages. . . .
> (2) I've always thought that underpopulated countries in Africa are vastly underpolluted; their air quality is probably vastly inefficiently low [sic] compared to Los Angeles or Mexico City. . . .
> (3) Concern over an agent [poison] that causes a one-in-a-million change in the odds of prostate cancer is obviously going to be much higher in a country where people survive to get prostate cancer than in a country where under-5 mortality is 200 per thousand. . . . (as quoted in Foster 1993, 10–11).

Because poor and minority communities in the United States and abroad are least able to resist, they are most likely to be on the receiving end of pollution. Yet these same people receive fewer economic benefits (jobs) from polluting industries than other communities receive. The people who benefit the least bear the highest burden.

For more information, look up the following subjects in InfoTrac College Edition:
Environmental justice
Hazardous waste sites

Or visit the following Web sites:
Center for Health, Environment, and Justice
http//www.chej.org/
Environmental Protection Agency, link on environmental justice
http//www.epa.gov/history/topics/justice/index.htm

Institutionalized racism occurs when the normal operation of apparently neutral processes systematically produces unequal results for majority and minority groups.

Double jeopardy means having low status on two different dimensions of stratification.

Another factor in the persistence of racial inequalities is the indirect inheritance model of social class described in Chapter 7, which implies that patterns of inequality established in past generations will persist. **Institutionalized racism** is the term used to describe the persistence. It means that apparently color-blind forces such as educational attainment produce systematically unequal results for members of majority and minority groups. This form of racism is an important dimension of contemporary racial and ethnic inequalities.

Combating Inequality: Race versus Class

In this chapter and chapter 7, we have shown how both social class, on the one hand, and race or ethnicity, on the other hand, affect one's life chances. When a person has a lower status on both of these dimensions, we speak of **double jeopardy.** This means that disadvantages snowball. For example, poor African American teenagers are more likely than poor white teenagers to be unemployed or to end up in prison.

Sociologists have hotly debated whether race or class is more important for understanding the structure of inequality in the United States today. The question most often asked is, "Is the status of lower-class African Americans due to the color-blind forces of class stratification, or is it due to class-blind racism?" A 1978 book titled *The Declining Significance of Race* set the tone for much of this debate. In it, African American sociologist W. J. Wilson argued that the status of African Americans had less to do with racism than with the simple inheritance of poverty and the changing nature of the U.S. economy. As well-paying factory jobs disappeared and as other forms of employment shifted from the inner cities to the suburbs, the position of the poorest third of the African American population has disintegrated. Joblessness is up, the number of female-headed households is up, rates of drug use are up, and so on. For this reason, Wilson argued that the solutions to racial inequality are primarily economic. By developing strategies that create full employment and better jobs for *all* Americans, Wilson (1987) believes that the serious problems faced by African Americans will largely be eliminated.

Most sociologists disagree. They doubt that policies based on social class alone will be enough to resolve the problem of racial inequality in the United States. True, there is an African American middle class (even an African American upper class), and it is a serious mistake to assume that racism keeps all racial minorities poor and powerless. Nevertheless, race continues to be a fundamental cleavage in U.S. society. Not only does being African American prove to be a handicap in social-class attainment, but it also continues to be important in social relationships. The finding that middle-class African Americans are much more likely than are middle-class whites to live with poor neighbors—both white and African American—suggests that the issue goes beyond class (Alba et al. 2000). Any successful strategy for combating inequality in the United States will have to simultaneously address issues of social class and of race and ethnicity.

Ethnic Relations in Comparative Perspectives

In 1991 in Brooklyn, New York, an African American child was accidentally struck and killed by a car driven by a Hasidic Jewish man. Charges that the slightly injured driver was treated while the child lay trapped beneath the car led to four days of violence in

■ The devastation produced by racial tensions and riots in cities such as Los Angeles and Miami differs only in degree from this scene of interethnic conflict and destruction in Sarajevo, capital of Bosnia/Herzegovina.

© Chandler/Corbis Sygma

the Crown Heights neighborhood, where poor African Americans and poor Hasidic Jews live side by side. In 1992, following the acquittal of four police officers charged with the videotaped beating of African American motorist Rodney King, Los Angeles erupted in riots that left 51 people dead.

Dramatic events such as these tend to obscure the more subtle nature of most ethnic conflict in the United States; they also may obscure the fact that the same processes deny minority groups their rights and opportunities in other nations, as well. Germany has witnessed a rising wave of neo-Nazi attacks on Turkish immigrants. In Rwanda, civil war between the Tutsi and Hutu tribes decimated the population. In Bosnia, tens of thousands of people were killed and millions displaced by the strife between Serbians and Croatians.

Fifty years ago, many social scientists believed that industrialization, the emergence of mass communications and education, and the development of impersonal bureaucratic forms of government would reduce interethnic antagonisms by creating new loyalties, based on a unified nation state. This has not happened, and the question is, of course, "Why not?" Ethnic conflict in the former Yugoslavia illustrates how majority/minority relations both vary across societies and are similar.

A Case Study: Bosnia

The violence that overwhelmed Bosnia in the 1990s had its roots many decades earlier.

Bosnia/Herzegovina was one of the five territories that were patched together into the nation of Yugoslavia in 1918, following the disruptions of World War I. Like its neighboring territory, Serbia, Bosnia/Herzegovina was predominantly made up of Christian peasants. But because of centuries spent under Turkish rule, both Serbia and Bosnia/Herzegovina also included many Moslem residents. In contrast, Slovenia and Croatia were more economically developed and overwhelmingly Christian, as was the far poorer fifth territory, Montenegro (Lydall 1989).

Although pressed into one country, residents of each territory considered themselves separate ethnic groups or even nationalities. So, too, did the ethnic Albanians scattered around the country. In addition, many Moslem Serbs and Croats considered themselves members of a separate Muslim ethnic or national group. And many Christian Serbs and Croats agreed that Moslems formed a separate group.

Despite Serbian political dominance between the two world wars, Croatia and Slovenia maintained a significant economic advantage over the other regions. Perhaps as a result of these political and economic tensions, Croatian Nazi collaborators, known as the Ustashi, killed hundreds of thousands of Serbs, Jews, and Gypsies during World War II. When the Ustashi surrendered in 1945, many thousands of Croatians were put to death by Serbian partisans (Bell-Fialkoff 1993).

In the wake of this conflict, Field Marshall Tito (Josip Broz) came to power determined to turn Yugoslavia into a modern, unified nation state. The system of government and of workplace management that Tito instituted was consistent with the accommodation model of ethnic relations discussed earlier in this chapter. The cultural integrity and equality of each ethnic group was to be maintained at the same time that industrialization and education were expected to forge a collective Yugoslav identity. Although there were periods of ethnic unrest, the postwar period did see a gradual increase in the number of people who identified as Yugoslavians rather than Serbs, Croatians, Muslims, Albanians, or some other ethnic group.

Why did Yugoslavia disintegrate, then, eventually erupting into enormous ethnic strife? Two factors seem critical. First, when Tito died in 1980, political power shifted from the central government to republic-level Communist parties, and local politicians had a great deal to gain from increasing the influence and power of their own region; their use of the news media to spread nationalist sentiments was particularly important in exacerbating ethnic intolerance (Massey, Hodson, & Sekulic 1999). Second, during the mid-1980s, the country experienced an economic crisis: living standards in Yugoslavia declined by at least one-fourth, with inflation reaching more than 2,500 percent in 1989 (Sekulic, Massey, & Hodson 1994). In this increasingly competitive and insecure environment and particularly in the highly segregated ethnic neighborhoods, ethnic differences became the basis for distrust, fear, and conflict (Massey et al.).

The situation became even more dangerous following the breakup of the Soviet Union in 1989. Because the Yugoslav economy depended heavily on trade with and support from the Soviet Union, once that country collapsed, so too did the Yugoslav economy. Horrific ethnic strife, centered in Bosnia, soon followed. Although all sides engaged in violence, most of the worst violence was committed by Christian Serbs who used systematic rape and murder to drive out Moslem Bosnians. This "ethnic cleansing," which continued throughout the early 1990s, resulted in the murder of about 250,000 Bosnians, out of a pre-war population of 4.4 million.

Although the political process that underlay the ethnic strife in Bosnia may be unique to this society, the economic processes appear to be typical of those accompanying ethnic conflict in societies throughout history: Racial and ethnic hostilities are most pronounced when economic resources are scarce and the majority group advantage is most threatened.

Where This Leaves Us

Racism and interethnic conflicts are problems worldwide. Croats, Serbs, Israelis, and Palestinians battle with guns and bombs; African Americans and white Americans battle in the courts and schoolyards.

This does not mean, though, that these conflicts cannot be lessened or even eliminated. Irish people no longer are refused employment as was common in the nineteenth century, and Jews no longer are prohibited from living in certain neighborhoods or belonging to certain clubs as was common until the 1960s. Ideas about race and ethnicity are social constructions that change as societies change. To combat prejudice and discrimination, we will need to combat subtle and institutionalized racism, and we will need to address the social class inequalities that support racial and ethnic inequalities. Doing so will be both especially difficult and especially crucial if economic hard times continue in the United States.

Summary

1. A race is a category of people treated as distinct due to the physical characteristics that have been given social importance. An ethnic group is a category whose members are thought to share a common origin and culture. Both race and ethnicity are socially constructed categories.

2. The United States is a semi-caste system, in which the population is stratified by both race and class.

3. The concepts of majority and minority groups provide a generic framework for examining structured inequalities by ascribed status. Interaction between majority- and minority-group members may take the form of conflict, assimilation, accommodation, or acculturation.

4. Prejudice and discrimination help create and maintain social distance. Segregation also helps reinforce differences and inequalities.

5. In the United States, white ethnicity is now largely a symbolic characteristic. Its main consequence is that it has become the "standard" American ethnicity against which other groups are judged.

6. On many fronts, African Americans have improved their position in U.S. society. Nevertheless, African American families continue to have a median income that is far lower than that of white families. Major areas of continued concern are high rates of female-headed households, unemployment, and housing segregation, and the growing divide between middle-class African Americans and poor African Americans living in segregated neighborhoods that lack employment opportunities.

7. Hispanics are the largest and one of the fastest growing minority groups in the United States; within a few years, they will replace African Americans as the largest minority group. Because many Hispanics are recent immigrants from less-developed countries, they generally have poor educations and low earnings. Studies show, however, that they face fewer barriers to achievement than African Americans.

8. Native Americans are the least prosperous and least assimilated group in the United States.

9. Asian Americans have used education as the road to social mobility. Even the newest immigrant groups far outstrip white Americans in their pursuit of higher education. Despite some discrimination, Asian Americans have higher family incomes than white Americans and very low levels of residential segregation.

10. Arab Americans are primarily middle class: well educated, with good jobs. Prejudice against Arab Americans is strong, but did not increase substantially following the 9/11 terrorist attacks.

11. Both race and class are important for understanding the experience of African Americans. Although the color-blind forces of the indirect inheritance model make it hard enough to erase economic disadvantage, prejudice and discrimination continue to pose additional handicaps.

12. From a comparative perspective, the key factor increasing ethnic and racial antagonisms appears to be an increase in competition for scarce resources.

Thinking Critically

1. Within the next 50 years or so, European Americans will be a numerical minority within the United States. In sociological terms, do you think they will be a minority group? What social, economic, or political changes do you expect as a result of changes in the relative size of the different U.S. racial and ethnic groups?

2. In thinking about the relationship between prejudice and discrimination, we generally assume that prejudice is the cause of discrimination. Can you think of a time or situation when the reverse might be true, that is, that prejudice would follow from discrimination?

3. Some scholars contend that the major cause of racial/ethnic inequality in the United States today is institutionalized, not individual, racism. If this is so, what recommendations would you offer to policy makers who wanted to reduce racial or ethnic differences in quality of life?

4. When the Smithsonian Institution mounted an exhibition showing the horrors that resulted from the bombing of Hiroshima during World War II, some American veterans protested that the exhibit paid too much attention to Japanese suffering and portrayed the United States as an unfeeling aggressor. How should a multicultural society like the United States deal with such issues? What would you have done if you were in charge of the exhibit? Why?

5. How do you think race and ethnic relationships within the United States affect our international relationships? What connection, if any, do you think there is between racial discrimination within the United States and racial conflict abroad?

Sociology on the Net

 The Wadsworth Sociology Resource Center: Virtual Society

> **http://sociology.wadsworth.com/**

The companion Web site for this book includes a range of enrichment material. Further your understanding of the chapter by accessing Online Practice Quizzes, Internet Exercises, InfoTrac College Edition Exercises, and many more compelling learning tools.

Chapter-Related Suggested Web Sites

> *American Studies Web*
> **http://www.georgetown.edu/crossroads/asw/**

> *The Race and Ethnicity Collection*
> **http://eserver.org/race/**

> *Racial Legacies and Learning*
> **http://www.pbs.org/adultlearning/als/race/**

 InfoTrac College Edition

> **http://www.infotrac-college.com/wadsworth**

Access the latest news and research articles online—updated daily and spanning four years. InfoTrac College Edition is an easy-to-use online database of reliable, full-length articles from hundreds of top academic journals and popular sources. Conduct an electronic search using the following key search terms:

> **Race relations**
>
> **Racial inequality**
>
> **Affirmative action**
>
> **Environmental racism**

 For more information related to this chapter, visit the Opposing Viewpoints Resource Center. Be sure you check all the options on the toolbar—viewpoints, references, statistics, and so on—and search under the following subjects:

> **Race relations**
>
> **Racism**
>
> **Affirmative action**

Suggested Readings

Feagin, Joe R., and Sikes, Melvin P. 1994. *Living with Racism: The Black Middle-Class Experience.* Boston: Beacon. Documentation of how discrimination and racism continue to pervade American culture and affect the lives of even middle-class African Americans.

Gourevitch, Philip. 1998. *We Wish to Inform You That Tomorrow We Will Be Killed With Our Families: Stories from Rwanda.* New York: Farrar, Straus & Giroux. A gripping account, written by a journalist, that describes the horrifying violence that decimated Rwanda's population. Gourevich analyzes how internal power struggles, colonial interventions, and economic devastation fostered unremitting violence between the Hutus and Tutsis, even though they had lived together peacefully for years.

Haizlip, Shirlee Taylor. 1994. *The Sweeter the Juice: A Family Memoir in Black and White.* New York: Simon & Schuster/Touchstone. A wonderfully readable personal story of the African American author's search to find her mother's sister and other relatives, who had left her when she was a child because they passed for whites. The book illustrates that racial categories are not accurate—and that our own racial genes may not be exactly what we think they are.

Loung, Ung. 2000. *First They Killed My Father: A Daughter of Cambodia Remembers.* New York: Harper-Collins. A riveting memoir of the almost unending brutality imposed by the Cambodian government on millions of its own citizens during the late 1970s.

Morales, Ed. 2003. *Living in Spanglish: The Search for Latino Identity in America.* New York: St. Martin's. Explores how Latinos are creating a new culture, blending their Latino roots with Anglo-American culture.

Takaki, Ronald. 1993. *A Different Mirror.* Boston: Little, Brown. A powerful retelling of America's distant and very recent past from the standpoint and often in the words of the many different ethnic groups usually left out of U.S. history books.

Temple-Raston, Dina. 2003. *A Death in Texas: A Story of Race, Murder, and a Small Town's Struggle for Redemption.* New York: Henry Holt. On June 7, 1998, three white men from Jasper, Texas chained an African American man to the back of a pickup truck and dragged him to his death. In this book Temple-Raston explores how the townspeople responded to news of this crime, in the process illuminating America's racist legacy.

Sex, Gender, and Sexuality

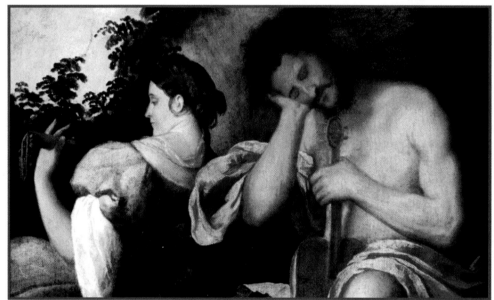

Photodisc Green/Getty Images

Sexual Differentiation

Men and women are different. Biology differentiates their physical structures, and cultural norms in every society differentiate their roles. In this chapter, we describe some of the major differences in men's and women's lives as they are socially structured in the United States. We will be particularly interested in the extent to which the ascribed characteristic of sex has been the basis for structured inequality.

Sex versus Gender

In understanding the social roles of men and women, it is helpful to make a distinction between gender and sex. **Sex** refers to the two biologically differentiated categories, male and female. It also refers to the sexual act that is closely related to this biological differentiation. **Gender,** on the other hand, refers to the normative dispositions and behaviors that cultures assign to each sex.

Although biology provides two distinct and universal sexes, cultures provide almost infinitely varied **gender roles.** Each man is pretty much like every other man in terms of sex—whether he is upper class or lower class, African American or white, Chinese or Apache. Gender, however, is a different matter. The rights, obligations, dispositions, and activities of the male gender are very different for a Chinese man than for an Apache man. Even within a given culture, gender roles vary by class, race, and subculture. In addition, of course, individuals differ in the way they act out their expected roles. Some males play an exaggerated version of the "manly man," whereas others display few of the expected characteristics.

Just how much of the difference between men and women in a particular culture is normative and how much is biological is a question of considerable interest to social and biological scientists. This question has led some biologists to investigate whether characteristics we typically think of as male and female also characterize non-human species. If they did, that would lend support to the idea that these male/female differences are biological. Results from these studies are decidedly mixed. Lions may be the kings of the jungle, but it is female lions who do all the hunting. Male baboons are certainly dominant over female baboons, but male marmosets (small monkeys) take care of the young, while whales and elephants live in matriarchal families. Female goby fish sport bright colors to attract male attention, but in most bird species, it is males who use their colors to attract the opposite sex.

For the most part, social scientists are more interested in gender than in sex. They want to know about the variety of roles that have been assigned to women and men and, more particularly, about the causes and consequences of this variation. Under what circumstances does each sex have more or less power and prestige? How does having more or less power affect women's and men's everyday lives? And what accounts for the recent changes that have occurred in gender roles in our society?

Gender Roles Across Cultures

A glance through *National Geographic* confirms that there is wide variability in gender roles across cultures. The behaviors we normally associate with being female and male are by no means universal. Among the Wodaabe, a nomadic tribe of

---

Sex is a biological characteristic, male or female.

Gender refers to the expected dispositions and behaviors that cultures assign to each sex.

Gender roles refer to the rights and obligations that are normative for men and women in a particular culture.

western Africa, boys carry mirrors with them from the time they can walk. Even when boys spend days alone in the bush herding cows, they begin each day by fixing their hair and putting on their jewelry, lipstick, mascara, and eye-liner. In contrast, because girls are primarily evaluated on their health and ability to work hard, they are expected to pay far less attention to their appearance than do boys. Wodaabe courtship mostly takes place during yearly dances, judged by women who select the winners based on the men's physical beauty and charm. Afterwards, the women openly approach the men they find most attractive to be their romantic partners.

Despite cross-cultural variations such as these, and despite the fact that women do substantial amounts of work in all societies (often providing more than half of household food), in almost all societies women have less power and less value than men (Kimmel 2000). A simple piece of evidence is parents' almost universal preference for male children (Sohoni 1994). This preference can be life threatening to girls. Demographers have determined that worldwide there are 100 million fewer women than there would be if boys and girls were equally valued. A 1992 Bombay study (cited in Sohoni 1994) found that of 8,000 abortions performed after parents had used amniocentesis to determine the sex of their fetuses, only one aborted fetus was male. Beyond selective abortion and female infanticide (killing of infant girls), boys often receive preferential treatment—for example, more food and medical care—while girls are more likely to be neglected and to die as a result. The preference for boys is less strong in modern industrial nations, but parents in the United States still prefer their first child to be a boy by a two-to-one margin (Holloway 1994; Sohoni 1994).

One result of female power disadvantage is widespread violence toward women. According to the respected international organization, Human Rights Watch, "Abuses against women are relentless, systematic, and widely tolerated, if not explicitly condoned. Violence and discrimination against women are global social epidemics" (Human Rights Watch 2004). For example:

- In the United States during 1996, more than 840,000 women were murdered, raped, assaulted, or robbed by an intimate (spouse, ex-spouse, or boyfriend). That number is more than five times the number of male victims of intimate violence for the same year (U.S. Department of Justice 1998). Violence between intimates, or domestic violence, is further discussed in Chapter 12.
- In Uganda, Sierra Leone, Bosnia, and elsewhere, armies have used rape both as a systematic tool to subjugate the population and as a form of "sport" for soldiers.
- More than 100 million women, mostly in African countries but also in Asia, South America, and Europe, have undergone the ritual of genital mutilation—removal of some or all of the clitoris and surrounding genitalia in order to control women's sexual desire and behavior (World Health Organization 2000).
- In India, according to the U.S. State Department's 1999 Country Report on Human Rights Practices, 6,006 new brides were known to have been murdered in 1997 by their husbands or in-laws; many more such murders go unreported. These "dowry deaths" occur when the husband and his family believe that the new wife brought an inadequate dowry (sum of money and goods) to them. The murderers are rarely punished.

At home and abroad, violence against women results from the lower status accorded to women. In growing numbers, women around the world are demanding equal rights. The 1995 World Conference on Women in Beijing, China, may well have marked a turning point for the international women's movement. In developing nations, of course, the primary question arising from the conference is how to pay for the programs that are necessary to improve women's lives. In the United States and the rest of the developed world, the questions are somewhat more subtle: How are gendered identities developed, and what are the institutional forces that maintain inequality, even in the absence of overt violence and discrimination?

Perspectives on Gender Inequality

Women rather than men bear children because of physical differences between the sexes. Most of the differences in men's and women's life chances, however, are socially structured. Different sociological theories offer different explanations for the persistence of this structured gender inequality.

Structural-Functional Theory: Division of Labor

The structural-functional explanation of gender inequality is based on the premise that a division of labor is often the most efficient way to get a job done. In the traditional sex-based division of labor, the man does the work outside the family and the woman does the work at home. According to this argument, a gendered division of labor is functional because specialization will (1) increase the expertise of each sex in its own tasks, (2) prevent competition between men and women that might damage the family, and (3) strengthen family bonds by forcing men and women to depend on each other.

Of course, as Marx and Engels noted, any division of labor has the potential for domination and control. In this case, the division of labor has a built-in disadvantage for women because by specializing in the family, women have fewer contacts, less information, and fewer independent resources. Because this division of labor contributes to family continuity, however, structural functionalists have seen it as necessary and desirable.

Conflict Theory: Segmented Labor Markets and Sexism

According to conflict theorists, women's disadvantage is not an historical accident. It is designed to benefit men. In addition, contemporary sex differences are designed to benefit the capitalist class. It is easy to see how women's lower status can benefit men, but how can it benefit capitalism?

In precapitalist societies, where economic activities required little physical strength and could be carried on while caring for children, women made major contributions to providing for their families' basic needs (Chafetz 1984). Even today, it is estimated that in societies based on gathering and simple hoe agriculture, women are responsible for 60 to 80 percent of a society's food needs (Blumberg 1978). Moreover, their economic activities make them an active part of their community and increase their power over community and group decisions.

The rise of capitalism and industrialization moved work away from the home and made it difficult for women to be economically productive while bearing and rearing children. As a result, the power of women declined. Capitalists benefited because having women at home who would cook, clean, and raise children made it easier for men to function as workers in the industrial economy.

The Environment

Ecofeminism

In 1962, Rachel Carson published a book called *Silent Spring*. It changed the world. Raising an impassioned, scientific voice against environmental destruction, Carson predicted, among other things, that many bird species would soon become extinct because of unrestricted use of pesticides. One result of the book was a ban on the use of the pesticide DDT in the United States. Today, we can thank Carson for saving our national emblem—the bald eagle—from the brink of extinction and for helping to launch the environmental movement.

Carson died in 1964, before the modern feminist movement fully emerged. With the growth of the feminist and environmental movements in the 1970s, many feminists came to conclude that women should play a central role in protecting the environment. Dissatisfied that the environmental movement of the time lacked a feminist perspective and that the feminist movement largely ignored issues of nature and the environment, a new way of thinking about politics, spirituality, and nature emerged. This feminist perspective on the environment and ecology is called *ecofeminism*.

Politics

Issues of power and subordination are important to ecofeminism. According to this perspective, human attempts to control and dominate (and thereby degrade) nature are culturally and politically related to the patriarchal systems that allow men to exert power and control over women. Believing that there is a fundamental connection between social justice and respect for the environment, ecofeminists are typically concerned with a whole range of social issues, including animal, civil, gender, and gay/lesbian rights.

Spirituality

Ecofeminists hope to develop new cultures that can live *with* the Earth rather than master it. Consequently, many have derived inspiration from ancient goddess myths, in which the creator is female, the Earth is revered, and spirituality is centered on oneness with nature. Within these spiritual traditions, the Earth is seen as sacred in and of itself, and all of nature's elements—rivers, forests, and creatures—have value simply because they exist, not just because of the uses to which they can be put. Ecofeminists recognize that humans must use some of the Earth's resources to survive, but believe that we also must protect the environment for future generations (Adams 1994; Diamond & Orenstein 1990).

Nature

The spiritual traditions noted above influence ecofeminists' views about nature, as well as their environmental activism. Ecofeminism includes individuals who believe that change is promoted by front-line social activism as well as those who believe that social change will be created through dedicated spiritual practice. Ecofeminism embraces communities and individuals who are uncompromisingly radical as well as those who operate within a more conventional political framework. Whether dedicated to tree-planting, alternative healing, organic food, performance art, benevolent witchcraft, animal rights, or new forms of activism based on ancient myths, ecofeminists are held together by a common belief that the future of the planet depends on our ability to develop ways of interacting with the environment and with one another that depend less on power over others than on harmony with others (Adams 1994; Diamond & Orenstein 1990).

For more information, look up the following subjects in InfoTrac College Edition:
Ecofeminism

Or visit the following Web sites:
Ecofeminism
http://www-rcf.usc.edu/~orenstei/ecofem/
http://lib.colostate.edu/research/english/ecofem.html#eco

Over the last few decades, however, the shift toward having fewer children has allowed women to leave the household and to increase their political role and participation in the paid workforce; Map 10.1 shows the percentage of women legislators in countries around the world. As a result of these changes, the status of women is improving, if slowly. Nevertheless, men remain substantially advantaged in prestige and power.

MAP 10.1
Women in Political Office, 2002
Percentage of women legislators in single or lower chamber of congress or parliament
SOURCE: United Nations Statistics Division 2003.

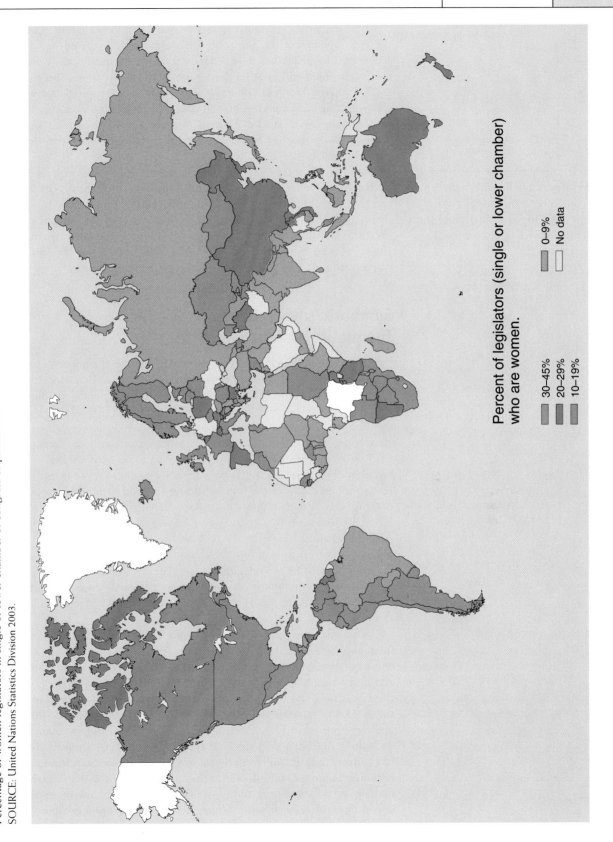

Percent of legislators (single or lower chamber) who are women.

30–45%
20–29%
10–19%
0–9%
No data

One explanation conflict theorists give for this continuing gender inequality is the development of the segmented labor market (discussed in Chapter 14). The segmented labor market is two-tiered and sharply gendered. It creates one set of jobs for women and another, better set, with higher wages, for men (Catanzarite 2003; Cohen & Huffman 2003a, 2003b). In this way, capitalists divide the working class, coaxing working-class men into a coalition against working-class women. Further, female labor can be used to provide a cushion against economic cycles. Women can be laid off in slack times and hired back when employment demands are up (Bonacich 1972).

One of the most important mechanisms for keeping women in their place is **sexism**—the belief that women and men have biologically different capacities and that these differences form a legitimate basis for unequal treatment. Conflict theorists explain sexism as an ideology that is part of the general strategy of stratification. If others can be categorically excluded, the need to compete individually is reduced. Sexism, then, is a means by which men reduce women's access to scarce resources and keep those resources for themselves.

Symbolic Interactionism: Gender Inequality in Everyday Life

Symbolic interactionist theory is particularly useful for understanding the sources and consequences of sexism in everyday interactions. Two recent studies illustrate this perspective. Karin Martin (1998) was interested in understanding how boys and girls learn gender-normative ways of moving, using physical space, and comporting themselves. To answer this question, she studied 112 preschoolers in 5 different classrooms, at 2 different preschools, with 14 different teachers. She found that teachers routinely structure children's play and impose discipline in ways that reinforce gender differences. Little boys are actively discouraged from playing "dress-up" (even though many of them enjoy doing so), and little girls are discouraged from running, crawling, and lying on the floor. Whereas boys are allowed to express their fun and enthusiasm by shouting and moving about wildly, girls are disciplined to raise their hands, lower their voices, and restrict their movements. Likewise, boys' play was allowed to be more rough and tumble than that of girls. By the end of preschool, then, boys and girls are well on their way to learning the nonverbal behaviors and communication styles that are typical of, and seem so natural for, adult men and women. We will discuss these gendered differences in more detail later in this chapter.

A second study illustrating the symbolic interactionist perspective on gender inequality was based on observations conducted at a sleep-away camp during the course of one summer (McGuffey & Rich 1999). At this camp, high-ranking boys attained power and popularity primarily through athletic prowess. They bolstered their positions and won approval from other boys through acts of aggressiveness toward lower-ranking boys and through sexual harassment of girls. In addition, and most importantly, high-ranking boys led other boys in teasing, assaulting, or excluding any boys they deemed too "feminine" and any girls they deemed too "masculine." Interestingly, high-ranking boys had sufficient power that they could redefine "feminine" activities that they enjoyed (such as hand-clapping games) as masculine. In these ways, high-ranking boys could maintain their status and power over other boys, while almost all boys maintained greater status and power than girls.

Sexism is a belief that men and women have biologically different capacities and that these form a legitimate basis for unequal treatment

Gender as Social Construction and Social Structure

To sociologists, gender is not simply something that individuals have—a biological given—but rather is something that is constantly recreated in individual socialization, in medical and cultural practices, and in social interaction. Similarly, sociologists describe gender as an attribute not only of individuals but also of social structures.

Developing Gendered Identities

From the time they are born, girls are treated in one way and boys in another—wrapped in blue blankets or pink ones, encouraged to take up sports or sewing, described as cute or as strong before they are old enough to truly exhibit individual personalities. In these ways, as the symbolic interactionists studies illustrate, children learn their gender and gender role. As a result of this training, one of the first elements that youngsters distinguish as part of their self-concept is their gender identity. By the age of 24 to 30 months, they can correctly identify themselves and those with whom they come into contact by sex, and they have some ideas about what this means for appropriate behavior (Cahill 1983).

Young children's ideas about what it means to be a boy or a girl tend to be quite rigid. They develop strong stereotypes for two reasons. One is that the world they see is highly divided by sex: In their experience, women usually don't build bridges and men usually don't crochet. The other important determinant of stereotyping is how they themselves are treated. Substantial research shows that parents treat boys and girls differently. They give their children "gender-appropriate" toys, they respond negatively when their children play with cross-gender toys, they allow boys to be active and aggressive, and they encourage their daughters to play quietly and visit with adults (Orenstein 1994). When parents do not exhibit gender-stereotypic behavior and do not punish their children for cross-gender behavior, their children are less rigid in their gender stereotypes (Berk 1989).

As a result of this learning process, boys and girls develop strong ideas about what is appropriate for girls and what is appropriate for boys. Because males and male behaviors have higher status than female behaviors, boys are punished more than girls for exhibiting cross-gender behavior. Thus, little boys are especially rigid in their ideas of what girls and boys ought to do. Girls are freer to engage in cross-gender behavior, and by the time they enter school, many girls are experimenting with boyish behaviors.

Reinforcing Biological Differences

Because of gender socialization, girls and boys and men and women understand quite well what a "proper" male or female should be like. These ideas can become self-fulfilling prophecies: The belief that males and females are naturally different biologically keeps males and females different (Lorber 1994). Shari Dworkin (2003) researched this idea during two years of participant observation at two gyms. Trainers at both gyms told women customers that they could lift weights

Despite many changes in gender roles in the United States, boys and girls still tend to experience large doses of traditional gender socialization.

© Tony Freeman/PhotoEdit

without fear because only men can "bulk up." Nonetheless, 25 percent of women didn't lift at all because they feared developing "masculine" muscles. Another 65 percent restricted their weight lifting to shorter periods or lighter weights because they discovered that otherwise they did develop bigger muscles. By the end of two years training, these women remained relatively unmuscular. They lacked muscles not because they were inherently unable to develop them but because they chose not to do so, based on their beliefs about proper male/female differences.

Biological sex differences can also be reinforced by medical practices. Doctors sometimes prescribe hormones to keep girls from growing "too tall" and boys from being "too short." Doctors also offer plastic surgery to women with small breasts and men with small pectoral muscles. In this way, the very bodies we see around us come to reinforce social ideas about male/female differences.

Our belief in the naturalness of biological differences is also reinforced when we are, in essence, kept from seeing how similar males and females can be. Television offers far more coverage of female cheerleaders and male football teams than of male cheerleaders and female football teams, reinforcing the idea that it is impossible for women to play strenuous sports and that no "real men" would be interested in cheerleading. Similarly, Olympic games that evaluate female figure skaters on their grace and male skaters on their speed and power force female and male skaters to develop different skills and leave audiences believing that female and male skaters naturally have quite different abilities. The same is true for athletic rules that limit the size of the basketball court on which girls can play or that forbid male and female athletes from competing together.

"Doing Gender"

Gender differences are also reinforced when we "do gender." Sociologists use the term "doing gender" to refer to everyday activities that individuals engage in to affirm that they understand what is expected of them as male or female (West & Zimmerman 1987). Women who are professional body-lifters almost always wear long, blonde hair so that no one will question their femininity despite their muscles (Weitz 2004a); male nurses might talk about their athletic interests or sexual conquests to keep others from questioning their masculinity. Each of us does gender every day when we (whether male or female) choose to wear skirts or jeans, to speak softly or boldly, to get a butterfly tattoo or shark tattoo, and so on. In these ways we participate in the social construction of gender. Another way to think about this is that gender is not something that we innately have, but rather is something that we do.

Gender as Social Structure

Gender is also a social structure, a property of society (Risman 1998). Gender is built into social structure when workplaces don't provide day care; women don't receive equal pay; fathers don't receive paternity leave; basketballs, executive chairs, and power drills are sized to fit the average man; and husbands who share equally in the housework are subtly ridiculed by their friends. Importantly, this suggests that changing gender roles and attitudes will only produce social change if there are parallel changes in the social structure of gender. Equally important, when social structure changes, gender roles and attitudes change. For example, Barbara Risman found that men whose wives died or deserted them and their children quickly learned how to be good "mothers," nurturing their children as women would.

Connections

Personal Application

How are you doing gender right now? Are you sitting with your legs splayed apart or crossed at the ankles? Are you wearing any makeup? What kind, and for what purposes? What color clothes are you wearing? (If you are male, you are probably not wearing pink.) If you are snacking on a muffin while reading this book, did you apologize beforehand to anyone else who was present and explain how you know you need to lose weight? These are all examples of doing gender.

Differences in Life Chances by Sex

In terms of race and social class, women and men start out equal. The nurseries of the rich as well as the poor contain about 50 percent girls. After birth, however, different expectations for females and males result in their having very different life chances. This section examines some of the structural social inequalities that exist between women and men.

Health

Women are at a substantial disadvantage in most areas of conventional achievement; in informal as well as formal interactions, they have less power than men. But men, too, face some disadvantages from their traditional gender roles.

Perhaps the most important difference in life chances involves life itself. Boys born in 2005 can expect to live 74.9 years and girls 81.4 years (U.S. Bureau of the Census 2002b). On average, then, women live more than 6 years longer than men. Part of this difference is undoubtedly biological. But men's gender roles also contribute to their lower life expectancies (Waldron 1994).

A major way male gender roles endanger men is by encouraging men to "prove" their masculinity through dangerous activities. As a result, young men are more than twice as likely to die in motor vehicle accidents and six times more likely to be killed by guns (Minino 2002). Similarly, men are far more likely than women to earn their living through dangerous jobs, such as fishing and lumbering.

The sex differential in life expectancy is more complex than the greater risk taking of young men, however. Consider men's greater risk of dying from heart disease. Evidence suggests that men have this disadvantage not because they experience more stress than women but because, under the same levels of stress, they are more vulnerable to heart disease. Current thinking attributes men's greater risk of heart disease at least partly to the male gender role's low emphasis on nurturance and emotional relationships. Maintaining family and social relationships is usually viewed as women's work, and when men end up without women to do this work for them (never married, divorced, or widowed), they also frequently end up alone (Kessler & McLeod 1984; Stroebe & Stroebe 1983; Wallerstein & Kelly 1980). Ultimately, this lack of social support leaves men especially vulnerable to stress-related diseases and may explain why their suicide rate is four times higher than women's (Minino 2002; Nardi 1992).

Education

Fifty years ago, few young women went to college. Those who did were encouraged to focus on earning an "MRS." (i.e., getting married) rather than a B.A. These days, women and men are about equally represented among high school graduates and among those receiving bachelor's and master's degrees. It is not until the level of the Ph.D. or advanced professional degrees (law and medicine) that women are disadvantaged in quantity of education.

Probably more important than the differences in level of education are the differences in types of education. From about the fifth grade on, sex differences emerge in academic aptitudes and interests: Boys take more science and math, whereas girls more often excel in verbal skills and focus their efforts on language and literature. In large part, these sex differences in aptitudes and interests are socially created (Sadker

& Sadker 1994). In all subjects, but especially math and science, teachers typically assume that boys have a better chance of succeeding. One result is that teachers more often ask girls simple questions about facts and ask boys questions that require use of analytic skills. When boys have difficulty, teachers help them learn how to solve the problem, whereas when girls have difficulty, teachers often do the problem for them. By the time students arrive at college, girls often lack the necessary prerequisites and skills to major in physical sciences or engineering, even if they should develop an interest in them (Sadker & Sadker). As a result, women college graduates are over-represented in education and the humanities, and men are overrepresented in engineering and the physical sciences—fields that pay considerably higher salaries.

Table 10.1 shows the proportion of bachelor's degrees earned by women in various fields of study in 1971 and in 2000. You can see from the table that there were changes over this period. Women comprised a far higher proportion of business, math, social science, and pre-law graduates in 2000 than in 1971. In fact, women now comprise about half of all business graduates and three-quarters of all pre-law graduates.

Still, striking differences between men and women remain. In 2000, only 20 percent of graduates in engineering and 28 percent of graduates in computer sciences were women. Meanwhile, 76 percent of graduates in education and more than 80 percent in home economics, health sciences (mostly nursing), and library sciences were women (U.S. Department of Education 2000). Because engineers and computer scientists make a great deal more money than teachers, librarians, and nurses, these differences in college majors have implications for future economic well-being. This situation is an example of institutionalized sexism. (Recall that Chapter 9 discussed the parallel concept of institutionalized racism.)

TABLE 10.1
Bachelor's Degrees Earned, by Field, 1971 and 2000
Between 1971 and 2000, there was a substantial narrowing of the sex gap in educational focus. Nevertheless, engineering continues to be largely a male preserve, while education attracts a disproportionate number of women undergraduates. Because engineers earn roughly three times what teachers earn, this difference in educational direction is one reason why, on average, women earn less than men.

Field of Study	Percent Female	
	1971	2000
Business	9.1	49.7
Computer and information sciences	13.6	28.1
Education	74.5	75.8
Engineering	0.8	20.4
Health sciences	77.1	83.8
Home economics	97.3	87.9
Library and archival sciences	92.0	87.5
Pre-law	6.0	73.0
Mathematics	37.9	47.1
Social sciences and history	36.8	51.2

SOURCE: U.S. Bureau of the Census 1996; U.S. Department of Education 2000.

Work and Income

In 2001, 93 percent of men compared with 76 percent of women aged 25 to 44 were in the labor force (U.S. Bureau of the Census 2002b). This gap is far smaller than it used to be but probably larger than it will be in the future (Figure 10.1). Studies show no sex differences at all in the proportion of male and female college students who expect to be employed at age 25 or age 50 (Affleck, Morgan, & Hayes 1989). Although most young women expect to be mothers, they also expect to be full-time permanent members of the labor force.

Despite growing equality in labor-force involvement, major inequalities in the rewards of paid employment persist. In 2000, women who were full-time workers earned 76 percent as much as men (U.S. Bureau of Labor Statistics 2001). Overall, this percentage has not changed much since 1950. Why do women earn less than men? The answers fall into two categories: differences in the types of jobs men and women have and differences in earnings of men and women in the same types of jobs.

Different Occupations, Different Earnings

Women are often employed in different jobs than men, and the jobs women hold pay less than the jobs men hold. The major sex difference as shown in Figure 10.2 is that women dominate clerical and service jobs, whereas men dominate blue-collar jobs. The proportion of men and women in professional and managerial jobs is nearly equal. Generally, though, men professionals are doctors and women professionals are nurses; men manage steel plants and women manage dry cleaning outlets.

There are three major reasons why men and women have different jobs: gendered jobs, different qualifications, and discrimination.

1. *Gendered jobs.* Many jobs in today's labor market are regarded as either "women's work" or "men's work." With the exception of domestic labor, manual labor is almost exclusively men's work; nursing and typing are largely women's work. These occupations are so sex segregated that many men and women would feel uncomfortable working in a job where they were so clearly the "wrong" sex. Women in male-dominated occupations report experiencing subtle forms of discrimination such as exclusion from informal leadership and decision-making networks, sexual harassment, and other forms of hostility from male coworkers (Chetkovich 1998; Jacobs

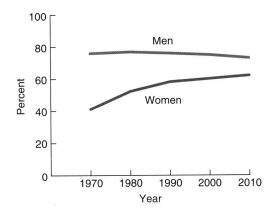

FIGURE 10.1
Labor-Force Participation Rates of Men and Women Aged 16 and Over, 1970 to 2010 (estimated)
In the 30 years between 1970 and 2000, women's participation in the labor force grew significantly, while men's barely changed. Sex differences in labor-force participation are expected to continue to decline slowly.
SOURCE: U.S. Bureau of the Census 2002b.

FIGURE 10.2

Differences in Occupation by Sex, 2000

Men and women continue to be employed in different occupations in the United States. The most striking differences are in technical, sales, and clerical work, where women predominate, and in various blue-collar jobs, held primarily by men. Within categories, men typically hold higher-status positions than women.

SOURCE: U.S. Bureau of the Census 2002b.

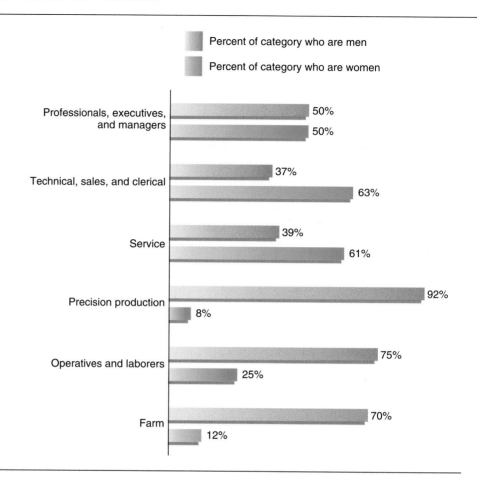

■ Percent of category who are men

■ Percent of category who are women

Professionals, executives, and managers — 50% / 50%

Technical, sales, and clerical — 37% / 63%

Service — 39% / 61%

Precision production — 92% / 8%

Operatives and laborers — 75% / 25%

Farm — 70% / 12%

1989b). These forms of informal discrimination serve as "glass ceilings"—invisible barriers to women's promotions in traditionally male careers (Freeman 1990). In contrast, men working in female-identified occupations report that negative stereotypes about men who do "women's work" form the major barrier to more men entering occupations such as nurse, elementary school teacher, librarian, or social worker. When men do work in nontraditional fields, however, they are likely to encounter "glass escalators"—rapid promotions "up" to more administrative positions and prestigious specialties (Williams 1992).

Although there has been little change in the number of men employed in female-dominated occupations, a growing number of jobs that used to be reserved for men—such as insurance adjusters, police officers, and bus drivers—have opened to women in the last two decades. Research shows, however, that this is not because of women's increased access to good jobs. Rather, women by and large are moving into jobs that men are abandoning because of deteriorating wages and working conditions (Reskin 1989).

2. *Different qualifications.* Although the differences are smaller than they used to be, women continue to major in fields of study that prepare them to work in relatively low-paying fields, such as education, while men are more likely to choose more lucrative fields. More important than these differences in educational qualifications are disparities in experience and on-the-job training. Because many women have taken time out of the labor force for child rearing, they have less job experience than men their own age (Marini 1989). Perceiving that women are more likely to be

short-term employees than men, employers have invested less in training and mentoring women (Tomaskovic-Devey & Skaggs 2002). As a result, women are less likely to be promoted to management positions.

3. *Discrimination.* Although men and women have somewhat different occupational preparation, a large share of occupational differences is due to discrimination by employers (Bielby & Baron 1986; Hesse-Biber & Carter 2000). Employers reserve some jobs for men and some for women based on their own gender-role stereotypes. Similarly, women are less likely than men to be promoted. One study of women engineers, for example, found that 5 to 10 years after completing school, women were already significantly behind their male peers, even though they had identical education and experience (Robinson & McIlwee 1989).

Same Job, Different Earnings

Not all jobs are substantially sex segregated. Some, such as flight attendant, teacher, and research analyst, contain considerable proportions of both men and women. Generally, however, men earn substantially more than women who do the same work (Table 10.2). There are three main reasons why women earn less than men in these circumstances.

1. *Different titles.* Very often, men and women who do the same tasks are given different titles—women will be maids or executive assistants, and men doing the same work will be janitors or assistant executives. Simply because one job category is considered "male" and is occupied by males it is paid a higher wage.

2. *Segmented labor markets.* Even when women and men have the same job titles, women tend to be paid less. One reason for this is that, within any given occupation, men tend to hold the more prestigious, better-paying positions (Hesse-Biber & Carter 2000; McBrier 2003). For example, men lawyers tend to work in large, high-paying firms, specializing in prestigious fields, while women tend to work in small, low-paying firms, specializing in less prestigious fields. These differences reflect what is called a segmented labor market, discussed in more detail in

Although sexism continues to have an impact, more and more women are finding employment in fields formerly open only to men.

TABLE 10.2
Sex Differences in Earnings from the Same Occupation
Even when women have the same occupation as men, they tend to earn substantially less money. In part because of their family responsibilities, women tend to be employed in smaller and lower-paying firms. They also experience substantial discrimination in both employment and promotional opportunities.

Occupation	Median Weekly Earnings		
	Males	Females	Male/Female Ratio
Accountants	$953	$690	1.4
Engineers	1,126	949	1.2
Natural scientists	1,007	726	1.4
Computer programmers	968	868	1.1
Lawyers	1,439	1,053	1.4

SOURCE: U.S. Bureau of Labor Statistics 2001.

focus on

Technology

How Is Technology an Equity Issue?

"Technology is an equity issue," writes Corlann Bush of Montana State University. It has "everything to do with who benefits and who suffers, whose opportunities increase and whose decrease, who creates and who accommodates" (Bush 1993, 206). To a significant extent, who benefits and who suffers depends on the social conditions in place when technological innovations appear (McGinn 1991). We can examine a few ways in which technology has affected African Americans and women in our society. In some cases, technology has apparently benefitted African Americans or women, whereas in other cases it assuredly has not.

Technology and African Americans

- Eli Whitney's 1793 invention of the cotton gin made large-scale plantations profitable and created the need for a large supply of cheap labor—a problem solved by importing many more African slaves (McGinn 1991, 118).

- Automobiles made it easier for people to live in neighborhoods distant from their workplaces, but the resultant suburbs have been primarily white, with African Americans concentrated in central cities. New expressways destroyed cohesive city neighborhoods and parks, further alienating racial minorities (Johnson 1993).

- In the 1950s and 1960s, television gave impetus to the civil rights movement as millions of Americans watched Dr. Martin Luther King's demands for racial justice and also saw police dogs and fire hoses unleashed on demonstrators (McGinn 1991, 120). Yet, television tends to depict African Americans comically, unfa-

© Michael S. Yamashita/Corbis

■ Software development and other high-paying computer jobs are less available to women than to men. Reasons include a lack of female role models in computer programming and an emphasis in video games and computer-based learning programs on activities of interest primarily to boys.

Chapter 14. The gist of this concept is that *where* you work may be as important as what you do in determining your income and opportunity.

3. *Family responsibilities.* Another important reason why women earn less than men within the same occupation is that women's ability to maximize their careers is curtailed by their family responsibilities (McBrier 2003). More often than men, women have to choose jobs close to home so that it is easy to drop junior off at the baby-sitter or to take a child to the doctor during the lunch hour. Similarly, women are much less likely than men to be able to uproot their families and move to advance their careers.

Recognizing these family claims on women employees, employers are less likely to hire them for jobs requiring long career tracks or geographical moves. As we have seen, employers are also less likely to invest in training women and less likely to

vorably, or in lesser roles—when it depicts them at all (Johnson 1993, 272).

- Good jobs in today's information society are disproportionately unlikely to go to African Americans (Wessells 1990). Largely because of economic disparities, only 43 percent of African American children have access to a computer at home, compared to 77 percent of white non-Hispanic children (U.S. Bureau of the Census 2002b).

Technology and Women

- Beginning with the development of the factory during the Industrial Revolution, women have increasingly engaged in paid employment—a situation that gradually liberated them from sole reliance on family roles as their only means of economic support.
- A proliferation in recent decades of new technological jobs in many sectors from dentistry to machine maintenance has meant expanding job options for women. According to some research findings, working-class women see new work technologies as generally improving their circumstances (Walshok 1993). Never-

theless, repetitive motion syndrome (also called repetitive stress injury or RSI), which can result from keyboarding on word processors, has disabled increasing numbers of women; many have difficulty getting compensation from their employers (Elmer-Dewitt 1994).

- Despite electronic "labor-saving" devices, women spend about as much time today in domestic work as they did 50 years ago (Walshok 1993). Partly, this situation is due to more rigorous cultural standards for cleanliness. Also, technology helped change housekeeping tasks such as clothes washing from communal or neighborhood activities to individual, isolated ones (Cowan 1993).
- Software development and programming as well as other high-paying computer jobs are currently less

available to women than to (white) men. Reasons include a lack of female role models in computer programming and an emphasis in video games and computer-based learning programs on activities of interest primarily to boys (Rosenberg 1994; Wessells 1990).

- Many women find much discussion on the Internet rude and insensitive (Rosenberg 1994). Furthermore, snuff porn (depicting women being mutilated, raped, and murdered) has infiltrated this technology (Elmer-Dewitt 1995), a situation of dire concern to many feminists and others.

In sum, technology has two faces. We all must become knowledgeable about its benefits and its dangers if we want to devise technology-related social policies that will create a more equal world (Bush 1993; Johnson 1993).

For more information, look up the following subjects in InfoTrac College Edition:
Internet and blacks
Digital divide

Or visit the following Web site:
The Reader's Companion to American History: Women and the Work Force
http://college.hmco.com/history/readerscomp/rcah/html/ah_093200_womenandthew.htm

promote women. This pattern affects all women, even those who intend to work for 40 years, intend to remain childless, or who are the sole support for themselves or their families (Marini 1989). Studies show, however, that discrimination is strongest against women who have children (Crittenden 2001).

Gender and Power

As Max Weber pointed out, differences in prestige and power are as important as differences in economic reward. When we turn to these rewards, we again find that women are systematically disadvantaged. In the family, business, the church, and elsewhere, women are less likely to be given positions of authority.

Unequal Power in Social Institutions

Women's subordinate position is built into most social institutions. The New Testament urges, "Wives, submit yourselves unto your own husbands" (Ephesians 5:22), the traditional marriage vows require women to promise to obey their husbands, and women were not allowed to vote in the United States until 1919.

In politics, prejudice against women leaders is declining—but it is still quite strong. In 2000, women comprised 56 percent of voters but only 22 percent of state legislators and 14 percent of U.S. Congress members (Center for American Women and Politics 2003). In every institution, traditional norms have specified that men should be the leaders and women the followers. Social scientists call this situation *patriarchy*.

Unequal Power in Interaction

As we noted in Chapter 4, even the informal exchanges of everyday life are governed by norms; that is, they are patterned regularities, occurring in similar ways again and again. Careful attention to the roles men and women play in these informal interactions shows clear differences—all of them associated with childhood socialization and with women's lower prestige and power.

An easy-to-see example is that women smile more (Goffman 1974). They smile to offer social support to others, and they smile to express humility. Studies of informal conversations also show that men regularly dominate women in verbal interaction (Tannen 1990). Men take up more of the speaking time, they interrupt women more often, and most important, they more often interrupt successfully. Finally, women are more placating and less assertive in conversation than men, and women are more likely to state their opinions as questions ("Don't you think the red one is nicer than the blue one?"). This pattern also appears in committee and business meetings—one reason women employees are less likely than men to get credit for their ideas (Tannen 1994).

Laboratory and other studies show that this male/female conversational division of labor is largely a result of differential status (Kollock, Blumstein, & Schwartz 1985; Ridgeway & Smith-Lovin 1999; Tannen 1990). When clear-cut situational factors, such as a student/teacher situation, give women more status in a conversation, they cease to exhibit low-status interaction styles. Nevertheless, a study of conversations between physicians and patients showed that patients are much more likely to interrupt when the doctor is a woman (West 1984). It would appear that the lower status accorded to women in U.S. society cannot be overturned merely by changes in occupational or political roles.

A Case Study: Sexual Harassment

Sexual harassment consists of unwelcome sexual advances, requests for sexual favors, or other verbal or physical conduct of a sexual nature.

The impact of women's relative lack of power becomes clear when we look at the topic of **sexual harassment**—unwelcome sexual advances, requests for sexual favors, and other unwanted verbal or physical conduct of a sexual nature. Although estimates vary widely depending on the definition and sample used, as many as half of all working women probably experience sexual harassment during their lifetime (Welsh 1998). Men also can be sexually harassed—by men as well as by women—but this occurs far less often.

There are two forms of sexual harassment (Shapiro 1994). By law, harassment exists either when an employer, teacher, or other supervisor expects sexual favors (from inappropriate touching to sexual intercourse) in exchange for something else: keeping one's job, getting a good grade or letter of recommendation, and so on. Sexual harassment also exists when an individual finds it impossible to do his or her job because of a hostile sexual climate, such as pornographic photographs posted in an office or constant sexist or sexual jokes.

Sexual harassment ranges from subtle hints about the rewards for being more friendly with the boss or teacher to rape. In the less-severe instances (and sometimes even in the more-severe instances), the subordinate may be reluctant to make, literally, a federal case of it. This pattern of reluctance to report the matter became an issue in 1991 when Anita Hill, a law professor, accused Supreme Court nominee Clarence Thomas of sexual harassment that allegedly had occurred years earlier.

Sexual harassment exists because women have less power than men. (Similarly, men are only harassed in situations where they have little power.) But sexual harassment not only *reflects* women's relative powerless social position, it also helps to *keep* them in that position. For example, women students in engineering classes or firms who experience sexual harassment are less likely to continue to pursue a career in engineering. They also may lose confidence in their abilities and their judgment, and may suffer long-lasting psychological troubles (Sadker & Sadker 1994).

Fighting Back: The Feminist Movement

To fight back against sexual harassment, woman battering, job discrimination, and the other problems discussed in this chapter, women—and men—have united in the feminist movement (Evans 2003; Freedman 2002). The first American feminist movement arose in the mid-nineteenth century. At the time, women's legal status was essentially that of property. Like slaves (male or female), women regardless of race could not own property, vote, make contracts, or testify in a court of law, and only two small colleges admitted women. Many women (both black and white) who were active in the movement to abolish slavery took from their experience the belief in equality and the organizing skills needed to start the feminist movement. By the end of the nineteenth century, the most egregious legal restrictions on

■ The feminist movement today is diverse. On the one hand, many business and professional women fight for equal rights in legislatures and courtrooms. Others take to the streets to protest the violence and inequality that affects women of all nationalities and social classes.

AP/Wide World Photos

Connections

Social Policy

Title IX of the federal Educational Amendments of 1972 is a product of liberal feminist activism. It prohibits sex discrimination in any educational institution or activity (from kindergarten through graduate school) that receives federal funding. Title IX applies to everything from financial aid and class offerings to athletics and health insurance, and has led to a dramatic rise in women's educational attainment and their athletic participation. Observers agree that the U.S. Women's Soccer Team now dominates the sport, and won the 1991 and 1999 World Cups, directly because of Title IX.

women's lives had been lifted, and a growing (though still small) list of colleges accepted women students. At this point, feminist activity shifted almost entirely to obtaining the vote (suffrage) for women. In 1920, Congress adopted the Nineteenth Amendment, which granted female suffrage.

After the passage of the Nineteenth Amendment, feminist activity declined precipitously. This changed in the 1960s, as two groups of women began pressing for change. The first group consisted of middle-aged, middle-class women who were brought together in mainstream political and professional organizations and who began to press for an end to job discrimination and to legal barriers that held women back. The second group consisted of young women activists brought together in the civil rights and anti-Vietnam War movements. These women quickly grew tired of making coffee and typing leaflets while male activists made all the decisions, and they quickly came to understand the parallels between racism and sexism.

At its core, the feminist movement holds that women and men are deserving of equal rights, and that women's lives, culture, and values are as important as are men's. Liberal feminists typically work to ensure that women have equal opportunity in all fields of life; because of feminist activism, academically select high schools and colleges no longer can refuse to admit female students. Radical feminists emphasize the need to change basic structures of society, not just ensure that women have equal access to existing structures. Radical feminists have, for example, led the battle against woman battering and for women's studies programs and research; 30 years ago, no sociology textbook would have included a chapter on sex and gender.

The Sociology of Sexuality

Like gender, sexuality is also a product of both biology and culture. Ideas about "proper" sexuality vary cross-culturally, and have varied historically. A hundred years ago, a woman who admitted to enjoying sexual pleasure could have been declared insane and locked in a mental hospital. Now, a woman who does not enjoy sexual pleasure is labeled frigid and advised to consider therapy. In ancient Greece, male youths were expected to engage in homosexual behavior with their adult male mentors; these days, adults (of either sex) who have sexual relations with minors can be imprisoned. In this section, we look at current sexual behavior in the United States.

Premarital Sexuality

In few areas of our lives are we free to improvise. Instead, we learn social scripts that direct us toward appropriate behaviors and away from inappropriate ones. Sex is no exception.

Some of the more important norms surrounding dating behavior are concerned with the amount of acceptable sexual contact. Premarital sexual activity has become increasingly common and accepted in the United States (Luker 1996). In the 1950s, about 40 percent of teenagers reported engaging in premarital sex, whereas in 1995, 63 percent of women younger than age 19 and 83 percent of those younger than age 25 did so (Abma et al. 1997). Perhaps more important, in the 1950s couples typically only had sex if they intended to marry. This is no longer the case. Changes in premarital sexual activity have been more pronounced for women than for men, with the result that men and women are now much more alike in sexual attitudes and experience (Laumann et al. 1994).

In recent years, however, concern about AIDS has fostered slightly more conservative attitudes toward sexual behavior. A longitudinal, national survey of single men between the ages of 17 and 19 found that both acceptance of premarital sex and the percentage who had ever had sexual intercourse rose between 1979 and 1988 but fell between 1988 and 1995 (Ku et al. 1998), although the lifetime number of sexual partners did not decrease over the 16-year time period. Increasing numbers of young people now use condoms the first few times they have sexual relations with a new partner. After that, though, most conclude that they know and can trust their partners and so abandon condom use. Women are especially likely to believe that their partner loves them and wouldn't hurt them; men are especially likely to believe that they are invulnerable and don't need to worry. Unfortunately, in most cases it is impossible to know if another individual has a sexually transmitted disease unless the individual admits it. But many individuals don't know they are infected, while others know but won't tell.

Marital Sexuality

In certain important ways, the sexual scripts followed by married couples have changed little over time. For example, frequency of sexual activity seems to have changed very little among married people over the years (Call, Sprecher, & Schwartz 1995; Laumann et al. 1994). And, now as in the past, most couples find that the frequency of intercourse declines steadily with the length of the marriage. The decline appears to be nearly universal and to occur regardless of the couple's age, education, or situation. After the first year, almost everything that happens—children, jobs, commuting, housework, finances—reduces the frequency of marital intercourse (Call et al. 1995). The overall conclusion drawn from research is that, after the first year of marriage, sex is of decreasing importance to most people. Nevertheless, satisfaction with both the quantity and the quality of one's sex life is essential to a good marriage (Blumstein & Schwartz 1983; Laumann et al. 1994).

Despite these historical continuities, the sexual scripts followed by married couples have undergone some important changes in recent decades. First, oral sex, a practice that was limited largely to unmarried sexual partners and the highly educated in earlier decades, is now more common. Second, women and men are now equally likely to have extramarital affairs. The double standard has disappeared in adultery, and recent studies suggest that as many as 50 percent of both men and women have had an extramarital sexual relationship (Laumann, Knoke, & Kim 1985).

Sexual Minorities

Although the majority of the population is heterosexual—preferring sex and romance with the opposite sex—significant minorities diverge from this script. This section discusses homosexuals and transgendered persons.

The largest of the sexual minorities is homosexuals (also known as gays and lesbians). Homosexuals are people who prefer sexual and romantic relationships with members of their own sex. On well-regarded surveys, somewhere between 2 to 6 percent of Americans admit recent homosexual activity or describe themselves as homosexual, with rates about twice as high among men as among women (Binson et al. 1995; Lauman et al. 1994). Considerably more report ever engaging in homosexual activity, and it is likely that many more are unwilling to honestly report their behavior and preferences.

■ In February 2004, the city of San Francisco began issuing marriage licenses to gay and lesbian couples. This couple received one of the first licenses.

© Kim Kulish/Corbis

Attitudes toward homosexuality have fluctuated greatly over time. During the last 50 years, however, American attitudes have become increasingly more positive, partly as a result of a growing gay rights movement. (This movement is described in more detail in Chapter 16). In a Gallup Poll conducted in 2003, 59 percent of surveyed Americans agreed that homosexual activity between consenting adults should be legal. Support for gay rights was highest among persons who were less religious, younger, urban dwellers, non-Southerners, more educated, and more liberal in general.

Transgendered persons are those whose gender and sexual identity is not indisputably male or female. There are two main types of transgendered people: hermaphrodites and transsexuals.

Hermaphrodites are persons who are born with ambiguous genitalia, such as a small penis as well as ovaries. Hermaphroditism is a naturally occurring, if rare, phenomenon. In the early stage of fetal development, all fetuses are sexually ambiguous. All fetuses (and adult humans) produce both male and female hormones (including estrogen and testosterone), and these hormones lead to sexual differentiation—the development of ovaries, penises, and so on—later in fetal development. Hermaphroditism occurs when that differentiation is incomplete. When such cases are identified, doctors typically use surgery or hormones to transform the individual's body into one that more closely matches our accepted ideas of what males or females should look like. As when plastic surgeons give women larger breasts, these medical interventions serve to reinforce social ideas about proper sexuality. In contrast, some other cultures recognize the existence of more than two sexes (Herdt 1994; Lorber 1994).

Unlike hermaphrodites, transsexuals' sex is not ambiguous: there are no observable differences between them and other heterosexual males or females. Instead, transsexuals are persons who psychologically feel that they are trapped in the body of the wrong sex. As with hermaphrodites, most doctors consider it appropriate to prescribe hormones or perform surgery (removing penises and constructing vaginas or vice versa) to give transsexuals the bodies they desire. Some observers, however, question the wisdom of these medical interventions (Meyerowitz 2002). They wonder whether, in a society that allowed both men and women more freedom, anyone would feel "trapped" in the wrong body, and they question whether there is really something so wrong with men who enjoy "chick flicks," taking care of children, and spending time on their appearance that they should require surgery. To these observers, the medical treatment of transsexuality is another example of the social construction of both gender and sexuality.

Where This Leaves Us

Gender roles have changed dramatically over the past 30 years, in ways that have affected us deeply. As structural-functionalists point out, traditional roles had their virtues. Everyone knew what was expected of them, and complementary male/female roles forced both sexes to depend on each other and held families together. In contrast, the decline in traditional gender roles has brought stress to many people—not only to men who lost rights and power and but also to women who found themselves caught between changing expectations.

But conflict theorists are also correct: Everyone did not benefit equally from traditional roles, and everyone paid some price for maintaining them. Women endured lower earnings, narrow educational and occupational opportunities, sexual harassment, sexist prejudice and discrimination, and, sometimes, physical violence. Men who held to traditional masculine gender roles experienced more stress, less nurturing relationships, and shorter lives.

Sex is a biological category, something we are born with. But sex, gender, and sexuality are also socially constructed. Doctors change physical bodies so that individuals' sex and gender better fit social expectations. Society, in general, continually evolves its ideas of what it means to be male and female, masculine and feminine, and all of us contribute to this process when we socialize our children, "do gender," and interact with each other. Creating a more just world will require that we change the social structure of gender and sexuality as well as its interpersonal aspects.

Summary

1. Although there is a universal biological base for sex differentiation, a great deal of variability exists in the roles and personalities assigned to men and women across societies. In almost all cultures, however, women have less power than men.

2. Structural-functional theorists argue that a division of labor between the sexes builds a stronger family and reduces competition. Conflict theorists stress that men and capitalists benefit from sexism and a segmented labor market that relegates women to lower-status positions. Symbolic interactionism does not address why gender inequality arose but does help us understand how it is perpetuated in interaction.

3. Sex stratification is maintained through socialization. From earliest childhood, females and males learn ideas about sex-appropriate behavior and integrate them into their self-identities. Biological differences between the sexes are reinforced by medical and social practices.

4. Gender is not simply an individual attribute. It is also a property built into social structures. Finally, it is

something we do. Sociologists use the term "doing gender" to refer to everyday activities that individuals engage in to affirm that they understand what is expected of them as male or female.

5. Men as well as women face disadvantages due to their gender roles. These include higher mortality, fewer intimate relationships, and higher stress.

6. Women and men are growing more similar in their educational aspirations and attainments and in the percentage of their lives that they will spend in the work force.

7. Women who are full-time, full-year workers earn 76 percent as much as men. This is because they have different (poorer-paying) jobs and because they earn less when they hold the same jobs. Causes include different aspirations and educational preparation, discrimination, and women's greater family obligations.

8. Women's subordinate position is built into family, religious, and political institutions. Although some of this has changed, men disproportionately occupy leadership positions in social institutions. They also dominate women in conversation.

9. The feminist movement has fought for almost two centuries to improve the position of American women, and has had many notable successes.

10. Sexual relations are a common aspect of contemporary courtship, although sexual behavior patterns have become slightly more conservative as a result of the AIDS epidemic.

11. Homosexuality is growing more accepted in the United States. Some sociologists question whether the medical treatment of transgendered persons reflects and reinforces traditional ideas about gender roles.

Thinking Critically

1. Suppose you want your daughter to consider science as a future profession. Short of coercion, how would you go about encouraging her to consider this career choice? As a member of the PTA at your daughter's school, what changes would you encourage her school to make in order to increase the chances of girls considering science as a profession?

2. Chapter 9 discussed institutionalized discrimination. Can you think of some specific examples of how institutionalized discrimination works against women in the workplace? Against men? How specifically might affirmative action programs (also discussed in Chapter 9) help to alleviate this discrimination?

3. If men have more power, why do they die earlier and have higher rates of heart disease, suicide, and alcoholism? As women gain power, should we expect them to have similar health problems? Why or why not?

4. In TV commercials, males predominate about nine to one as the authority figure, even when the products are aimed at women. To what extent do you think this is a reflection of the male/female differences in conversational styles described in this chapter? What other factors do you think are involved?

Sociology on the Net

 The Wadsworth Sociology Resource Center: Virtual Society

http://sociology.wadsworth.com/

The companion Web site for this book includes a range of enrichment material. Further your understanding of the chapter by accessing Online Practice Quizzes, Internet Exercises, InfoTrac College Edition Exercises, and many more compelling learning tools.

Chapter-Related Suggested Web Sites

Institute for Women's Policy Research
http://www.iwpr.org

Department of Labor: Women's Bureau
http://www.dol.gov/dol/wb/

The UN Internet Gateway on the Advancement and Empowerment of Women
http://www.un.org/womenwatch/

American Men's Studies Association
http://mensstudies.org

InfoTrac College Edition

http://www.infotrac-college.com/wadsworth

Access the latest news and research articles online—updated daily and spanning four years. InfoTrac College Edition is an easy-to-use online database of reliable, full-length articles from hundreds of top academic journals and popular sources. Conduct an electronic search using the following key search terms:

Gender discrimination

Gender roles

Glass ceiling

Sexual harassment

For more information related to this chapter, visit the Opposing Viewpoints Resource Center. Be sure you check all the options on the toolbar—viewpoints, references, statistics, and so on—and search under the following subjects:

Homosexuality

Sexual behavior

Teenage sexual behavior

Feminism

Suggested Readings

Kimmel, Michael. 1995. *Manhood in America: A Cultural History.* New York: Free Press. An authoritative, entertaining, and wide-ranging history of men in the United States by a respected social scientist in the field of men, manhood, and masculinity.

Orenstein, Peggy. 1994. *School Girls: Young Women, Self-Esteem, and the Confidence Gap.* New York: Doubleday. A very good book, written in association with the American Association of University Women, on young women in today's middle schools. The book is based on participant observation in two California schools.

Weitz, Rose. 2004. *Rapunzel's Daughters: What Women's Hair Tells Us About Women's Lives.* New York: Farrar, Straus and Giroux. Using interviews about their hair collected from a large and diverse sample of women, Weitz shows how young girls are socialized to see their appearance as central to their identities, how teenagers use their appearance to manipulate their identities, and how women find that their appearance continues to affect all aspects of their lives, including their careers and romances.

Age and Health Inequalities

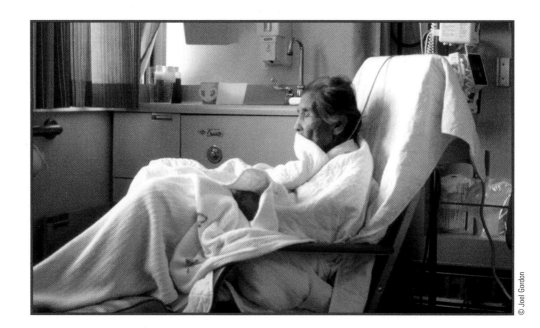

© Joel Gordon

The Structures of Aging and Health Care

In this chapter, we consider two related topics—aging and health. Because children are seldom able to perform difficult tasks as well as adults and because elderly persons experience declining strength and endurance as they age, all societies are differentiated by age. The physical and psychological changes associated with aging are not of primary interest to sociologists. Rather, we are concerned with the ways that age norms and roles structure the behavior and opportunities of people in different age categories. We want to consider whether the rights and privileges that are associated with age roles represent another system of structured inequality.

Although the vast majority of people older than age 65 live independently and are in good health, after age 85 health concerns and limitations become a major factor in daily life. But it is not just the elderly who are interested in health and who are affected by changes in the health-care system. More than 16 percent of the U.S. population has no health insurance, and health-related debt is a major cause of personal bankruptcy (Newman 1999b; Sullivan, Warren, & Westbrook 2000). Changes in medical technology have altered the way we think about the viability and the quality of life at both ends of the age structure—for both the premature infant and the very old. Furthermore, health is the single most important factor that influences overall quality of life. Thus, we want to consider why the U.S. health-care system has taken the particular form it has and who pays the bill for our health care.

Age Differentiation and Inequality

Society is run mainly by adults between the ages of 30 and 65. These adults control jobs, industry, education, and wealth. In the following sections, we examine the extent to which young and elderly persons can be considered structurally disadvantaged relative to this middle-aged group.

The Status of Young People

In terms of class, status, and power, youth rank significantly behind adults. To a large extent, this inequality represents stratification: The lower status of youth is justified by institutionalized norms applicable to an entire category of individuals. Young people are systematically excluded from participation in many of our society's institutions by both legal restrictions and economic forces.

Legal Restrictions

In law, people younger than 18 are not responsible for their contracts (thus, they usually are not allowed to make any), and they are considered less responsible than adults for their bad deeds. Just as legal statutes once declared women and minority members incapable of self-government, the law still declares the young to be incapable. As you may be painfully aware from personal experience, young people have few legal rights. Their rights to drink, drive, work, own property, and marry are limited. In some cities, curfew laws require them to be off the streets at night. They have to pay adult prices but cannot see adult movies. In church, school, and industry, they must pass age barriers before they are allowed full participation. Since 1970, 18-year-olds can vote, but they no longer can drink legally. We now face the irony of staffing our armed forces with young people we will not trust with a beer. Although raising the drinking age springs from the praiseworthy objective of reducing deaths from drinking and driving, it is nevertheless a reminder that many older people believe young people cannot make responsible decisions and need to be protected from themselves.

For the most part, these age-related rules reflect society's belief that age is a reasonable measure of competency; moreover, it is one that can be administered easily and efficiently. Thus, instead of devising some test of maturity that one must pass before obtaining voting privileges, society uses a simple rule of thumb: Persons under 18 are not mature enough for this purpose, but persons 18 and over are.

Although there is growing sentiment that neither ethnicity nor sex is a good guide to competency, no such change of sentiment has occurred about the utility of age for making relevant judgments about youth. Unlike the case of women and minority groups, there is no movement to reduce inequality for teenagers. In fact, discrimination against the young may be the last bastion of approved inequality before the law. For example, although it is illegal to discriminate in housing based on gender, race, or religion, court decisions have consistently affirmed the right of "adult" communities to keep out younger people. These age rules receive very high consensus approval from society; little pressure exists to repeal them.

Economic Status

Owing to underdeveloped work skills, limited experience, desire for part-time or flexible schedules, and simple discrimination, a very large percentage of youth in the United States earn minimum wage. Even these earnings are under assault. First, the law provides for a subminimum training wage to persons under age 20. Second, many youths cannot find any jobs at all. In 2001 when 4.8 percent of the civilian labor force was unemployed, the unemployment figure was 11.0 percent for those aged 16 to 24 (U.S. Bureau of the Census 2002b). Among African American young people, 19.1 percent are unemployed. As a consequence, families headed by people younger than 25 are more than three times as likely to be living in poverty as is the average family.

Social Integration

Failure to integrate young people fully into major social structures is reflected in several indices of social disorganization. Although youth are underrepresented among voters and workers, they are overrepresented in accident and crime statistics. In 2001, people between the ages of 15 and 24 represented only 14 percent of the population but accounted for 41 percent of all arrests (U.S. Department of Justice 2002).

As we pass into young adulthood, most of us eventually become integrated into mainstream society. We are hired for jobs, become married, finish school, have children, and generally become plugged into society. For the minority who never become integrated into society's economic and social structures—individuals who disproportionately belong to racial and ethnic minorities—the lower status of youth becomes permanent.

Connections

Social Policy

Whenever a new house is built in the suburbs, the suburban community gains yearly property tax dollars from the new owners. But those new owners also cost communities because they require new community-provided services. One of the most expensive services to provide is new schools, which these days are expected to include not only classrooms but also swimming pools, computer labs, and the like. As a result, some suburbs are now trying unofficially to keep out families with children. Two common policies are to give financial incentives to builders who develop adult-only communities and to refuse permits for building apartments with more than two bedrooms.

American Diversity

focus on

The "Graying" and "Browning" of America

With each passing year, fewer American babies are born and more Americans turn 65. At the same time, the white population is shrinking while the Hispanic and nonwhite populations are growing. What are the combined consequences of these two trends—the "graying" of America and the "browning" of America?

One obvious result of having more older Americans and more nonwhite Americans is that in future there will be more Americans who are both older *and* nonwhite. This will have many important consequences, for the experience of old age is substantially different for nonwhite compared to white Americans (Takamura 2002). Most importantly, minority elderly are more likely than others to be poor: 26 percent of African American elderly live below the poverty line, compared to 21 percent of Latino elderly and only 8 percent of white non-Hispanic elderly. This has serious implications for health, health care, and social services, because poorer persons are both more likely to need services and less able to pay out of pocket for any services not covered by Medicare.

But the problems extend beyond those who live in poverty. Even when

incomes are equivalent, and after controlling for education, age, sex, marital status, and urban residence, minority elderly are still more likely than white elderly to lack health insurance. As a result, they find it much more difficult to get the medical treatment they need, and their health problems are much more likely to spiral out of control, making them more difficult (and expensive) to treat in the long run.

Finally, even if they are able to obtain health care, cultural barriers may make that health care less effective than it would otherwise be (Capitman 2002; Hayes-Bautista, Hsu, & Perez 2002; Takamura 2002). When health-care workers and health-care clients do not speak the same language, communication will necessarily be poor. Doctors and nurses may not understand what their patients need, and patients may not understand what their health-care

providers want them to do. In these circumstances, patients easily become dissatisfied, do not follow instructions, and do not return for follow-up care. In turn, care providers may come to regard patients as "noncompliant," and conclude that the patients are unintelligent or unmotivated. This is obviously a problem in the current situation, in which most doctors are white non-Hispanics but increasing numbers of patients are nonwhite. But it is also a problem for elderly white patients who now live in nursing homes. Although most doctors and nurses are white, day-to-day care in nursing homes is primarily left to nurse's aides, virtually all of whom are nonwhite and most of whom are immigrants. For all these reasons, policy makers will need to pay close attention to both the graying and the browning of America.

For more information, look up the following subjects in InfoTrac College Edition:
Minority elderly
Cultural competence
Cultural competency

Or visit the following Web sites:
American Medical Students Association (two web pages):
Cultural Competency in Medicine
http://www.amsa.org/programs/gpit/cultural.cfm
The Minority Elderly Population
http://www.amsa.org/adv/mac/aging.cfm

The Status of People 65 and over

Although they may have less power than they used to, elderly persons in our own society are not particularly disadvantaged—as long as they keep their health. For these relatively healthy older persons, their social status is more socially than physically determined.

Health, Life Expectancy, and Changing Definitions of Old

Perhaps the most important change in the social structure of the older population is that being older than 65 is now a common stage in life—and often a long one. The average person, on reaching age 65, can now expect to live 18 more years—19 more if she is a

■ As more Americans live into old age—and actively—you can expect to see a few more runners and water-skiers like these "old old" athletes.

white woman and 16 more if he is a white man; for African Americans, the figures are 17 and 14 years, respectively (U.S. Bureau of the Census 2002b). Because more and more people in the United States are living into their 80s and beyond, demographers now tend to divide the over-65 population into two categories: the "old old" (age 75 and above) and the relatively "young old" (ages 65 to 74) (Treas 1995).

For the majority of us, most of the "young old" and even many of the "old old" years will be healthy years. Certainly most people experience some loss of energy and physical stamina at age 65 or 70, and most older people report having at least one chronic health problem, such as arthritis, hypertension, or hearing loss (Federal Interagency Forum on Aging Related Statistics 2000). But a remarkably large proportion experience no major health limitations. Demographers estimate that three-fourths of the years of life remaining after age 70 will be spent in good enough health to permit independent living in the community (Crimmins, Hayward, & Saito 1994). Race and ethnicity, however, significantly affect one's chances of good health in old age: 74 percent of non-Hispanic whites aged 65 or older consider themselves healthy, compared to 65 percent of Hispanics and only 59 percent of African Americans (Federal Interagency Forum on Aging Related Statistics 2000).

Economic Status

The economic condition of older people has improved sharply over the last 35 years. In 1959, 35 percent of the population older than 65 had incomes below the poverty level; in 2000, this figure was only 10.2 percent (U.S. Bureau of Census 2002b). This improvement is directly related to increases in Social Security benefits, expanded coverage by private pension plans, and a more comprehensive system of benefits (including Medicare benefits) for the elderly. As a result, poverty rates are lower among those above age 65 than among those below that age.

Although only 10.2 percent of the population older than 65 live below the poverty level, most of the rest are not wealthy. The average older person experiences only a minor income loss in the years immediately following retirement, but there are three dark spots in this picture. First, more than one-quarter of the population experience an income drop of more than 50 percent after retirement. Those people who were living close to the economic margin during middle age—who accumulated few or no assets, who didn't pay off a home mortgage, and who contributed relatively little to either Social Security or a pension fund—find that retirement brings poverty or near poverty. These people are disproportionately women and minorities, people whose work-life earnings were low (Federal Interagency Forum on Aging Related Statistics 2000). Second, widowhood produces a severe economic blow for many older women, who find that a substantial portion of their pensions died along with their husbands (Holden, Burkhauser, & Feaster 1988). Third, a significant proportion of the "old old" outlive their physical and financial resources and find themselves poor, ill, or disabled. This group is disproportionately female and minority.

Discrimination

In the past, mandatory retirement regulations forced some people out of the labor force regardless of their physical ability, economic need, or desire to work. In 1986, however, federal legislation outlawed mandatory retirement except for a few jobs (such as police officer) for which age-related abilities are considered legitimate bases for discrimination.

A far more important problem than mandatory retirement is age discrimination. This discrimination begins during middle age, when women and men older than 40 are considered by some to be too old to learn new skills or take new jobs. Age-based discrimination is particularly hard on people who need a new job after age 40. One study, for example, found that it took workers older than 55 twice as long to find a new job after a plant closed as it did for workers younger than 45 (Love & Torrence 1989). The study also found that the new jobs of older workers were substantially worse than the new jobs of younger workers.

Honor and Esteem

Societal rewards distributed by stratification systems go beyond income and power. They also include prestige, esteem, and social honor. It is in this regard that the status of older people has been most seriously at issue. Our idealized version of the past holds that elderly people used to be highly regarded and respected. Careful examinations of cross-cultural and historical data, however, suggest that older people seldom have as much prestige as younger adults (Stearn 1976); historically, our attitude toward them has been ambivalent at best (Achenbaum 1985). Hags, crones, dirty old men, and other negative stereotypes of older people crowd our literature. Whether viewed as shrews and curmudgeons or as perfect grandparents, older people are seldom regarded with complete respect or admiration (Cool & McCabe 1983; Hummert et al., 1994, 1995; Levin 1988).

Stereotypes about older people—as inflexible or less competent, for example—form the basis of **ageism,** the belief that age determines individuals' abilities and is a legitimate basis for unequal treatment. In U.S. society, many circumstances work to reduce the esteem in which we hold older people. Age is associated with reductions in vigor and physical beauty, both of which are highly prized in U.S. society (Thorson 1995). Moreover, because of rapid expansion of the educational system within the last several decades, older people are substantially less well educated than younger adults. Only 69 percent of Americans older than 65 have completed high school, compared with 84 percent of all Americans older than age 25 (U.S. Bureau of the Census 2002b). In a society in which knowledge and education are taken as indications of intelligence and worth, older people are considered less worthy than others. Then, too, the life experience of older people is considered less valuable in a society that is changing rapidly, as U.S. society is today. Young people no longer need old people to teach them to plant fields or knit clothes, but old people often need young people to teach

Ageism is the belief that age determines individuals' abilities and is a legitimate basis for unequal treatment.

them how to run a computer or program a videotape recorder. Finally, because people older than 65 (or 70) are generally outside the mainstream of economic competition, they seldom receive the prestige of active earners (Thorson 1995). Together, these factors mean that, although individual older people may be politically active and have good educations, incomes, and health, they often find themselves being patronized and condescended to by younger individuals.

Social Integration

We have defined adulthood as a period of maximum integration into social institutions. This involvement is gradually reduced during old age. Active parenting is the first role to drop off, followed by the work role. Many, especially women, also experience the loss of the marital role through the death of their spouses. Nevertheless, they often remain active in the world of church, family and friends, and community, at least until the advent of old, old age, when health limitations may restrict activities.

Studies of older people demonstrate that one of the most important predictors of life satisfaction is relationships with close friends. These are usually age peers and often, at the end of the life course, brothers and sisters. Interestingly, close ties with their children are not a uniform blessing. Especially when age starts to become a serious handicap, ties with children become tinged with dependency and ambivalence (Thorson 1995). Because ties with age peers usually lack this element, they tend to be more gratifying.

Youth versus Age: The Battle over the Public Purse

Over the last few decades, the American population has grown progressively older; just between 1980 and 2000, the median age increased by 5.3 years (U.S. Bureau of the Census 2002b). In numerical terms, then, the relative power of young people has declined substantially. Even if numbers were the only criterion, however, we would still expect youth to retain advantages over age. But older people vote and children

■ Since 1970, poverty among the elderly has consistently declined, while poverty among children has increased. These children live in a rural Vermont town where unemployment is high due to the decline in industrial jobs.

do not. Although children have various representatives to act in their interests (parents, social workers, and so on), they are basically disfranchised. One result is that, increasingly, children lose out in the battle for scarce public resources.

According to a report produced by the U.S. Congressional Budget Office (2004), federal expenditures for mandatory spending on child-related programs (Aid to Families with Dependent Children, Head Start, food stamps, child nutrition, child health, and aid to education) were only one-tenth the size of federal expenditures on programs primarily for the elderly (Social Security, Medicare, and retirement benefits). This disparity is gradually increasing because cutting programs for the elderly, especially Social Security benefits, has become too politically risky. In contrast, child-related programs have been the targets of major cutbacks in federal spending for the last 30 years.

A consequence of these changes is that the condition of children in the United States has deteriorated, whereas the condition of the elderly has improved. Figure 11.1 compares changes in poverty rates between 1970 and 2000. These data show that poverty has increased among children and decreased among the elderly. Within 22 years, we have gone from a society in which elderly persons were much *more* likely than children to be poor to a society in which elderly persons are much *less* likely to be poor.

Why have children become less advantaged than the elderly? Several explanations have been suggested. First, although everybody has parents, fewer and fewer people have children living at home. Some childless people and some whose children have grown have little sympathy with the problems of providing for children. Second, the rise in unwed motherhood and divorce means that growing numbers of children receive support from only one parent (almost always the mother). Only about half of all parents pay the full amount of child support they owe, and about one-quarter make none of their required payments (U.S. Bureau of the Census 2002b). Finally, older persons are more likely to vote than other age groups. This makes them a potentially powerful group, although cleavages of race, class, and sex work against such unity. Nevertheless, older people have increasingly united—in the

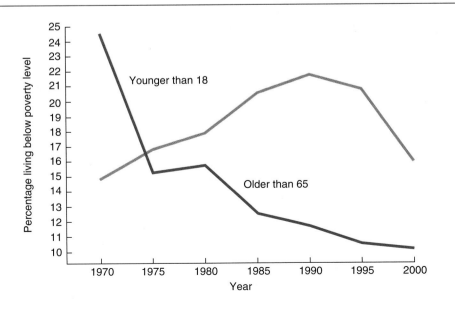

FIGURE 11.1
Changing Levels of Poverty among Youth and the Elderly, 1970–2000
Since 1970, poverty rates have increased steadily among Americans under the age of 18 and decreased among those older than 65.
SOURCE: U.S. Bureau of the Census 1982, 1993a, 2002b; www.census.gov/hhes/poverty/pov95/povest1.html.

so-called "gray lobby"—to protect existing Social Security and Medicare benefits and, to a more limited extent, to fight for new health benefits such as prescription drug coverage (Quadagno 2002).

Explanations for Age Stratification

Most people, most of the time, assume age stratification is caused by physiological age differences. The young and the old have less status because they are less competent and less productive. To some extent, this explanation is correct. It does not, however, explain cross-cultural or historical variations in the status of age groups. Conflict and structural-functional perspectives, as well as modernization theory, furnish insightful explanations about why age stratification exists and why it varies across societies. Symbolic interactionism helps us understand how ideas about youth and old age are socially constructed and reinforced.

Structural-Functional Perspective

The structural-functional perspective focuses on the ways in which age stratification helps fulfill societal functions. From this perspective, the restricted status of youth is beneficial because it frees young people from responsibility for their own support and gives them time to learn the complex skills necessary for operating in society.

At the other end of the age scale, the functionalist perspective is the basis for **disengagement theory.** The central argument of this theory is that older people voluntarily disengage themselves from active social participation, gradually dropping roles in production, family, church, and community even before disability connected with age requires it. This disengagement is functional for society because it allows younger people, with new ideas and skills appropriate to a fast-changing society, to take their places. Disengagement also makes possible an orderly tran-

Disengagement theory, a functionalist theory of aging, argues that older people voluntarily disengage themselves from active social participation.

Concept Summary

A Comparison of Three Explanations of Age Stratification

	Major Assumptions	Conclusions About Youth	Conclusions About Old Age	Overall Evaluation
Structural-Functional Theory	Age groups cooperate for common good	Young people's exclusion from full social participation is for good of self and society	Older people disengage voluntarily; good for self and society	Currently adequate to explain status of older people but not compelling to explain status of young people
Conflict Theory	Age groups compete for scarce resources	Young people are excluded so that others may benefit	Older people are excluded so that senior positions open up for younger adults	Useful to explain status of young people but not to explain today's older people
Modernization Theory	Changes in institutions alter the value of special resources that age groups hold (land, labor, and knowledge)	Unspecified; by implication, status goes down because labor not necessary	Status of older people has decreased because traditional bases of power have eroded	Useful to explain low social honor of older people and young people

sition from one generation to the next, avoiding the dislocation caused by people dropping dead in their tracks or dragging down the entire organization by decreased performance. It is functional for the individual because it reduces the shame of declining ability and provides a rest for the weary. In sum, according to functionalists, the lack of participation of older people is agreed on by older people and others and benefits all (Hendricks & Hendricks 1981).

Conflict Perspective

The conflict view of age stratification produces a picture of disengagement that is far less benign—a picture of rejection and discrimination springing from competition for scarce resources. These resources are primarily jobs but also include power within the family. Conflict theorists suggest that barring young and older people from the labor market is a means of categorically eliminating some groups from competition and improving the prospects of workers between the ages of 25 and 65.

Empirical evidence supports this view. Early in U.S. history, few people retired. Not only could they not afford to, but their labor was still needed. As immigration provided cheap and plentiful labor, the need for elderly workers decreased. In addition, as unions established seniority as a criterion for higher wages, older workers became more expensive than younger ones. In response to these trends, management instituted compulsory retirement to get rid of older workers. The mandatory retirement rules occurred long before Social Security, in an era when few employees had regular pension plans. Thus, compulsory retirement usually meant poverty for older people (Atchley 1982).

It was not until 1965 that social structures such as Social Security and private pensions began to make retirement a desirable personal alternative for some. At present, retirement often suits both the aging worker and the economic system. As supporting the growing elderly population—paying for Medicare, Social Security, and other benefits—becomes an increasing burden on a shrinking working-age population, however, conflict may once again emerge between the generations over economic interests.

In preindustrial societies such as Papua New Guinea, aged persons are revered because they are responsible for teaching traditional beliefs and important survival skills to the young.

Modernization Theory of Aging

According to the **modernization theory of aging**, older people have low status in modern societies because the value of their traditional resources has eroded. This is a result of three simultaneous events: the decline in importance of land (disproportionately owned by older people) as a means of production, the increasing productivity of society, and a more rapid rate of social change (Cowgill 1986). Next, we look at each of these in turn.

First, when land is the most vital means of production, those who own it have high status and power. In many traditional societies, land ownership is passed from father to son. This gives fathers a great deal of power even if they live to an age when they are physically much less able than their sons. This explanation, of course, applies only in a society where wealth resides in transferable property, either land or animals.

Second, in societies with low levels of productivity, it is not feasible to exclude either young or elderly persons from productive activity. Everyone's labor is needed. In industrial societies, however, productivity is so high that many people can be freed

The **modernization theory of aging** argues that older people have low status in modern societies because the value of their traditional resources has eroded.

from direct production. They can study, they can do research, they can write novels, or they can do nothing at all. In such a society, the labor of young and elderly persons becomes expendable: Society doesn't need it anymore.

Third, technological knowledge has grown at an ever-accelerating pace. Because most of us learn the bulk of our technological skills when we are young, this rapid change produces an increasing disadvantage for older workers. Their technical skills become outdated. Thus, rapid social change works to the disadvantage of older people (Thorson 1995).

Symbolic Interactionist Theory

As we have seen in other chapters, symbolic interactionism addresses a somewhat different question than either structural-functional, conflict, or in the case of aging, modernization theory. Whereas each of the other three major perspectives on aging focuses on questions related to age differences in status and access to resources, symbolic interactionist theory is more concerned with how young, middle-aged, and elderly persons feel about their lives and how they interact with others. It is also more concerned with the ways our ideas about aging and different age groups are socially constructed. With respect to our earlier discussion of ageism, symbolic interactionists are interested in the ways the media reflect and reinforce age-related stereotypes through their portrayals of older persons (and children) in magazines and on TV. Similarly, symbolic interactionists are interested in understanding how the language we use to describe older persons—geezers, fogeys, coots, old bags—also reinforce ageism. Finally, symbolic interactionists focus on the ways we constantly remind one another to "act our age." We may worry about the young girl who dresses "too provocatively" for her age or laugh at the middle-aged man or woman who dresses "like a teenager." The fact that media, language, and interaction must continuously reinforce age-related expectations suggests that the link between age, behavior, and personality is not as natural as it often seems (Laz 1998). Symbolic interactionists highlight how ageism—like racism, sexism, and other forms of prejudice—can limit an individual's life chances.

Evaluating the Theories

Although both young people and old people have lower status than adults in midlife, their experiences are rather different. Thus, theories that explain the status of young people may not be as effective in explaining the status of older people (see the Concept Summary).

Overall, it would appear that structural-functional, conflict, and modernization theories can contribute to understanding the status of the young. Certainly, as structural-functional theorists suggest, when people are protected from full responsibilities while very young, both the young people and society benefit. Nevertheless, the continued disadvantage of young adults and their subsequent poverty, lack of social integration, and deviance are hardly functional for them or society. In this case, it seems appropriate to attribute their low status to the systematic disadvantages they face in competing for scarce resources controlled by an older generation. Part of this disadvantage stems, as modernization theory suggests, from devaluation of their traditional resource: the capacity for low-skill, physically demanding work.

The status of older people is harder to account for. Modernization theory is not really applicable to their rising economic status, although it may explain their low social honor. Nor does conflict theory seem entirely adequate to explain the eco-

The quality of life one experiences in old age probably has less to do with age than with social class, sex, and race or ethnicity. Those who are most disadvantaged in old age are those who lived close to the poverty line during their earning years. Even among the relatively affluent, widowhood often means a sharp reduction in income.

© Karim Shamsi-Basha/The Image Works

nomic status of older people; rather, at this point in history, the disengagement of the older worker seems mutually attractive to younger and older people. But systems of inequality often overlap. Thus, the status of both young and old varies for men and women, for rich and poor, and for majority and minority group members.

Symbolic interaction theory attacks these questions from a different angle. Although it cannot on its own explain the status of young or old people, it does help us understand what it is like to be a youth, to be middle-aged, or to be an older person. It also helps explain how these statuses are reinforced over time, and suggests how we might change the social constructions of youth and old age through actions such as changing media portrayals.

Health and Health Care

Health and health care are basic concerns for most Americans. Over the last quarter century, Americans have increasingly tried to quit smoking, cut down on cholesterol, and increase their exercise, as medical scientists, the media, and our culture in general have proclaimed it a moral duty to take care of one's health (Weitz 2004b, 118 125). Of course, taking these precautions makes good sense because health status is the single most important factor influencing overall quality of life. For many Americans, however, worry over medical expenses is the companion to concern about living a healthy lifestyle. These worries are realistic; medical costs have become one of the leading causes of personal bankruptcy in the United States.

The sociological study of health, then, is central to understanding almost every facet of life; it is relevant to sociologists who study fertility and mortality and socialization and self-identity, to those concerned with the family, and to those interested in studying inequalities based in class, race, gender, and age. It is also of concern to those who study the changing nature of work. Nowhere are the themes of professionalization and an expanding service-based economy more visible than in the arena of health care. Nowhere is the impact of technology more vivid or the political and economic stakes higher. In this section, we want to consider health and the political economy of health care: How does structured inequality affect health; why has the U.S. health-care system taken the particular form it has; and who pays the bill for America's health care?

Good health is not simply a matter of taking care of yourself and having good genes. Although both elements play important parts in health, good health is related to social statuses such as gender, social class, and race or ethnicity. The study of how social statuses relate to the distribution of illness and mortality (i.e., death) in a population is called **social epidemiology**. In this section, we provide an overview of social epidemiology in the United States and then briefly examine how changes in social structure have influenced life expectancy in Russia and Eastern Europe.

Social epidemiology is the study of how social statuses relate to the distribution of illness and mortality.

Social Epidemiology

In the United States, the average newborn can look forward to 76.7 years of life (U.S. Bureau of the Census 2002b). Although some will die young, the average person in the United States now lives to be a senior citizen (Table 11.1). This is a remarkable achievement given that life expectancy was less than 50 years at the beginning of the twentieth century. Not everyone benefited equally, however: Men, African Americans, and poorer people on average die younger than women, whites, and more affluent individuals.

There is a great deal more to health, of course, than just avoiding death. The distribution of illness and disability is at least as important as the distribution of mor-

TABLE 11.1

Health and Life Expectancy by Sex, Race, and Family Income, United States, 2001
Generally, people of higher status report better health: Men report better health than women, white Americans report better health than black Americans, and those with higher incomes report better health than those with lower incomes. For the most part, differentials in life expectancy parallel these differentials in health. The exception is sex: Despite their better health, men have lower life expectancy than women.

	Life Expectancy at Birth	Percentage Reporting Fair or Poor Health
Total	77.2	9
Sex		
Male	74.4	9
Female	79.8	10
Race		
White	77.2	8
Black	72.2	15
Family Income		
Poor	NA	21
Near Poor	NA	16
Not Poor	NA	6

SOURCE: U.S. Department of Health and Human Services 2003.

tality in evaluating a population's overall well-being. Although only 1 out of every 100 people dies each year in the United States, well over half of these 100 people experience some sort of serious illness that affects the quality of their lives and their ability to hold jobs or maintain social relationships. In the following sections, we consider why and how gender, social class, and race/ethnicity are related to illness and mortality.

Gender

On average, U.S. women live more than 5 years longer than U.S. men (Table 11.1). Yet although women live longer, they also report significantly worse health than men at all ages: more high blood pressure, arthritis, asthma, diabetes, cataracts, corns, and hemorrhoids (Lane & Cibula 2000; Waldron 1994). These differences mean that women more often than men find their last years plagued by disabilities and discomfort. Because of this combination of longer lives and more illnesses, women are considerably more likely than men to eventually enter a nursing home.

Social Class

The higher one's income, and thus one's social class, the longer one's life expectancy and the better one's health (Table 11.1). The effects of social class are complex (Robert & House 2000). They are partially attributable to poorer people's inability to afford expensive medical care, but environmental, economic, and psychosocial factors play a considerable role. Lower-income people, for example, are more likely to live in unhealthy conditions, near air-polluting factories, or in substandard housing. Low-income people are also less likely to have control over the world around them than those who are better off. As a result, they are likely to experience higher levels of stress and to cope with that stress through self-destructive drinking, smoking, and other risky behaviors.

Race and Ethnicity

Because a higher proportion of African Americans, Hispanics, and Native Americans than non-Hispanic whites are poor, the impacts of low socioeconomic status on health disproportionately affect these minorities. Because of lower incomes, non-whites are significantly more likely than whites to lack health insurance (Mills 2001). Because they are poor, racial and ethnic minorities are also 47 percent more likely than others to live near a hazardous waste facility, which may emit toxins into the surrounding ground, air, or water (Ember 1994).

But even after we control for income, racial and ethnic minorities face obstacles to good health simply because they are minorities. A language barrier, for instance, often separates Hispanic patients from health-care professionals (Vega & Amero 1994). Furthermore, regardless of income, minority group members experience prejudice and discrimination that raise their risk of physical and psychological distress (Ulbrich, Warheit, & Zimmerman 1989). In addition, as a result of continued segregation, even middle-class African Americans are more likely than whites to live in neighborhoods where violence and pollution pose constant threats to health.

A Case Study: Declining Life Expectancy in Eastern Europe

The single most important social factor affecting mortality is the standard of living—access to good nutrition, safe drinking water, adequate housing free from exposure to environmental hazards, and decent medical care. Differences in living standards help to explain why African American infants in the United States are more than twice as likely as white infants to die in their first year of life and why the average life expectancy of African American men is 7 years less than that of the average white non-Hispanic male. Differences in living standards also help to explain why, on average, Americans can expect to live 42 years longer than citizens of Sierra Leone. Throughout the world, improvements in living standards have been accompanied by increased life expectancy. Consequently, the shocking decline in life expectancy in Eastern Europe is one of the most surprising current developments in world health. Although there is some question about the reliability of Soviet health statistics, it is quite clear that Russian men, for example, now can expect to live almost 8 years less than their Soviet counterparts did in 1960, and Russian women can expect to live one-and-a-half years less (Cockerham 1997). Why? The answer to this question probably rests on four basic aspects of social structure.

First, as was discussed in Chapter 8, Eastern Europe is plagued by extensive environmental pollution. Pollution is strongly implicated in the onset of life-threatening illnesses, such as cancer and respiratory diseases. The decline in life expectancy appears to be greatest among individuals with the least education living in the most developed, industrialized regions (Cockerham 1997). According to some estimates, pollution-related diseases were responsible for about 12 percent of the increase in deaths in Russia (Haub 1994).

Second, some analysts have blamed the decrease in life expectancy in republics of the former Soviet Union on problems within the Soviet health-care system. Funded only by the money left over after defense and industrial needs were addressed and relying heavily on physician assistants and poorly equipped hospitals, the Soviet health-care system was designed to prevent the spread of infectious disease more than it was to deal with the treatment of chronic health problems. As long as the major causes of death were related to epidemics of contagious disease, the Soviet health-care system made substantial progress in improving life expectancy. Once these problems were controlled, however, the Soviet system proved ineffective in dealing with the chronic illnesses, such as heart disease, that now constitute the major causes of death in industrial societies. Most important, even though Soviet-style medicine guaranteed access to health care, it did not end differences in the *quality* of care provided to political elites and to the less-privileged classes (Cockerham 1997).

Third, some medical researchers have suggested that increased stress is the single most important variable in the health of large populations. Thus, as affluence in Japan increased and its standing in the world community moved up, Japanese life expectancy increased; likewise, as the status and standard of living in Eastern Europe moved down, the stress caused by these changes produced a decrease in life expectancy (Hertzman, Frank, & Evans 1994).

Finally, unhealthy lifestyles certainly contributed to increasing mortality in Eastern Europe (Cockerham 1997). Eastern Europeans are among the world's heaviest drinkers, smokers, and consumers of fat in the diet. These behaviors reflect social conditions, not just individual choices. In Eastern Europe, where fresh fruits and vegetables are often absent from store shelves and heavy alcohol consumption is encouraged by workplace norms, external constraints place sharp limits on the range of health behaviors from which one might choose. There as here, then, unequal access to education, employment, safe environments, and other social rewards have life-and-death consequences.

The U.S. Health-Care System

Medicine is a social institution. It has a complex and enduring status network, and the relationships among actors are guided by shared norms and roles. Most of us occupy the status of patient in this institution. However, there are dozens of other statuses. Approximately 10 million people in the United States are employed in health-related institutions. They include phlebotomists and X-ray technicians, aides, pharmacists, and hospital administrators. We will focus on just two of these statuses: physicians and nurses.

Physicians

Less than 5 percent of the medical workforce consists of physicians. Yet they are central to understanding the medical institution. Physicians are responsible both for defining ill health and for treating it. They define what is appropriate for those with the status of patient, and they play a crucial role in setting hospital standards and in directing the behavior of the nurses, technicians, and auxiliary personnel who provide direct care.

As will be described in Chapter 14, a profession is a special kind of occupation that demands specialized skills and permits creative freedom. No occupation better fits this definition than that of physician. Until about 100 years ago, however, almost anyone could claim the title of physician; training and procedures were highly variable and mostly bad (Starr 1982). About the only professional characteristic early practitioners shared was that their product was highly unstandard! With the establishment of the American Medical Association in 1848, however, the process of professionalization began; the process was virtually complete by 1910, at which point strict medical training and licensing standards were adopted.

Learning the Physician Role

Most people who enter the medical profession have high ideals about helping people. Studies of the medical school experience, however, suggest that the strenuous training schedule of student physicians deals a temporary blow to these ideals. On crowded days, patients come to be defined as enemies who create unnecessary work. Thus, interns and residents learn to "GROP" (get rid of patients) by referring patients elsewhere, treating them hastily, and discharging them as soon as possible (Mizrahi 1986). For a patient to actually get a busy intern's attention, the patient usually must have a really interesting, curable disease; be intelligent, cooperative, and appreciative; and not have lost the intern's sympathy by causing his or her illness through self-destructive behaviors.

Medical education is a grueling experience. There is a huge amount of technical information to learn, and the biggest rewards in medicine go to those who have the most technical and least personalized types of skills and practices. Once they survive the rigors of medical school, many physicians renew their commitment to helping people, but nowhere in the medical profession does the structure of rewards encourage personal care (Weitz 2004b).

Understanding Physicians' Income and Prestige

The medical profession provides an example of stratification theories. Why are physicians predominantly male and nurses predominantly female? Why are physicians among the highest-paid and highest-status professionals in the United States? In 1998

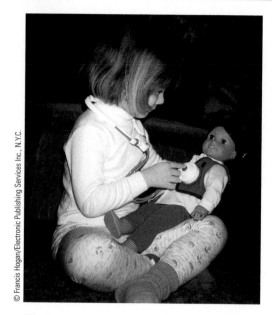

© Francis Hogan/Electronic Publishing Services Inc., N.Y.C.

■ The popularity of doctor and nurse play sets and the large number of children who aspire to medical professions demonstrate the continuing prestige of doctors in contemporary society.

(the latest data available as of 2003), the median income for all physicians was $160,000, after paying for deductible professional expenses. General and family practitioners earned a median net salary of $130,000 and general surgeons earned an average of $240,000 (U.S. Bureau of Labor Statistics 2003).

According to structural-functionalists, there is a short supply of persons who have the talent and ability to become physicians and an even shorter supply of those who can be surgeons. Moreover, physicians must undergo long and arduous periods of training. Consequently, high rewards must be offered to motivate the few who can do this work to devote themselves to it. The conflict perspective, on the other hand, argues that the high income and prestige accorded physicians have more to do with physicians' use of power to promote their self-interest than with what is best for society.

Central to the debate on whether physicians' privileges are deserved or are the result of calculated pursuit of self-interest is the role of the American Medical Association (AMA). The AMA sets the standards for admitting physicians to practice, punishes physicians who violate the standards, and lobbies to protect physicians' interests in policy decisions. Although less than half of all physicians belong to the AMA, it has enormous power. One of its major objectives is to ensure the continuance of the free market model of medical care, in which the physician remains an independent provider of medical care on a fee-for-service basis. In pursuit of this objective, the AMA has consistently opposed all legislation designed to create national health insurance, including Medicare and Medicaid. It has also tried to ban or control a variety of alternative medical practices such as midwifery, osteopathy, and acupuncture (Weitz 2004b).

The Changing Status of Physicians

Up to 30 or 40 years ago, the physician was an independent provider who had substantial freedom to determine his or her conditions of work and who was regarded as a nearly godlike source of knowledge and help by patients. Much of this is changing. The many signs of changes include the following (Coburn & Willis 2000; Weitz 2004b):

- A growing proportion of physicians work in incorporated group practices, where fees, procedures, and working hours are determined by others. As a result, physicians have lost a significant amount of their independence. These bureaucratized structures are also more likely to have profit rather than service as a dominant goal.
- The public has grown increasingly critical of physicians. Getting a second opinion is now general practice, and malpractice suits are about as common as unquestioning admiration. Patients are critical consumers of health care rather than passive recipients.
- Fees and treatments are increasingly regulated by government agencies and managed care organizations (insurance companies) that are concerned about keeping costs down. The vast number of patients whose bills are paid by insurance companies or government agencies gives these groups some control over what treatments will be funded and at what fees.

Being a physician is still a very good job, offering high income and high prestige. But it is also part of an increasingly regulated industry that is receiving more critical scrutiny than ever before.

Nurses

Of the nearly 10 million people employed in health care, the largest category includes the 1.8 million who are registered nurses. Despite their great importance to
the health-care system, their status remains relatively low. Why is this, and why have
attempts to improve nurses' status achieved only modest success?

Nurses' Current Status

Nurses play a critical role in health care, but they have relatively little independence. Although the nurse usually has much more contact with the patient than
does the physician, the nurse has only limited authority over patient care. Nurses
are subordinate to physicians both in their day-to-day work and in their training. Physicians determine the training standards that nurses must meet, and they
enforce these standards through licensing boards. On the job, physicians give
instructions and supervise. Because the majority of physicians are male and the
majority of nurses are female, the income and power differences between doctors
and nurses parallel the gender differences in other institutions (Table 11.2). This
makes the hospital a major arena in the battle for gender equality. In part, women
have fought the battle by joining physicians rather than beating them. Many
women who previously would have become nurses are now raising their aspirations: These days, almost half of first-year medical students are female (Barzansky
& Etzel 2002).

Changing the Status of Nurses

To improve the status and position of nurses, nursing's leadership has worked for
decades to raise educational levels (Weitz 2004b). Until the 1960s, the standard nursing credential was an RN (registered nurse) diploma, obtained through a hospital-
based training program. Now, almost all RNs hold 2- or 4-year nursing degrees from
community colleges or universities. In addition, a small percentage of nurses obtain
graduate degrees and become "advanced practice nurses," such as nurse practitioners
or nurse-midwives. Advanced practice nurses enjoy considerably more autonomy, status, and financial rewards than do other nurses, including the right (in most states) to
prescribe certain medications.

The drive to increase nurses' education and thus their status has succeeded only
partially. Because many hospitals believe that associate-degree nurses receive the
best practical training and make the best employees, associate-degree programs have

TABLE 11.2
Physicians and Registered Nurses: Income, Sex, and Race, 2000
Nurses earn less than a quarter of what physicians earn. Critics wonder whether this reflects
real differences in training and responsibility or whether it is another instance of traditional
women's jobs being evaluated as less worthy than traditional men's jobs.

	Physicians	Registered Nurses
Median net income	$160,000*	$44,840
Percentage female	29.3%	93.1%
Percentage African American	5.6%	9.9%
*1998 data		

SOURCE: U.S. Bureau of the Census 2002b; U.S. Bureau of Labor Statistics 2002.

■ One of the most striking changes in U.S. medicine is the growth in the number of women physicians.

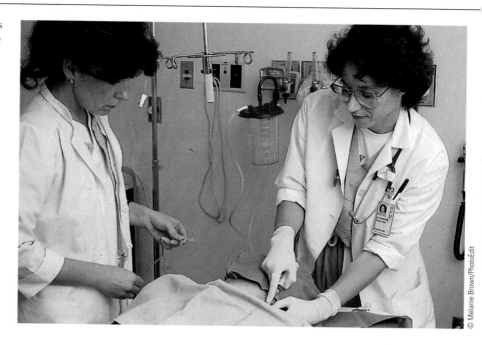

remained more popular than higher-level training. Meanwhile, to control costs, hospitals have shifted many services to outpatient clinics where fewer RNs are needed, salaries for nursing are lower, and nursing jobs are less interesting and prestigious (Brannon 1996; Norrish & Rundall 2001). At the same time, now that women have wider occupational opportunities, fewer choose a job, such as nursing, that requires weekend and midnight shift work, makes intense emotional demands, includes relatively little independence, and has a short career ladder. As a result, the nursing profession is having difficulty attracting members of the quality that it used to. Finally, although more men now work as nurses, the field is still considered a "woman's profession," and for that reason, salaries and status remain relatively low.

The High Cost of Medical Care: Who Pays?

Medical care is the fastest-rising part of the cost of living. In 1970, Americans spent an average of $340 per person on doctors' services and hospital treatments. By 2000, they spent an average of $2,278 (U.S. Bureau of the Census 2002b).

Underlying many of the analyses of health care is one question: "Who pays?" There are three primary modes of financing health care in the United States: paying out of pocket, private insurance, and government programs. The cost of health care is so high that only the very rich can afford to pay out of pocket. Most Americans must rely on private insurance or government programs. The remainder have no insurance, and often are unable to pay for health care.

Private Insurance

Approximately 70 percent of people in the United States have health insurance coverage from a private plan. Almost all of these obtain their insurance through their employer. Such insurance tends to be limited to employed adults (and their families) who work for large corporations or government agencies. Many jobs in small businesses and most minimum-wage jobs do not provide insurance benefits.

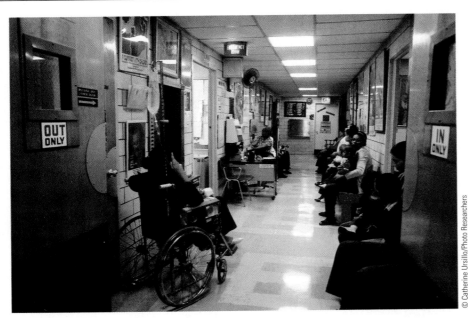

People who have no health insurance are particularly likely to receive their medical care in emergency-room settings where waits are long, care is costly, and follow-up treatment is rare.

© Catherine Ursillo/Photo Researchers

Government Programs

The government has several programs that support medical care. The federal government provides some health care through its Veterans Administration hospitals, but its two largest programs are Medicaid and Medicare. In addition, local governments provide medical care through public health agencies and public hospitals.

Medicare is a government-sponsored health insurance policy for citizens over 65; premiums are deducted from Social Security checks. The enactment of this program in 1965 did a great deal to improve the quality of health care for the elderly. More than 95 percent of the elderly are now covered by health insurance. This is not a cheap program, however; in 2000, the government paid more than $224 billion in Medicare benefits (U.S. Bureau of the Census 2002b).

Medicaid is a federal cost-sharing program that provides federal matching funds to states that provide medical services to the very poor. The eligibility of individuals, however, and the services available are determined by states. As a result, some states offer much more generous medical care than others. States are only required to offer Medicaid to individuals whose family incomes, as of 2001, are less than $19,019 (for a family of three) and who additionally are either under age 6 or pregnant.

The Uninsured

A significant portion of the U.S. population—14 percent—has no medical coverage (Table 11.3). Thanks to Medicare, nearly 100 percent of the elderly are insured; those who fall through the cracks are children, the unemployed, the working poor, and those who work for small businesses. Uninsured persons are more likely than others to postpone needed medical care. As a result, they are more likely to require hospitalization for otherwise preventable health problems (Weitz 2004b). Every county in the United States does provide public hospital-based emergency care to the so-called "medically indigent." Once they seek care, however, uninsured individuals are often kept waiting for several hours by overworked

Connections

Example

How does lack of health insurance lead to illness or even death? Consider what happens to people with diabetes. Some diabetics can control their disease with exercise and diet, but the rest must rely on injections of insulin, which their body has problems producing. Those who lack health insurance may have trouble affording the regular medical check-ups needed to know how much insulin to take. They also might not be able to afford the insulin injections. As a result, their diabetes can spiral out of control, leading to strokes, gangrene and foot amputations, blindness, comas, kidney failure, and even death—all avoidable with affordable health care.

TABLE 11.3

Americans' Health Insurance Coverage, 2000

Approximately 14 percent of the U.S. population has no health insurance. These people, generally the unemployed and the working poor, are concentrated among those who are younger than age 65 and have family incomes of less than $20,000.

	Persons without Health Insurance	
	Percent	No. of Persons (in millions)
Sex		
Male	14.9	20.2
Female	13.1	18.5
Age		
Younger than 18	11.6	8.5
18–24	27.3	7.3
25–34	21.2	7.9
35–44	15.5	6.9
45–64	12.6	7.9
65 or older	0.2	0.7
Race and Hispanic origin		
White	12.9	29.3
African American	18.5	6.6
Hispanic origin*	32.0	10.8
Asian	21	2
Household income		
<$25,000	22.7	13.9
$25,000–49,999	18.8	12.8
$50,000–74,999	11.0	6.5
$75,000 or more	6.9	5.6

*Persons of Hispanic origin may be of any race.

SOURCE: Mills 2001.

and underpaid hospital staff before seeing a doctor. Once seen, they are more likely than others to receive substandard care. As a result of all these factors, uninsured persons are more likely than others with similar conditions and backgrounds to die once admitted to a hospital (Weitz 2004b).

Why Doesn't the United States Have National Health Insurance?

Health care in the United States is available on a fee-for-service basis. Like dry cleaning, you get what you can afford. If you cannot afford it, you might not get any. The United States is the only industrialized nation that does not guarantee health care to all of its citizens. In the rest of the industrialized world, medical care is like education— regarded as a good that should be available to all regardless of ability to pay.

Why is the United States alone among industrialized nations in having no national health insurance? Certainly, the AMA has strongly opposed national health

MAP 11.1
Percent of Persons without Health-Care Coverage, 2000
SOURCE: Mills 2001.

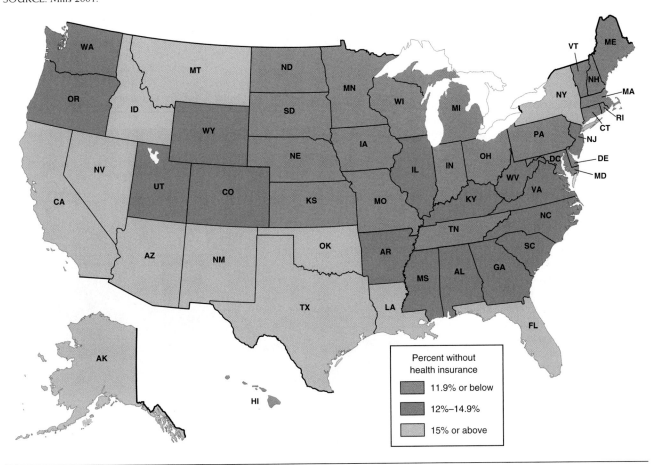

Percent without
health insurance

11.9% or below

12%–14.9%

15% or above

insurance, but that cannot be the whole reason. It opposed Medicaid and Medicare too, and those programs have existed since 1965. Nor is public opposition the reason. Polls consistently find that about two-thirds of the U.S. public believes it is the federal government's responsibility to make sure that all Americans have health care coverage (Gallup Poll 2002).

Despite this growing public support, the forces that led to the rise of national health insurance in other countries have not (yet) come together in the United States (Rothman 1993). In other countries, conservative political parties supported national health insurance as a way of reducing popular ferment that might otherwise have led to support for socialist parties. Here, socialist parties never really took root. In other countries, labor unions fought for national health insurance. Here, unions were able to gain members by offering health insurance as a benefit of membership, and so did not press for national health insurance. In other countries, the citizenry defined the responsibilities of government broadly, and so supported national health plans. Here, citizens traditionally have distrusted "big government." In other countries, all social classes believed they would benefit from national health insurance. Here, the middle and upper classes by and large have access to health insurance and so have little

focus on Technology

Technology and the Meaning of Life

In 1997, scientists announced a very important new arrival—Dolly. The lamb named Dolly was the first animal cloned from an adult mammal. Dolly is just one example of how technological change can push society to rethink its norms and values. Through advances in medical technology, babies born at 5 to 6 months' gestation and weighing less than a pound are now surviving. Developments in human genetics make it possible to determine in advance whether adults and fetuses carry genes that predispose them to cancer, Alzheimer's, deafness, and other conditions. Medical interventions can prolong the lives of the terminally ill, and medical technologies can sustain the human body even after the heart and brain stop functioning.

Developments in medical technology pose challenges that no society or culture is prepared to address. They raise questions, for instance, about what constitutes a good life and about how much we ought to be willing to pay for it.

Computerized axial tomography (CAT) and magnetic resonance imaging (MRI) are techniques that have enormous potential for diagnosing tumors, cancer, heart murmurs, and other conditions. Yet, the minimum installation fee for equipment to perform either

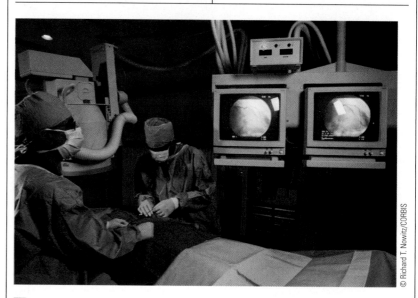

© Richard T. Nowitz/CORBIS

Open-heart surgery saves lives—but at very high costs per patient. Should we spend our health-care dollars on intensive care for the few or on more basic care for the many?

© Leong Ka Tai/Material World

China's medical practices have effectively reduced the death rate at a relatively low cost.

interest in paying taxes to support health care for the poor. If Americans get national health insurance in the future, it will be because the middle class, labor unions, and corporations all find it increasingly difficult to pay their health care bills.

An Alternative Model

Modern medical technology has enhanced our ability to extend and save lives. It is, however, extraordinarily expensive. In China, a largely rural society with a population in excess of 1 billion, Western-style medicine has taken a back seat to prevention. The focus, instead, has been on community organization for health-care delivery; improved sanitation, housing, and food; and traditional healing practices that Western physicians are only now coming to appreciate. China's medical practices have effectively reduced the death rate at a relatively low cost. Throughout the less developed nations, in countries such as Costa Rica, Sri Lanka, Cuba, and Vietnam, the Chinese model of health care has proven

one is $1 million. Should all hospitals have this equipment? If not, will the cardiac patient in a small rural hospital or the indigent patient in a central city emergency room receive the same quality of service that the more affluent patient receives at a major teaching hospital?

One billion dollars will buy 500 people a liver transplant or fund the Arkansas public school system for 1 year. Can we have it all and, if not, which should we choose? Could the millions of dollars spent on saving the lives of extremely premature infants be better spent on vaccinations and proper nutrition for America's poor children? The drugs used to prolong the lives of AIDS patients can cost $18,000 annually for one person. Who should pay the cost?

In the wake of medical advances, societies find themselves struggling not only with questions about medical costs and what constitutes quality of life; they also find themselves struggling with questions regarding the very meaning and nature of life. If human beings can create life in a laboratory and can alter the very forms that life will take through genetic interventions, where does God fit in? Who will have the power and authority to decide who or what gets cloned? Is it ethical to clone humans solely to "harvest" their organs for transplantation? For that matter, is it ethical to purchase organs from living donors?

As noted in Chapter 2, sociologist William T. Ogburn (1922) pointed out the period of cultural lag between the arrival of a new technology and a society's adaptation to it. Structural and cultural adaptations to medical technology are only beginning. In Oregon, a state board prioritizes all medical services each year and decides which will be paid for under Medicaid. Federal officials continue to struggle with questions regarding who should have access to individuals' genetic information, the appropriateness of DNA sampling of people released from prison, and how to protect people from discrimination when they are found to have a genetic defect (Martindale 2001). What other structural and cultural adaptations will have to be made? Who do you think should be involved in making these decisions? Who do you think will be?

For more information, look up the following subjects in InfoTrac College Edition:
Human cloning
Genetic screening

Or visit the following Web sites:
Medical Ethics: Tough Choices Project
http://net.unl.edu/newsFeat/med_eth/me_index.html
BMC Medical Ethics
http://www.biomedcentral.com/bmcmedethics

more effective in reducing mortality than building high-tech hospitals. Critics of the U.S. system have likewise argued that U.S. tax dollars would be more wisely spent on nutrition and preventive health-care programs than on expensive new biomedical technologies.

Where This Leaves Us

Sociological analysis suggests that health and illness are socially structured. To paraphrase C. Wright Mills again, when one person dies too young from stress or bad habits or inadequate health care, that is a personal trouble, and for its remedy we properly look to the character of the individual. When whole classes, races, or sexes consistently suffer significant disadvantage in health and health care, then this is a social problem. The correct statement of the problem and the search for solutions require us to look beyond individuals to consider how social structures and institutions have fostered these patterns. The sociological imagination suggests that significant improvements in the nation's health will require changes in social institutions—increasing education, reducing poverty and discrimination, improving access to good quality housing and food, and so on. Equalizing access to health care will also help, but is considerably less important than making these social changes.

Summary

1. Young people suffer many structured inequalities and are not well integrated into society's institutions. Among the consequences of this are a high crime rate and a high level of poverty among families headed by young adults.

2. The population over 65 is perhaps better off now than it ever has been. Although the older population may suffer from low esteem, there have been sharp improvements in economic conditions. The disadvantages of aging are most pronounced for women, for the "old old," and for those who were living on the economic margin during their working years.

3. As a group, children are poorer than older people. This is due to three factors: declining public willingness to support young people out of the public purse, the increasing proportion of children whose fathers do not help support them, and greater political activism by older people.

4. Structural-functional theory (disengagement theory) is most useful for understanding the improved status of older people, while conflict theory is most useful for understanding the disadvantages of young people. Modernization theory is most useful for understanding why youth and old people have low social prestige in U.S. society, and symbolic interactionism helps us understand how age norms and age inequalities are reproduced through media, language, and in social interaction.

5. Gender, social class, and race/ethnicity all help explain the social epidemiology (i.e., distribution) of health in the United States. Men, racial and ethnic minorities, and those with lower socioeconomic status have higher mortality rates.

6. The health disadvantage associated with lower socio-economic status goes far beyond a simple inability to afford health care. Lower social class is associated with lower standards of living, more stress, lower education, polluted environments, and poorer coping strategies, all of which increase the likelihood that individuals will have poor health.

7. Physicians are professionals; they have a high degree of control not only over their own work but also over all others in the medical institution. Structural function-alists argue that physicians earn so much because of scarce talents and abilities, but conflict theorists argue that high salaries are due to an effective union (the AMA). Physicians have less independence than they used to have.

8. Nurses comprise the largest single occupational group in the health-care industry. Nurses earn much less than physicians, have less prestige, take orders instead of give them, and are predominantly female.

9. Most people in the United States have private health insurance coverage. Medicare insures almost all senior citizens, but 14 percent of Americans are uninsured. Uninsured persons are more likely than others to postpone getting needed health care, to become ill, and to die if they become ill.

10. The United States is the only industrialized nation that does not make medical care available regardless of the patient's ability to pay. We do not have national health insurance because of lack of pressure from a socialist party, labor unions, or the middle and upper classes, and because of Americans' fear of "big government."

Thinking Critically

1. Should U.S. society treat all people, regardless of age, as individuals? Do age requirements for voting, driving, and so on assume, for example, that all 18-year-olds are more responsible and better informed than any 17-year-old? Isn't this stereotyping? Is there any justification for it?

2. Society has greater tolerance for the inequalities of the young than for those of the old. We more often think that the disadvantages of old age are "unfair." Why do you suppose society is so tolerant of the disadvantages of young people?

3. Achievement, efficiency, progress, and individual freedom are basic values in the dominant U.S. culture. Based on what you know about modernization theory, how might these basic values affect the prestige of older people in the United States?

4. Nurses earn less than physicians, but they earn more than many other jobs with equal training requirements. Why do so few males enter nursing? What could change this gender gap?

Sociology on the Net

 The Wadsworth Sociology Resource Center: Virtual Society

http://sociology.wadsworth.com/

The companion Web site for this book includes a range of enrichment material. Further your understanding of the chapter by accessing Online Practice Quizzes, Internet Exercises, InfoTrac College Edition Exercises, and many more compelling learning tools.

Chapter-Related Suggested Web Sites

National Institutes of Health
http://www.nih.gov

National Institute of Aging
http://www.nih.gov/nia/

Section on Aging and the Life Course of the American Sociological Association
http://www.asanet.org/sectionaging/

American Association of Retired Persons
http://www.aarp.org

 InfoTrac College Edition

http://www.infotrac-college.com/wadsworth

Access the latest news and research articles online—updated daily and spanning four years. InfoTrac College Edition is an easy-to-use online database of reliable, full-length articles from hundreds of top academic journals and popular sources. Conduct an electronic search using the following key search terms:

Health care reform

Age discrimination

Life expectancy

Aging

 For more information related to this chapter, visit the Opposing Viewpoints Resource Center. Be sure you check all the options on the toolbar—viewpoints, references, statistics, and so on—and search under the following subjects:

Medically uninsured persons

Elderly

Suggested Readings

Callahan, Daniel. 1998. *False Hopes: Why America's Quest for Perfect Health Is a Recipe for Failure.* NY: Simon and Schuster. One of America's foremost ethicists argues that we need to emphasize providing primary care to all rather than high-technology care to a few.

Friedan, Betty. 1993. *The Fountain of Age.* New York: Simon and Schuster. Most well known for her book, *The Feminine Mystique,* published in 1963, Friedan now has tackled the issue of aging in our society, particularly as aging affects women in the United States.

Leach, Penelope. 1994. *Children First.* New York: Knopf. A policy book by an author who has become a television personality. The book argues that the industrialized world is forgetting its children and that we need to put children first in our family and national political decisions.

Weitz, Rose. 2004. *The Sociology of Health, Illness, and Health Care: A Critical Approach.* 3rd edition. Belmont, CA: Wadsworth. An excellent overview of how social factors affect who becomes ill or disabled, the experiences of illness and disability, and the health-care world.

Cross-Cutting Statuses: Race, Sex, and Age

Intersections

In the past three chapters on stratification, we have dealt with unequal life chance by race and ethnicity, sex, and age. For each of these characteristics, we have been able to demonstrate that there is a hierarchy of access to the good things in life and that some groups are substantially disadvantaged. We introduced the concept of double jeopardy in Chapter 9 to show how disadvantage snowballs. African American women earn less than African American men and white women, African American teenagers are twice as likely as white teenagers to be unemployed, and old women are more likely than old men to be poor. In this section we briefly review the special problems of three cases of cross-cutting statuses.

Race and Gender

Nonwhite women face a two-pronged dilemma. First, they have not benefited from the sheltered position of traditional women's roles. Nonwhite women have always worked outside the home: For example, married African American women were six times more likely to be employed at the turn of the century than married white women (Goldin 1992). Although they worked, they still had to face the economic and civic penalties of being women. Consequently, minority women traditionally have had less to lose and more to gain from abandoning conventional gender roles. On the other hand, nonwhite women face a potential conflict of interest: Is racism or sexism their chief oppressor? Should they work for an end to racism or an end to sexism? If they choose to work for women's rights, they may be seen as working against men of their own racial and ethnic group.

Current income figures indicate that sex is more important than race in determining women's earnings: The difference in earnings among Hispanic, African American, and European American women is relatively small compared to the difference between women and men. This suggests that fighting sex discrimination should be more important than fighting racial discrimination. But this conclusion overlooks the dependence of women and children on the earnings of their husbands and fathers. For example, because of the low earnings and limited employment opportunities of African American men, African American women and children are three times more likely than their white counterparts to live below the poverty level. As a result, nonwhite women have much to gain by fighting racism as well as sexism.

The dilemma remains. The women's rights movement is often seen as a middle-class white social movement; racial and ethnic movements have been seen as men's movements. Nevertheless, minority women have a long history of resistance to both forms of discrimination, racism and sexism.

Aging and Gender

Aging poses special problems for women. First is the problem of the double standard of aging: The signs of age—wrinkles, loose skin, and gray hair—are considered more damaging for women than for men. Thus, age is associated with greater decreases in prestige and esteem for women than for men.

The life expectancy gap between men and women also makes the experience of old age very different for women than for men. On average, women live about 5 years longer than men. Taken together with the fact that women are usually 2 years or so younger than their husbands, this works out to a 7-year gap between when a woman's husband dies and when she dies. This mortality difference has enormous consequences for the quality of life. First, it means that most men will spend their old age married, have the care of a spouse during illness, and be able to spend their last years at home being cared for by a spouse. The average woman, on the other hand, will spend the last years of her life unmarried, will spend most of these years living alone with no one to care for her, and is more likely to be cared for in a nursing home during her last illness.

Sex differences in mortality mean that old age is disproportionately a society of women. Above age 75, there are nearly twice as many women as men. The decreasing availability of men in older age groups has consequences for social roles. It means that, as women get older, heterosexual contacts, whether through marriage or in bridge groups, are less important for structuring social life. It also means that fewer elderly women are married and more of them live alone (Federal Interagency Forum on Aging Related Statistics 2000). The feminization of old age is an increasing component of the poverty of old age. Female-headed households are poorer than male-headed households at all ages, and older people are no exception. As the elderly population rose out of poverty over the last three decades, female-headed households were disproportionately left behind.

Race and Age

Ethnic minority groups (other than Asians) earn substantially less during their peak working years than do non-Hispanic whites, and this means that they are less likely to have accumulated assets such as home ownership to cushion income loss during retirement. In 1999, for instance, the median net worth of older African American households was $13,000 compared to a median net worth of $181,000 for older white households (Federal Interagency Forum on Aging Related Statistics 2000). But the most significant link between minority status and aging is that members of minority groups are less likely to live to experience old age! Whereas 75 percent of white males can expect to survive until they are 65, only 58 percent of African American males will live until retirement age. For white and African American women, the figures are 85 and 75 percent, respectively.

An issue that applies to minority aging in today's increasingly diverse society concerns how elderly immigrants experience U.S. culture. Many immigrants, such as those from Asian or Latino countries, have internalized their native culture's belief that an old person deserves respect and prestige simply because she or he has lived a long time. However, U.S. culture places high value on activity, productivity, and individual achievement—values that are inconsistent with high prestige for old people. The assimilation of elderly immigrants and their children and grandchildren requires mutual adjustments in age roles and expectations (Lin & Liu 1993; Paz 1993).

MicroCase Online Exercise

For an interactive exercise using MicroCase data sets, go to the text companion website at http://sociology.wadsworth.com/brinkerhoff/essentials6e. Choose Chapter 11 and then select "Intersection Exercise" from the left navigation bar.

Family

© Eastcott-Momatiuk/The Image Works

Marriage and Family: Basic Institutions of Society

Recent decades have seen many changes in American family life. Birth rates have declined sharply, divorce rates have reached record levels, the proportion of single-parent families has increased, and for the first time, women with small children have entered the paid labor force in large numbers. In addition to these statistical trends, major shifts in attitudes and values have occurred. Homosexuality, premarital sex, and extramarital sex have all become more acceptable. Related to many of these changes are the dramatic changes in the roles of women in our society.

These changes in family life have been felt, either directly or indirectly, by all of us. Is the family a dying institution, or is it simply a changing one? In this chapter, we examine the question from the perspective of sociology. We begin with a broad description of marriage and the family as basic social institutions.

To place the changes in the U.S. family into perspective, it is useful to look at the variety of family forms across the world. What is it that is really essential about the family?

Universal Aspects

In every culture, the family has been assigned major responsibilities, typically including the following (Murdock 1949; Seccombe & Warner 2004):

1. Replacement of the population through reproduction.
2. Regulation of sexual behavior.
3. Economic responsibility for dependents—children, the elderly, the ill, and the handicapped.
4. Socialization of the young.
5. Ascription of status.
6. Provision of intimacy, belongingness, and emotional support.

Because these activities are important for individual development and the continuity of society, every society provides some institutionalized pattern for meeting them. No society leaves them to individual initiative. Although it is possible to imagine a society in which these responsibilities are handled by religious or educational institutions, most societies have found it convenient to assign them to the family.

Unlike most social structures, the family is a biological as well as a social group. The **family** is a relatively permanent group of persons linked together in social roles by ties of blood, adoption, marriage, or marriage-like commitments who live together and cooperate economically and in the rearing of children. This definition is very broad; it would include a mother living alone with her child as well as a man living with several wives. The important criteria for families are that their members assume responsibility for each other and are bound together—if not by blood, then by some cultural ceremony such as marriage or adoption that ties them to each other relatively permanently.

Marriage is an institutionalized social structure that provides an enduring framework for regulating sexual behavior and childbearing. Many cultures tolerate

The **family** is a relatively permanent group of persons linked together in social roles by ties of blood, marriage, or adoption who live together and cooperate economically and in the rearing of children.

Marriage is an institutionalized social structure that provides an enduring framework for regulating sexual behavior and childbearing.

■ Some modern American families, like these fundamentalist Mormons, live a polygamous life despite legal and social opposition from most of their fellow citizens and from most other Mormons.

AP/Wide World Photos

other kinds of sexual encounters—premarital, extramarital, or homosexual—but most cultures discourage childbearing outside marriage. In some cultures, the sanctions are severe, and almost all sexual relationships are confined to marriage; in others, nonmarital sexuality incurs relatively little punishment.

Marriage is also a legal contract, specifying the obligations of each spouse. Until very recently, those obligations were sharply divided by sex: by law, husbands were obligated to financially support their wives, and wives were obligated to provide domestic services and sexual access to their husbands. These sex-specific obligations only started changing with the rise of the modern feminist movement in the 1970s.

Marriage is important for childbearing because it imposes socially sanctioned roles on parents and the kin group. When a child is born, parents, grandparents, and aunts and uncles are automatically assigned certain normative obligations to the child. This network represents a ready-made social structure designed to organize and stabilize the responsibility for children. Children born outside marriage, by contrast, are more vulnerable. The number of people normatively responsible for their care is smaller, and, even in the case of the mother, the norms are less well enforced. One consequence is higher infant mortality for children born outside of marriage in almost all societies, including our own.

Marriage and family are among the most basic and enduring patterns of social relationships. Although blood ties are important, the family is best understood as a social structure defined and enforced by cultural norms.

Cross-Cultural Variations

Families universally are expected to regulate sexual behavior, care for dependents, socialize the young, ascribe status, and offer emotional and financial security. The importance of these tasks, however, varies across societies. Status ascription is a greater responsibility in societies in which social position is largely inherited; regulation of sexual behavior is more important in cultures without contraception. In our own society, we have seen the priorities assigned to these family responsibilities change substantially over time. In colonial America, economic responsibility and replacement through reproduction were the family's primary functions; the provision of emotional support was a secondary consideration. More recently, however, some of the responsibility for socializing the young has been transferred to schools and day-care centers; financial responsibility for dependent elderly persons has been partially shifted to the government. At the same time, intimacy has taken on increased importance as a dimension of marital relationships.

Although all families share the same basic functions, hundreds of different family forms can satisfy these needs. This section reviews some of the most important ways cultures have fulfilled family functions.

Family Patterns

The basic unit of the American family is the wife-husband pair and their children. When the couple and their children form an independent household living apart from other kin, we call them a **nuclear family.** When they live with other kin, such as the wife's or husband's parents or siblings, we refer to them as an **extended family.**

Extended families are found in all types of societies, although they are defined as the ideal family form only in some premodern, nonindustrialized societies. Where extended families are seen in the United States, they are often the result of financial hardship.

Marriage Patterns

In the United States and much of the Western world, a marriage form called **monogamy** is practiced; each man may have only one wife (at a time), and each woman may have only one husband. Many cultures, however, practice some form of **polygamy**—marriage in which a person may have more than one spouse at a time. The most frequent pattern (practiced by the nineteenth-century Mormons, for example), is to allow a man to have more than one wife at a time (**polygyny**). Less frequently, the form is **polyandry,** in which a woman may have more than one husband at a time.

Viewed cross-culturally, polygyny has been the most popular marriage pattern. In a study of 250 cultures, Murdock (1949, 1957) found that 75 percent prefer polygyny, 24 percent prefer monogamy, and only 1 percent prefer polyandry. In the contemporary world, polygyny is most common in African societies, where approximately 24 percent of married men have more than one wife (Ingoldsby & Smith 1995). Since there are nearly equal numbers of men and women in society, the practice of polygyny has limits; if some men in society have more than one wife (typically those with the most wealth and status), other men have to do without. Consequently, even in societies where polygyny is the preferred marriage pattern, the majority of men living there (and in the world at large) actually practice monogamy.

A **nuclear family** is a family in which the couple and their children form an independent household living apart from other kin.

An **extended family** is a family in which a couple and their children live with other kin, such as the wife's or husband's parents or siblings.

Monogamy is marriage in which there is only one wife and one husband.

Polygamy is any form of marriage in which a person may have more than one spouse at a time.

Polygyny is a form of marriage in which one man may have more than one wife at a time.

Polyandry is a form of marriage in which one woman may have more than one husband at a time.

The U.S. Family over the Life Course

Family relationships play an important role in every stage of our lives. As we consider our lives from birth to death, we tend to think of ourselves in family roles. Being a youngster usually means growing up in a family; being an adult usually means having a family; being elderly often means being a grandparent. Family ties and family roles are an important part of the developmental process from birth to death (Juster & Vinovskis 1987).

Because of the close tie between family roles and individual development, we have organized this description of the U.S. family into a life course perspective. This means that we will approach the family by looking at age-related transitions in family roles.

Childhood

U.S. norms specify that childhood should be a sheltered time. Children's only responsibilities are to accomplish developmental tasks such as learning independence and self-control and mastering the school curriculum. Norms also specify that children should be protected from labor, physical abuse, and the cruder, more unpleasant aspects of life.

Childhood, however, is seldom the oasis that our norms specify. A sizable number of children are physically or emotionally abused by their parents; current estimates suggest that 1 of 10 girls experiences rape or attempted rape during childhood and many more experience other forms of abuse (Tjaden & Thoennes 1998). In addition, nearly one-fifth of all American children grow up in poverty. Although the proportion of children living in poverty in the United States has fallen slightly since 1993, the gap between rich and poor children appears to be growing and is now second highest among industrialized nations (Bennett & Lu 2000).

An important change in the social structure of the child's world is the sharp increase in the proportion of children who grow up in single-parent households: 33 percent of children are now born to single mothers (U.S. Bureau of the Census 2002b). Many more experience the divorce of their parents and sometimes a second divorce between their parents and stepparents (Coleman, Ganong, & Fine 2000). Perhaps because single parents cannot provide as much money or time as two parents, studies show that, on the average, children whose parents divorce have poorer self-esteem, academic performance, and social relationships than other children. These differences are slight, however, and stem primarily not from

As increasing numbers of U.S. women, including those with infants younger than age 1, have entered the labor force, day-care centers have become much more important aspects of early childhood socialization.

the divorce itself, but from the poverty and parental conflicts that precede or follow it (Coontz 1997; Demo & Cox 2000; Lamanna & Riedmann 2000). Consequently, some of these children would not have been any better off if their parents had remained married.

The increasing participation of women in the labor force has added another social structure to the experience of young children: the day-care center. In 2001, about two-thirds of mothers of children younger than age 6 were employed (U.S. Bureau of the Census 2002b). Approximately half of all preschoolers are cared for by relatives while their mothers are at work, with half of these cared for by their fathers (Casper 1997). About one-quarter of all preschool children with an employed mother are enrolled in a day-care center. Research is divided on the effects of day care on children. Some research suggests that day care can increase children's stress levels and behavioral problems (National Institute of Child Health and Human Development, 2003; Watamura et al. 2003). Other research suggests that any negative effects of child care may be limited to certain types of children, families, or programs, and that attending quality day care can increase children's school achievement and reduce their anxiety levels when they begin formal schooling (Love et al. 2003). In the United States, however, high-quality programs are hard to find; few day-care centers have a large, stable, well-trained staff. Because low-quality day-care centers are also less expensive, children from low-income families are more likely to experience the disadvantages of low-quality day care. Nonetheless, many families have to place children in day care. For these families, day care is often far superior to leaving children unattended or with distracted or unqualified family members.

Adolescence

Contemporary social structures make adolescence a difficult period. Because society has little need for the contributions of youth, it encourages young people to become preoccupied with trivialities—such as concern over personal appearance or the latest music. Yet, because adolescence is a temporary state, the adolescent is under constant pressure. Questions such as "What are you going to do when you finish school?", "What are you going to major in?", "What went wrong in Friday night's game?", and "How serious are you about that boy [girl]?" can create strain. That strain can be particularly high for gay and lesbian youth, who may find themselves interested in someone of the "wrong" sex, confused about their own feelings, and fearful over how their family might react.

Adolescents are supposed to become independent from their parents, acquiring adult skills and their own values. They are supposed to shift from the family to peer groups as a source of self-esteem. They must learn how to impress new people and, last but not least, they are supposed to have fun (Gullotta, Adams, & Markstrom 2000). The average adolescent begins to date at about age 14, and an adolescent who is far behind may find that parents and friends are concerned. Thus, although society does not appear to expect much from them, adolescents experience a great deal of role strain. Many adults believe adolescence was the worst rather than the best time of their lives.

The Transition to Adulthood

Some societies have **rites of passage,** formal rituals that mark the end of one age status and the beginning of another. In our own society, there is no clear point at which we can say a person has become an adult.

Rites of passage are formal rituals that mark the end of one age status and the beginning of another.

In our society, rituals such as graduations and weddings continue to have symbolic significance. Nevertheless, when as many as half of all couples cohabit before their weddings and many people don't graduate from college until their 30s or later, the transition to adulthood becomes somewhat fuzzy.

Propinquity is spatial nearness.

Homogamy is the tendency to choose a mate similar in status to oneself.

Endogamy is the practice of choosing a mate from within one's own racial, ethnic, or religious group.

Although expectations about adulthood vary greatly by sex, in the United States adulthood usually means that a person adopts at least some of the following roles: being employed and supporting oneself and one's dependents, being out of school, voting, marrying, and having children. Some of these social roles are optional, and people may be considered adults who never vote, marry, or, in the case of women, hold a paid job (Hogan & Astone 1986). Nevertheless, the exit from adolescence always entails "escaping" from dependence on parents and family.

The normative and most common transition sequence is to finish or leave school, get a job, marry, and have children in that order, but major changes have taken place in this sequence over the last two decades. In the United States there is a great deal of fluidity and reversibility in late adolescence and early adult years (Rindfuss, Swicogood, & Rosenfeld 1987). Youths may leave home and return several times before becoming independent (White 1994).

Early Adulthood: Dating and Mate Selection

Nearly all Americans marry. In fact, the United States is the "marryingest" of industrialized nations. By the time they reach 30, a very high proportion of people in the United States have been married at least once. This strong cultural emphasis on marriage is one of the reasons that so many gays and lesbians also want to marry their life partners.

At first glance, it appears as if all persons are on their own in the search for a suitable spouse; few Americans (outside of certain religious and ethnic communities) rely on matchmakers or arranged marriages. On further reflection, however, it is clear that parents, schools, and churches are all engaged in the process of helping young people find suitable partners. Schools hold dances designed to encourage heterosexual relationships, churches have youth groups partly to encourage members to date and marry within their church, parents and friends introduce somebody "we'd like you to meet." Although dating may be fun, it is also an obligatory form of social behavior—it is normative.

The Changing Age of First Marriage

In the 1950s, teenagers dated in order to find a spouse. Many did so very quickly, and more than 50 percent of U.S. women were married before their 21st birthday. Times have changed. Teenagers no longer date with the expectation of settling down early. Because people are marrying later and more people are marrying for a second or even third time, dating is no longer an activity restricted to the teen years. Forty percent of women and 30 percent of men have never been married by ages 25 to 29. Although not all of these people are looking for a spouse, most are looking for at least a temporary partner. Courtship and dating are activities of individuals who are 28 or 35 years old as well as of teens.

Sorting through the Marriage Market

Over the course of one's single life, one probably meets thousands of potential marriage partners. How do we narrow down the marital field?

Obviously, you are unlikely to meet, much less marry, someone who lives in another community or another state. In the initial stage of attraction, **propinquity,** or spatial nearness, operates in this and a much more subtle fashion, by increasing

■ Interracial marriage and dating have become far more common and socially accepted over the last few decades.

the opportunity for continued interaction. It is no accident that so many people end up marrying coworkers or fellow students. The more you interact with others, the more positive your attitudes toward them become—and positive attitudes may ripen into love (Homans 1950).

Spatial closeness is also often a sign of similarity. People with common interests and values tend to find themselves in similar places, and research indicates that we are drawn to others like ourselves. Of course, there are exceptions, but faced with a wide range of choices, most people choose a mate of similar status. This pattern is called **homogamy** (Kalmijn 1998). Most people also marry within their racial, ethnic, or religious group, a practice known as **endogamy.** Intermarriages can only occur when individuals have contact with persons from other groups and accept those others as more or less equal. Although intermarriage remains rare, it has increased substantially in the last few decades, especially among Jews, Asian Americans, and more educated Americans (Kalmijn).

Physical attractiveness may not be as important as advertisers have made it out to be, but studies do show that appearance is important in gaining initial attention (Sullivan 2001). Its importance normally recedes after the first meeting.

Dating is likely to progress toward a serious consideration of marriage (or cohabitation) if the couple discover similar interests, aspirations, anxieties, and values (Kalmijn 1998; Seccombe & Warner 2004). When dating starts to get serious, couples begin checking to see if they share values such as the desire for children and how to divide household labor. If he wants her to do all the housework and she thinks that idea went out with the hula hoop, they will probably back away from marriage.

Responding to Narrow Marriage Markets

Whether dating leads to marriage also depends on the local supply of "economically attractive" men. As early as 1987, William Julius Wilson noted that one of the reasons African American women were much less likely to marry than white women was the shrinking pool of well-educated African American men with good

Connections
Personal Application

Your college education is likely to affect who you marry. Many people find a spouse in college classrooms or activities. If you attend a college where students are overwhelmingly from one religion, race, or ethnic group, you are more likely to marry within that group. If college throws you into contact with many others whose backgrounds are different from your own, you will be more likely to "marry out." If you don't marry during college, your college education will still likely affect who you marry by affecting the kind of work you do and the kinds of people you meet on the job.

jobs and earnings. The "male marriageable pool index" is the tool that researchers have developed to assess this hypothesis. Results are clear; local marriage markets do matter. A shortage of males employed in good jobs with adequate earnings sharply reduces the likelihood that a woman will marry or cohabit (Lichter et al.1992; Raley 1996; Teachman, Tedrow, & Crowder 2000). In fact, differences in the availability of marriageable men account for at least 40 percent of the race difference in overall marriage rates. Similarly, local marriage markets affect the rates of intermarriage: members of racial and ethnic minority groups have significantly higher rates of intermarriage when they live in areas where members of their group are rare (Qian 1999).

In an interesting sidebar, researchers have found that "economically attractive" women are also more likely to marry. Their greater attractiveness to potential male partners apparently more than makes up for the fact that women with full-time employment and higher earnings tend to be choosier about the men they date and marry (Lichter et al. 1992).

Middle Age

The busiest part of most adult lives is the time between the ages of 20 and 45. There are often children in the home and marriages and careers to be established. This period of life is frequently marked by role overload simply because so much is going on at one time. Middle age, that period roughly between 45 and 65, is by contrast a quieter and often more prosperous period. Studies show that both men and women tend to greet the empty nest and then retirement with relief rather than regret (Goudy et al. 1980; White & Edwards 1990).

Many middle-aged couples, however, look forward to the empty nest period only to find that they are sandwiched between the demands of their parents and their children. This generation squeeze has been fostered by two distinct trends. The first is rising life expectancy. Reflecting this trend, a 1996 national survey found that 24 percent of U.S. households include someone who is providing care for a relative older than age 65 (National Alliance for Caregiving 1997). At the same time, difficulties their children face in establishing themselves economically, rising age at marriage, and increasing rates of separation and divorce have increased the likelihood that middle-aged adults still have children living at home. In 2000, 56 percent of all men aged 18 to 24 were still living at home. After children move out the first time, nearly half move back home at least once (Goldscheider & Goldscheider 1994). As a result, even at ages 30 to 34, 17 percent of all divorced men live with their parents (Casper & Bryson 1998). As a result of these pressures from both sides, many middle-aged people do not find the relief from family responsibility that they had hoped for.

The average parent copes quite well with children's continued coresidence in the parental home. When a large national sample of parents was asked to rate how well it worked for them to have an adult child in their home, nearly 70 percent gave a positive reply. Nevertheless, a full quarter of parents reported having disagreements once a week or more (Aquilino & Supple 1991). As a result, it is not surprising that research shows parents experience an increase in feelings of well-being when their children leave home (White & Edwards 1990). This feeling of well-being, however, depends on their children's having established themselves successfully in an alternative living arrangement and continued positive association between parents and children.

Age 65 and Beyond

One of the most important changes in the social structure of old age is that it is now a common stage in the life course—and often a long one. Almost all of us can count on living to age 65. Furthermore, if you live to age 65, you can expect to live an average of 16 more years. Most of these years will be healthy ones. Although most people age 70 and older experience some loss of stamina, 75 percent experience no major health limitations (Federal Interagency Forum on Aging Related Statistics 2000).

Family roles continue to be critical in old age. Having spouses, children, grandchildren, and brothers and sisters all contribute to well-being. Marriage is an especially important relationship, one that provides higher income, live-in help, and companionship. Because of men's shorter life expectancy and their tendency to marry younger women, however, marriage is unequally available: 79 percent of men aged 65 to 74 are still married compared to only 55 percent of women that age.

Whether older people are married or not, relationships with children and grandchildren are an important factor in most of their lives. Most grandparents visit grandchildren every month and report very good relationships with them (American Association of Retired Persons 1999). In fact, as of 2003, 2½ million grandparents are the primary caregivers for their grandchildren (U.S. Bureau of the Census 2002b). Many children and families would have great difficulty without the help of grandparents, and many grandparents consider involvement with their grandchildren an important source of personal satisfaction (Allen, Blieszner, & Roberto 2000).

The nature of intergenerational relationships depends substantially on the ages of the generations. When the older generation falls into the "young old" category, they are generally still providing more help for their children than their children are for them (Hogan, Eggebeen, & Clogg 1993). They are helping with down payments and grandchildren's college educations or providing temporary living space for children who have divorced or lost their jobs.

Both grandparents and grandchildren gain great personal satisfaction from their relationships.

© Eastcott/Momatiuk/Woodfin Camp & Associates

As the senior generation moves into the "old old" category, however, relationships must be renegotiated (Mutran & Reitzes 1984). Even in the "old old" category, most people continue to be largely self-sufficient, but eventually will need help of some kind—for shopping, home repairs, and social support. Although these services are available from community agencies, most older people rely heavily on their families, especially their daughters (Kemper 1992; Lye 1996). Understandably, though, both older people and their children are happiest when these relationships are free of dependency. Elderly persons much prefer to live alone rather than with their children (Bayer & Harper 2000).

Roles and Relationships in Marriage

Marriage is one of the major role transitions to adulthood, and most people marry at least once. In fiction, the story ends with the wedding, and we are told that the couple lived happily ever after. In real life, though, the work has just begun. Marriage means the acquisition of a whole new set of duties and responsibilities, as well as a few rights. What are they and what is marriage like?

Gender Roles in Marriage

Marriage is a sharply gendered relationship. Both normatively and in actual practice, husbands and wives and mothers and fathers have different responsibilities. Although many things have changed, U.S. norms specify that the husband *ought* to work outside the home; it is still considered his responsibility to be the primary provider for his family—even though this is no longer true for a significant minority of families. Similarly, although most Americans now believe that husbands and wives should share in household labor, most still expect that the wife will do the larger share. In fact, women currently perform two-thirds to three-quarters of household labor and most women as well as men regard this as a fair arrangement (Coltrane 2000). Although women who work outside the home typically do less housework than other women, this still leaves many of them—especially those with young children—subject to severe cases of role overload, or role strain. One adaptation women make to this overload is to lower their standards for cleanliness, meals, and other domestic services. They let their family eat at McDonald's and let the iron gather dust. Another adaptation women make is to hire other women to perform domestic tasks. In this way, domestic labor remains a woman's job and the idea that women are responsible for this work is reinforced. In addition, since most employers are white and middle class and employees are nonwhite and working class, paid domestic labor also reinforces class and race divisions within society (Hondagneu-Sotelo 2001; Parreñas 2000; Wrigley 1995).

The Parental Role: A Leap of Faith

The decision to become a parent is a momentous one. Children are extremely costly, both financially and in terms of emotional wear and tear. It currently costs about $10,000 per year to raise a middle-class child—not including college expenses (U.S. Bureau of the Census 2002b). Parenthood, however, is one of life's biggest adventures. Few other undertakings require such a large commitment of time and money on so uncertain a return. The list of disadvantages is long and certain: It costs a lot of money, takes an enormous amount of time, disrupts usual activities, and causes at least occasional stress and worry. Also, once you've started, there is no backing out; it is a lifetime commitment. What are the returns? You hope for love and a sense of family, but you know all around you are parents whose children cause them heartaches and headaches. Parenthood is really the biggest gamble most people will ever take. In spite of this, or maybe because of it, the majority of people want and have children.

Mothering versus Fathering

Despite some major changes, the parenting roles assumed by men and women still differ considerably (Cancian & Oliker 2000). Mothers are the ones most likely to drop out of the labor force to care for infants and young children; they are the ones most likely to care for sick children and to go to school conferences (Cancian & Oliker). Fathers, on the other hand, are the ones likely to carry the major burden of providing for their families. Generally, parenthood increases men's attachment to the workforce at the same time as it decreases that of women (Thompson & Walker 1989); child rearing also reduces the size of women's social support networks and temporarily increases the number of kin included in men's (Munch, McPherson, & Smith-Lovin 1997).

The overwhelming proportion of mothers who are employed— around 80 percent as of 2001—has exerted pressure for fathers to increase their role in child care. Although research still finds that fathers "help" rather than "take responsibility" and that they are more likely to play with children than change diapers, fathers have increased their role in child care. A growing proportion of fathers, however, do not live with their children. Among nonresidential fathers, contact tends to be low and child care virtually nonexistent.

■ Although fathers now take more responsibility for child care and household tasks than they did in previous generations, mothers still bear far more of these burdens, leaving many feeling overworked and underappreciated.

Stepparenting

During 2000, 9 percent of American children lived with a biological parent and a step-parent (U.S. Bureau of the Census 2002a). About one-third of U.S. children will live in a stepfamily before they are 18—most often with a mother and a stepfather (Coleman, Ganong, & Fine 2000). If parenting is difficult, stepparenting is more so (Coleman, Ganong, & Fine). Often stepparents are unsure what role they should take in their stepchildren's lives, and often their spouses and stepchildren are equally ambivalent. Older children, especially, are likely to reject advances from stepparents and to discourage warm relationships, although others eventually develop close relationships with their stepparents. Stepmothers typically face more difficulties than stepfathers because stepmothers typically are more involved in their stepchildren's lives. In addition, stepmothers face more competition for the children's affections, since noncustodial biological mothers are far more likely to remain involved in their children's lives than are noncustodial biological fathers.

Marital Happiness

Recent studies suggest that marital happiness peaks during the honeymoon, drops sharply over the next few years, and then continues to drop more slowly afterwards (Bradbury, Fincham, & Beach 2000). The presence of children in the home—especially teenagers—seems particularly likely to reduce marital happiness.

focus on

Technology

Kinkeeping, Gender Roles, and the Internet

One of the most frequently cited outcomes of the Industrial Revolution has been the rise of geographic mobility. Instead of young families living near or even with their parents and other kin, families now move around the country in search of the best jobs and prospects. What are the consequences of this change for family ties?

Decades of research convincingly demonstrate that older people in the United States have not been abandoned by their children. In a national survey conducted in 2000, 73 percent of adults reported telephoning their parents at least once in the last month, and 49 percent had done so five or more times (Maritz Poll on American Habits #83, Feb. 2000). Grandchildren keep in touch with their grandparents, and siblings maintain contact with each other. The fact that family members remain strongly connected throughout the life course, even though they often live many miles apart, has led some sociologists to propose that the typical family form in contemporary society is not the nu-

clear family but one that might more appropriately be called the modified extended family (Litwak 1960).

Technology has been a major factor in the creation of this new family form; rather than relying on close physical proximity to ensure continuing contact and support, the modified extended family uses planes, trains, automobiles, the telephone, and the Internet to link nuclear family members to their more extended kin (Litwak & Kulis 1988). Across the Internet, family members now plan celebrations, share information and photos, solve problems, and generally keep track of one another's "comings and goings" via the family home page. How might this use of the Web create important changes in the family roles of men and women?

Although anyone can use a phone or drive a car, studies show that, at

least until now, women are overwhelmingly the ones to play the role of "kinkeeper." They send the birthday cards, organize family reunions, provide most of the emotional support, and generally keep the family in touch with one another. One result of this gender-based division of labor is that female relatives are usually closer to each other than are male relatives. On the other hand, many men now use the Internet to stay in contact easily with friends and relatives. This raises the question: Will men take on a larger share of the kinkeeping role as families come to use the Web more frequently to communicate and connect with one another? If so, will the bonds between male relatives begin to take on the closeness and intimacy now associated with the relationships between female family members? What do you think?

 For more information, look up the following subjects in InfoTrac College Edition:
Internet and social (as keywords)
Internet and family (as keywords)

 Or visit the following Web site:
Pew Internet and American Life Project
http://www.pewinternet.org/reports/reports.asp?Report=55&Section=ReportLevel2&Field=Level2ID&ID=398

Contemporary Family Choices

As discussed in Chapter 2, U.S. norms have changed over time to permit much wider variation in the way that people achieve core values. Although a happy family life remains at the center of things people in the United States value, the ways that individuals choose to meet this value have changed considerably. An increasing number of people will find themselves making three choices about marriage and the family: to marry or cohabit; to have children, either within or outside of marriage, or remain childless; and to place work ahead of family or vice versa.

Marriage or Cohabitation

During the last 30 years, the chances that an individual will *ever* live in a cohabiting relationship has increased more than 400 percent for men and 1,200 percent for women. Cohabitation is also an increasingly common stage in the courtship process, with approximately half of all recently married couples cohabiting before they were married (Smock 2000).

Although cohabitation is often a prelude to marriage—55 percent of married couples cohabited before marriage in recent years—there is good evidence that much of the decline in marriage and remarriage rates in the United States is linked to the increasing numbers of individuals who prefer to live with one another outside of legal marriage. Even if they plan to marry, most cohabitors simply do not believe that anything would change if they did so. Probably the most important exception to this rule is that nearly one-third of men, but only about one-sixth of women, report that their "freedom to do what they want" would be diminished if they were married (Bumpass & Sweet 1991).

It is clear that normative pressures to marry are not as high as they once were. About one-fifth of cohabiting persons do not expect ever to marry or to marry again. Among all unmarried persons, only about one-third agree that it is better to marry than to go through life single, and one-third say it would be all right for them to have a child without ever marrying (Bumpass & Sweet 1991). Because of cohabitation, however, being unmarried does not necessarily mean being single or living alone.

Having Children . . . or Not

Although most people in the United States plan to have children, increasingly they choose to do so outside of marriage. Many will choose to postpone parenthood, and increasing numbers will choose to remain childless.

Nonmarital Births

As of 2002, 33.8 percent of all births in the United States are to unmarried women (Hamilton, Martin, & Sutton 2003). Most of these births (about three-quarters) are to women 20 years of age and older. Now that most women participate in the labor force, many have the economic and psychological resources to tackle the tough job of parenting on their own. Some of these will choose to become pregnant or adopt even if they are not married. For the same reasons, births to unmarried women also have increased in Europe (Table 12.1).

Nonmarital childbearing among teenagers raises special concern, although fortunately the rate has declined steadily over the last decade (Hamilton, Martin, & Sutton 2003). Because teenage mothers are less likely to complete both high school and college, they are also likely to suffer economic hardship. In many instances, however, teenage pregnancy stems from poverty as well as causing it (Luker 1996). Girls who face bleak futures sometimes conclude that single motherhood is a reasonable way to seek love and happiness. Other girls become pregnant because they fear that using contraceptives would suggest to a new boyfriend that they are "loose." Still others lack the power to insist that contraception be used.

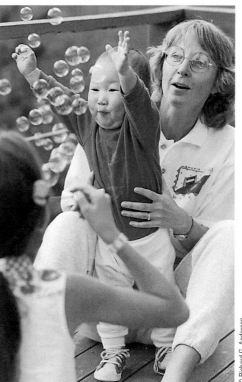

■ Many single women now have the financial and psychological resources needed to raise a child on their own by choice.

TABLE 12.1
Percentage Births to Unmarried Women, Selected Countries: 1970–1999

	1970	1980	1990	1999
United States	11	18	28	32
Bulgaria	9	11	12	35
Denmark	11	33	46	45
Italy	2	4	6	9
Netherlands	2	4	11	23
Sweden	18	40	47	55
Switzerland	4	5	6	10
United Kingdom	8	12	28	39

SOURCE: Council of Europe 2000; U.S. Bureau of the Census 2002b.

Connections

Historical Note

Although the rate of childbearing among unmarried teenagers has risen sharply, the rate of *pregnancy* has not. The difference is that now more pregnant teenagers keep their babies and don't get married. Prior to the 1970s, this was an extremely rare choice, for the shame of raising a child alone was overwhelming. Girls who found themselves pregnant had three choices: getting an abortion (usually illegal and sometimes life-threatening), having a "shotgun" wedding, or giving up their babies. Those who chose the last option usually left their home towns and stayed in special institutions, where they could hide their pregnancies until their babies were born and given up for adoption.

Regardless of the mother's age, having a child outside marriage does not necessarily mean raising a child alone. About 40 percent of nonmarital childbirths are to women who live with a partner (Smock 2000). Many of these women will eventually marry the fathers of their babies or another man; others will continue to share parenting outside of marriage.

Delayed Childbearing

Many married women are choosing to postpone childbearing until 5 or even 10 years after their first marriages. Today, 26 percent of U.S. women aged 30 to 34 are still childless as are 18 percent of those aged 40 to 44 (U.S. Bureau of the Census 2000b). Although many still intend to have children eventually, childbearing is no longer seen as an inevitable consequence of marriage.

Currently, the average number of children born to U.S. women is approximately two. This small family size is due largely to changes in the role of women and

■ About one-third of all births in the United States are to unmarried women, and most of these are age 20 or over.

Mark Peterson/Corbis

American Diversity

Is the Gay Family a Contradiction in Terms?

In a landmark 1986 decision upholding a Georgia law that criminalized same-sex sexual activity, Supreme Court Justice White argued that laws granting privacy rights to families did not apply to homosexuals. There was, he said, "No connection between family, marriage, or procreation on the one hand and homosexual activity on the other. . ." (cited in Weston 1991, 208). Is Justice White right? Are heterosexuals the only ones with families?

Because homosexuals have parents, siblings, and cousins, the definition of family at issue in this decision is obviously a narrow one that centers around marriage and children. On the question of marriage, the justice is correct by definition. In a few jurisdictions, gay or lesbian partners may gain some legal protection by registering their union " as a "domestic partnership." Although some states and municipalities have permitted marriage between gay and lesbian partners, provisions permitting such marriages are under close scrutiny. In addition, the 1996 federal Defense of Marriage Act prohibits any federal recognition of gay marriage (such as veterans benefits or Social Security benefits to surviving spouses) and relieves states of any obligation to recognize gay marriages legally contracted elsewhere.

On the other hand, although same-sex lovers obviously cannot produce children from their union, they can have children in a variety of other ways: Many lesbians and gay men can (and do) have children from previous heterosexual relationships; in some states they can adopt children; and lesbian women can have children through artificial insemination. As a result, somewhere between 4 and 14 million children have homosexual parents (Bozett 1987).

The issue of gay families has become a public issue in both the gay and straight communities (Eskridge 1996). Should gays be allowed to adopt children? Should being gay be sufficient by itself to make a parent unfit for child custody or visitation rights? Should lesbians be able to use artificial insemination? Should gay and lesbian partners be allowed to marry legally so that their health insurance and Social Security benefits could be shared by their partners?

These are questions that go to the heart of the family. The traditional view is that homosexual unions are both unnatural and sinful. Others define the family by long-term commitment, and they are willing to tolerate and encourage a variety of family forms—including gay and lesbian families—as long as they contribute to stable and nurturing environments for adults and children. In fact, research consistently finds that growing up with gay or lesbian parents has no measurable effect on children's sexual identity, personality, or social development (Fitzgerald 1999; Patterson 2000).

There is no question that both homosexual activity and gay marriage are regarded more favorably now than in the past. A national survey conducted in 2003 found that 38 percent of all Americans approve of gay marriages—and 52 percent of Americans ages 18 to 29 approve (Pew Research Center 2003b). That same year, the Supreme Court declared unconstitutional all laws criminalizing homosexual activity, and some parts of Canada legalized gay marriage.

Contrary to Justice White's opinion, homosexuals obviously do have families. The question society must address is whether such families should receive the same legal recognition and protection as other families.

© Shelley Gazin/Corbis

■ Growing numbers of children in the United States are being raised and nurtured by gay or lesbian parents and their same-sex partners.

For more information, look up the following subjects in InfoTrac College Edition:
Gay families
Gay marriage

Or visit the following Web site:
http://www.ngltf.org/library/index.cfm

changes in the security of family roles. Although the average woman does not yet place career roles over family roles, women today desire economic security and more personal freedom, both of which are adversely affected by taking time out for having children. If the divorce rate remains high and if women's participation in the labor force rises—both of which appear likely—then having children will become even less attractive.

Choosing Childlessness

Already, increasing numbers of women—and men—have decided that they are uninterested in having children. Of course, this choice depends on access to effective contraception. But it also reflects social changes. As more and more women find satisfaction in their work and other aspects of their lives, they are less likely to believe that they need children to have a good life. They may also conclude that having children would make it difficult for them to pursue their careers. If they meet men who feel the same way, they may decide against having children. These decisions are bolstered by the belief—backed by research—that having children reduces marital satisfaction. Childlessness is also particularly common among women who were the eldest daughters in large families; these women often feel that they already raised several children, and have no interest in doing so again.

Work versus Family

Several developments in the labor force have squeezed the time that people have available to spend at home. First, 76 percent of married women aged 25 to 34 are now in the labor force (U.S. Bureau of the Census 2002b). Second, for many of the nation's workers, workdays and work weeks are growing longer. Hourly employees are often coerced or bribed into working overtime hours—even double shifts; others must do so to make ends meet. Professionals have to work early, late, and on weekends and take work home in order to demonstrate that they are serious players. As a result, parents experience a time bind at home. Family meals are increasingly rare, and time at home becomes rigidly scheduled as parents try to get themselves to work, their laundry done and cleaning picked up, and their children to school or other activities.

This time bind is often explained as the inevitable result of decreasing real wages, global competitiveness in the workplace, and the growing taste for expensive consumer goods. In an influential study, however, sociologist Arlie Hochschild (1997) argues that many middle-class parents are choosing to spend more time at work because they find work more rewarding than being at home with their family. The more hectic it gets at home, the nicer the job looks. Bosses and coworkers hardly ever spill their juice, dirty their diapers, cry, or slam out of the house because they cannot use the car. Compared to home, the workplace tends to be relatively quiet and orderly and the work rewarding. For many, work rather than home is the place where you can put your feet up and drink a quiet cup of coffee, work is the place where you can get advice on your meddlesome mother-in-law or crumbling marriage, and work is the place where employers notice that you're under a lot of stress and provide free professional counseling. Plus, of course, at work there are paychecks, promotion opportunities, and recognition ceremonies.

Much of the horror expressed over parents finding their jobs less work than their families can be traced to anxiety over changing gender roles. It has always been supposed that men would find work both satisfying and preferable to staying at home with the kids. Now that women report the same preferences, it is defined as a problem. One way that people handle this dilemma is simply by having fewer children, and the time bind is surely one reason why women in many European countries are only having one child (see Chapter 15). Another solution is to turn the care of children—day care, after-school care, swimming lessons, birthday parties, psychological counseling, summer vacations—over to paid professionals (Wrigley 1995). These solutions, however, are simply means to help parents work even longer hours with less interference from their families. Real solutions would require a reduction in overtime work, a reduction in the hours of the work week, and a cultural shift that valued raising children as much as careers. For the time being, it seems most likely that the family will continue to lose the time war, with men, women, and children spending increasingly long hours away from home and away from each other.

Problems in the American Family

There are couples who swear that they never have an argument and never disagree. These people are certainly in the minority, however, for most intimate relationships involve some stress and strain. We become concerned when these stresses and strains affect the mental and physical health of the individuals and when they affect the stability of society. In this section, we cover two problems in the U.S. family: violence and divorce.

Violence

Child abuse is nothing new, nor is wife battering. These forms of family violence, however, didn't receive much attention until recent years. In a celebrated court case in 1871, a social worker had to invoke laws against cruelty to animals in order to remove a child from a violent home. There were laws specifying how to treat your animals, but no restrictions on how wives and children were to be treated. In recent years, however, we have become both more aware and less tolerant of violence in the home.

The incidence of child abuse is particularly hard to measure, since it is difficult to obtain permission to interview children outside of their parents' presence. Surveys of child protective services professionals give us at least a starting point for estimating abuse. These surveys suggest that each year, 1.5 million children are known to be sexually, physically, or emotionally abused by their parents or caregivers, with about one-third of these receiving serious physical injuries (Sedlak & Broadhurst 1996). This figure is obviously an underestimate, as it does not include those whose abuse remains hidden. For this reason, the best data currently available come from a survey of 16,000 Americans conducted for the National Institute of Justice (Tjaden & Thoennes 1998). Of women interviewed, 10 percent reported experiencing rape or attempted rape during their childhood, primarily at the hands of family members. These figures, too, are likely to be substantial underestimates, since they do not include those adults who refused to talk about their experiences.

The same survey gives us our best measure of the extent of violence between adults in families (Tjaden & Thoennes 2000). The survey found that 22 percent of women and 7 percent of men have been physically assaulted by a spouse or cohabitant

TABLE 12.2
Violence between Spouses and Cohabitants

	Percentage who have been assaulted
Men by female spouses or cohabitants	7%
Women by male spouses or cohabitants	22
Women by female cohabitants	11
Men by male cohabitants	23

SOURCE: Tjaden & Thoennes 2000.

of the opposite sex (Table 12.2). Women were twice as likely as men to have required medical care after being assaulted. Violence was almost as common in male homosexual couples as in heterosexual couples, but was much rarer among lesbians.

Recently, concern also has been raised about violence directed at dependent, vulnerable, elderly parents by their adult children. Research has found, however, that most victims of elder abuse have been attacked by their spouses rather than by their children (Bergen 1998).

Family violence is not restricted to any class or race (Johnson & Ferraro 2000). It occurs in the homes of lawyers as well as the homes of welfare mothers. Violence is most likely to occur when individuals feel they are losing control, whether over their spouse or over other aspects of their lives. One reason men are more likely than women to beat their spouses is because they are more likely to believe that they *should* control their spouses (Johnson & Ferraro).

Solutions to family violence are complex. The first step, however, is to make it clear that violence is inappropriate and illegal. New laws against spousal rape and other forms of family violence may clarify what used to be rather fuzzy norms about whether family violence was appropriate (Straus & Gelles 1986).

Divorce

The **divorce rate** is calculated as the number of divorces each year per 1,000 married women.

Lifetime divorce probability is the estimated probability that a marriage will ever end in divorce.

In the United States, more than 2 million adults and approximately 1 million children are affected annually by divorce. The **divorce rate,** calculated as the number of divorces each year per 1,000 married women, has risen steadily in the post–World War II period. In 2001, it stood at 20—that is, 20 of 1,000, or 2 percent of all married women in the United States divorce annually. Another way of looking at divorce is to calculate the probability that a marriage will *ever* end in divorce—the **lifetime divorce probability.** Of marriages begun in 1890, for example, the proportion eventually ending in divorce was approximately 10 percent (Cherlin 1981). Today, however, experts estimate that about half of first marriages will end in divorce (Figure 12.1) (Bumpass, Raley, & Sweet 1995).

What are the factors that make a marriage more likely to fail? Table 12.3 displays some of the predictors of marital failure within the first 5 years of marriage. Research consistently finds six factors especially important (Bramlett & Mosher 2002; Teachman 2002):

- *Age at marriage.* Probably the best predictor of divorce is a youthful age at marriage. Marrying as a teenager or even in the early 20s doubles chances for divorce compared with those who marry later (Table 12.3).

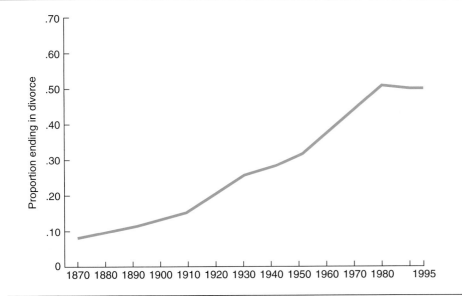

FIGURE 12.1
Changing Probability of Divorce, 1870–1995
There has been a dramatic increase in the likelihood that marriages will end in divorce. Half of recent first marriages are expected to end in divorce.
SOURCE: Adapted from Cherlin 1981 (reprinted by permission, Harvard University Press); Martin and Bumpass 1989; Bumpass, Raley, and Sweet 1995.

- *Parental divorce.* People whose parents divorced are more likely to divorce themselves.
- *Premarital childbearing.* Having a child before marriage reduces the stability of subsequent marriages. Premarital conception followed by a postmarital birth, however, does not seem to increase the likelihood of divorce.
- *Education.* The higher one's education, the less likely one's marriage is to end in divorce. Part of this is because people with higher educations are more likely to come from two-parent families, avoid premarital childbearing, and marry later. Independent of these other factors, however, higher education does reduce the chances of divorce.
- *Race.* African Americans are substantially more likely than whites, Hispanics, or Asians to get divorced, although the difference has declined over time (Teachman 2002).
- *Religion.* Catholics are significantly less likely than others to get divorced, even after a variety of other demographic variables are taken into account.

Societal-Level Factors

Age at marriage, parental divorce, premarital childbearing, education, race, and religion affect whether a particular marriage succeeds or fails. These personal characteristics, however, cannot account for the fact that slightly more than 50 percent of all marriages now fail. The shift from a lifetime divorce probability of 10 percent to one of 50 percent within the last century is a social problem, not a personal trouble, and to explain it we need to look at social structure.

The change in marital relationships is probably most clearly associated with changes in economic institutions. Rising divorce rates are not unique to the United States (Table 12.4). Although divorce has always been more prevalent in the United States than elsewhere, most industrialized nations have experienced substantial increases in divorce. In France, Greece, and the United Kingdom, for instance, the divorce rate tripled between 1970 and 2000. In these countries and in the United States, the shift from an agricultural to an industrial economy and then to a service

TABLE 12.3
Probability of First Marriage Breaking up within the First 10 Years
The probability that a first marriage will ever end in divorce is about 50 percent. If we limit our focus to the first 10 years of marriage, 23 percent of all first marriages end in this period. Divorce is more likely for African Americans, children of divorced parents, and persons who marry young, have limited education, or have a child before marriage.

	% Ending in Divorce
Total	23%
Age at Marriage	
<18	48
18–19	40
20–24	29
25 or over	24
Education	
Less than 12 years	42
12 years	36
13 years or more	29
Children before marriage	
No	31
Yes	50
Race	
White	32
African American	47
Hispanic	34
Asian	20
Children of divorced parents	
Yes	43
No	29

SOURCE: Bramlett & Mosher 2002.

economy, changes detailed in Chapter 14, has revolutionized the technologies and relationships essential to production. One result of this revolution is that a middle-class wage earner's chief economic asset is education and experience. You can walk away from a marriage and take these assets along; the same is not true with land, which is often tied up in family relationships. Another result is the increased opportunity for women to support themselves outside marriage. On the other hand, changes in the economy have made it more difficult for lower-class men to support themselves or a family. The resulting economic hardships cause enormous stress within relationships, often resulting in divorce.

Because of these economic changes, women and men are now less impelled or less able to marry or to stay married. Because no new incentive for marriage has proven to be more effective than economic need, marriages have less institutional support than before. Nevertheless, it is important to note that most people whose marriages end in divorce eventually remarry, with remarriage especially common for young people, whites, and men (Clarke 1995; Coleman, Ganong, & Fine 2000).

TABLE 12.4
Divorce Rates for Selected Countries, 1970–2000

Country	Divorces per 100 Marriages				
	1970	1980	1990	1995	2000
Czech Republic	26	31	38	38	43
France	12	22	32	38	38**
Germany*	17	25	29	33	39**
Greece	5	10	9	17	16
Sweden	23	42	45	52	55
United Kingdom	16	35	38	40	43**

*German data for 1970–1990 are for West Germany only.
**1998 or 1999 data

SOURCE: Sardon 2002; U.S. Bureau of the Census 1995, 1998.

How Serious Are These Problems?

In Chapter 4, we noted that some people view institutions as constraints that force people into uncomfortable and perhaps oppressive relationships; others see institutions as providing the stability and comfort frequently associated with old shoes. This difference of views is nowhere more present than in the case of the family.

■ Despite the very real problems many families face, family relationships continue to be a major source of satisfaction for most Americans throughout their lives.

Theoretical viewpoints sharply influence perceptions of the health of the modern family (Adams 1985). If the family is an oppressive institution, then divorce is a form of liberation; if the family is the source of individual and community strength, then divorce undermines society. In this section, we briefly review some of the major criticisms of the contemporary family and conclude with a perspective on the future.

Loss of Commitment

Some critics argue that a major problem with the U.S. family—and they would further argue, with U.S. culture—is an accent on individual growth at the expense of commitment (Bellah et al. 1985, 1991). This criticism is most likely to come from structural functionalists, who traditionally stress the subordination of individual to community needs, but it also comes from symbolic interactionists concerned with stable self-identity.

The most prominent symptom of the alleged emphasis on individual happiness and growth at the expense of long-term commitment is the rapid rise in the number of women who are raising children on their own. Because men don't want to be tied down by wives and children and because wives don't want to be tied down by husbands, fathers have walked away and mothers have let them, even encouraged them. Following divorce, many fathers all but sever ties to their children (Lye 1996; Seltzer 1994). The result is that the basic family unit, in the sense of long-term commitment to sharing and support, has become the mother-child pair. Husband-wife and father-child relationships are increasingly viewed as temporary and even optional. To many critics, this increasingly voluntary nature of family ties is dysfunctional, reducing the stability of the family and decreasing its ability to perform one of its major tasks: caring for children. Among the ill consequences they note is the increasing proportion of women and children living in poverty.

Inequality

Sociologists from a conflict perspective are more apt to criticize the family for its oppression of women and children. They point to the lower power of women and children relative to men and to the number of women and children who suffer abuse in the family. For example, Denzin (1984, 487–488) declares, "A patriarchal, capitalist society which promotes the ownership of firearms, women, and children; which makes homes men's castles; and which sanctions societal and interpersonal violence in the forms of wars, athletic contests, and mass media fiction (and news) should not be surprised to find violence in its homes. Violence directed toward children and women is a pervasive feature of sexual divisions of labor which place females and children in subordinate positions to adult males."

From this perspective, the family has not really changed for the worse: It has traditionally been an oppressive institution maintained by force and fraud. These critics do note the development of greater egalitarianism in the contemporary family, but generally find that these changes are occurring too slowly.

One major developing problem from this perspective is that women's and children's independence from husbands and fathers is coming before their independence in the marketplace. The solution they recommend is equal opportunity and equal pay for women and state support for children in the form of family allowances. They note that in Scandinavian countries—where women have near equal opportunity in the job market and the state amply supports day care, education, and health care—single motherhood is not linked to poverty.

Another major problem noted by these critics is that even though women are increasingly encouraged to adopt what had formerly been viewed as "masculine" roles, the traditionally "female" role remains undervalued. As a result, few men are willing to fully share parenting or housework and many will abandon their children if their marriages break up. Furthermore, neither women nor men who wish to remain home with their children feel that they receive respect or support for doing so. From this perspective, true solutions to problems within the family will require not only economic change but also change in the way we think about gender.

Where This Leaves Us

Some of the functions performed by the traditional family, such as care of dependents, are not as important today as they used to be. Nevertheless, the family continues to perform vital functions for society and individuals. Despite social welfare programs, families remain the major source of economic support for children and of social support for people of all ages.

The family is also an important arena in which we develop our self-concept, learn to interact with others, and internalize society's norms. Without the strong bonds of love and affection that characterize family ties, these developmental tasks are difficult if not impossible. Thus, the family is essential for the production of socialized members, people who can fit in and play a productive part in society.

The family is important not just in childhood but throughout the life course. To cite just a few examples, people with close ties to their kin report greater satisfaction with their life, their marriages, and their health; they are less likely to abuse their children or get divorced; they are also more likely to be able to ride out personal and family crises (Lavee, McCubbin, & Patterson 1985; McGhee 1985; Rosenthal 1985).

Given these benefits that the family gives to both the individual and society, it would seem to be a reasonable goal to support the family, while simultaneously reducing some of its more oppressive features. This goal is not impossible. Despite current rates of divorce, illegitimacy, childlessness, and domestic abuse, there are signs of health in the family: the durability of the mother-child bond, the frequency of remarriage, the number of stepfathers who willingly support other men's children, and the frequency with which elderly persons rely on and get help from their children.

There is no doubt that the family is changing. When you ask a young man what his father did when he was growing up, you are increasingly likely to hear, "What father?" or "Which father?" These recent changes must be viewed as at least potentially troublesome. At present we have no institutionalized mechanisms comparable to the family for giving individuals social support or for caring for children. The importance of these tasks suggests that the needs of families and especially children must be moved closer to the top of the national agenda.

Summary

1. Marriage and family are the most basic institutions found in society. In all societies, these institutions meet universal needs such as regulation of sexual behavior, replacement through reproduction, child care, and socialization.

2. Individual development over the life course is closely paralleled by changing family roles. The transition to adulthood is closely linked to adoption of marital and parental roles.

3. High rates of divorce and increases in the labor-force participation of women have led to major changes in the social structure of childhood. Nowadays, many U.S. children spend some time in a single-parent

household before they are 18. About one-quarter of preschoolers with employed mothers attend day-care centers. Although high-quality day-care programs can benefit children, low-quality programs are more common and may be detrimental.

4. Because of postponement of marriage and high rates of divorce, dating is no longer just a teen activity. Mate selection depends on love, but also on propinquity, homogamy, endogamy, and value consensus.

5. Intergenerational bonds are important to individuals. Studies show that parent-child bonds remain strong over the life course and that sibling bonds remain important even in old age. The one exception is that at all ages, noncustodial fathers often have only weak ties to their children.

6. The increasing participation of wives in the breadwinning role is a major change in family roles. The change has not been accompanied by a significant increase in husbands' housekeeping or child-care work.

7. Cohabitation is now a common stage of courtship. Many cohabiting couples eventually marry, although a sizable minority of male cohabitors believe that marriage would limit their ability and freedom to do the things they want.

8. Growing numbers of women now choose to delay childbearing or to forego having children altogether. One-third of all births in the United States now occur outside of marriage. Teenage pregnancy stems from poverty, a lack of easily available contraception, and a lack of other ways to find meaning and personal satisfaction.

9. Violence against both children and intimate partners is relatively common in U.S. homes. In both homosexual and heterosexual relationships, men are more likely than women to batter their partners. Family violence is strongly related to multiple family problems.

10. It is estimated that 50 percent of first marriages will end in divorce. Factors associated with divorce include age at marriage, parental divorce, premarital childbearing, education, race, and religion. Reduced economic dependence on marriage underlies many of these trends.

11. Perceptions of the health of the family depend on the theoretical orientation of the viewer. Two problems are loss of commitment and inequality within the family.

Thinking Critically

1. Why do you think the United States has by and large adopted a nuclear family form? What functions does it serve for individuals and society? What are its major dysfunctions?

2. Analyze the mate selection processes that you (or someone close to you) have undergone. Show how propinquity, homogamy, endogamy, and appearance were or were not involved. What role did parents play?

3. Do you know anyone who is taking care of an elderly parent or grandparent? Why do you think that person rather than some other family member has assumed that responsibility? What personal characteristics and what relational characteristics are involved?

4. How many children do you plan to have? What do you think the advantages and disadvantages will be? How do you think you and your significant other will decide whose child-care responsibilities are whose?

5. How would you test the hypothesis that unemployment increases child abuse? What ethical issues would your research design raise?

Sociology on the Net

The Wadsworth Sociology Resource Center: Virtual Society

http://sociology.wadsworth.com/

The companion Web site for this book includes a range of enrichment material. Further your understanding of the chapter by accessing Online Practice Quizzes, Internet Exercises, InfoTrac College Edition Exercises, and many more compelling learning tools.

Chapter-Related Suggested Web Sites

National Council on Family Relations
http://www.ncfr.org

Administration for Children and Families
http://www.acf.dhhs.gov

American Association for Marriage and Family Therapy
http://www.aamft.org

InfoTrac College Edition

http://www.infotrac-college.com/wadsworth

Access the latest news and research articles online—updated daily and spanning four years. InfoTrac College Edition is an easy-to-use online database of reliable, full-length articles from hundreds of top academic journals and popular sources. Conduct an electronic search using the following key search terms:

Marriage

Cohabitation

Divorce

Domestic violence

 For more information related to this chapter, visit the Opposing Viewpoints Resource Center. Be sure you check all the options on the toolbar—viewpoints, references, statistics, and so on—and search under the following subjects:

Divorce

Domestic violence

Teenage pregnancy

Suggested Readings

Cherlin, Andrew. 1999. *Public and Private Families: An Introduction* (2nd ed.). New York: McGraw-Hill. A textbook treatment of families that provides a sophisticated and coherent overview of the major historical patterns in the family as well as the major current issues. Cherlin is a widely respected research scholar who offers a well-balanced and insightful analysis of families.

Cosaro, William A. 1997. *The Sociology of Childhood.* Newbury Park, Calif.: Sage. A synthesis of theoretical and empirical work on children and how they shape and are shaped by their families and society. Also included are chapters on children as social problems and on the social problems of children.

Rubin, Lillian B. 1994. *Families on the Fault Line: America's Working Class Speaks about the Family, the Economy, Race,* *and Ethnicity.* New York: HarperCollins. A book based on interviews conducted with 162 working-class and lower-middle-class families of various ethnic and racial backgrounds during a period in which the economy had contracted and shifted. The author examines the impact of this change on family life and the stress it caused.

Weston, Kath. 1991. *Families We Choose: Lesbians, Gays, Kinship.* New York: Columbia University Press. A book based on the premise that we create our own families: families being those persons whom we actively choose to include as members. This theme is specifically applied to lesbians and gay men as they create families through involved decision making.

CHAPTER 13

Education and Religion

© Will & Deni McIntyre/CORBIS

Educational and Religious Institutions

This chapter examines two institutions, education and religion. Both are central components of our cultural heritage and have profound effects on us as individuals and on the norms and values of our society. Most Americans are directly and personally affected by these institutions: 99 percent of people in the United States have been to school, and 66 percent belong to a church or synagogue (Gallup Poll 2001). Even those who do not go to school or church live in a society that is shaped by the norms and values of these two institutions.

The Sociological Study of Education: Theoretical Views

The **educational institution** is the social structure concerned with the formal transmission of knowledge. It is one of our most enduring and familiar institutions. Nearly 3 of every 10 people in the United States are involved in education on a daily basis as students or staff. As former students, parents, or taxpayers, all of us are involved in education in one way or another.

The **educational institution** is the social structure concerned with the formal transmission of knowledge.

What purposes are served by this institution? Who benefits? Structural-functional and conflict theories offer two different perspectives on these questions.

Structural-Functional Theory: Functions of Education

A structural-functional analysis of education is concerned with the consequences of educational institutions for the maintenance of society. It points out how education contributes to the maintenance of society and how education can be a force for change and conflict.

Structural-functionalists point out that the educational system has been designed to meet multiple needs. Major manifest (intended) functions of education include cultural reproduction, social control, assimilation, training and development, selection and allocation, and promotion of change:

• *Cultural reproduction.* Schools transmit society's culture from one generation to the next by teaching the ideas, customs, and standards of the culture. We learn to read and write our language, we learn the Pledge of Allegiance, and we learn history. In this sense, education builds on the past and conserves traditions.

• *Social control.* Second only to the family, schools are responsible for socializing the young into patterns of conformity. By emphasizing a common culture and instilling habits of discipline, obedience, cooperation, and punctuality, the schools are an important agent for encouraging conformity.

• *Assimilation.* One goal of schools is to assimilate persons from diverse backgrounds into a unified society. By exposing students from all ethnic backgrounds, all

■ In all societies, education is an important means of reproducing culture. In addition to neutral skills such as reading and writing, children learn many of the dominant cultural values. In America, this means learning about George Washington and the American Revolution (although not necessarily about Chief Red Cloud or Sojourner Truth). In Japan, it means learning about the devastation caused when the United States dropped the atom bomb on Hiroshima (although not necessarily about Pearl Harbor) (Gordon 1993).

regions of the country, and all social backgrounds to a more or less common curriculum, schools help create and maintain a common cultural base.

• *Training and development.* In addition to teaching cultural ideas and behavioral patterns, schools transmit knowledge and teach skills like reading, writing, and arithmetic.

• *Selection and allocation.* Schools are like gardeners; they sift, weed, sort, and cultivate their products, determining which students will be allowed to go on and which will not. Standards of achievement are used as criteria to channel students into different programs on the basis of their measured abilities. Ideally, the school system ensures the best use of the best minds. The public school system is a vital element of our commitment to equal opportunity.

• *Promotion of change.* Schools also act as change agents. Although we do not stop learning after we leave school, new knowledge and technology are usually aimed at schoolchildren rather than at the adult population. In addition, schools can promote change by encouraging critical and analytic skills and skepticism. Schools, particularly colleges and universities, are also expected to produce new knowledge.

Conflict Theory: Education and the Perpetuation of Inequality

Conflict theorists agree that education reproduces culture, socializes young people into patterns of conformity, and sifts and sorts. But because conflict theorists see the social structure as a system of inequality designed to benefit the powerful—whites, men, and the wealthy—at the expense of the rest of us, naturally they see any institution that reproduces this culture in a negative light. Three of the major conflict arguments are summarized.

Education as a Capitalist Tool

Some conflict theorists argue that mass education developed because it benefited the interests of the capitalist class (Bowles & Gintis 1976). To support this argument, these theorists point to the schools' **hidden curriculum,** which socializes young people into obedience and conformity. This curriculum—learning to wait your turn, follow the rules, be punctual, and show respect—prepares young people for life in the industrial working class (Gatto 2002). Conflict theorists also note that schools teach young people to expect unequal rewards on the basis of differential achievement, and so teach young people to accept inequality.

This argument has generated a great deal of controversy. Certainly, capitalists and industry have tried to affect the content of schooling in a variety of ways supportive of their interests. Most recently this has taken the form of demanding more emphasis on basic skills—better reading, writing, and arithmetic skills—and less emphasis on creativity and critical thinking.

The **hidden curriculum** socializes young people into obedience and conformity.

Education as a Cultural Tool

In all societies, education is an important means of reproducing culture. In addition to neutral skills such as reading and writing, children learn many of the dominant cultural values. In the United States, this means two things: First, school curricula have typically focused on the history, art, literature, and scientific contributions of Western civilization; second, the way that students are taught has generally been designed to accommodate middle-class European American values. In a pluralistic society such as ours, the cultural heritage children learn at home may not be consistent with the values and history they learn at school. For example, Native American children, taught by their parents never to embarrass a peer, are suddenly expected to correct one another publicly in the classroom. U.S. history texts may describe Indian-Anglo violence but make little or no mention of the factors that provoked Native American nations to warfare. Similarly, textbooks may give little or no coverage to the waves of anti-Chinese violence in the United States in the late nineteenth century, the removal of Japanese Americans to relocation camps during World War II, the rich cultural and economic presence of Mexicans in the American Southwest, or the many contributions of African Americans to U.S. history and culture. The cultural disconnect between minority students and the American school system may partly explain the disproportionately high dropout rates characteristic of U.S. minority students.

Credentialism

One supposed outcome of free public education is that merit will triumph over origins, that hard work and ability will be allowed to rise to the top. Conflict theorists, however, argue that the emphasis on educational credentials as a means of deciding who will get the best jobs and highest status has done little to equalize economic opportunity. Instead, a subtle shift has taken place. Instead of inquiring who your parents are, prospective employers ask what kind of education you have and where you got it. Because people from affluent families tend to end up with the best educational credentials, the emphasis on credentials serves to keep "undesirables" out. Conflict theorists argue that educational credentials are mere window dressing; apparently based on merit and achievement, credentials are often a surrogate for race, gender, and social class. The use of educational credentials to measure individuals' worth is called **credentialism** (Collins 1979).

Credentialism is the use of educational credentials to measure an individual's worth.

Education, Symbolic Interactionism, and Self-Fulfilling Prophecies

In the modern world, the elite cannot directly ensure that their children remain members of the elite. To pass their status attainments on to their children, they must provide their children with appropriate educational credentials (Robinson 1984). To an impressive extent, they are able to do so: Students' educational achievements are very closely related to their parents' social status.

How does this happen? Do schools discriminate? Or, in a fair and open competition, do students with the most resources win? To answer these questions, we need to look at the processes that take place within the schools. This is where a symbolic-interactionist perspective is useful, for it encourages us to look at two of these processes: interpersonal interactions and self-fulfilling prophecies.

Interactions with Teachers

Children from disadvantaged or minority backgrounds are likely to experience multiple hardships that work against them in all their daily experiences in school. They are less likely to have a set of encyclopedias and a computer at home. Furthermore, it is more likely that their parents will be too caught up in the struggles of day-to-day living to have the time or energy to help them with their studies.

■ Children enter school with very unequal backgrounds. Some have been taken to the library and read to while others have been abandoned to the company of the television. Some have been playing learning-readiness games on their computers while others have been playing in dirty stairwells. These inequalities pose a huge obstacle to educational opportunity. As a result, the benefits of education go disproportionately to those who originally had the advantages.

Paying the bills may be more important than trying to improve their children's SAT scores. For these reasons, they begin their interactions with teachers at a decided disadvantage.

But the disadvantages faced by poor and minority students go beyond simple economics. Poorly educated parents, for example, are likely to give more attention to conformity than to independent thinking when socializing their preschoolers (Alexander et al. 1987). Further, they are less likely to attend parent/teacher conferences, are less comfortable talking to their children's teachers, and are less able and sometimes less willing to help their children with their schoolwork. Minority parents, too, may be less comfortable talking with teachers, especially if their English-speaking skills are weak or if teachers treat them disrespectfully.

In addition, white children of the middle and upper-middle classes have more of what has been called **cultural capital**—familiarity and identification with elite culture (Bourdieu 1984). They are more likely to have been introduced at home to the sort of art, music, and books that middle-class teachers value. It also means that they will dress and behave in a socially approved manner. This cultural capital will help them in their interactions with teachers (DiMaggio & Mohr 1985; Farkas et al. 1990; Kalmijn & Kraaykamp 1996; Teachman 1987), and it will influence how effectively they are able to translate their education into occupational success.

Self-Fulfilling Prophecies

One of the major processes that takes place in schools is, of course, that students learn. When they graduate from high school, many can type, write essays with three-part theses, and even do calculus. In addition to learning specific skills, they also undergo a process of cognitive development in which their mental skills grow and expand. In the ideal case, they learn to think critically, to weigh evidence, and to develop independent judgment.

An impressive set of studies demonstrates that cognitive development during the school years is enhanced by complex and demanding work without close supervision and by high teacher expectations. Teachers and curricula that furnish this setting produce students who have greater intellectual flexibility and higher achievement scores. Such students are also more open to new ideas, are less authoritarian, and are less prone to blind conformity (Miller, Kohn, & Schooler 1985, 1986).

Unfortunately, the availability of these ideal learning conditions varies by students' social class, race, and ethnicity. Studies show that teachers demand the most from students who share their backgrounds: The greater the difference between their own background and that of their pupils, the more rigidly they structure their classrooms and the fewer the demands they place on their students (Alexander, Entwisle, & Thompson 1987). As a result, students learn less when they are from a lower social class or different race/ethnicity than is their teacher.

This process is a perfect example of a self-fulfilling prophecy: When teachers assume that certain students cannot succeed, they give those students less opportunity to do so. Similarly, girls don't get taught calculus, boys are discouraged from enrolling in cooking classes, African Americans are encouraged at music but not at science, and so on. This process helps to keep disadvantaged students disadvantaged.

Connections

Personal Application

What social-class advantages or disadvantages did you bring with you to college? Did you grow up around parents who read the *New York Times*, or around parents who could not read? Did your parents pay for you to receive extra tutoring, music lessons, theater tickets, a computer of your own, or a junior year abroad? Or did your parents need you to work in order to help them pay the household bills? Did your high school have all the latest facilities, or a leaky roof and out-of-date textbooks? These advantages and disadvantages will continue to affect you as you go through college.

Cultural capital refers to familiarity and identification with elite culture.

Current Controversies in American Education

In recent years, various proposals have emerged to improve the quality of education in the United States and to give young Americans the tools needed to be more competitive in an increasingly global job market. Three proposals that have met widespread adoption are tracking, high-stakes testing, and school choice.

Tracking

Tracking occurs when evaluations made relatively early in a child's career determine the educational programs the child will be encouraged to follow.

Tracking is the use of early evaluations to determine the educational programs a child will be encouraged to follow. When students enter first grade, they are sorted into reading groups on the basis of ability. By the time they are out of elementary school, some students will be directed into college preparatory tracks, others into general education (sometimes called vocational education), and still others into remedial classes. At all levels, and regardless of their actual abilities, minority and less affluent students are more likely to be put into lower tracks (Gamoran & Mare 1989).

Ideally, tracking is supposed to benefit both the gifted and the slow learners. By having classes that are geared to their levels, both should learn faster and should benefit from increased teacher attention. In addition, classes should run more smoothly and effectively when students are at a similar level. In some ways, this is indeed true. Nevertheless, one of the most consistent findings from educational research is that students are generally helped academically by assignment to high-ability groups but hurt if they are put in low-ability groups (Hallinan & Sorenson 1986; Shavit 1984).

An important reason students assigned to low-ability groups learn less is because they are taught less. They are exposed to less material, asked to do less homework, and,

■ Because students who are assigned to lower-track classes receive few rewards for their academic efforts from either parents or teachers, many will quit school and seek their rewards from delinquent peers.

© AP/Wide World Photos

in general, are not given the same opportunities to learn. One study found that the average student in high-track English classes was asked to do an average of 42 minutes of homework a night; in low-track English classes, the average was 13 minutes (Oakes 1985). Because teachers expect low-track students to do poorly, the students find themselves in a situation where they cannot succeed—a perfect self-fulfilling prophecy.

Less formal processes also operate. Students who are assigned to high-ability groups, for instance, receive strong affirmation of their academic identity; they find school rewarding, have better attendance records, cooperate better with teachers, and develop higher aspirations. The opposite occurs with students placed in low-ability tracks. They receive fewer rewards from their efforts, their parents and teachers have low expectations for them, and there is little incentive to work hard. Many will cut their losses and look for self-esteem through other avenues such as athletics or delinquency (Rosenberg, Schooler, & Schoenbach 1989).

These negative effects of tracking diminish in schools where mobility between tracks is encouraged, teachers are optimistic about the potential for student improvement, and schools continue to place academic demands on students in noncollege tracks (Gamoran 1992; Hallinan 1994).

High-Stakes Testing

Both federal and many local laws now require schools to measure student performance using standardized achievement tests. In many school districts, students must now pass these "high stakes" tests before they can move on to a higher grade. Moreover, not only students, but teachers and schools increasingly are evaluated, punished, or rewarded based on results from standardized examinations.

The emphasis on documenting school achievement through standardized test performance has pressed schools to pay more attention to the quality of the education their students receive, and has encouraged them to make sure that all students receive good training in basic skills such as reading, writing, and arithmetic.

But high-stakes testing also has had unanticipated negative consequences (Berliner & Biddle 1995). Because few schools have received additional resources to meet these new goals, they have had to drop courses in subjects such as art, music, and foreign languages and use these teachers for classes in reading, writing, and arithmetic—even though the teachers lack the training to teach these subjects (Berliner & Biddle). Furthermore, teachers can afford to spend time only on teaching those aspects of the subjects that appear on the tests. In addition, teachers now must devote time simply to teaching test-taking skills. Meanwhile, the testing process itself costs school districts considerable time, energy, and money.

High-stakes testing also means that some students will be held back a grade and thereby stigmatized as failures. At the end of the 2002/2003 school year, for example, 23 percent of Florida third graders were held back because they failed to score high enough on the state reading test (Winerip 2003). Yet research suggests that holding students back can *reduce* their long-term academic performance and *increase* their chances of dropping out. Moreover, those who fail will disproportionately be lower class and minority, due to the cultural biases inherent in any standardized tests. Similarly, when standardized achievement exams are used to determine who should graduate, be admitted to college, or receive financial aid, they are likely to increase rather than reduce inequality between races and social classes (McDill, Natriello, & Pallas 1986). Finally, there is some evidence that, to artificially improve their schools' rankings on high-stakes tests, schools are encouraging or even forcing low-performing students to leave school—turning potential dropouts into "push-outs" (Lewin & Medina 2003).

American Diversity

focus on

What Do IQ Tests Measure?

- How many legs does a Kaffir have?
- Who wrote *Great Expectations*?
- Which word is out of place?
  sanctuary—nave—altar—attic—apse
- If you throw the dice and 7 is showing on top, what is facing down? 7—snake eyes—boxcars—little joe's—11

If you answered two, Dickens, attic, and 7, then you get the highest possible score on this test. What does that mean? Does it mean that you have genetically superior mental ability, that you read a lot, or that you shoot craps? What could you safely conclude about a person who got only two questions right?

The standardized test is one of the most familiar aspects of life in American schools. Whether it is the California or the Iowa Achievement Test, the SAT or the ACT, students are constantly being evaluated. Most of these tests are truly achievement tests; they measure what has been learned and make no pretense of measuring the capacity to learn. IQ tests, however, are supposed to measure the innate capacity to learn—mental ability. On these tests, African American, Hispanic, and Native American students consistently score below white students, and working-class students score substantially below middle-class students. The obvious question is whether these tests are fair measures. Do African American, Hispanic, Native American, and working-class youths

have lower mental ability than middle-class or white youths?

Before we can answer this question, we must first ask another: What is mental ability? It is an aspect of personality, "the capacity of the individual to act purposefully, to think rationally and to deal effectively with his environment" (Wechsler 1958, 7).

Do questions such as those that opened this section measure any of these things? No, they do not. We can all imagine people who act purposefully, think rationally, and deal effectively with the environment but do not know who wrote *Great Expectations* and are ignorant about dice or church architecture. These people may be foreigners, they may have lacked the opportunity to go to school, or they may have come from a subculture where dice, churches, and nineteenth-century English literature are not important.

For this reason, good IQ tests try to measure the ability to think and reason independently of formal education. Examples of items from such a test are reproduced in Figure 13.1. Do these tests achieve their intention? Do they measure the ability to reason independently of years in school, subcultural

background, or language difficulties? Again, the answer seems to be no.

There are two ways in which these tests are not culture-free. The first is that they reflect not only reasoning and knowledge but also competitiveness, familiarity with and acceptance of timed tests, rapport with the examiner, and achievement aspiration. Students who lack these characteristics may do poorly even though their ability to reason is well developed.

The more serious fault with such nonverbal tests is their underlying assumption. Reasoning ability is not independent of learning opportunities. How we reason, as well as what we know, depends on our prior experiences. The deprivation studies of monkeys and hospitalized orphans (see Chapter 3) demonstrate that mental and social retardation occur as a result of sensory deprivation. Just as the body does not develop fully without exercise, neither does the mind. Thus, reasoning capacity is not culture-free; it is determined by the opportunities to develop it. For this reason, there will probably never be an IQ test that will not reflect the prior cultural experiences of the test taker.

 For more information, look up the following subjects in InfoTrac College Edition:
Intelligence tests
Intellect

 Or visit the following Web sites:
National Center for Fair and Open Testing
http://www.fairtest.org
Association for Childhood Education International
http://www.udel.edu/bateman/acei, see link on standardized testing

---

School choice refers to a range of options (vouchers, tax credits, magnet and charter schools, home schooling) that enable families to choose where their children go to school.

School Choice

Concern about the quality of American public education has led to a variety of proposals and programs for increasing school choice. **School choice** refers to a range of options (including tuition vouchers, tax credits, magnet schools, charter schools, and home schooling) that enable families to choose where their children go to school.

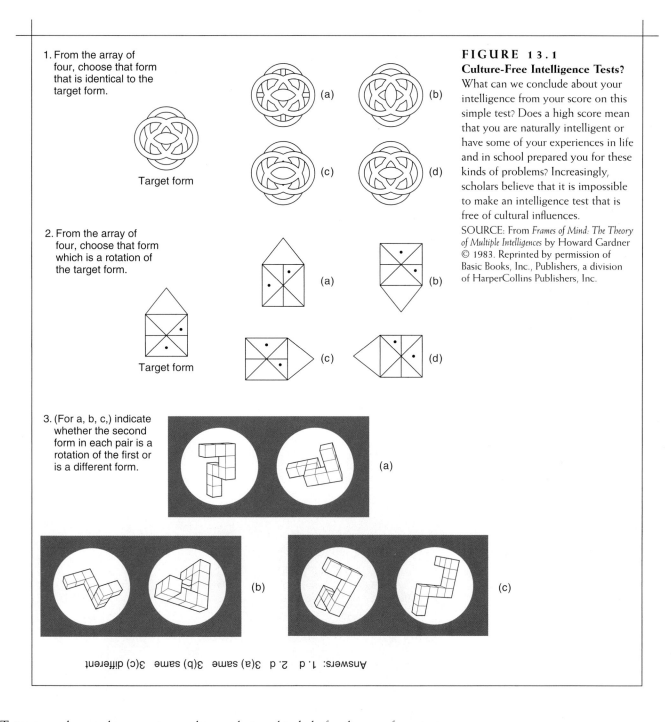

1. From the array of four, choose that form that is identical to the target form.

Target form

(a) (b) (c) (d)

2. From the array of four, choose that form which is a rotation of the target form.

Target form

(a) (b) (c) (d)

3. (For a, b, c,) indicate whether the second form in each pair is a rotation of the first or is a different form.

(a)

(b) (c)

Answers: 1. d 2. d 3(a) same 3(b) same 3(c) different

FIGURE 13.1
Culture-Free Intelligence Tests?
What can we conclude about your intelligence from your score on this simple test? Does a high score mean that you are naturally intelligent or have some of your experiences in life and in school prepared you for these kinds of problems? Increasingly, scholars believe that it is impossible to make an intelligence test that is free of cultural influences.

SOURCE: From *Frames of Mind: The Theory of Multiple Intelligences* by Howard Gardner © 1983. Reprinted by permission of Basic Books, Inc., Publishers, a division of HarperCollins Publishers, Inc.

Tuition vouchers and income tax credits are designed to help families pay for private (and, in some cases, religious) schools. Magnet schools are public schools that try to attract students through offering high-quality special programs or approaches; most commonly these schools emphasize either basic skills, language immersion, arts, or math and science. Charter schools are similar to magnet schools but are privately controlled. Charter schools receive some public funding and are subject to some public oversight, such as requirements that they offer certain courses and that their students meet specified measures of academic performance.

Proponents of school choice argue that when schools compete with each other for students, they will provide better quality services, in the same way that Ford and Chevrolet compete to provide better cars (Chubb & Moe 1990; Schneider, Teske, & Marschall 2000). The school choice movement reflects the animosity toward "big government" that has been building in the United States for the last quarter century, and is part of a broader movement toward **privatization**: the process of taking goods and services out of governmental control and instead treating them like any other marketable commodity. School choice has found supporters on the left as well as the right: black separatists, liberal believers in free-form "alternative schools," and Evangelical Christians all may prefer that their children attend schools where their own values are reinforced.

Although there is some merit to the arguments for school choice, it is difficult to scientifically document its benefits. The problem is that students who participate in school choice programs differ from other students from the beginning. Their parents are often more educated than other parents. More importantly, by definition their parents are committed to seeking out the best education for their children, knowledgeable about the options available, and willing to invest time and effort in obtaining the best options for their children. As a result, no matter what schools their children attend, they will likely do well.

Opponents of school choice identify several unintended negative consequences of these programs. First, these programs reinforce social inequality. Because tuition vouchers and tax credits do not cover the full cost of tuition and transportation, only middle- and upper-income children can afford to use them. Second, because white and affluent parents typically prefer not to send their children to schools with many poor or nonwhite students, school choice programs unintentionally increase segregation (Saporito 2003). Third, school choice programs reduce Americans' commitment to public education and to maintaining high-quality schools in all neighborhoods. Yet all of us suffer when the quality of education the next generation receives is reduced.

Privatization is the process of taking goods and services out of governmental control and instead treating them like any other marketable commodity—something to be bought and sold in a competitive market.

■ Teachers and curricula that couple complex, demanding schoolwork with opportunities for independent learning, extracurricular activities, and ample formal and informal rewards produce students with greater intellectual flexibility and higher achievement test scores.

© Jeffry W. Myers/Corbis

College: Who Goes and What Does It Do for Them?

If your grandparents attended college, they were part of the educational elite. The period following World War II saw a tremendous growth in high school and college education, and today almost half of recent high school graduates ages 18 to 21 are enrolled in college.

Who Goes?

As Figure 13.2 shows, all segments of the population have been affected by this expansion in education, but significant differences still remain. African Americans are still slightly less likely than whites to graduate high school and Hispanics are considerably less likely to do so, partly because many Hispanics emigrated here as adults.

Until just a few years ago, non-Hispanic white males were the group most likely to be enrolled in college, but this has changed (Figure 13.3). Today, white females are the group most likely to be in college, although white men are still most likely

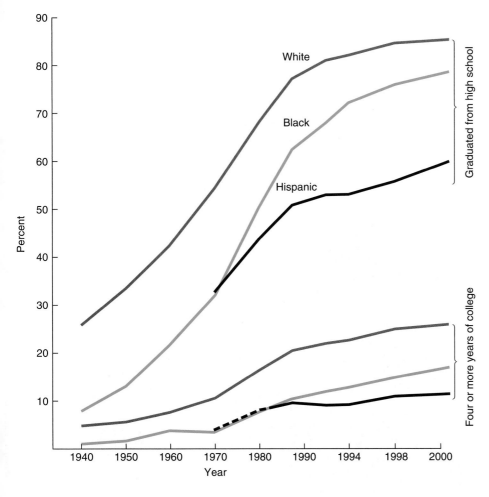

FIGURE 13.2
Educational Achievement of Persons 25 and Older by Race and Ethnicity, 1950–2000
Among whites, the proportion of adults graduating from high school has tripled in the last 50 years; among African Americans, the increase in education has been even more dramatic. Nevertheless, African Americans and Hispanics continue to be disadvantaged in terms of quantity of education.
SOURCE: U.S. Bureau of the Census 1975b, 380; 1989a, 1989b, 1989c; 1993a, 2002b.

FIGURE 13.3
Percentage of High School Graduates Ages 18 to 21 Enrolled in College, by Race, Ethnicity, and Sex, 1975 and 2000
Comparisons by sex, race, and ethnicity show increasing similarity in the likelihood that high school graduates from each category will attend college.
SOURCE: U.S. Bureau of the Census 2002b.

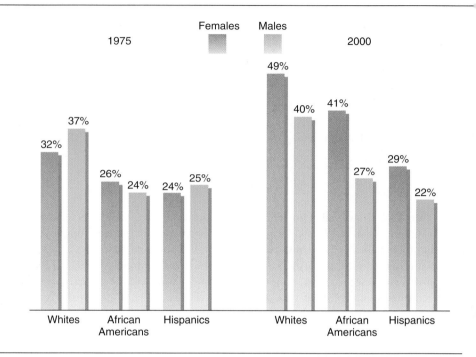

to receive professional and doctoral degrees. African American women are now as likely as white men and almost as likely as white women to enroll in college. African American men, Hispanic women, and Hispanic men continue to have significantly lower rates of college attendance (U.S. Bureau of the Census 2002b).

The Bottom Line—What Will College Do for Me?

A major incentive for college attendance is the belief that it will pay off economically. The evidence clearly supports this. College graduates are more likely to get satisfying professional jobs with good benefits and are less likely to be unemployed.

■ At its best, college encourages creative and critical thinking and broadens one's view of the world.

© Mark Harmel/FPG

TABLE 13.1
Economic Returns of Additional Education, 2001
For all groups, there is a substantial economic payoff for educational attainment after high school. Full-time, full-year workers who completed college annually earned nearly twice as much as high school graduates.

	Median Annual Income				
	4 Years High School	4 Years College	Master's Degree	% Income Increase from High School to Bachelor's	% Income Increase from Bachelor's to Master's
All	$19,900	$37,203	$49,324	87%	33%
Male	28,342	49,984	61,959	76	24
Female	15,664	30,972	40,744	98	32
White	20,294	37,600	49,804	85	33
African American	17,384	35,510	42,505	105	20
Hispanic	17,483	31,235	42,899	79	37

SOURCE: U.S. Bureau of the Census 2002c.

They also earn nearly double the income of high school graduates. But as Table 13.1 shows, the economic impact of education varies by sex and race. At every educational level, earnings of males exceed those of females and earnings of whites exceed those of nonwhites.

Finally, research shows that college graduates are more knowledgeable about the world around them, more active in public and community affairs, less traditional in their religious and gender-role beliefs, and more open to new ideas than those who don't have a college degree (Funk & Willits 1987; Weil 1985); they also lead healthier lives and live longer (Ross & Mirowsky 1999). When one considers these nonmonetary rewards as well as the monetary rewards, the benefits of college become obvious.

The Sociological Study of Religion: Theoretical Views

Unlike education, which we are forced by law to take part in, we have a choice about participating in religious organizations. Nevertheless, most people in the United States choose to participate, and religion is an important part of social life. It is intertwined with politics and culture, and it is intimately concerned with integration and conflict. At the micro level, sociologists examine the consequences of religious belief and involvement for the lives of ordinary individuals. At the macro level, sociologists examine how society affects religion and how religion affects society. Of particular concern is the contribution of religion to social order and social change.

What Is Religion?

How can we define *religion* so that our definition includes the contemplative meditation of the Buddhist monk, the speaking in tongues of a modern Pentecostal, the worship of nature in Native American cultures, and the formal ceremonies of the Catholic

■ Ritual occasions such as bar mitzvahs and first communions bring people together as a community of believers, helping them to reaffirm their shared values.

© Bill Aaron/PhotoEdit

Religion is a system of beliefs and practices related to sacred things that unites believers into a moral community.

church? Sociologists define **religion** as a system of beliefs and practices related to sacred things that unites believers into a moral community (Durkheim [1915] 1961, 62). This emphasis on the sacred allows us to include belief systems that invoke supernatural forces as explanations of earthly struggles (Stark & Bainbridge 1979, 121), as well as those that attribute personality and will to nature. It does not include, however, belief systems such as Marxism or science that do not emphasize the sacred.

Sociologists who study religion treat it as a set of values. They do not, however, ask whether the values are true or false: whether God exists, whether salvation is really possible, or which is the true religion. Rather sociologists examine the ways in which culture, society, and class relationships affect religion and the ways in which religion affects individuals and social structure.

Why Religion?

Religion is a fundamental feature of all societies; Figure 13.4 shows the distribution of the world's population into the major religions as well as the distribution of religions in the United States. Whether premodern or industrialized, each society has forms of religious activity and expressions of religious behavior. Why? One answer is that every individual and every society must struggle to find explanations for, and meaning in, events and experiences that go beyond personal experience. The poor man suffers in a land of plenty and wonders, "Why me?" The woman whose child dies wonders, "Why mine?" The community struck by flood or tornado wonders, "Why us?" Beyond these personal dilemmas, people may wonder why the sun comes up every morning, why there is a rainbow in the sky, and what happens after death.

The **profane** represents all that is routine and taken for granted in the everyday world, things that are known and familiar and that we can control, understand, and manipulate.

Religion helps us interpret and cope with events that are beyond our control and understanding; tornadoes, droughts, and plagues become meaningful when they are attributed to the workings of some greater force (Spilka, Shaver, & Kirkpatrick 1985). Beliefs and rituals develop as a way to control or appease this greater force, and eventually they become patterned responses to the unknown.

The **sacred** consists of events and things that we hold in awe and reverence—what we can neither understand nor control.

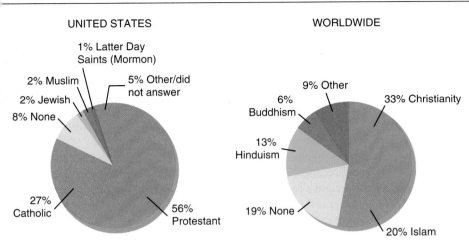

UNITED STATES

WORLDWIDE

1% Latter Day
Saints (Mormon)

2% Muslim

2% Jewish

8% None

27%
Catholic

5% Other/did
not answer

56%
Protestant

9% Other

6%
Buddhism

13%
Hinduism

19% None

33% Christianity

20% Islam

FIGURE 13.4
Religious Affiliation in the United States and Worldwide, 2000
83 percent of Americans are Protestant or Catholic. Worldwide, Christianity remains the largest religion, but it is shrinking while Islam is growing.
Note: Judaism, Sikhism, Jainism, Shinto, Zoroastrianism, and Confucianism each account for less than 1% of world population.
SOURCE: www.gallup.com/poll/indicators/indreligion3.asp, www.religioustolerance.org, U.S. Department of the Census 2002b.

Rain dances may not bring rain, and prayers may not lead to good harvests, but both provide a familiar and comforting context in which people can confront otherwise mysterious and inexplicable events. Regardless of whether they are right or wrong, religious beliefs and rituals help people cope with the extraordinary events they experience.

Within this general sociological approach to religion, there are three distinct theoretical perspectives. Structural-functionalists, not surprisingly, focus on the functions that religion serves for both individuals and societies. Conflict theorists focus on how religion can foster or repress social conflict. The third perspective, associated with the work of Max Weber, combines elements from these first two perspectives.

Durkheim: Structural-Functional Theory of Religion

The structural-functional study of religion begins, most importantly, with the work of Emile Durkheim. Durkheim began his analysis of religion by identifying the three elements shared by all religions, which he called the elementary forms of religion ([1915] 1961).

First, all religions divide human experience into the sacred and the profane. The **profane** represents all that is routine and taken for granted in the everyday world, things that are known and familiar, that we can control, understand, and manipulate. The **sacred,** by contrast, consists of the events and things that we hold in awe and reverence—what we can neither understand nor control.

Second, all religions hold a set of beliefs about the supernatural that help people explain and cope with the uncertainties associated with birth, death, creation, success, failure, and crisis. These beliefs form the basis for official religious doctrines.

Third, all religions have rituals. In contemporary Christianity, rituals are used to mark such events as births, deaths, weddings, Jesus's birth, and the resurrection. In earlier eras, many Christian rituals were closely tied to planting and harvest; these are still important ritual occasions in many religions.

AP/Wide World Photos

Religious rituals help individuals cope with events that are beyond human understanding, such as death, illness, drought, and famine.

The Functions of Religion

Durkheim argued that religion is universal, so it must serve important functions for society. At the societal level, the major function of religion is that it gives tradition a moral imperative. Most of the central values and norms of any culture are reinforced through its religions. These values and norms cease to be merely the *usual* way of doing things and become perceived as the *only* moral way of doing them. They become sacred. When a tradition is sacred, it is continually affirmed through ritual and practice and is largely immune to change.

For individuals, Durkheim argued that the beliefs and rituals of religion offer support, consolation, and reconciliation in times of need. On ordinary occasions, many people find satisfaction and a feeling of belongingness in religious participation. This feeling of belongingness is the moral community, or community of believers, that is part of the definition of religion.

Conflict Theory: Marx and Beyond

Like Durkheim, Marx saw religion as a supporter of tradition. This support ranges from injunctions that the poor and oppressed should endure rather than revolt (blessed be the poor, blessed be the meek, and so on) and that everyone should pay taxes (give unto Caesar) all the way to the endorsement of inequality implied by a belief in the divine right of kings.

Marx differed from Durkheim by interpreting the support for tradition in a negative light. Marx, an atheist, saw religion as the "opiate of the masses"—a way the elite kept the eyes of the downtrodden happily focused on the afterlife so that the poor would not notice their earthly oppression. This position is hardly value-free, and much more obviously than structural-functional theory, it does make a statement about the truth or falsity of religious doctrine.

Modern conflict theory goes beyond Marx's view. Its major contribution is in identifying the role that religions can play in fostering or repressing conflict between social groups. Religion has certainly *contributed* to the conflict between Serbian Orthodox Christians, Croatian Catholics, and Bosnian Muslims, as well as between Protestant and Catholic Irish. On the other hand, it has also served to *repress* conflict when Muslim, Christian, Hindu, and other clergy have taught impoverished people to accept their fate as God's will. In both instances, religion can and has served as a tool for groups to use in their struggles for power. Interestingly, although Marx believed that religion would always help those who have power to maintain it, recent theorists have identified cases in which religion has helped those who *lack* power to improve their position. One example of this is the powerful role the African American church and leaders such as the Reverend Martin Luther King, Jr., played in fighting for civil rights in the United States.

Another contribution of conflict theory to the analysis of religion is the idea of the dialectic—that contradictions build up between existing institutions and that these contradictions lead to change. Specifically, conflict theorists suggest that social change can foster change in religion.

Weber: Religion as an Independent Force

Max Weber's influential theory of religion combines elements of structural-functionalism and of conflict theory. Like Durkheim and other structural-functionalists, Weber was interested in the forms of religion and their consequences for individuals and society. Further, Weber was also concerned with the processes through which religious answers are developed and how their content affects society. In his focus on the relationship between social and religious change, Weber shared interests similar to those of the conflict theorists. Whereas conflict theorists typically focus on the ways that social conflict leads to religious change, Weber was most concerned with the way that changes in religious ideology could promote other types of social change.

For most people, religion is a matter of following tradition; people worship as their parents did before them. To Weber, however, the essence of religion is the search for knowledge about the unknown. In this sense, religion is similar to science: It is a way of coming to understand the world around us. And as with science, the answers religion provides may challenge the status quo as well as support it.

Where do people find the answers to questions of ultimate meaning? Often they turn to a charismatic religious leader. **Charisma** refers to extraordinary personal qualities that set the individual apart from ordinary mortals. Because these extraordinary characteristics are often thought to be supernatural in origin, charismatic leaders can become agents for dramatic social change. Charismatic leaders include Christ, Muhammad, and, more recently, Joseph Smith (Mormonism), and David Koresh (Branch Davidians). Such individuals give answers that often disagree with traditional answers. Thus, Weber sees religious inquiry as a potential source of instability and change in society.

In viewing religion as a process, Weber gave it a much more active role than did Durkheim. This is most apparent in Weber's analysis of the Protestant Reformation.

> **Charisma** refers to extraordinary personal qualities that set an individual apart from ordinary mortals.

The Protestant Ethic and the Spirit of Capitalism

In a classic analysis of the influence of religious ideas on other social institutions, Weber ([1904–1905] 1958) argued that the Protestant Reformation paved the way for capitalism. The values of early Protestant religion, which Weber called the Protestant ethic, included the belief that work, rationalism, and plain living are moral virtues, whereas idleness and indulgence are sinful. What happens to a person who follows this ethic—who works hard, makes business decisions on rational rather than emotional criteria (for example, firing inefficient though needy employees), and is frugal rather than self-indulgent? Such a person will grow rich. According to Weber, it was not long before wealth became an end in itself. At this point, the moral values underlying early Protestantism became the moral values underlying early capitalism.

In the century since Weber's analysis, other scholars have explored the same issues, and many have come to somewhat different conclusions. Nevertheless, this research has not changed Weber's major contribution to the sociology of religion: that religious ideas can be the source of tension and change in social institutions.

Tension Between Religion and Society

Each religion is confronted with two contradictory yet complementary tendencies: the tendency to reject the world and the tendency to compromise with the world (Troeltsch 1931). When a religion denounces adultery, homosexuality, and fornication, does the church categorically exclude adulterers, homosexuals, and fornicators, or does it adjust its expectations to take common human frailties into account? If "it is easier for a camel to go through the eye of a needle than for a rich man to enter

■ This young American belongs to the Hindu cult commonly known as Hare Krishna.

the kingdom of God," must the church require that all its members forsake their worldly belongings?

How religions resolve these dilemmas is central to their eventual form and character. Scholars distinguish two general types of religious organizations: church and sect. The *church* represents the successful compromisers, and the *sect* represents the virtuous outsiders (Sherkat & Ellison 1999).

In everyday language, we use the term "church" to refer to Christian religious organizations or places of worship. Sociologists, on the other hand, use the term **church** to refer to any religious organization that has become institutionalized. Churches have endured for generations, are supported by and support society's norms and values, and have become an active part of society. Their involvement in society does not necessarily mean that they have compromised essential values. They still retain the ability to protest injustice and immorality. From the abolition movement of the 1850s to the Civil Rights struggle of the 1960s to the demonstrations against U.S. military intervention in Iraq, churchmen and women have been in the forefront of social protest. Nevertheless, churches are generally committed to working with society. They may wish to improve it, but they have no wish to abandon it.

Sects are religious organizations that reject the social environment in which they exist (Stark & Bainbridge 1985, 1987). Religions that reject sexual relations (Shakerism), automobiles (the Amish), or monogamy (nineteenth-century Mormonism) are examples of sects that differ so much from society's norms that their relationships with the larger society are often hostile. They reject major elements of the larger culture and are in turn rejected by it.

The categories of church and sect are what Weber referred to as *ideal* types. The distinguishing characteristics of each type are summarized in the Concept Summary. Although no church or sect may have all of these characteristics, the ideal types serve as useful benchmarks against which to examine actual religious organizations.

Churches are religious organizations that have become institutionalized. They have endured for generations, are supported by and support society's norms and values, and have become an active part of society.

Sects are religious organizations that reject the social environment in which they exist.

Distinctions Between Churches and Sects

Church and sect are ideal types against which we can assess actual religious organizations. Many religious organizations combine some characteristics of both. Nevertheless, Catholicism and Lutheranism are obviously churches, whereas the Nation of Islam has many of the characteristics of a sect.

	Churches	**Sects**
Degree of tension with society	Low	High
Attitude toward other institutions and religions	Tolerant	Intolerant; rejecting
Type of authority	Traditional	Charismatic
Organization	Bureaucratic	Informal
Membership	Establishment	Alienated
Examples	Catholics, Lutherans	Jehovah's Witnesses, Amish, Nation of Islam

Churchlike Religions

Although all churches are institutionalized religious organizations, some are more institutionalized than others. In some societies, a state church automatically includes virtually every member of the society. The fate of the church and the fate of the nation are wrapped up in each other, and the church is vitally involved in supporting the dominant institutions of society. Examples include the Roman Catholic church in Europe during the Middle Ages and Islam in Afghanistan during the Taliban's rule. In other societies, religious organizations accommodate both to society and to other religions. In the United States, Jewish, Catholic, Lutheran, Methodist, and Episcopalian clergy meet together in ecumenical councils, pray together at commencements, and generally adopt a live-and-let-live policy toward one another.

Structure and Function of Churchlike Religions

Churchlike religions tend to be formal bureaucratic structures with hierarchical positions, specialization, and official creeds specifying their religious beliefs. Leadership is provided by a professional staff of ministers, rabbis, imams, or priests, who have received formal training at specialized schools. These leaders are usually arranged in a hierarchy from the local to the district to the state and even the international level. Religious services almost always prescribe formal and detailed rituals, repeated in much the same way from generation to generation. Congregations often function more as audiences than as active participants. They are expected to stand up, sit down, and sing on cue, but the service is guided by ceremony rather than by the emotional interaction of participants.

Generally, people are born into churchlike religions rather than converting to them. People who do change churches often do so for practical rather than emotional reasons: They marry somebody of another faith, another church is nearer, or their friends go to another church. Individuals also might change churches when their social status rises above that of most members of their church, so that Baptists might become Methodists and Methodists might become Episcopalians (Sherkat &

Ellison 1999). Most individuals who change churches have relatively weak ties to their initial religion. Nevertheless, few make large changes: Orthodox Jews become Conservative Jews and members of one small Baptist church join a different small Baptist church (Stark & Finke 2000).

Churchlike religions tend to be large and to have well-established facilities, financial security, and a predominantly middle-class membership. As part of their accommodation to the larger society, churchlike religions usually allow the scriptures to be interpreted in ways that are relevant to modern culture. Because of these characteristics, these religions are frequently referred to as *mainline churches*, a term denoting their centrality in society.

Sectlike Religions

Within the category of religious organizations that have greater tension with society, there is a great deal of variability. We can distinguish three levels of tension. First are cults, with the greatest tension, then sects, and finally established sects. The last begin to approach institutionalization.

Cults

A **cult** is a sectlike religious organization that is independent of the religious traditions of society.

A religious organization that is independent of the religious traditions of society is a **cult** (Stark & Bainbridge 1979). Examples of cults in the United States are the Church of Scientology, the Unification Church (Moonies), and the Hare Krishna. Each of these religions is rather foreign to the Judeo-Christian tradition of the United States: They have a different God or Gods or no God at all; they don't use the Old Testament as a text.

Cults tend to arise in times of societal stress and change, when established religions do not seem adequate for explaining the upheavals that individuals experience. Because they are so alien to society's institutions, however, cults are of little assistance in helping people cope with everyday lives in mainstream society. Instead, they often urge their members to alter their lives radically and to withdraw from society altogether. Because of the radical changes they demand, cults generally remain small, and many fail to survive more than a few years.

An **established sect** is a sect that has adapted to its institutional environment.

Sects

Within the general category of sectlike religions, those called sects occupy a medium position. They reject the social world in which they live, but they embrace the religious heritage of the surrounding society. The Amish are an excellent example: They base their lives on a strict reading of the Bible and remain aloof from the contemporary world.

Sects often view themselves as restoring true faith, which has been mislaid by religious institutions too eager to compromise with society. They see themselves as preservers of religious tradition rather than innovators. Like the Reformation churches of Calvin and Luther, they believe they are cleansing the church of its secular associations. However offbeat it may be in comparison to mainline churches, if a religious group in the United States uses the Bible as its source of inspiration and guidance, then it is probably a sect rather than a cult.

Established Sects

A sect that has adapted to its institutional environment has become an **established sect.** An established sect often retains the belief that it is the one true church, but it is less antagonistic to other faiths than are sects.

Whereas sects often withdraw from the world in order to preserve their spiritual purity, established sects are active participants. Frequently, the motivation for this participation in the world is to spread their message, to make converts, and to change social institutions. To the extent that they successfully meet these goals, they reduce the tension between themselves and society. To effect social change, they must have lobby groups and participate in political, economic, and educational institutions. To spread their message and make converts, they must associate with many outsiders.

The Mormons are a classic example of an established sect (Arrington & Bitton 1992). Over time, they have increased their accommodation to the larger society: As an official church position, they have abandoned polygamy, allowed African Americans to become priests, left the seclusion of a virtual state church in Utah, and spread throughout the world seeking converts. Mormons nevertheless retain many characteristics of sectlike religions, including lack of a paid clergy, an emphasis on conversion, and the belief that certain texts unique to Mormonism are at least as important as the Bible. These characteristic organizational features of sect-like religions are covered in the next section.

Structure and Function of Sectlike Religions

The hundreds of cults and sects in the United States exhibit varying degrees of tension with society, but all are opposed to some basic societal institutions. Not surprisingly, these organizations tend to be particularly attractive to people who are left out of or estranged from society's basic institutions—the poor, the underprivileged, the handicapped, and the alienated. For this reason, sects have been called "the church of the disinherited" (Niebuhr [1929] 1957). Not all of the people who convert to sectlike religions are poor or oppressed. Many are middle-class people who are spiritually rather than materially deprived. They are individuals who find established churches too bureaucratic; they seek a moral community that will offer them a feeling of belongingness and emotional commitment (Barker 1986). Others join sectlike religions such as Hasidic Judaism or Christian fundamentalist groups because they want to hold on to traditional norms and values that seem to have fallen from favor (Davidman 1991).

Sect membership is often the result of conversion or an emotional experience. Members do not merely follow their parents into the church; they are reborn or born again. Religious services are more informal than those of churches. Leadership remains largely unspecialized, and there is little, if any, professional training for the calling. The religious doctrines emphasize otherworldly rewards, and the scriptures are considered to be of divine origin and therefore subject to literal interpretation.

Sects and cults share many of the characteristics of primary groups: small size, informality, and loyalty. They are closely knit groups that emphasize conformity and maintain significant control over their members.

Many, if not all, of the churches in the world today started out as sects. Over the centuries, they grew and became part of the institutional structures of society. Not all sects, of course, adjust and become assimilated in this way. Some remain established sects, antagonistic to many institutions in the general society; others suffer eventual extinction.

A Case Study: Islam

Islam was founded in the seventh century A.D. by an Arab prophet named Muhammad, near Mecca, in what is now Saudi Arabia. It is the fastest-growing religion in the world, encompassing one-fifth of the world's population ("Islam" 2001). (In contrast, Christianity is shrinking—as is the number who claim no religion.) No matter where they are

Connections

Example

Unlike mainstream Mormonism, fundamentalist Mormonism remains a full-fledged (rather than established) sect, functioning outside of the law as well as mainstream religious traditions. Members of various breakaway Mormon groups believe that God desires them to engage in polygamy, and that the mainstream Mormon church moved away from God's will when it abandoned polygamy. Thousands are believed to practice polygamy in Utah and nearby states; most live in isolated communities, under close control of a charismatic leader. Law enforcement officers in Arizona and Utah have begun actively prosecuting polygamists, primarily because of cases in which middle-aged and elderly men married girls as young as 13. The mainstream Mormon church excommunicates identified polygamists.

Muslims throughout the world believe that, if possible, they should make at least one pilgrimage to the Sacred Mosque in Mecca. Aside from this commitment and a shared belief in the four other pillars of Islamic faith, however, there is considerable variation both within and between countries in the way that Islam is practiced and in the relationship between the church and the state.

found, Muslims share a set of common beliefs. All Muslims believe in a single all-powerful God whose word is revealed to the faithful in the Koran, a book similar to the Christian Bible or the Jewish Torah. All Muslims must follow the Five Pillars of Islam. They must (1) profess faith in one almighty God and Muhammad his prophet, (2) pray five times daily, (3) make charitable donations to the Muslim community and the poor, (4) fast during daylight hours during the month of Ramadan, the time when the Koran was revealed to Muhammad, and (5) if possible, make at least one pilgrimage to Mecca. Prayer is usually in a mosque (an Islamic house of worship) and is lead by an imam (a religious scholar). There is no formal central authority. Apart from this basic doctrine, however, there is considerable variation across countries in the relationship between Islamic clergy and the government and in the interaction between followers of the faith and members of the larger community. In some nations Islam more closely resembles a church, and in others it more closely resembles a sect.

Islam as a Sectlike Religion: The United States

Islam in the United States can be traced back to the importation of African Muslim slaves in the eighteenth and nineteenth centuries. In the late 1800s, new Muslim immigrants arrived and began to settle in the Midwest, especially North Dakota and Iowa. Today there are an estimated 6 million Muslims in the United States, about 40 percent of whom are African Americans (Smith 1999).

Of the many varieties of Islam practiced in the United States, perhaps the most widely recognized is that of the Nation of Islam, popularly known as "Black Muslims." Although the religious beliefs of this group differ markedly from those of traditional Islam, the Nation of Islam has become more like other mainstream Islamic groups in recent years (Smith 1999). The Nation of Islam functions as an established sect, actively participating in the broader social environment without antagonism to Christianity, the dominant American religion. (However, deeply-entrenched antisemitism remains a significant issue.) Membership in the Nation of Islam is growing most rapidly among poor and disenfranchised African American inner city residents. For these individuals, Islam can provide a sense of hope, community, identity, and freedom from the white-dominated world around them. In addition, the Muslim emphasis on community activism—antidrug campaigns and economic development—and on discipline and modest dress provide the sense of order and belonging commonly provided by all sects.

Islam as a Churchlike Religion: Afghanistan and Egypt

Until the collapse of the fundamentalist Islamic Taliban government following the U.S. invasion of Afghanistan in 2001, Afghanistan offered a good example of a modern-day Muslim state religion. Church and state were intertwined, with every citizen bound by religious law. For instance, all Afghan women, regardless of religion or nationality, could receive up to 80 lashes with a whip if they failed to cover their whole bodies and could be executed if they had sex outside of marriage. So long as Islamic clergy held all political power, there was little tension revealed between religion and the larger society.

In contrast, Egypt has a more or less secular government even though 90 percent of its population belongs to the Islamic church (Rubin 2002). Although there has been a recent upsurge in antigovernment violence by radical fundamentalist sects, Islam in Egypt is still more like a churchlike religion than a sect. Christians and

Muslims share the streets and businesses in relative peace. The government tacitly allows more moderate Islamic groups, like the Islamic Brotherhood, to exist in spite of official bans. Recently, the Egyptian government has also increased public discussion of moderate Islamic values and promised to improve services to the less fortunate, suggesting a willingness of church and state to work together in supportive roles (Murphy 1994). At the same time, the government also uses terror and repression to keep radical fundamentalists under control.

Islamic Fundamentalism

Recent years have seen a worldwide increase in Islamic fundamentalism. As is true of Christian fundamentalist churches, fundamentalist Muslim sects appeal especially to individuals who lack economic and political power. But, as we saw in Chapter 8, Islamic fundamentalism also appeals to educated Muslims who despair of Western political and cultural domination and of the decline in traditional morality that they believe accompanied that domination (Amanat 2001; Barber 2001; Jacquard 2002). Islamic fundamentalists call for a rejection of the excesses and corruption of modern, secular culture and a return to "true" religious principles. Only the most radical Islamic fundamentalist sects advocate violence as a means of restoring religious values and law. Most Muslims, in fact, say the concept of *jihad*—holy war—refers not to offensive warfare but rather to the need to defend social justice, through spiritual, economic, political, and, only when absolutely necessary, military means (Lawrence 1998). Islamic fundamentalism does not uniformly endorse violence nor is it a single unified movement or sect. It varies from country to country and includes both religious and political elements; acts of violence do not derive from core Islamic fundamentalist beliefs but from the problematic behavior of a limited number of individuals, sects, and political regimes (Gordon 1993).

Religion in the United States

When asked what religion they belong to, only 6 percent of Americans say they belong to no church. Most people identify themselves not only as religious but also as members of some particular religion. Most (56 percent) call themselves Protestants, but 27 percent are Catholics and 2 percent are Jews (see Figure 13.4). Despite their differences, these three religions embrace a common Judeo-Christian heritage. They accept the Old Testament, and they worship the same God. They rely on a similar moral tradition (the Ten Commandments, for example), which reinforces common values. This common religious heritage supplies an overarching sense of unity and character to U.S. society—providing a framework for the expression of our most crucial values concerning family, politics, economics, and education.

U.S. Civil Religion

Americans also share what has been called a civil religion (Bellah 1974, 29; Bellah et al. 1985). **Civil religion** is a set of institutionalized rituals, beliefs, and symbols sacred to U.S. citizens. These include reciting the Pledge of Allegiance and singing the national anthem, as well as folding and displaying the flag in ways that protect it from desecration. In many U.S. homes, the flag and/or a picture of the president is displayed along with a crucifix or a picture of the Last Supper.

Civil religion has the same functions as religion in general: It is a source of unity and integration, providing a sacred context for understanding the nation's history and

Civil religion is the set of institutionalized rituals, beliefs, and symbols sacred to the U.S. nation.

current responsibilities (Wald 1987). For example, shortly after the American colonies declared their independence from Britain, George Washington was declared commander of the U.S. army. With little military experience or charisma, Washington's major qualification for the job was that he didn't want it. Within weeks, he became an object of near worship. Why did this cult of Washington develop? It emerged, in part, because Washington symbolized the fledgling nation's unity and, in part, because his disdain for power made him a hero. In worshiping Washington, the colonists were worshiping their nation and the virtues they believed it embodied (Schwartz 1983). Since then, we have made liberty, justice, and freedom sacred principles. We believe the American way is not merely the usual way of doing things here but also the only

Within weeks of his appointment as commander of the army, Washington became an object of near worship. (ART WORK: "Memorial to George Washington," painting on glass.)

moral way of doing them, a way of life blessed by God. The motto on our currency, our Pledge of Allegiance, and our national anthem all bear testimony to the belief that the United States operates "under God" with God's direct blessing.

Trends and Differentials in Religiosity

Until the 1970s, many scholars implicitly assumed that religion would decline in importance as science and technology increased society's ability to explain and control previously mysterious events. As a result, they assumed that **secularization**—the process of transferring things, ideas, or events from sacred authority (the clergy) to profane authority (the state, medicine, and so on)—would gradually increase.

It is true that science has given us physical rather than supernatural explanations for more and more phenomena, but the rise over the last 30 years of fundamentalist Christian, Jewish, and Muslim groups has demonstrated that modernization and science do not necessarily undermine religious commitment (Sherkat & Ellison 1999; Stark & Finke 2000). Some theorists argue that continued religious commitment reflects a *rational* decision on the part of individuals who conclude that the costs of commitment (in time and money) are outweighed by the benefits: explanations for otherwise inexplicable events, the promise of supernatural rewards, integration into a community of like-minded individuals, the lending of supernatural authority to traditional values and practices, and so on (Stark & Finke).

If anything, religion plays a stronger role in American life now than in the past. More than two-thirds (68 percent) of U.S. residents now belong to a church or synagogue compared to no more than 12 percent in 1776 (Finke & Stark 1992). Moreover, conservative religious groups have grown more rapidly than liberal groups in recent years (Sherkat & Ellison 1999). One intriguing explanation for this is offered by sociologists Rodney Stark and Roger Finke (2000). They argue that church membership and religious participation are higher in the United States than in most other industrialized nations because of our highly developed, competitive, and unregulated **religious economy.** According to this perspective, there are so many religious organizations in this country that they continually compete with one another to provide better "consumer products," thereby generating greater "market demand" for them.

Although church membership in the United States appears to have increased, individual religiosity has declined (Table 13.2). *Religion* refers to a set of beliefs.

Secularization is the process of transferring things, ideas, or events from the sacred realm to the profane.

Religious economy refers to the competition between religious organizations to provide better "consumer products," thereby creating greater "market demand" for their own religions.

TABLE 13.2
Changing Religious Commitment, 1962–2001
During the last 40 years, there has been little decline in outward religious observance. There has, however, been a drop in the proportion who say that religion is very important to their lives, and *a sharp drop* in the proportion who think that the Bible is the actual word of God.

	1962–65	2000–01
Belong to a church or synagogue	73%	66%
Attended church last week	46	41
Have no religion	2	8
Religion is very important to their own lives	70	60
Believe Bible is actual word of God, to be taken literally word for word	65	33

SOURCE: Gallup Poll 2001; U.S. Bureau of the Census 2002b.

Many observers, including Marx, have thought that religion ought to appeal disproportionately to the poor and disadvantaged. Although the poor and disadvantaged are more likely than others to be fervent in their beliefs, studies show that churchgoing is correlated more strongly with being conventional than with being disadvantaged.

Religiosity is an individual's level of commitment to religious beliefs and to acting on those beliefs.

Religiosity refers to an individual's level of commitment to religious beliefs and to acting on those beliefs. Rates of church attendance have changed very little over the last several decades, but the proportion of Americans who say religion is very important in their lives has declined and the proportion who believe the Bible is the literal word of God has declined sharply. Although almost everybody believes in God, some people are more involved in religion than others (Table 13.3). The most striking differences in religiosity are related to age and sex. Older people, women, Southerners, and African Americans are more likely than others to attend religious services regularly (Sherkat & Ellison 1999).

One interesting question is the relationship between income, education, and religiosity. In the past, many scholars assumed that religion would appeal disproportionately to the poor, who were in greater need of hope, and to the uneducated, who were more likely to lack "scientific" explanations for natural and human events. As the data in Table 13.3 indicate, however, people with a college education are as likely to attend church as people who did not attend college. Although those with a college education are less likely, overall, to say that religion is important to them, those who are religious are as likely as their less educated counterparts to hold conservative religious beliefs (Sherkat & Ellison 1999). Moreover, those who work in the physical and life sciences, such as physics and biology, are more likely to be religious than those who work in the social and behavioral sciences, such as sociology and psychology (Stark & Finke 2000). In general, churchgoing appears to be more strongly associated with being conventional than with being disadvantaged. It is a characteristic of people who are involved in their communities, belong to other voluntary associations, and hold traditional values.

Consequences of Religiosity

Because religion teaches and reinforces values, it has consequences for attitudes and behaviors. People who are more religious tend to be friendlier, more cooperative, healthier, happier, and more satisfied with their lives and marriages (Ellison 1991, 1992; Sherkat & Ellison 1999; Thomas & Cornwall 1990). These benefits in large part stem from the social support and sense of belonging that individuals receive from their religious communities.

Persons who are more religious tend to have more conservative attitudes on sexuality and personal honesty; they also may have more conservative attitudes about family life, being more likely, for instance, to support the use of corporal punishment in disciplining children (Ellison, Bartkowski, & Segal 1996). Not surprisingly, some conservative religious groups have played significant roles in supporting conservative political movements, such as the antiabortion movement and certain right-wing hate groups.

Yet, we should not assume that church members necessarily adopt the conservative attitudes of their churches. Primarily because of immigration, for example, the membership in the Roman Catholic church in the United States has increased by about 25 percent over the past 40 years (Monroe 1995, 383). However, these new American Catholics aren't as likely to obey church leaders or to attend Sunday Mass as Catholics used to be. Although the Pope sees artificial birth control as sinful, 82 percent of U.S. Catholics disagree; 64 percent disagree that abortion is always wrong. Just one-third agree that only males should be priests. Nevertheless, 86 percent say they approve of the Pope (Sheler 1995), a sign, perhaps, that in the American Catholic church, members take what they like and leave the rest.

Moreover, even though religious training generally teaches and reinforces conventional behavior, religion and the church can be forces that promote social change. In the United States, African American churches and clergy played a significant role in the Civil Rights movement of the 1950s and 1960s; religion also played a supportive role in the struggle of Appalachian coal miners to unionize in the 1920s and 1930s (Billings 1990). In Latin America, liberation theology aims at the creation of democratic Christian socialism that eliminates poverty, inequality, and political oppression (Smith 1991).

The Religious Right

One of the most striking changes in religion in the United States today is the vitality and growth of the fundamentalist churches compared with the relative decline of mainline churches (Woodberry & Smith 1998). Fundamentalists can be found in all religions; there are fundamentalist Catholics, Baptists, Presbyterians, Lutherans, and Jews. Their common aim is to restore what they believe is the proper role of religion in a civil society. One outgrowth of this is the development of the Religious Right.

The Religious Right is a loose coalition of Protestant, and on some issues Catholic, fundamentalists who believe that the U.S. government and social institutions must be made to operate according to traditional Christian principles and practices (Woodberry & Smith 1998). They believe that it is a Christian obligation to be active politically in making the United States a Christian nation. They are united by opposition to abortion, liberal sexual mores, gay and lesbian rights, and the strict separation of church and state, among other things. The Religious Right is politically most powerful in the South and West.

The Religious Right is best understood as a political rather than a religious movement. It aims to influence public policy through normal political processes—lobbying, campaign contributions, and getting out the vote. In doing so, of course, it is building onto the existing foundations of civil religion.

Despite this link to established values and despite the growth in fundamentalist religions, the movement has had more publicity than power (Woodberry & Smith 1998). First, its power is limited because many fundamentalists believe that churches should save their energies and resources solely for religious endeavors. Second, fundamentalists clearly differ among themselves on matters of religious conviction; they are also divided on political issues such as abortion, gun control, military defense, and pornography (Moore & Whitt 1986). Interestingly, conservative Protestants do not differ from more liberal religious groups in their attitudes toward racial equality and in many ways hold more progressive positions on economic issues (Davis & Robinson 1996). Furthermore, with the exception of abortion, studies show that during the last 15 years, the social attitudes of evangelical Protestants have become more like those of Roman Catholics, Jews, and the more moderate to liberal Protestant denominations (DiMaggio, Evans, & Bryson 1996). Finally, the power of the Religious Right is limited because its leaders do not hold the kinds of positions in the media and elsewhere that would allow them to control how their views are presented to the broader public (Woodberry & Smith 1998). The net result is that although the Religious Right exercises "veto power" over certain issues within the Republican Party (such as ensuring that no presidential candidate will actively support increased access to abortion), its overall power is relatively limited.

TABLE 13.3
Percentage of Americans Who Attend Religious Services Regularly, 2002
Older people, women, Southerners, and African Americans are more likely than other Americans to attend religious services regularly.

Total	40%
Region	
Midwest	43
South	47
East	34
West	37
Age	
Below 29	33
30–49	41
50 and older	52
Sex	
Male	32
Female	47
Education	
No College	39
College Incomplete	41
College Graduate	42
Postgraduate	46
Race*	
White	36
Black	50
Other	36

*Data from 2000.

SOURCE: General Social Survey 2002.

Technology

The Electronic Church

Along with others around the world, Americans can take part in religion's electronic church with the touch of a dial. Programs from radio and television ministries are aired 24 hours a day every day on network, cable, and satellite stations. People who tune in to these religious broadcasts come from the same groups as those who watch television and attend church more often than average: They are older, mostly female, Protestant, and disproportionately from rural areas, the South, and the Midwest. They also tend to have less education and lower socioeconomic status than the average American (Alexander 1994; Bruce 1990).

Because their audience is neither affluent nor well-educated and because the electronic ministry actively solicits donations—according to one calculation up to 21 percent of airtime is used to raise funds (Abelman 1990; Hoover 1990)—critics of televangelism have argued that electronic preachers are using their persuasive powers to bilk naive audiences.

This criticism, however, has come almost entirely from nonviewers and at least some research suggests that the charge is largely unfounded. Studies show, for instance, that the average contribution to electronic ministries is $32 per month and that most of that money comes from middle-income viewers. In addition, some ministries such as Pat Robertson's 700 Club raise most of their funds from a small group of wealthy supporters (Hoover 1990).

Critics have also charged that televangelism does not fulfill one of the major functions of religion: bringing people together into a moral community. Because most people who listen to televangelists also attend church, this criticism has relatively little merit. Nevertheless, the electronic church does reach two sizable groups of non-churchgoers: older people and people with physical disabilities who are unable to attend regular church services. The cost of airtime makes televangelism enormously expensive, and the religious needs of these people could not, perhaps, be met if the electronic church did not devote considerable time to fund raising.

For more information, look up the following subjects in InfoTrac College Edition:
Religious television programs
Religious broadcasting

Or visit the following Web sites:
Religious Broadcasting
http://religiousbroadcasting.lib.virginia.edu/home.html
Wikipedia
http//wikipedia.org/wiki/televangelism

Where This Leaves Us

Structural-functional theory and conflict theory are both right. On the one hand, schools and churches are preservers of tradition. Both institutions socialize young people to understand and accept traditional cultural values and to find their place in society. Occasionally schools and churches teach people to think for themselves, but more often both stress unquestioning acceptance of authority and of contemporary social arrangements, including social inequalities. On the other hand, schools and churches are in the forefront of social change. Nowhere are the battles over oppression in the least-developed nations, abortion, or homosexuality fought more bitterly than in the councils of our major churches. Nowhere are the battles over race relations, sex and class equity, and clashing cultural values fought more bitterly than on school boards. Even if you are not religious and even after you finish your education, you must not ignore the vital roles education and religion play in creating or impeding social change.

Summary

1. The structural-functional model of education suggests that education meets multiple social needs: Education contributes to cultural reproduction, social control, assimilation, the teaching of specific skills, the selection of students for future adult roles, and the promotion of change.

2. Conflict theory suggests that education helps to maintain and reproduce the stratification structure through three mechanisms: training a docile labor force that accepts inequality (the hidden curriculum); emphasizing credentials to save the best jobs for the children of the elite; and perpetuating the dominant culture.

3. Symbolic interactionists explore some of the processes through which education can reproduce inequality. Key elements of this process are differences in cultural capital that children enter school with and self-fulfilling prophecies on the part of teachers that keep disadvantaged students from improving their lot.

4. Three current controversies in education are tracking, high-stakes testing, and school choice. Tracking generally helps students who are put in high-ability groups but hurt those put into low-ability groups. High-stakes testing has encouraged schools to pay more attention to the quality of the education their students receive, but has also made it difficult for schools to fund many essential programs and has forced schools to pay as much attention to test-taking skills as to actual teaching. School choice gives parents and students options but can reinforce inequality and reduce support for public education.

5. About half of U.S. high school graduates between 16 and 24 are enrolled in college. There are few sex differences in college enrollment, and enrollment of minorities is increasing, with the notable exception of African American men.

6. Education pays off handsomely in terms of increased income, better jobs, and lower unemployment. It also offers nonmonetary benefits such as the likelihood of a longer life.

7. The sociological study of religion concerns itself with the consequences of religious affiliation for individuals and with the interrelationships of religion and other social institutions. It is not concerned with evaluating the truth of particular religious beliefs.

8. Durkheim argued that religion is functional because it provides support for the traditional practices of a society and is a force for continuity and stability. Weber argued that religion generates new ideas and thus can change social institutions. In contrast, Marx argued that religion serves as a conservative force to protect the status quo. More recent conflict theorists have explored the role that religion can play in either fostering or repressing social conflict.

9. All religions are confronted with a dilemma: the tendency to reject the secular world and the tendency to compromise with it. The way a religion resolves this question determines its form and character. Those that make adaptations to the world are called churches, whereas those that reject the world are called sects.

10. The primary distinction between a cult and a sect is that a cult is outside a society's religious heritage whereas a sect often sees itself as restoring the true faith of a society. Both tend to be primary groups characterized by small size, intense "we-feeling," and informal leadership.

11. U.S. civil religion is an important source of unity for the U.S. people. It is composed of a set of beliefs (that God guides the country), symbols (the flag), and rituals (the Pledge of Allegiance) that many people of the United States of all faiths hold sacred.

12. Despite earlier predictions, secularization has not increased significantly in recent years in the United States. Rather, mainstream religious organizations remain strong and fundamentalist groups are growing in popularity. Nevertheless, the proportion of people in the United States who say religion is very important to them and who believe the Bible is the literal word of God is declining. Women and older people are most likely to be religious; education has little impact on religiosity.

13. A major development in the contemporary church is the growth of fundamentalism and its political arm, the Religious Right.

Thinking Critically

1. Do you think colleges and universities should require all students to take at least one course dealing with multiculturalism or diversity? Why or why not?

2. Given the same levels of funding now available for public education, how would you organize elementary and secondary classrooms to best meet the needs of all students?

3. Can you foresee a time in your own life when at least some aspects of religion might be more important or less important than they now are? Why do you say so?

4. Some scholars have worried that the end result of the Religious Right may be a government that places the Bible before the Constitution. Do you agree with this forecast? If the Religious Right were to gain power, what changes would you expect to occur? Do you think they would be good for the United States? Why or why not?

Sociology on the Net

The Wadsworth Sociology Resource Center: Virtual Society

http://sociology.wadsworth.com/

The companion Web site for this book includes a range of enrichment material. Further your understanding of the chapter by accessing Online Practice Quizzes, Internet Exercises, InfoTrac College Edition Exercises, and many more compelling learning tools.

Chapter-Related Suggested Web Sites

U.S. Department of Education
http://www.ed.gov

The American Religion Data Archive
http://www.arda.tm

Academic Info—Religion
http://academicinfo.net/religindex.html

InfoTrac College Edition

http://www.infotrac-college.com/wadsworth

Access the latest news and research articles online—updated daily and spanning four years. InfoTrac College Edition is an easy-to-use online database of reliable, full-length articles from hundreds of top academic journals and popular sources. Conduct an electronic search using the following key search terms:

Ability grouping (tracking)

Standardized tests

Church and state

Religiosity

Christian right

For more information related to this chapter, visit the Opposing Viewpoints Resource Center. Be sure you check all the options on the toolbar—viewpoints, references, statistics, and so on—and search under the following subjects:

Religion

School choice

Islamic fundamentalism

Suggested Readings

Davidman, Lynn. 1991. *Tradition in a Rootless World: Women Turn to Orthodox Judaism.* Berkeley: University of California Press. A fascinating ethnographic account of the lives of modern women who convert to strict religious communities in part to find husbands who will support the women's desire to stay home and raise large families.

Kozol, Jonathan. 1991. *Savage Inequalities: Children in America's Schools.* New York: Crown. A comparison with detailed examples of schools and school districts across the United States that illustrates serious inequalities, with minority children generally attending the U.S.'s most crowded and least well-equipped schools.

Smith, Jane I. 1999. *Islam in America.* New York: Columbia University Press. An excellent overview of the history, religious beliefs, and social life of American Muslims.

CHAPTER 14

Politics and the Economy

© Eduardo Garcia/Taxi/Getty Images

Political and Economic Institutions

Political and economic institutions are so closely related that we sometimes refer to them as a single institution—the political economy. Although we can treat them separately to some extent, we will see that many issues, such as the political power of organized labor, cut across both institutions. Throughout the chapter, a primary concern is to provide a sociological perspective that will help you to interpret your own political and economic experiences as well as the headlines in the national news.

Power and Political Institutions

Coercion is the exercise of power through force or the threat of force.

Authority is power supported by norms and values that legitimate its use.

Connections

Social Policy

Although states have a legitimate monopoly on coercion, not all coercion engaged in by states is legitimate. According to the nonprofit organization Amnesty International, torture is still widely used in Mexico, Russia, China, and many other countries. Persons accused of ordinary crimes, as well as political activists, are sometimes beaten, held under water until they almost drown, given electrical shocks to their genitalia, raped, and more. To fight against torture, the United Nations has established an International Criminal Court that has the authority to prosecute political leaders who authorize torture. In addition, it has established a body of independent observers authorized to assess the treatment of prisoners throughout the world and to prevent the use of torture in prisons.

Lisa wants to watch *Sex and the City,* while John wants to watch MTV's *Jackass;* fundamentalists want prayer in the schools and the American Civil Liberties Union wants it out; state employees want higher salaries and the citizens want lower taxes. Who decides?

Whether the decision maker is Mom or the Supreme Court, those who are able to make and enforce decisions have power. As we discussed in Chapter 7, power is the ability to direct others' behavior, even against their wishes. Here we will describe two kinds of power: coercion and authority. Although both mothers and courts have power, there are obvious differences in the basis of their power, the breadth of their jurisdiction, and the means they have to compel obedience. The social structure most centrally involved with the exercise of power is the state, and that is the focus of this section.

Coercion

The exercise of power through force or the threat of force is **coercion.** The threat may be physical, financial, or social injury. The key is that we do as we have been told only because we are afraid not to. We may be afraid that we will be injured, but we may also be afraid of a fine or of rejection.

Authority

Threats are sometimes quite effective means of making people follow your orders, but they tend to create conflict and animosity. It would be much easier if people would just agree that they were supposed to do whatever it was you told them. This is not as rare as you might suppose. This kind of power is called **authority;** it refers to power that is supported by norms and values that legitimate its use. When you have authority, your subordinates agree that, in this matter at least, you have the right to make decisions and they have a duty to obey. This does not mean that the decision will always be obeyed or even that each and every subordinate will agree that the distribution of power is legitimate. Rather, it means that society's norms and values legitimate the inequality in power. For example, if a parent tells her teenagers to be in at midnight, they may come in later. They may even argue that she has no right to run their lives. Nevertheless, most people, including children, would agree that the parent does have the right.

Because authority is supported by shared norms and values, it can usually be exercised without conflict. Ultimately, however, authority rests on the ability to back up commands with coercion. Parents may back up their authority over teenagers with threats to ground them or take the car keys away. Employers can fire or demote workers. Thus, authority rests on a legitimization of coercion (Wrong 1979).

In a classic analysis of power, Weber distinguished three bases on which individuals or groups can be accepted as legitimate authorities: tradition, extraordinary personal qualities (charisma), and legal rules.

Traditional Authority

When the right to make decisions is based on the sanctity of time-honored routines, it is called **traditional authority** (Weber [1910] 1970c, 296). Monarchies and patriarchies are classic examples of this type of authority. For example, a half century ago, the majority of women and men in our society believed that husbands ought to make all the major decisions in the family; husbands had authority. Today, most of that authority has disappeared. Traditional authority, according to Weber, is not based on reason; it is based on a reverence for the past.

Charismatic Authority

An individual who is given the right to make decisions because of perceived extraordinary personal characteristics is exercising **charismatic authority** (Weber [1910] 1970c, 295). These characteristics (often an assumed direct link to God) put the bearer of charisma on a different level from subordinates. Gandhi's authority was of this form. He held neither political office nor hereditary position, yet he was able to mold national policy in India.

AP/Wide World Photos

Traditional authority, like that enjoyed by King Mohamed VI of Morocco, exists when an individual's right to make decisions for others is widely accepted based on time-honored beliefs.

Rational-Legal Authority

When decision-making rights are allocated on the basis of rationally established rules, we speak of **rational-legal authority.** This ranges all the way from a decision to take turns to a decision to adopt a constitution. An essential element of rational-legal authority is that it is impersonal. You do not need to like or admire or even agree with the person in authority; you simply follow the rules.

Rational-legal authority is the kind on which our government is based. When we want to know whether the president or the Congress has a right to make certain decisions, we simply check our rule book: the Constitution. As long as they follow the rules, most of us agree that they have the right to make decisions and we have a duty to obey.

Combining Bases of Authority

Analytically, we can make clear distinctions among these three types of authority. In practice, the successful exercise of authority usually combines two or more types (Wrong 1979). An elected official who adds charisma to the rational-legal authority stipulated by the law will have more power; the successful charismatic leader will soon establish a bureaucratic system of rational-legal authority to help manage and direct followers.

All types of authority, however, rest on the agreement of subordinates that someone has the right to make a decision about them and that they have a duty to obey it.

Traditional authority is the right to make decisions for others that is based on the sanctity of time-honored routines.

Charismatic authority is the right to make decisions that are based on perceived extraordinary personal characteristics.

Rational-legal authority is the right to make decisions that is based on rationally established rules.

Power

Concept Summary

Concept	Definition	Example From Family
Power	Ability to get others to act as one wishes in spite of their resistance; includes coercion and authority	"I know you don't want to mow the lawn, but you have to do it anyway."
Coercion	Exercise of power through force or threat of force	"Do it or else."
Authority	Power supported by norms and values	"It is your duty to mow the lawn."
Traditional authority	Authority based on sanctity of time-honored routines	"I'm your father, and I told you to mow the lawn."
Charismatic authority	Authority based on extra-ordinary personal charac-teristics of leader	"I know you've been won-dering how you might serve me, . . . "
Rational-legal authority	Authority based on submis-sion to a set of rationally established rules	"It is your turn to mow the lawn; I did it last week."

Political Institutions

Power inequalities are built into almost all social institutions. In institutions as varied as the school and the family, roles associated with status pairs such as student/teacher and parent/child specify unequal power relationships as the normal and desirable standard.

In a very general sense, **political institutions** are all those institutions con-cerned with the social structure of power, including the family, the workplace, the school, and even the church or synagogue. The most prominent political institution, however, is the state.

Political institutions are institu-tions concerned with the social structure of power; the most prominent political institution is the state.

The **state** is the social structure that successfully claims a monop-oly on the legitimate use of coer-cion and physical force within a territory.

Power and the State

The **state** is the social structure that successfully claims a monopoly on the legiti-mate use of coercion and physical force within a territory. It is usually distinguished from other political institutions by two characteristics: (1) its jurisdiction for legiti-mate decision making is broader than that of other institutions, and (2) it controls the use of coercion in society.

Jurisdiction

Whereas the other political institutions of society have rather narrow jurisdictions (over church members or over family members, for example), the state exercises power over the society as a whole.

Generally, states are responsible for gathering resources (taxes, draftees, and so on) to meet collective goals, arbitrating relationships among the parts of soci-

It is the responsibility of the government in Washington to see to it that people have help in paying for doctors and hospital bills.	52%*

The government should reduce the income differences between the rich and the poor, perhaps by raising the taxes of wealthy families or by giving income assistance to the poor.	44%*

The federal government should provide assistance to arts organizations (art museums, dance, operas, theater groups, and symphony orchestras) if they need financial assistance to operate.	50%**

FIGURE 14.1
Americans' Perceptions of Government Responsibilities, Percentage Who Agree with the Statment

*2000 data
**1998 data
SOURCE: General Social Survey, 2002.

ety, and maintaining relationships with other societies. As societies have become larger and more complex, the state's responsibilities have grown. Recent polls (Figure 14.1) indicate that about half of all Americans think the U.S. government is also responsible for supporting the arts, aiding the poor, and helping those who lack access to health care.

State Coercion

The state claims a monopoly on the legitimate use of coercion. To the extent that other institutions use coercion (for example, the family or the school), they do so with the approval of the state, and as the state gives, the state also takes away. Thus, the state has withdrawn approval from husbands who beat their wives, parents who beat their children, and teachers who hit their students.

The state uses three primary types of coercion. First, the state uses its police power to claim a monopoly on the legitimate use of physical force. It is empowered to imprison and even to kill people in certain circumstances. This claim to a monopoly on legitimate physical coercion has been strengthened in recent years by the declining legitimacy of coercion in other institutions, such as the home and the school. Second, the state uses taxation, a form of legitimated confiscation. Finally, the state is the only unit in society that can legally maintain an armed force and that is empowered to deal with foreign powers.

A variety of social structures can be devised to fulfill these functions of the state. Most basically, states can be categorized into two basic political forms: authoritarian systems and democracies.

Authoritarian Systems

Most people in most times have lived under **authoritarian systems.** Authoritarian governments go by a lot of other names: dictatorships, military juntas, despotisms, monarchies, theocracies, and so on. What they have in common is that the leadership was not selected by the people and cannot be changed by them. In some of these countries, elections may be held, but the elections are rigged so that only certain individuals can win. Afghanistan under the Taliban was an authoritarian system, as is Cuba under Castro.

Authoritarian structures vary in the extent to which they attempt to control people's lives, the extent to which they use terror and coercion to maintain power, and the purposes for which they exercise control. Some authoritarian governments,

Authoritarian systems are political systems in which the leadership is not selected by the people and legally cannot be changed by them.

such as monarchies and theocracies, govern through traditional authority; others have no legitimate authority and rest their power almost exclusively on coercion.

Democracies

Democracies are political systems that provide regular, constitutional opportunities for a change in leadership according to the will of the majority.

There are several forms of **democracies,** many of them rather different from that of the United States. All democracies, however, share two characteristics: there are regular, constitutional procedures for changing leaders, and these leadership changes reflect the will of the majority.

In a democracy, two basic groups exist: the group in power and one or more legal opposition groups that are trying to get into power. The rules of the game call for sportsmanship on both sides. The losers have to accept their loss and wait until the next constitutional opportunity to try again, and the winners have to refrain from eliminating or punishing the losers. Finally, there has to be public participation in choosing among the competing groups.

Conditions for Democracy

Why are some societies governed by democracies and others by authoritarian systems? The answer appears to have less to do with virtue than with economics. Democracy is found primarily in the wealthier nations of the world. A large and relatively affluent middle class is especially important. Members of the middle class usually have sufficient social and economic resources to organize effectively; their economic power and organization enable them to hold the government accountable. The key factor, however, is not the overall wealth of the nation, or even the size of the middle class, but the way the wealth is distributed: Democracy is also found in poorer nations that do not have extremes of income inequality, such as Costa Rica and Sri Lanka (Simpson 1990). Still, democracy is possible even in the absence of these conditions. The largest democracy in the world, for instance, is India, which has a relatively small middle class and tremendous income inequality.

Democracy may also flourish in societies with many competing groups, each of which comprises less than a majority. In such a situation, no single group can win a

■ Democracy is now taking root in South Africa, where the financial and political power of the white minority has been counterbalanced by the sheer numbers and political determination of the black majority.

majority of voters without negotiating with other groups; because each group is a minority, safeguarding minority political groups protects everybody.

Although democratic stability depends on competing interest groups, two additional conditions must be met. First, if minority political groups are so divided or ineffective that there is little chance they can win an election, the public may become disillusioned with the democratic process (Weil 1989). Second, if competing interest groups do not share the same basic values and interests, they are not likely to be able to abide by the rules of the game. The repeated failure of peace talks and eruptions of violence between Israelis and Palestinians demonstrate how fundamental differences in values and interests can make it difficult for democracy to exist.

Who Governs? Models of U.S. Democracy

Everyone agrees that the United States is a democracy. Political parties that have at least moderately different economic and social agendas vie for public support, and every 2, 4, or 6 years there are opportunities to "turn the rascals out" and replace the leadership. There is substantial debate, however, about whether the decisions made by our leaders really reflect the will of the majority. This section outlines three sociological models of how these decisions are made: pluralist, power-elite, and state autonomy. Although there are some important differences among the three major models of decision making in the United States (see the Concept Summary), a notable feature of all three is that the key actors are organized entities—businesses, unions, political action committees, or government agencies—rather than individuals.

Structural-Functional Theory: The Pluralist Model

The pluralist model focuses on the processes of checks and balances within the U.S. government and on coalition and competition among governmental and nongovernmental groups. This model argues that the system of checks and balances built into the U.S. Constitution makes it nearly impossible for either the judicial,

Comparison of Three Models of American Political Power

Concept Summary	Pluralist	Power-Elite	State Autonomy
Basic units of analysis	Interest groups	Power elites	Government bureaucraccy
Source of power	Situational; depends on issue	Inherited and positional; top positions in key economic and social institutions	Control of personnel and budget of government
Distribution of power	Dispersed among competing diverse groups	Concentrated in relatively homogeneous elite	Held by bureaucrats
Limits of power	Limited by shifting and cross-cutting loyalties	Potentially limited when other groups can unite in opposition	Limited if elite is unified
Role of the state	Arena where interest groups compete	One of several sources of power	A major source of power

legislative, or executive branch of government to force its will on the other branches. Similarly, the model argues that different groups with competing vested interests hold power in different sectors of American life. Some groups have economic power, some have political power, and some have cultural power.

A vital part of this model is the hypothesis of shifting allegiances. According to pluralist theorists, different coalitions of interest groups arise for each decision. For example, labor unions will ally themselves with automakers in favoring tariffs against Japanese imports; when it comes to domestic issues such as wages, however, these two groups will oppose each other. This pattern of shifting allegiances keeps any interest groups from consistently being on the winning side and keeps political alliances fluid and temporary rather than allowing them to harden into permanent and unified cliques (Dahl 1961, 1971). As a result of these processes, pluralists see the decision-making process as relatively inefficient but also relatively free of conflict, a process in which competition among interest groups (each of which has some sources of power) keeps any single group from gaining significant advantage over the others.

Research suggests the limits of the pluralist model. Typically, the power elite stick together, while other groups lack the resources to successfully challenge the elite (Burris & Salt 1990; Clawson & Su 1990; Korpi 1989). In the United States, programs designed to share wealth or access to opportunities more equitably—such as civil rights or Social Security—have had the best chance of success when a sense of crisis caused the elite to favor the change or when the elite disagreed among themselves (Jenkins & Brent 1989).

Conflict Theory: The Power-Elite Model

The **power elite** comprises the people who occupy the top positions in three bureaucracies—the military, industry, and the executive branch of government—and who are thought to act together to run the United States in their own interests.

In contrast to the pluralist model, the power-elite model contends that a relatively unified elite group makes all major decisions—in its own interests (Domhoff 1998). In his classic work, *The Power Elite*, C. Wright Mills (1956) defined the **power elite** as the people who occupy the top positions in three bureaucracies: the military, industry, and the executive branch of government. Through a complex set of overlapping cliques, these people share decisions having at least national consequences (Mills, 18).

There is no question that the power elite has become more diverse since Mills's day. The independent power of the military has declined, whereas that of the cultural elite—which includes both movie stars and religious leaders—has increased. Increasing numbers of African Americans, Hispanics, and women hold high corporate positions and elected office, especially at local levels. Nevertheless, the percentages remain low. Moreover, most "outsiders" who become part of the power elite came from at least middle-class homes, attended elite schools, and are willing and able to fit in: light-skinned minorities, Jews who marry Christians, and women who learn to play golf and even to smoke cigars (Zweigenhaft & Domhoff 1998).

The power-elite theory argues that individuals have power by virtue of the positions they hold in key institutions. If the interests of these individuals and institutions were in competition with one another, this model would not be significantly different from the pluralist model. The key factor in elite theory is the unity of purpose and outlook that top position holders have as a result of common membership in the upper class.

In sum, the power-elite model shares with the pluralist model a belief that different groups with different types of power can come together in shifting allegiances to create social change. The models differ in that the pluralist model assumes that power is generally distributed in a relatively equitable fashion and, as a result, that virtually all groups can change their situations if they so desire. The power elite model, on the other hand, argues that power is very unevenly distributed and therefore creating meaningful social change is difficult, unless people organize together in unions, social movements, and the like.

A Third Perspective: The State Autonomy Model

Some scholars argue that the government bureaucracy is a powerful, independent actor in political decisions. The federal government employs 3 million people directly. In addition, its policies indirectly determine the employment of tens of millions of people who work for national defense contractors, state and local governments, schools, and social welfare agencies. Each year, the federal government collects and spends more than $1 trillion. It seems only common sense that the state (meaning the federal bureaucracy) is in a powerful position to get what it wants. Its agenda is linked not to class, as in the power-elite model, but to the maintenance and extension of bureaucratic power.

A good example of this approach is Hooks' (1990) analysis of the profound effect that Pentagon and Defense Department policies have had on the development of the microelectronics and aeronautics industries. Hooks' findings show that the competitiveness of U.S. high-technology firms has been seriously jeopardized by defense policy. Because the Pentagon's goals have been strategic rather than commercial, the state has pursued its own interests at the expense of the economic elite. Only when the elite are unified can they counterbalance the power of state autonomy.

Individual Participation in U.S. Government

Democracy is a political system that explicitly includes a large proportion of adults as political actors. Yet it is easy to overlook the role of individual citizens while concentrating on leaders and organized interests. This section describes the U.S. political structure and process from the viewpoint of the individual citizen.

Who Votes?

The average citizen is not politically oriented. About one-third of the voting-age population does not even register to vote, and almost half (45 percent in 2000) does not vote in national elections (U.S. Bureau of the Census 2002b). An astonishing 75 to 80 percent do not vote in typical local elections.

This low level of political participation poses a crucial question about the structure of power in U.S. democracy. Who participates? If they are not a random sample of citizens, then some groups probably have more influence than others. Studies show that voters differ from nonvoters by social class and age, but not by race or ethnicity.

Social Class

One of the firmest findings in social science is that political participation (indeed, participation of any sort) is strongly related to social class. Whether we define participation as voting or letter writing, people with more education, more income, and more prestigious jobs are more likely

© Paul Conklin/PhotoEdit

■ Although all U.S. citizens over the age of 18 have the right to vote, middle-aged, better off, and better educated citizens are most likely to do so.

TABLE 14.1

Participation in the 2000 Presidential Election, Among Voting-Age Population

The likelihood of registering to vote and actually voting is greater among people who are older, better educated, employed, and non-Hispanic.

	Percentage Registered	Percentage Who Voted
Total	64%	55%
Education		
8 years or less	36	27
Some high school	46	34
High school graduate	60	49
Some college	70	60
College graduate or more	77	72
Race/ethnicity		
White	66	56
African American	64	54
Hispanic	35	28
Age		
18–20	41	28
21–24	49	35
25–34	55	44
35–44	64	55
45–64	71	64
65	76	68
Employment status		
Employed	65	56
Unemployed	46	35

SOURCE: U.S. Bureau of the Census 2002b.

to be politically active. They know more about the issues, have stronger opinions, are much more likely to believe they can influence political decisions, and thus are more likely to try to do so. This conclusion is supported by data on voting patterns from the 2000 presidential election (Table 14.1). The higher the level of education, the greater the likelihood of voting—those who have graduated from college are twice as likely to vote as those who have not completed high school.

Age

Another significant determinant of political participation is age. Political interest, knowledge, opinion, and participation steadily increase with age: 60 percent of all voters in the 2002 election were 45 or older (U.S. Bureau of the Census 2002b). Even in the turbulent years of the Vietnam War, when young antiwar demonstrators were so visible, young adults were significantly less likely to vote than were middle-aged individuals. In that period, many young adults engaged in other forms of political participation that did, in fact, influence political decisions. In most time periods, however, the low participation of younger people at the polls is a fair measure of their overall participation.

TABLE 14.2
Percentage Voting Democrat or Republican in 1996 Presidential Election*
Although American parties are not closely tied to social class, better-off individuals tend to vote Republican, and less-educated and nonwhite individuals tend to vote Democrat.

	Democrat	Republican
Total	58	42
Race/ethnicity		
White	54	46
African American	99	1
Education		
Grade school	82	18
High school	60	40
College	49	51
Sex		
Male	51	50
Female	65	35
Age		
18–37	58	42
38–53	58	42
54–69	56	44
70–85	64	36
86 and over	57	43

*Most recent data on party affiliation available as of 2003.

SOURCE: U.S. Bureau of the Census, 2000.

Race and Ethnicity

Racial differences in political participation have virtually disappeared. In fact, after social class has been taken into consideration, it seems likely that being African American increases political participation. African Americans are more apt than whites to want changes made in the system, and they turn to political participation as a means to effect these changes (Guterbock & London 1983, 440). Low Hispanic participation is partly traceable to low socioeconomic status. Mostly, however, it is simply due to the fact that one-third of Hispanic adults are not citizens and so are not eligible to vote.

Which Party?

Although in theory the United States could have many political parties, in practice political power in this country is divided between two parties, the Democratic Party and the Republican Party. Although both major political parties in the United States are basically centrist, there are philosophical distinctions between them. For the last century, the Democratic Party has been more likely to support the interests of the poor, the working class, and minorities, and the Republican Party has been more likely to favor policies supporting economic growth. Because of these characteristics, the Democrats tend to attract more female, younger, minority, and less-educated voters than do the Republicans (Table 14.2).

A growing proportion of voters align themselves with neither party but declare an intention to vote on the basis of issues rather than party loyalty. When the 10 percent (or more) of voters who call themselves independent go to the polls, however, they usually have to choose between a Democratic and a Republican candidate.

Why So Few Voters?

The United States prides itself on its democratic traditions. Yet U.S. citizens are only half as likely to vote as are citizens of other Western nations, and rates of voting in the U.S. have declined steadily for the last century. How can we explain this low and declining rate of political participation? As we've seen, persons are most likely to vote if they are older, more affluent, more educated, employed, and non-Hispanic. These data have led some scholars to conclude that people are most likely to vote if they are educated enough to understand the political process and if they feel they have a stake in maintaining the existing system (Nie, Verba, & Petrocik 1976). Yet this theory cannot explain why voting rates have declined steadily even though both income and educational achievement have increased. As a result, some now argue that political participation has declined because a growing number of even well-off Americans believe that the political process is corrupt, that the Democrats and Republicans are more similar than different, and that it makes little difference who gets elected (Southwell & Everest 1998).

Voting rates may also be low at least in part because politicians (sometimes intentionally, sometimes unintentionally) have made it difficult for people to vote (Piven & Cloward 1988, 2000). Until only a few years ago, both registering to vote and voting were more cumbersome in the United States than in any other Western democracy. In many states, individuals had to register annually, pass literacy tests, or pay special taxes. They also had to both register and vote in specific locations during specific limited hours, which was especially difficult for persons who held working-class jobs.

Voter registration has increased significantly since passage of the National Voter Registration Act of 1993 but, as was evident in the 2000 presidential election, barriers to voting still remain. In St. Louis, Missouri and elsewhere throughout the country, potential voters were hampered by transportation difficulties and by polling places that closed too early to accommodate their vote. Perhaps the major reason that voting turnout has been so low, however, is that no major political party has sought to involve marginalized Americans nor to actively address their concerns; voting rates have been highest when social movements have energized these constituencies to believe that they can make a difference and when political parties have responded by attempting to reach out to them (Winders 1999).

A Case Study: Felon Disenfranchisement

As we've seen, a surprising number of Americans choose not to participate in the democratic process. An even more surprising number of Americans are *kept* from participating. It is estimated that approximately 4.7 million individuals are barred from voting—disenfranchised—because they were once convicted of a felony (Ugger & Manza 2002). In some states, only those still in prison are forbidden from voting; in other states, a felony conviction brings lifelong **felon disenfranchisement.** Because the United States has both a high rate of felony convictions (primarily for drug-related crimes) and unusually restrictive laws on the voting rights of ex-felons, the United States has a higher rate of felon disenfranchisement than almost any other

Felon disenfranchisement is the loss of voting privileges suffered by those who have been convicted of a felony. In some states, felon disenfranchisement applies only to those in prison; in other states, it is lifelong.

country. In essence, the very possibility of rehabilitation is ignored: Someone convicted at age 20 of selling marijuana, for example, might be ineligible to vote for the rest of his or her life, even if he or she never again commits a crime and becomes a successful worker, parent, and community citizen.

Importantly, because poverty sometimes pushes individuals to commit crimes, and because the criminal justice system more often convicts poor criminals than equally guilty wealthy criminals, those subject to felon disenfranchisement overwhelmingly are poor. The number of disenfranchised poor people is high enough to significantly decrease the chances of electing politicians who favor helping the poor (Ugger & Manza 2002).

Modern Economic Systems

Our description of the U.S. political process has crossed over into discussions of economic processes again and again. From the role of the working class to the role of the power elite, we find that understanding government requires understanding the economic relationships that underlie it. At this point, we turn to an explicit assessment of economic relationships and how they affect the individual worker and the political economy.

Economic institutions are social structures concerned with the production and distribution of goods and services. Issues such as scarcity or abundance, guns or butter, and craftwork or assembly lines are all part of the production side of economic institutions. Issues of distribution include what proportion goes to the worker versus the manager, who is responsible for supporting nonworkers, and how much of society's production is distributed on the basis of need rather than effort or ability. The distribution aspect of economic institutions intimately touches the family, stratification systems, education, and government.

In the modern world, there are basically two types of economic systems: capitalism and socialism. Because economic systems must adapt to different political and natural environments, however, we find few instances of pure capitalism or pure socialism. Most modern economic systems represent some variation on the two and often combine elements of both.

Economic institutions are social structures concerned with the production and distribution of goods and services.

Capitalism is the economic system in which most wealth (land, capital, and labor) is private property, to be used by its owners to maximize their own gain; this economic system is based on competition.

Capitalism

The economic system in which most wealth (land, capital, and labor) is private property, to be used by its owners to maximize their own gain, is **capitalism.** This economic system is based on competition. Each of us seeks to maximize his or her own profits by working harder or devising more efficient ways to produce goods. Such a system encourages hard work, technical innovation, and a sharp eye for trends in consumer demand. Because self-interest is a powerful spur, such economies can be very productive.

Even when it is very productive, though, a capitalist economy has drawbacks. These drawbacks all center around problems in the distribution of resources. First, the capitalist system at its most ideal represents a competitive bargain between labor (workers) and capital (owners of industries), both of whom control a necessary resource. But this is not a bargain between equals: almost always, capital has more power to obtain the bargain it wants than does labor. As a result, workers earn only a fraction of what capitalists earn. Second, those who have neither labor nor capital

A Global Perspective

Democratic Socialism in Sweden

focus on

What would it be like to live and work in Sweden? You would have guaranteed access to quality public transportation; guaranteed income if you were ill, disabled, or elderly; guaranteed access to comfortable housing; and free education all the way through college, graduate school, or professional training. After you or your partner gave birth to or adopted a baby, you would be entitled to a full year of paid parental leave. Once you went back to work, you could use a free, high-quality, state-funded day-care center. In exchange for these benefits, you would pay about 25 percent of your paycheck in federal income taxes and almost as much in local taxes.

Sweden is a democratic socialist society with an economy that mixes corporate capitalism with significant welfare benefits for workers and non-

workers alike. Because Sweden is a democracy, the majority of Swedes have voted to receive these benefits and to pay high taxes for them. But Sweden's economy wasn't always arranged this way.

Sweden owes its economic organization in part to the rise of a strong labor movement (Koblik 1975). As industrialization began in Sweden in the 1870s, labor union members worked to create the Social Democratic Party, a

Swedish day-care center.

to bargain with (children, stay-at-home moms, the elderly, the disabled, and workers whose jobs have disappeared) always lose out, for with nothing to exchange, they are outside the market. They must rely on aid from others, which is not always forthcoming. Third, because public goods such as streets, watersheds, sewers, or defense offer no profit, pure capitalism has no interest in providing them. Yet society cannot function without these services. Thus capitalist systems must have some means of distribution other than the market.

Socialism

If capitalism is an economic system that maximizes production at the expense of distribution, socialism is a system that stresses distribution at the expense of production. As an ideal, **socialism** is an economic structure in which productive tools are owned and managed by the workers and used for the collective good.

In theory, socialism has several major advantages over capitalism. First, societal resources can be used for the benefit of society as a whole rather than for individuals. This advantage is most apparent in regard to common goods such as protecting the environment. A related advantage is that of central planning. Because resources are controlled by the group, they can be deployed to help reach group goals. This may mean diverting them from profitable industries (say, for example, those making

Socialism is an economic structure in which productive tools (land, labor, and capital) are owned and managed by the workers and used for the collective good.

political party dedicated to equitable wages, job security, and welfare programs for the entire society. While Communists in Russia were fighting and winning the Russian Revolution in 1914–1917, members of Sweden's Social Democratic Party were politicking for seats in parliament. After holding power on and off during the 1920s, the Social Democratic Party won an important election in 1932 and then retained virtually uninterrupted political power until the present day. This has allowed the welfare state established by the party to develop deep roots.

The welfare state's emergence and success also reflects the deeply held Swedish belief in the responsibility of the community to look out for all its members. This attitude, in turn, has been fostered by the cultural homogeneity of the Swedish population. Until about 25 years ago, the population of Sweden was overwhelmingly ethnically Swedish. Currently, however, foreign immigrants and individuals who have at least one immigrant parent comprise close to 20 percent of the Swedish population. As a result, many question whether support for the welfare state will decline if ethnic Swedes become unwilling to extend the benefits of their social system to immigrants, and if immigrants bring with them a philosophy of individual rather than social responsibility.

Not everyone in Sweden is a member of the Social Democratic Party, of course. Conservative groups favor a freer market economy. Furthermore, Social Democrats today are worried about whether Sweden's welfare society can survive the current international economic downturn (Olsen 1996). Controlling Sweden's transnational corporations so that they do not export jobs and continue to pay high taxes at home is proving to be more and more difficult. Some economists are beginning to point out that Sweden's market socialism is based on an inherent irony: Strong and profitable capitalist businesses are necessary so that workers can be employed and taxes for welfare benefits can be collected. But insisting on generous worker benefits and full employment eats into capitalist profits (Olsen).

 For more information, look up the following subjects in InfoTrac College Edition:
Sweden welfare

 Or visit the following Web sites:
Sweden.se
http://www.sweden.se/
The Swedish Institute
www.si.se

---

bicycles, televisions, or compact discs) to industries that are viewed as more likely to benefit society in the long run, such as education, agriculture, or steel. The major advantage claimed for socialism, however, is that it produces equitable (although not necessarily equal) distribution.

The creed of pure socialism is *from each according to ability, to each according to need.* An explicit goal of socialism is to eliminate unequal reward as the major incentive to labor. Workers are expected to be motivated by loyalty to their community and their comrades. Unfortunately, the childless woman is not likely to be motivated to do her best when the incompetent worker next to her takes home a larger paycheck simply because she has several children and thus a greater need. Nor is the farmer as likely to make the extra effort to save the harvest from rain or drought if his rewards are unrelated to either effort or productivity. Because of this factor, production is usually lower in socialist economies than in capitalist economies.

The Political Economy

Both capitalism and socialism can coexist with either authoritarian or democratic political systems. Socialism and democracy are combined in many Western European nations, such as the United Kingdom and Sweden. Other nations, such as China and Cuba, combine socialism with an authoritarian political system. We often use (and misuse) the term "communist" to refer to societies in which a socialist economy is guided by a political elite and enforced by a military elite. The goals of socialism (equality and efficiency) are still there, but the political form is authoritarian rather than democratic.

Likewise, some capitalist nations are democratic and some are authoritarian. The United States and Japan have both capitalist economies and democratic political systems. Singapore and Saudi Arabia, on the other hand, have capitalist economies but autocratic political systems, in which elections are either nonexistent or virtually meaningless. These examples remind us that both capitalism and socialism can coexist with authoritarian regimes.

Mixed Economies

Most Western societies today represent a mixture of both capitalist and socialist economic structures within a more-or-less democratic framework. In many nations, services such as the mail and the railroads and key industries such as steel and energy are socialized. This socialism rarely results from pure idealism. Rather, public ownership is often seen as the best way to ensure continuation of vital services that are not profitable enough to attract private enterprise. Other services—for example, health care and education—have been partially socialized because societies have judged it unethical to deny these services to the poor and too inefficient to provide them on the open market.

In the case of many socialized services, general availability and progressive tax rates have gone far toward meeting the maxim from each according to ability, to each according to need. There are still inequalities in education and health care, but many fewer than there would be if these services were available on a strictly cash basis. The United States has done the least among major Western powers toward creating a mixed economy, and our future direction is unclear. By and large, the Republican Party has pushed to reduce government provision of social services and the Democratic Party has pushed to increase such services. The future mix of socialist and capitalist principles will reflect political rather than strictly economic conditions.

The U.S. Economic System

In the late twentieth century, the United States has what is called a postindustrial economy. To understand this concept, a brief historical review of economic forms is required.

The Postindustrial Economy

Primary production is extracting raw materials from the environment.

Secondary production is the processing of raw materials.

In a preindustrial economy, the vast majority of the labor force is engaged in **primary production,** extracting raw materials from the environment. Prominent among primary production activities are farming, herding, fishing, forestry, hunting, and mining. Preindustrial economic structures were characteristic of Europe until 500 years ago and are still typical of many societies.

Industrialization meant a shift from primary to **secondary production,** the processing of raw materials. For example, ore, cotton, and wood are processed by the steel, textile, and lumber industries, respectively; other secondary industries turn these materials into automobiles, clothing, and furniture. The shift from primary to secondary production is characterized by enormous increases in the standard of living.

■ Primary production involves direct contact with natural resources—fishing, hunting, farming, forestry, and mining.

Postindustrial development rests on a third stage of productivity, **tertiary production.** This stage is the production of services. The tertiary sector includes a wide variety of occupations: physicians, schoolteachers, hotel maids, short-order cooks, and police officers. It includes everyone who works for hospitals, governments, airlines, banks, hotels, schools, or grocery stores. None of these organizations produces tangible goods; they all provide service to others. They count their production not in barrels or tons but in numbers of satisfied customers. The tertiary sector has grown very rapidly in the last half-century and is projected to grow still more. As Figure 14.2 illustrates, only 19 percent of the labor force was involved in tertiary production in 1920; by 1956, the figure had grown to 49 percent, and by 2000 it was estimated to include 78 percent of the labor force. Simultaneously, the portion of the labor force employed in primary production has almost disappeared, and the proportion employed in secondary production has halved. These shifts do not mean that primary and secondary production are no longer important. A large service sector depends on primary and secondary sectors that are so efficiently productive that large numbers of people are freed from the necessity of direct production.

As we saw in Chapter 8, a major change in the last 20 years is that America's postindustrial economy is increasingly interwoven with that of other nations. This globalization is likely to increase in the near future.

Tertiary production is the production of services.

The Dual Economy

The U.S. economic system can be viewed as a **dual economy.** Its two parts are the complex giants of the industrial core and the small, competitive organizations that form the periphery. They are distinguished from each other on two dimensions: the

A **dual economy** consists of the complex giants of the industrial core and the small, competitive organizations that form the periphery.

FIGURE 14.2
Changing Labor Force in the United States
Since 1820, the labor force in the United States has changed drastically. The proportion of workers engaged in primary production has declined sharply while the proportion engaged in service work has expanded greatly.
SOURCE: U.S. Bureau of Labor Statistics, 2002.

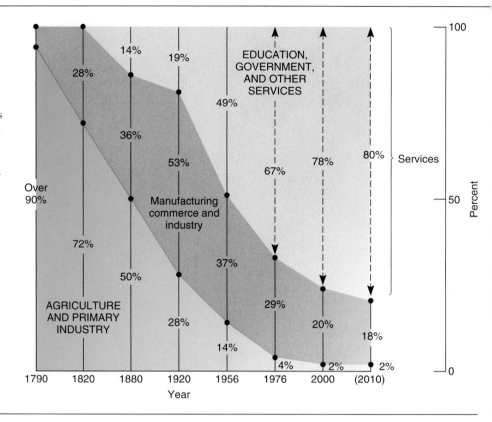

complexity of their organizational forms and the degree to which they dominate their economic environment (Baron & Bielby 1984).

The Industrial Core: Corporate Capitalism

Although there are more than 250,000 businesses in the United States, most of the nation's capital and labor are tied up in a few transnational giants that form the industrial core. The top 20 U.S. companies are large bureaucracies that control billions of dollars of assets and employ thousands of individuals. These giants loom large on both the national and international scene.

At the local level, we are familiar with the situation in which a region's one major employer holds city and county government hostage and bargains for tax advantages and favorable zoning regulations in exchange for increasing or retaining jobs. The growing size and interdependence of firms in the industrial core are now causing this scene to be reenacted at the federal level.

Wealthy capitalists are linked to each other by shared ownership of large firms; large firms are linked to one another by having common members on their boards of directors and through their dealings with the same financial institutions. As a result of this interdependence, relations among large firms have become more cooperative than competitive. Although decreased competition reduces productivity and efficiency, it increases joint political influence (Mizruchi 1989, 1990). For example, as the proportion of the nation's assets held by the top 100 firms increased, their political power increased and the taxes they were required to pay decreased—even while individual income tax rose (Jacobs 1988).

Transnational corporations constantly seek new access to natural resources and new markets for their products.

That power can extend to influencing U.S. foreign policy. A desire to protect the interest of transnational companies like Dole and United Fruit certainly played a role in the U.S. decision to lend support to repressive regimes in Guatemala, Honduras, and other Latin American countries at various points during the twentieth century. Similarly, to protect transnational oil companies, the United States covertly orchestrated the 1951 coup of democratically elected Iranian Prime Minister Mohammad Mossadegh and subsequently propped up the authoritarian regime of the Shah of Iran (Kinzer 2003). (Popular resentment of the Shah's repressive regime eventually led to the Islamic Iranian revolution in 1979, which stimulated Islamic fundamentalism worldwide.) Some observers suspect that the U.S. conflicts with Iraq in 1991 and 2003 had more to do with protecting U.S. oil interests than with fighting against terrorism.

The Periphery: Small Business

The periphery of the U.S. economy is made up largely of small businesses that are owned and run by a family or small group of partners. They are usually characterized by few employees, economic uncertainty, and relatively little bureaucratization of management and authority. The periphery can be divided into a formal and an informal economy.

The Formal Economy The formal economy consists of small banks, farms, retail stores, restaurants, and repair services. Some small manufacturing companies are also part of this sector. This segment of the periphery is what Marx called the petit (pronounced petty) bourgeois. The **petit bourgeois** are those who use their own modest capital to establish small enterprises in which they and their family provide the primary labor (Bechhofer & Elliott 1985).

Although this class is smaller than it was 100 years ago, it continues to furnish jobs for a substantial portion of the population and to be an important part of both the industrial and service sectors (Bechhofer & Elliott 1985). It has been

The **petit bourgeois** are those persons who use their own modest capital to establish small enterprises in which they and their family provide the primary labor.

■ The small businesses of the periphery provide important economic opportunities for minority Americans, such as these Korean American grocers.

an especially important avenue of opportunity for minorities in the United States. Koreans, Hispanics, and African Americans who face discriminatory barriers in the corporate world can sometimes achieve moderate prosperity by operating neighborhood grocery stores, laundries, and fast-food franchises.

Informal Economy   An important sector of the periphery is the underground or **informal economy.** This is the part of the economy that escapes the record keeping and regulation of the state. It includes illegal activities such as prostitution, smuggling immigrants, and running numbers, but it also includes a large variety of legal but unofficial enterprises such as home repairs, house cleaning, and garment subcontracting. Often referred to disparagingly as "fly-by-night" businesses, enterprises in the informal sector are nevertheless an important source of employment. This is especially true for those segments of the population who would like to avoid federal record keeping: illegal aliens, foreign students, senior citizens and welfare recipients who don't want their earnings to reduce their benefit levels, adolescents too young to meet work requirements, and many others (Portes & Sassen-Koob 1987).

The Segmented Labor Market

Parallel to the dual economy is a dual labor market, generally referred to as a **segmented labor market,** in which hiring, advancement, and benefits vary systematically between the industrial core (the corporate sector) and the periphery (the competitive sector).

In the corporate sector, firms generally rely on what are called internal labor markets. Almost all hiring is done at the entry level, and upper-level positions are filled from below. At all levels, credentials are critical for hiring and promotion. Within core firms, there are predictable career paths for both blue- and white-collar workers. Although both job security and benefits may have declined over the last 20 years (Newman 1999a), employment remains generally secure, and benefits still are relatively good. Wages and benefits are best in the very largest firms (Villemez & Bridges 1988).

Connections

Example

The illegal drug trade is a major part of the informal economy, worth many billions of dollars. Like other parts of the economy, the drug trade includes producers of raw goods (such as poppy growers), manufacturers (such as those who refine heroin from poppy), distributors, buyers, and sellers, at both retail and wholesale levels. Like any other business, the drug trade tries to keep its costs down. One way it does so is by not paying any taxes. For those in the drug trade, ordinary business expenses include occasional prison terms, as well as the price of weapons and bribes.

In the competitive sector, on the other hand, credentials are less important, career paths are short and unpredictable, security is minimal, and benefits are low or nonexistent. At the same time, there is less bureaucratization and red tape, and both workers and managers have more freedom in their work.

The competitive sector offers a haven of employment for those who do not meet the demands of the corporate sector: those who do not have the required credentials, who have spotty work records, who want to work part-time, or who have been "down-sized" from the corporate world. As a result, a disproportionate number of youths, minorities, and women work in the competitive sector. Because keeping a job and getting a promotion are governed almost exclusively by personal factors rather than by seniority or even ability, however, this sector is less likely to promote minorities or women; there is no affirmative-action officer at Joe's Café. As a result, the gender gap in wages is significantly larger in the competitive sector than it is in core industries (Coverdill 1988).

The **informal economy** is the part of the economy that escapes the record keeping and regulation of the state; also known as the underground economy.

The **segmented labor market** parallels the dual economy. Hiring, advancement, and benefits vary systematically between the industry core and the periphery.

Professions are occupations that demand specialized skills and creative freedom.

Work in the United States

From the individual's point of view, economic institutions mean jobs. For some, jobs are just jobs; for others, they are careers. But 40 years of involvement in the world of work is central to most people's lives.

Occupations

Aside from the simplest consequences of working (income and filling up much of your time), what you do at work is probably as important as whether you work. Here some of the important differences between the professions and white- and blue-collar work are described.

Professions

Occupations that demand specialized skills and creative freedom are **professions.** Their distinctive characteristics include (1) the production of an unstandardized product, (2) a high degree of personal involvement, (3) a wide knowledge of a specialized technique, (4) a sense of obligation to one's art, (5) a sense of group identity, and (6) a significant service to society (Gross 1958). The definition of professions was originally developed for the so-called learned professions (law, medicine, and college teaching). It applies equally well, however, to actors, dancers, and potters.

The rewards that professionals achieve vary considerably. Physicians and lawyers can receive very high incomes while dancers and potters typically earn very little. The major reward that all professionals share, however, is substantial freedom from supervision. Because their work is nonroutine and requires personal judgments, professionals have been able to demand—and get—the right to work their own hours, do things their own way, and arrange their own work lives.

Freedom from supervision remains the most outstanding reward of professional work, but it is a reward that is being eroded. Increasingly, people in the professions work for others within bureaucratic structures that constrain many of the most characteristic aspects of professional life.

© Peter Menzel

■ Like most at the upper-end of the white collar spectrum, this professional has a job that requires a college education, the ability to think independently, and excellent communication skills.

What Color is Your Collar?

Fifty years ago, the color of your collar was a pretty good indication of the status of your job and your gender. Men who worked with their hands wore blue (or brown or flannel) collars; women who worked with their hands as beauticians, maids, or waitresses often wore pink. Managers and others who worked in clean offices wore white collars. Those days are past. The labor force is far more diversified, and some of the old guidelines no longer work. The bagger at Safeway wears a white shirt and tie; the librarian wears blue jeans and sandals. Yet the librarian is a white-collar worker and the bagger is not.

Traditional white-collar workers are managers, professionals, typists, and salespeople—those who work in clean offices and are expected to be able to think independently. Blue-collar workers are people in primary and secondary industry who work with their hands; they farm, assemble telephones, build houses, and weld joints. Although some blue-collar workers earn more than some white-collar workers, blue-collar jobs are characterized by lower incomes, lower status, lower security, closer supervision, and more routine.

Fifty years ago, this simple division of the labor force included most workers. These days it leaves out a growing category of low-skilled, low-status workers who often appear in company-supplied brown polyester suits or turquoise jackets and who fry hamburgers, stock K-Mart shelves, and collect money at the "U-Serv" gas station. Characteristically, these workers hold jobs that have a short or nonexistent career ladder, and they earn the minimum wage or close to it.

FIGURE 14.3
The Shifting Job Market: Projected Changes Between 2000 and 2010
The demand for labor is expected to grow between 2000 and 2010. But many of the jobs with the largest growth rates are low-skill and low-wage jobs.
SOURCE: U.S. Bureau of Labor Statistics, 2003.

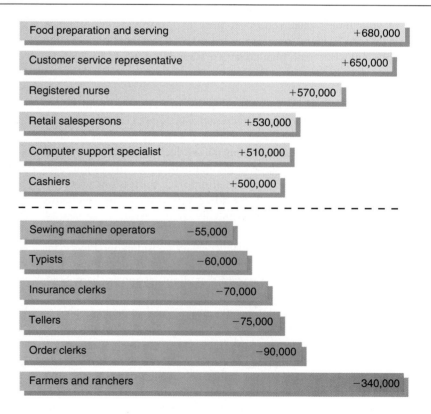

Food preparation and serving	+680,000
Customer service representative	+650,000
Registered nurse	+570,000
Retail salespersons	+530,000
Computer support specialist	+510,000
Cashiers	+500,000
Sewing machine operators	-55,000
Typists	-60,000
Insurance clerks	-70,000
Tellers	-75,000
Order clerks	-90,000
Farmers and ranchers	-340,000

Occupational Outlook

As the graph in Figure 14.2 indicated, the outlook for the future is for greater expansion of the tertiary sector and even more reductions in employment in secondary and primary production. What will this mean for the kinds of jobs that are available in the future? Some of the changes projected between 2000 and 2010 are illustrated in Figure 14.3.

Some traditional occupational categories are expected to suffer major declines. The occupations with the largest projected decreases include both blue- and white-collar jobs: Typists and word processors will have a harder time finding jobs, as will sewing-machine operators and farmers. The declining opportunities in these occupations reflect a variety of factors: changing age structure, loss of U.S. jobs due to migration of industry overseas, and new technology.

The most controversial issue is what kind of new jobs the economy will offer. Optimistic observers note that executive and professional jobs are growing faster than average and suggest that the high quality and good pay of these new jobs indicate what awaits today's college graduates. Others focus on the rapid increase in what one critic has called "McJobs" (Ritzer 1996). Although not all these jobs entail selling hamburgers, many are low-status jobs with low wages and no benefits: health aides, personal and home-care aides, and cashiers. Figure 14.3 shows the jobs that are expected to grow the most, and those that are shrinking fastest.

Both the optimists and the critics are correct in their expectations for the future. Good jobs for college graduates and those with technical training—computer engineers and scientists, registered nurses, and system analysts—are growing rapidly. At the same time, however, bad jobs, often traditionally done by females and paying very poorly—for home health aides, waiters and waitresses, and personal care assistants—are also growing rapidly (James, Grant, & Cranford 2000). Thus, the fastest-growing occupations require either years of advanced education or almost no skill at all, and the

The fastest growing jobs in the United States today are minimum-wage service jobs that offer few benefits and fewer prospects for advancement.

© Seth Resnick/Gamma Liaison

latter offer very little reward. The big losers in the transformation of the labor market are likely to be the traditional working class: men and women who did skilled manual labor. If they do not go to college, few young people from working-class homes will be able to find secure, well-paying, unionized jobs like those their parents held.

The Meaning of Work

For most people, work is essential as the means to earn a livelihood. As noted in Chapter 7, one's work is often the most important determinant of one's position in the stratification structure and, consequently, of one's health, happiness, and lifestyle.

Work is more than this, however. It is also the major means that most of us use to structure our lives. It determines what time we get up, what we do all day, who we do it with, and how much time we have left for leisure. Thus, the nature of our work and our attitude toward it can have a tremendous impact on whether we view our lives as fulfilling or painful. If we are good at it, if it gives us a chance to demonstrate competence, and if it is meaningful and socially valued, then it can be a major contributor to life satisfaction.

Work Satisfaction

U.S. surveys consistently find that the large majority (80 percent) of workers report satisfaction with their work. Although such a report may represent an acceptance of one's lot rather than real enthusiasm, it is remarkable that so few report dissatisfaction.

Studies of job satisfaction concentrate on two kinds of rewards that are available from work. **Intrinsic rewards** arise from the process of work; you experience them when you enjoy the people you work with and feel pride in your creativity and accomplishments. **Extrinsic rewards** are more tangible benefits, such as income and security; if you hate your job but love your paycheck, you are experiencing extrinsic rewards.

Generally, the most-satisfied workers are those in the learned professions, people such as lawyers, doctors, and professors. These people have considerable freedom to plan their own work, to express their talents and creativity, and to work with others; furthermore, their extrinsic rewards are substantial. The least-satisfied workers are those who work on factory production lines. Although their extrinsic rewards are good, their work is almost completely without intrinsic reward; they have no control over the pace or content of the work and are generally unable to interact with coworkers. In between these extremes, professionals and skilled workers generally demonstrate the greatest satisfaction; semiskilled, unskilled, and clerical workers indicate lower levels of satisfaction. Nevertheless, even those who hold highly routine, physically demanding jobs such as cashiers and cooks at fast-food restaurants often enjoy the satisfactions that come from doing a job well, earning a steady paycheck, and socializing with fellow workers (Newman 1999b).

Alienation

Another dimension of the quality of work life is alienation. **Alienation** occurs when workers have no control over their labor. Workers are alienated when they do work that they think is immoral (build bombs) or meaningless (push papers or brooms, or put together small pieces with no understanding of how those small pieces will become part of some larger whole). Work is also alienating when it takes physical and emotional energy without giving any intrinsic rewards in return. Alienated workers feel *used*.

Intrinsic rewards are rewards that arise from the process of work; they include enjoying the people you work with and pride in your creativity and accomplishments.

Extrinsic rewards are tangible benefits such as income and security.

Alienation occurs when workers have no control over the work process or the product of their labor.

The concept of alienation was first developed by Karl Marx to describe the factory system of the mid-nineteenth century. In 1863, a mother gave the following testimony to a committee investigating child labor:

> When he was seven years old I used tó carry him [to work] on my back to and fro through the snow, and he used to work 16 hours a day. . . . I have often knelt down to feed him, as he stood by the machine, for he could not leave it or stop. (as quoted in Hochschild 1985, 3)

This child was truly an instrument of labor. He was being used, just as a hammer or a shovel is, to create a product that would belong to someone else.

Although few of us work on assembly lines any more, modern work can also be alienating. Service work, in fact, has its own forms of alienation, known as **emotional labor**. In occupations from nursing to teaching to working as flight attendants, not merely our bodies but also our emotions become instruments of labor. To turn out satisfied customers, we must smile and be cheerful in the face of ill humor, rudeness, or actual abuse. Studies of individuals in these occupations show that many have trouble with this emotional component of their work. After smiling for 8 hours a day for pay, they feel that their smiles have no meaning at home. They lose touch with their emotions and feel alienated from themselves (Hochschild 1985). This is especially true when workers feel that they have no control over their job conditions (Bulan, Erickson, & Wharton 1997).

Alienation is not the same as job dissatisfaction (Erickson 1986). Alienation occurs when workers lack control. It is perfectly possible that workers with no control, but with high wages and a pleasant work environment, will express high job satisfaction.

> **Emotional labor** refers to the work of smiling, appearing happy, or in other ways suggesting that one enjoys providing a service.

Technology and the Future of Work

The productivity of workers and the quality of their work experience are often tied directly to the technologies they work with. Some work technologies, such as the assembly line, increase alienation while they increase productivity. Others, such as the photocopier, appear to be unmixed blessings. Although computerization and automation have increased productivity per worker, many people argue that the new technology is inescapably antilabor. These critics point out three negative effects of technology on labor: de-skilling, displacing workers, and greater supervision.

1. *De-skilling.* Many observers believe that increased mechanization has reduced the skill level needed for many jobs to the point where it is difficult to take pride in craft or a job well done. The process of automating a job so that it takes much less skill than it used to is called *de-skilling.* De-skilling occurs at all levels of labor, not just on the assembly line. For example, in the days before word processors, photocopiers, and self-correcting typewriters, a good typist could take pride in the work. With the new technologies, almost anyone can turn out decent-looking copy.

An important element of the de-skilling process is that it reduces the scope for individual judgment. In hundreds of jobs across the occupational spectrum, computers make decisions for us. In the sawmill industry, for example, a computer now assesses the shape of a log and decides how it should be cut to get the most board feet of lumber from it. An important element of skill and judgment honed from years of experience is now made worthless. According to the chief proponent of this argument, the central process in de-skilling is the separation of mind and hand (Braverman 1974).

2. *Displacement of the labor force.* One of the most critical complaints about automation is that it replaces people with machines. A few examples suffice. Computerization in grocery stores has resulted in sharp reductions in employment by eliminating inventory clerks and pricing/repricing personnel, as well as reducing the skill level needed for cashiering to the point that an average 15-year-old could do it. In the automobile industry, a robot can replace 1.5 humans per shift—and the robot can work three shifts a day (U.S. Department of Labor 1985, 1986).

In industry after industry, more sophisticated technology has made sharp inroads into the number of hours of labor necessary to produce goods and services. Many have concluded that fear of job loss is one of the reasons employees seldom complain about the de-skilling aspects of their jobs. If they still have a job, they are happy (Vallas & Yarrow 1987).

3. *Greater supervision.* Computerization and automation give management more control over the production process. More aspects of the production process are determined by management through its computerized instructions, and fewer aspects are determined by the employees. Computers also keep more complex records on employees. For example, the scanner machines used in grocery stores do more than keep inventory records and add up your grocery bill. They also keep tabs on the checker by producing statistics such as number of corrections made per hour, number of items run through per hour, and average length of time per customer. It is not surprising, therefore, that studies show that computers have increased work alienation among the cashiers and typists who use them (Vallas 1987).

Whether new technologies are an enemy of labor may depend on which laborer we ask. Those persons, often women and less-skilled workers, whose jobs are being replaced by new technologies are unlikely to see anything wonderful about them. On the other hand, most professionals in the knowledge industries (education and communications) regard these technologies as a boon. Computers have expanded their job opportunities and enhanced their lives. Even these workers, however, may occasionally wonder if they really benefit when technologies such as fax machines, cell phones, and e-mail allow—or even require—that they work at home, expanding work into a 24-hour-a-day job.

One scholar has argued that technology by itself is a neutral force: It can aid management or it can aid the workers (Davies 1986). Which technologies are implemented and the way they are implemented reflect a struggle between labor and management, and this struggle, not the technology itself, will determine the outcome.

Globalization and the Future of Work

The consequences of "de-laborization"—loss of jobs—are substantial. Not only individual workers but also entire communities are impoverished as new technologies facilitate corporate decisions to move factories to other parts of the world where labor is much cheaper. In fact, globalization is leading our national economy through a process of reverse development: Like a least-developed country, we export raw materials such as logs and wheat and import manufactured products such as VCRs and automobiles. People in Mexico, Japan, and Korea have jobs manufacturing products for the U.S. market while U.S. workers are making hamburgers.

The loss of jobs is not affecting only blue-collar workers. Increasingly, white-collar jobs like computer programming also are being moved overseas by corpora-

Technology

focus on

The Luddites: Down with Machines

Since the dawn of the industrial era, there has been tension between labor and technology. In 1675, weavers rioted against the introduction of looms that could allegedly do the work of 20 people; in 1768, sawyers in London destroyed a mechanized sawmill. The most widely known revolt of labor against machinery, however, was the Luddite uprising in England between 1811 and 1816 (Thomis 1970).

Wool was a major part of the English economy in the early nineteenth century. Wool-making was largely a home industry; and in Lancashire, nearly every home was engaged in it. The work was tedious and difficult. Particularly difficult was the last stage, in which a worker wielding 50-pound shears finished the fabric by cutting off all the nubs. Being able to handle these shears for 88 hours a week (the standard work week) required great strength and skill. It was an esteemed occupation.

In 1811, finishing machines were introduced to do this work. Each machine replaced six men. Not only were the men out of a job, but also the skills developed over a lifetime were made worthless. As use of the machines spread, large numbers of men were thrown out of work, and their families starved. On the horizon were more machines to take over other phases of wool production. Added to this, England was engaged in the Napoleonic Wars, and associated trade embargoes made the price of food high. The classic ingredients for insurrection were in place. The focus of the workers' anger was the machines, and their response was to destroy them.

In 1811, a young man named Ned Ludd, or Lud, or maybe Ludlam, is alleged to have broken up his father's hosiery loom because he resented a rebuke. The incident, which may have been imaginary, coincided with the eruption of machine-breaking demonstrations, and the labor movement came to be called the Luddites.

In the early nineteenth century, any organization of labor was illegal. Nevertheless, laborers met secretly to plan well-organized attacks. A body of men with blackened faces would break into a shop and destroy all the machinery. As the movement progressed, it became less disciplined, and owners too were assaulted and their homes looted.

The government was uncertain how to respond, although a few liberals were sympathetic with labor. Lord Byron, for example, wrote that "however much we may rejoice in any improvement in ye arts which may be beneficial to mankind; we must not allow mankind to be sacrificed to improvements in Mechanism" (as quoted in Reid 1986). Finally, however, the government sent troops to restore order. Leaders and alleged leaders of the Luddites were hanged or deported to the far corners of the empire.

The Luddite movement caused hardly a pause in the increasing use of machines to replace workers. Now, every year many new technologies are developed that throw more people out of work. Although we have unemployment insurance, early retirement schemes, and welfare to cushion the blows, the process still causes human misery as valuable skills are debased and employment is lost. If we use the term *Luddite* to include all of those who question whether humanity benefits from the adoption of new technologies, there are probably plenty of people who remain Luddites in spirit.

For more information, look up the following subjects in InfoTrac College Edition:
Luddites

Or visit the following Web site:
Wikipedia: The Free Encyclopedia
http://www.wikipedia.org/wiki/Luddite

---

tions eager to find cheaper workers. Countries like India, Russia, and the Philippines offer highly skilled workers, fluent in English, who are willing to work for far less than will U.S. workers. A significant proportion of telephone call centers and data-processing operations for U.S. corporations now are located overseas. During 2003, about 400,000 white-collar jobs moved from the United States to other countries, and it is estimated that 3.3 million jobs yearly will be moving overseas by 2015, with one-third of those jobs in information technology (Greenhouse 2003).

What can public policy do to protect jobs in the United States? There are three general policy options: the conservative free-market option, new industrial policies, and the social welfare option (Hooks 1984).

The Conservative Free-Market Approach

Generally, business leaders and conservatives argue that the way to keep jobs in the United States is to reduce wages and benefits. If labor is cheap, they argue, business will have less incentive to automate or to move assembly plants to Mexico or Indonesia.

By default, this policy has been implemented. In communities across the nation, management has used threats of plant closings to force wage concessions and reduce benefits. The power of labor unions is reduced to negotiating benefit protection in the face of wage reductions. Because so many workers have been afraid of losing their jobs, organized labor's power has been sharply reduced. Thus, one result of "de-laborization" is the reduced economic circumstances of workers who still have jobs.

New Industrial Policies

Liberals argue that private profit should not be the only goal of economic activity and that the state should see to it that economic decisions protect communities' and workers' interests (Genovese 1988). Among the specific policies recommended are (1) federal trade policies that make U.S.-made goods more competitive in international markets and that reduce the advantage that foreign-made products have in the United States, (2) vigorous state investment in industries that will provide the largest number of decent jobs, (3) government oversight of mergers and plant closings to make sure plants behave responsibly, and (4) state support for worker efforts to buy and manage their own industries.

Social Welfare Policies

New industrial policies are designed to keep people working; social welfare policies aim at protecting those who are thrown out of work. Among the policies recommended are (1) 6-month notification of plant closings, (2) paid leave for soon-to-be displaced employees to look for jobs, (3) retraining programs for displaced workers, (4) relocation assistance for displaced workers, and (5) substantially more generous unemployment benefits (Blakely & Shapiro 1984). Of course, such suggestions are open to the same concerns that a number of Western European nations are now experiencing in regard to their welfare economies (see the Focus on a Global Perspective box in this chapter).

Where This Leaves Us

As you sit in the classroom to prepare yourself for a good position in the labor force, the economy itself is changing. Indeed, it is changing so fast that you may need to retool your skills several times before your work life is complete. Rapid developments in technology have dramatically changed the workplace from what it was 20 or even 10 years ago. The globalization of the economy has further changed the job situation for Americans. Although the specter of unemployment still haunts ethnic minorities and blue-collar workers the most, middle-class workers are also experiencing the pangs of job insecurity and alienation, as corporations strive to cut costs and jobs move overseas.

One political approach to the changing job situation involves social welfare policies that would better protect American workers and their families against both overseas competition and corporations focused on profitability and stock prices. Yet in an era of political conservatism, when relatively few Americans vote—and voting is especially uncommon among those who need the government's help the most—it seems unlikely that the United States will expand its protections for average citizens. The growing concentration of money and power also works against any meaningful changes. In the meantime, American workers will increasingly need to learn new skills and be creative to find and keep jobs.

Summary

1. Power may be exercised through coercion or through authority. Authority may be traditional, charismatic, or rational-legal.

2. Any ongoing social structure with institutionalized power relationships can be referred to as a political institution. The most prominent political institution is the state. It is distinguished from other political institutions because it claims a monopoly on the legitimate use of coercion and it has power over a broader array of issues.

3. Democracy is most likely to flourish in societies that have vibrant, competing interest groups, large middle classes, and relatively little income inequality.

4. Three major models are used in describing the U.S. political process: the pluralist model, the power-elite model, and the state autonomy model. None of them suggests that the average voter has much power to influence events.

5. Political participation is rather low in the United States; fewer than half of the people of voting age vote in most national elections, and fewer yet take an active role in politics. Political participation is greater among those with high social status and among middle-aged and older people—establishment types who are more likely to support the status quo.

6. Although the Democratic Party tends to attract working-class and minority voters and the Republican Party tends to attract better-off voters, both U.S. political parties tend to have middle-of-the-road platforms with broad appeal.

7. Some sociologists believe that voting rates are lower in the United States than in other industrialized countries because potential voters are politically alienated. Others believe that rates are low because government policies make it difficult for Americans to vote and because no major political party has actively sought to involve marginalized groups.

8. Capitalism is an economic system that maximizes productivity but tends to neglect aspects of distribution; socialism emphasizes distribution and neglects aspects of production. Both capitalism and socialism can occur in either democratic or autocratic political structures.

9. Changes from preindustrial to industrial to postindustrial economies have had profound effects on social organization. The tertiary sector of the economy occupies about three-quarters of the U.S. labor force; it includes highly paid professional occupations as well as maids and waitresses.

10. The United States has a dual economy containing two distinct parts: the industrial core and the periphery (or competitive sector). These are paralleled by a segmented labor market.

11. Economic projections show substantial changes in occupations in just the next 10 years. The largest number of new jobs will be low-status, low-wage service positions. The major losers will be those who have occupied traditional blue-collar jobs.

12. Although most U.S. workers report satisfaction with their work, some scholars argue that they are nevertheless alienated because they are estranged from the products of their labor or from their emotions.

13. Critics argue that automation and computerization have had three ill effects on labor: de-skilling jobs, reducing the number of jobs, and increasing control over workers. Nevertheless, some occupations have grown or been made easier through new technology.

Thinking Critically

1. The family and the classroom are more often authoritarian than democratic. Try to explain this in sociological terms.

2. What impact, if any, do you think the most recent presidential election will have on future voter participation rates? Discuss in terms of the factors identified in the text as being associated both with voting and with voter nonparticipation.

3. As an employee, what would you like about working in Sweden? What would you dislike? As an employer, what would you like and dislike about doing business in Sweden? How can Sweden's democratic socialist government continue to resolve these differences?

4. How has technology affected your schoolwork in the last 10 years? Has it given you new tools or robbed you of old skills or simply given your instructors an excuse to demand more of you?

5. How do you think a postindustrial economy will affect your working and economic future? In what ways is a postindustrialized economy a global economy? Which of the three general policy options outlined in the text do you think the United States will follow and why? In your opinion, which would be the best one to follow and why?

Sociology on the Net

The Wadsworth Sociology Resource Center: Virtual Society

http://sociology.wadsworth.com/

The companion Web site for this book includes a range of enrichment material. Further your understanding of the chapter by accessing Online Practice Quizzes, Internet Exercises, InfoTrac College Edition Exercises, and many more compelling learning tools.

Chapter-Related Suggested Web Sites

The Federal Election Commission
http://www.fec.gov

U.S. Department of Labor
http://www.dol.gov

U.S. Bureau of Labor Statistics
http://stats.bls.gov

InfoTrac College Edition

http://www.infotrac-college.com/wadsworth

Access the latest news and research articles online—updated daily and spanning four years. InfoTrac College Edition is an easy-to-use online database of reliable, full-length articles

from hundreds of top academic journals and popular sources. Conduct an electronic search using the following key search terms:

Democracy

Capitalism

Socialism

Political participation

Employment

Equal opportunity

 For more information related to this chapter, visit the Opposing Viewpoints Resource Center. Be sure you check all the options on the toolbar—viewpoints, references, statistics, and so on—and search under the following subjects:

Employment

Poverty

Suggested Readings

Domhoff, G. William. 1998. *Who Rules America: Power and Politics in the Year 2000* (3rd ed.). Mountain View, Calif.: Mayfield. A fully updated edition of a classic text on the power elite.

Hochschild, Arlie R. 1985. *The Managed Heart: The Commercialization of Human Feelings.* Berkeley: University of California Press. A study of alienation in service occupations, with a detailed examination of how flight attendants handle emotional work. This book is becoming a classic.

Mills, C. Wright. 1956. *The Power Elite.* New York: Oxford University Press. One of the most important discussions of the relationship among three areas of power in the United States: government, business, and the military. Mills, a conflict theorist, argues that the members of a small elite make the decisions that control U.S. society.

Newman, Katherine S. 1999. *No Shame in My Game: The Working Poor in the Inner City.* New York: Knopf. A finely detailed portrait of the lives of the urban working poor that documents both the difficulties and the rewards of low-wage work.

Piven, Frances Fox, and Cloward, Richard A. 2000. *Why Americans Still Don't Vote: And Why Politicians Want It That Way.* Boston: Beacon. Scholar-activists Piven and Cloward, who played key roles in passage of the "motor voter" National Voter Registration Act of 1993, argue that liberal politicians have done little to encourage Americans to vote while conservative politicians have worked tirelessly to ensure that the poor, minorities, and others *don't* vote.

Rubin, Beth A. 1996. *Shifts in the Social Contract: Understanding Change in American Society.* Thousand Oaks, Calif.: Pine Forge Press. A thoughtful analysis of how changes in the U.S. and world economies have resulted in changes in politics, in the institution of family, and in the way we can expect to live our everyday lives.

Politics, Religion, and the Culture War

Over the last decade, it has become commonplace for political commentators, theologians, scholars, and the popular press to lament the increasing polarization of American society. As examples of this polarization they cite:

- Heated debates among Episcopalians about the proper church stance on homosexual clergy
- Strongly divergent views toward abortion held by Right to Life and Freedom to Choose groups
- Conflict between those who support and oppose capital punishment
- The controversy regarding the appropriateness of providing government support to parochial schools and faith-based social organizations

The evidence seems to be mounting that Americans are no longer as tolerant of ideological differences as they once were. Some scholars have argued that the United States is increasingly split into two broad and antagonistic groups, each vying for control of American culture. The general public also appears to believe that this is true. A 1995 *Newsweek* poll found, for instance, that 86 percent of those surveyed agreed that "there was a time when people in this country felt that they had more in common and shared more values than Americans do today" (DiMaggio, Evans, & Bryson 1996). This divisiveness over important social values has appeared so extreme that it is frequently referred to as the "culture wars" (Hunter 1991; Wuthnow 1988). But what evidence is there that Americans have actually become much more deeply split in their attitudes?

Using data from the General Social Survey and the National Election Survey, researchers have found very little evidence that attitudes have become more polarized since the 1970s. Although attitudes toward abortion and (to a lesser extent) toward the poor have become more deeply divided, Americans have, in fact, become more unified in a number of respects—their support for racial integration and the rights of women to participate in public life, for example, as well as their tough stance on crime (DiMaggio et al. 1996). Differences in attitudes between young and old, African Americans and whites, Southerners and Northerners, high school dropouts and college graduates, and Protestants, Catholic, and Jews persist, to be sure. But the polarization of attitudes has steadily declined among all of these groups, with the one exception being a widening gap between Republicans and Democrats (DiMaggio et al.).

If attitudes have not generally become more hostile and divisive, why do so many observers believe they have? According to Miller and Hoffmann (1999), the answer lies in a series of political events in the 1970s and 1980s that made the terms *liberal* and *conservative* much more value-laden.

The 1970s saw the emergence of many conservative Christian political groups. In 1974, Third Century Publishers began publishing a wide range of Christian literature aimed at political issues. In 1978, Pat Robertson founded the Christian Broadcasting Network and in 1979 Jerry Falwell founded the Moral Majority. As we discussed earlier, these groups have had only marginal electoral success. There is no doubt, however, that they have been effective in increasing the tendency to label certain positions on moral issues as either wholly "liberal" or "conservative." Also, during the 1980s, the Reagan administration successfully used the term *conservative* to promote a social and economic agenda based on a smaller federal government and lower taxes. As such, it enjoyed considerable public support.

The rise of these conservative political groups was countered by the political activities of the large group of 1960s protesters who were reaching early and middle adulthood at the same time. As youths, these individuals had advocated the progressive social and political agenda that fostered the gay rights movement, political debates over the Equal Rights Amendment, and legal access to abortion. By the 1980s they were established adults, some of whom remained politically active. As a result, by this time not only were there more Christian political organizations with more visible and clearly defined conservative platforms, but competing organizations with alternative, liberal views also had a greater presence and greater visibility. The emergence of these groups and the political wrangling of their high-profile leaders no doubt fueled the increasingly negative view of liberals held by conservatives and vice versa (Miller & Hoffmann 1999; Wuthnow 1996).

Miller and Hoffmann (1999) argue that the end result of these historical, political events has been not only a polarization of the categories "liberal" and "conservative" but also an increased tendency for individuals to adopt whatever label is promoted by their religious denomination (or political party), no matter what their attitudes might be on any specific issue. If Miller and Hoffmann are right, the culture wars do not stem from fundamental differences in our attitudes but rather from a series of heated, high-profile, political campaigns.

MicroCase Online Exercise

For an interactive exercise using MicroCase data sets, go to the text companion website at http://sociology.wadsworth.com/brinkerhoff/essentials6e. Choose Chapter 14 and then select "Intersection Exercise" from the left navigation bar.

Population and Urban Life

© Reuters/CORBIS

Populations, Large and Small

Birth and death—nothing in our lives quite matches the importance of these two events. Naturally, each of us is most intimately concerned with our own birth and death, but to an important extent, our lives are also influenced by the births and deaths of those around us. Do we live in large or small families, large or small communities? Is life predictably long or are families, relationships, and communities periodically and unpredictably shattered by death?

In this chapter we take a historical and cross-cultural perspective on the relationship between social structures and population. The study of population is known as **demography,** and those who study it are known as demographers. Demographers focus on three processes: **fertility** (childbearing), **mortality** (deaths), and migration. Here we will look at these three demographic processes and also at the effect of population size on social relationships within communities. We are interested in questions such as how birthrates or community size affect social structures and, conversely, how changing social structures affect birthrates and community size.

Currently, the world population is 6.3 billion, give or take a couple hundred million. This is two and a half times as many people as lived in 1950. World population has grown because fertility has increased while life spans have lengthened and mortality has decreased. In part because of this growth, millions are poor, underfed, and undereducated; pollution is widespread; and the planet's natural resources have been ransacked.

These problems are among the causes of migration, which has caused some villages, cities, and nations to grow in size while others shrink. Because often some people enter an area just as others leave it, demographers focus on **net migration:** the number of people who move into an area minus the number who move out. Some migrants, known as *immigrants,* move from one country to another while others, known as *internal migrants,* move within a nation.

Migration, in turn, leads to another set of social concerns, as nations wrestle with how to respond to the newcomers in their midst. Some nations, like the United States, allow immigrants to eventually become citizens. Other nations, like Germany, refuse citizenship to almost all immigrants as well as to descendants of immigrants who have lived their whole lives in these nations. Immigration has substantial consequences, then, not only for population growth and economic development, but also for issues such as the meaning of citizenship and nationality.

In sum, population size and change is vitally linked to many of our era's crises. The next section examines the process by which current world population was reached.

Demography is the study of population—its size, growth, and composition.

Fertility is the incidence of childbearing.

Mortality is the incidence of death.

Net migration is the number of people who move into an area minus the number who move out.

Understanding Population Growth

Although population is concerned with intimate human experiences such as birth and death, the big picture of population growth and change can be understood only if we use statistical summaries of human experience. Three measures are especially important: the crude birthrate, the crude deathrate, and the rate of natural increase.

TABLE 15.1
The World Population Picture, 2003
In 2003, the world population was 6.3 billion and growing at a rate of 1.3 percent per year. Growth was uneven, however; the less developed areas of the world were growing much more rapidly than the more developed areas. As a result, most of the additions to the world's population were in poor nations.

Area	Crude Birthrate (Births/1,000 Persons)	Crude Deathrate (Deaths/1,000 Persons)	Rate of Natural Population Increase (in millions)
World	22	9	1.3
More developed nations	11	10	0.1
Less developed nations	24	8	1.6

SOURCE: Population Reference Bureau 2003.

The **crude birthrate** is the number of live births per 1,000 persons.

The **crude deathrate** is the number of deaths per 1,000 persons.

The **rate of natural increase** is the crude birthrate minus the crude deathrate, calculated as a percentage.

The **crude birthrate** is the number of live births per 1,000 persons, and the **crude deathrate** is the number of deaths per 1,000 persons. To find the **rate of natural increase,** we subtract the number of deaths from the number of births and then calculate the result as a percentage.

Table 15.1 shows these rates in 2003. Worldwide, the crude birthrate in 2003 was 22 births per 1,000 population; the crude deathrate was a much lower 9 per 1,000. Because the number of births exceeded the number of deaths by 13 per 1,000, the rate of natural increase of the world's population was 1.3 percent. If your savings were growing at the rate of 1.3 percent per year, you would undoubtedly think that the growth rate was very low. A growth rate of 1.3 percent in population, however, translates into a doubling time of 51 years. Although the rate of increase has actually been decreasing since the 1960s, it remains true that if the current rate of increase continues, the population will double to 12.2 billion by the year 2054.

The frightening prospect of welcoming another 6.3 billion people in our lifetime is complicated by the fact that the growth is uneven. As Table 15.1 shows, growth rates are startlingly different across the areas of the world. Africa is the world's fastest-growing continent. With a rate of increase of 2.4 percent per year, it will double its population size in only 29 years. In Europe, by contrast, deaths actually exceed births.

These differentials in growth are of tremendous importance. Almost all the additions to world population in the next several decades will take place in the least-developed nations. As a result, the world is likely to be poorer in 2050 than it is now. How did these different population patterns evolve?

Population in Former Times

For most of human history, fertility (childbearing) was barely able to keep up with mortality (death), and the population grew little or not at all. Historical demographers estimate that in the long period before population growth exploded, both the birth- and deathrates hovered around 40 to 50 per 1,000. (Birthrates are still almost 40 per 1,000 in Africa.) Translated into personal terms, this means that the average woman spent most of the years between the ages of 20 and 45 either pregnant or nursing. If both she and her husband survived until they were 45, she would produce an average of 6 to 10 children. The average life expectancy was perhaps 30 or 35 years.

Such a low life expectancy was largely due to very high infant mortality: perhaps one-quarter to one-third of all babies died before they reached their first birthday. Both birth and death were frequent occurrences in most preindustrial households.

The Demographic Transition in the West

The industrial revolution set in motion a series of events that revolutionized population in the West. First, mortality dropped; then, after a period of rapid population growth, fertility declines followed. Because studies of population are called demography, this process is called the **demographic transition.**

Decline in Mortality

General malnutrition was an important factor underlying high levels of mortality. Although few died of outright starvation, poor nutrition increased the susceptibility of the population to disease. Improvements in nutrition were the first major cause of the decline in mortality, beginning in the 1700s and continuing into the early twentieth century. New crop varieties from the Americas (corn and potatoes especially), new agricultural methods and equipment, and increased trade all helped improve nutrition in Europe and the United States. The second major cause of the decline in mortality was a general increase in the standard of living, as improved shelter and clothing left people healthier and better able to ward off disease. Changes in hygiene were vital in reducing communicable disease, especially those affecting young children, such as typhoid fever and diarrhea (Kiple 1993).

In the late nineteenth century, public-health engineering led to further reductions in communicable disease by providing clean drinking water and adequate treatment of sewage. For example, over the course of the twentieth century, the life expectancy of white Americans increased from 47 to 77 and the life expectancy of African Americans increased from 33 to 70 (U.S. Bureau of the Census 1975a, 1999a). Thus, although life expectancy has been increasing gradually since about 1600, the fastest increases have occurred in the twentieth century. Medical advances probably account for no more than one-sixth of this overall rise in life expectancy (Bunker et al. 1994). Instead, public-health initiatives and steady progress in eliminating deaths caused by poor nutrition and an inadequate standard of living are largely responsible for the decrease (McKinlay & McKinlay 1977; Weitz 2004b). Interestingly, once the standard of living in a nation reaches a certain point—approximately $6,400 per capita income—further increases in life expectancy depend less on increasing income than on reducing the income gap between rich and poor (Wilkinson 1996). This is one major reason why, compared to Americans, on average Cubans live almost as long and Swedes live longer.

Decline in Fertility

The Industrial Revolution also affected fertility, although less directly. The reduction in fertility was not a response to the drop in mortality or even a direct response to industrialization itself. Rather, it appears to have been a response to changed values and aspirations triggered by the whole transformation of life (Coale 1973).

Industrialization meant increasing urbanization, greater education, and the real possibility of getting ahead in an expanding economy. Pensions and other social benefits became more common with industrialization, so people didn't need to have many children to care for them in their old age (Friedlander & Okun 1996). Perhaps even more important, industrialization created an awareness of the possibility of doing things differently from how they had been done by previous generations. As a result, the idea of controlling family size to satisfy individual goals spread even to areas that had not

The **demographic transition** is the process of moving from the traditional balance of high birth- and deathrates to a new balance of low birth- and deathrates.

Connections

Personal Application

Have you ever traveled to a less developed country? If you did, the odds are that you got a nasty stomach virus for a day or two, but otherwise suffered no health problems. Yet malaria, cholera, dysentery, and the like kill millions in these countries each year. Why are American tourists virtually immune? Vaccinations, antibiotics, and access to soap and water help. But the most important reason is that they start out healthy, well-nourished, and well-clothed. As a result, even if they come in contact with dangerous germs, their bodies most likely will be able to fight against infection.

© Lauren Goodsmith/The Image Works

■ Many families in Africa, especially polygamous families, have numerous children, and overpopulation is a cause for some concern.

experienced industrialization, so that by the end of the nineteenth century, the idea of family limitation had gained widespread popularity (van de Walle & Knodel 1980). Currently in Europe and North America, birth- and deathrates are about even, and there is little population growth.

The Demographic Transition in the Non-West

In the less-developed nations of the non-West, Africa especially, birth- and deathrates remained at roughly preindustrial levels until the first decades of the twentieth century. After that, in some areas such as Latin America, Taiwan, Singapore, and South Korea, economic development and improvements in standard of living caused both death- and birthrates to plummet, much as they had previously done in the West. It took a few more decades for deathrates to begin falling in Africa, following improvements in public-health engineering and the introduction of very basic medical interventions (primarily childhood vaccinations and treatment for childhood diarrhea).

Unlike the mortality decline in the West and in the developed countries of Asia, in other countries decreasing deathrates were not gradual. Nor were they caused by changes in social structure or standard of living. Instead, deathrates dropped because of sudden change brought in from outside. Consequently, the changes in social structure and culture that are necessary to cause fertility decline are occurring more slowly, and populations continue to grow steeply. These patterns, however, are being affected dramatically by the AIDS epidemic. In African countries like Botswana, Swaziland, and Zimbabwe, where between 30 and 40 percent of the population is infected with HIV, deathrates have soared and population growth has slowed considerably (UNAIDS/WHO 2002).

Population and Social Structure: Two Examples

In this section we explore contemporary relationships between social structure and population in two societies: Ghana, where fertility is high, and Europe, where fertility is low.

Ghana: Is Fertility Too High?

Ghana is an example of a society in which traditional social structures encourage high fertility. It is also an example of a society in which high fertility may ensure continuing traditionalism—and poverty.

The Effects of Social Roles on Fertility

Fertility rates have declined in Ghana in recent years but remain high. Ghana still has a crude birthrate of 31 per 1,000 population. Mortality, however, is down to 10 per 1,000. This means that the rate of natural increase in Ghana is 2.1 percent per year (Population Reference Bureau 2003). If that rate continues, the population will double in less than 30 years. An aggressive family-planning program is unlikely to reduce this growth substantially because Ghanaian women, on average, still want 4.3 children while having 4.6 children (Ghana Statistical Service 1999).

One of the most important reasons for this high fertility is women's roles. In Ghana, children are an important—perhaps the only—source of esteem and power open to women. Women who cannot bear children risk divorce or abandonment. This is especially true for the 23 percent of Ghanaian women who live in polygamous unions. The number of children a woman has—especially the number of sons— strongly affects her position relative to that of her co-wives. Moreover, because infant mortality remains relatively high, Ghanaian women believe they must have four or more children to ensure that a couple survive to adulthood.

Another important cause of high fertility is the need for economic security. Most Ghanaians work in subsistence agriculture. To survive, families need children as well as adults to work in the fields. In addition, when children grow up and marry, they can add to the family's economic and political security by creating political and social allegiances to other families. Finally, children are the only form of old-age insurance available to Ghanaians, for parents who grow old or ill must rely on their children to support them. Conversely, having children is relatively inexpensive: no expensive medical treatment for children is available, schooling is either inexpensive or unaffordable, and children don't expect to own designer jeans or $150 tennis shoes. With a cost/benefit ratio of this sort, it is not surprising the Ghanaians desire many children.

The Effects of High Fertility on Society

Although high fertility may appear to be in the best interests of individual women, its consequences for society are less beneficial. Ghana's population is now doubling every 29 years. This means that just to maintain current levels of support for education, highways, agriculture, and the like, the country must double its budget.

Thus, a decision that is rational on the individual level turns out to be less wise on the societal level. Occasionally, people in the West make remarks of the sort: "Are they stupid? Can't they figure out they would be better off if they had fewer children?" Unfortunately for the argument, nations don't have children; women do. High fertility continues to be a rational choice for individual Ghanaians.

Policy Responses

To reduce its population growth, Ghana has established an excellent family-planning program that makes contraception available, convenient, and affordable to women who want it. When women want several children, however, access to contraception has limited impact. Currently, only 13 percent of all married women in Ghana use modern contraceptive methods (Population Reference Bureau 2003). Contraceptive use is considerably higher among younger, better educated, urban women. Study after study has found that the best way to reduce fertility is to combine access to contraception with educational and economic opportunities for women.

Europe: Is Fertility Too Low?

In a world reeling from the impact of doubling populations in the less-developed world, it is ironic that many developed countries are worried that fertility is too low.

■ In Europe, many families have only one child, and underpopulation is increasingly a cause for concern.

TABLE 15.2

Fertility and Population Growth in Europe, 2003

Overall deaths are now slightly exceeding births in Europe. Thus, some nations are already experiencing population decline. The last column in the table shows the combined impact of births, deaths, and migration into and out of a country.

Area	Crude Birth- rate	Crude Death- rate	Rate of Natural Increase	Average # of Children per Woman	Projected Population Change 2003–2050*
Europe, total	10	12	−.2%	1.4	−9%
Austria	9	9	.0	1.3	+1
Denmark	12	11	.1	1.7	+8
Germany	9	10	−.1	1.3	−18
Hungary	10	13	−.4	1.3	−25
Italy	9	9	−.1	1.2	−9
Romania	10	12	−.3	1.2	−21
Spain	10	10	.1	1.2	0
United Kingdom	11	10	.1	1.6	+8

*Includes the impact of net migration plus net natural growth.

SOURCE: Population Reference Bureau 2003.

The Effects of Social Roles on Fertility

Replacement level fertility requires that a women bear an average of 2.1 children—one to replace themselves, one to replace their partners, and a little extra to cover childhood mortality.

With modern levels of mortality, fertility must average 2.1 children per woman if the population is to replace itself: two children so that the woman and her partner are replaced and a little extra to cover unavoidable childhood mortality. This is called **replacement level fertility.** If fertility is less than this, the next generation will be smaller than the current one.

Currently in Europe, the average woman is having 1.4 children (Population Reference Bureau 2003). In Latvia, Greece, the Czech Republic, Italy, and other countries, the average woman is having only 1.1 to 1.2 children (Table 15.2). This means that the next generation of Europeans will be much smaller than previous ones, unless these countries absorb many new immigrants.

Why is fertility so low in Europe? In essence, the situation in Europe is the reverse of that in Ghana. Most women are educated and many hold paying jobs outside of the home. Women's social status is close to that of men (especially in the Scandinavian countries), so women do not need to have children to have a purpose in life or to assure their social standing. Because few Europeans work in agriculture, and all children are expected to be in school, having children doesn't add to a family's labor pool. Finally, European governments provide good safety nets in the form of disability insurance, health care, old-age pensions, and the like, which means that couples do not need to have children to take care of them in sickness or old age.

The Effects of Low Fertility on Society

Given the serious worldwide dilemmas posed by population growth and the very high density of many European nations, why should we consider low fertility a problem? There are two main concerns: the large numbers of old people compared to

Il paraît que je suis
un phénomène socio-culturel.

LA FRANCE
A BESOIN
D'ENFANTS.

CAMPAGNE RÉALISÉE PAR AVENIR.DAUPHIN GIRAUDY

© Mareschal/Image Bank and Population Reference Bureau, Inc.

■ Countries such as France are attempting to increase fertility above replacement levels through billboards such as this one, which declares "France Needs Children."

young people, and rising nationalistic fears resulting from the importing of immigrant labor.

Very low fertility creates an age structure in which the older generation is almost as large as the younger generation on whom it relies for support. As a result, it is increasingly difficult for European nations to fill all the occupations needed to keep a nation running, from taxi-drivers to doctors, and to compete in the global market. At the same time, the cost of paying for old-age pensions and health care is growing rapidly. (The same is true of Social Security in the United States.) By 2030, the most-industrialized nations will need to spend an additional 9 to 16 percent annually of their national net incomes, just to continue current pension and health benefits for older persons (Peterson 1999).

To counteract this problem, European countries have been importing workers from the Middle East and the Mediterranean. This has led to nationalist fears of cultural dilution. A survey conducted in 2003 found that 59 percent of Germans and 51 percent of French people felt Arabic immigrants were bad for their countries (Pew Research Center 2003a). An astounding 80 percent of Italians felt the same about Albanian immigrants. These feelings, in turn, have provoked anti-immigrant violence throughout Europe and have led many European nations to clamp down on immigration.

Policy Responses

In response to these concerns, many European nations have established incentives to encourage fertility. Among them are paid maternity leave, cash bonuses and housing subsidies for having more children, longer vacations for mothers, and graduated family allowances (Gautier 1999). Nevertheless, the costs of raising children far outstrip these benefits. As a result, while these incentive plans have kept birthrates from falling drastically, they have not helped to raise birthrates in countries where women have attractive alternatives outside the home (Gautier & Hatzius 1997).

TABLE 15.3
Fertility Decline in World Regions, 1950-2003
In the last half century, the average number of children per woman has declined worldwide.

Region	Average Number of Children per Woman	
	1950	2003
Africa	6.6	5.2
Asia	5.9	2.6
Europe	2.6	1.4
Latin America	5.9	2.7
North America	3.5	2.0
Oceania	3.8	2.4

SOURCE: Population Bulletin 1999; Population Reference Bureau 2003.

Population and Social Problems: Two Examples

Analysis of world population growth reveals a good news/bad news situation. The good news is that fertility is declining in every part of the world (Table 15.3). The bad news is that the population of the world will double within 50 or so years anyway. The reason for this gloomy prediction lies in the age structure of the current population. The next generation of mothers is already born—and there are a lot of them. Thus, we must plan for a world that will soon hold 12 to 13 billion people.

Population pressures can contribute to numerous social problems. In this section, we address two of them—environmental devastation and poverty.

Environmental Devastation: A Population Problem?

All around the world, there are signs of enormous environmental destruction: In the developed world, we have acid rain and oil spills; in Africa, desert environments are spreading rapidly due to deforestation and overgrazing. Both of these pose serious threats to the environment, but only the latter is truly a population problem.

It is estimated that the United States, which contains only 5 percent of the world's population, consumes one-quarter of the world's resources and produces nearly three-quarters of the world's hazardous waste (Ashford 1995). Our affluent, throwaway lifestyle requires large amounts of petroleum and other natural resources. Obtaining these resources results in the destruction of wilderness, the loss of agricultural lands, and the pollution of oceans. Using these resources causes illness-inducing air pollution, acid rain, and smog that are killing our forests. Although these problems would be less severe if there were half as many of us (and hence half as many cars,

Deforestation is devastating tropical rainforests in Brazil, the Philippines, and elsewhere.

© Harold Castro/FPG

factories, and Styrofoam cups), they are not really population problems. They stem from our way of life rather than our numbers.

In sub-Saharan Africa, however, population pressure is a major culprit in environmental destruction. In rural areas, the typical scenario runs like this: Population pressure forces farmers to try to plow marginal land and to plant high-yielding crops in quick succession without soil-enhancing rotations or fallow periods. The marginal lands and the overworked soils produce less and less food, forcing farmers to push the land even harder. They cut down forests and windbreaks to free more land for production. Soon, water and wind erosion becomes so pervasive that the topsoil is borne off entirely, and the tillable land is replaced by desert or barren rock. This cycle of environmental destruction—which destroys forests, topsoil, and the plant and animal species that depend upon them—is characteristic of high population growth in combination with poverty. When one's children are starving, it is hard to make long-term decisions that will protect the environment for future generations.

In sum, reducing population growth would reduce future pressure on natural resources, but it would not solve the current problem. The solution rests in an international moral and financial commitment to reducing rural poverty, improving farming practices, reducing the foreign debt of the less- and least-developed nations, *and* curbing wasteful and destructive practices in the developed nations.

Poverty in the Least-Developed World

Perhaps 500 million people around the world are seriously undernourished, and each year outbreaks of famine and starvation occur in Africa and Asia; a billion more are poorly nourished, poorly educated, and poorly sheltered. These people live in the same nations that have high population growth.

Some observers blame poverty in the developing nations on high fertility. Yet high fertility is not the only or even the primary cause of this poverty. Poverty and malnutrition result primarily from war, corruption, and inequality in nondemocratic countries and from a world economic system that extracts raw goods and profits from poorer countries (Chase-Dunn 1989; Dreze & Sen 1989; Sen 1999). It is a terrible irony that most poor countries export more food than they import (Lappé, Collins, & Rosset 1998). Cuba, for example, has become poorer in the 1990s not because of population growth but because its authoritarian government failed to develop true economic growth and instead relied heavily on subsidies from the now-defunct Soviet Union. People in the Democratic Republic of the Congo, meanwhile, are dying of starvation because of war rather than because of high fertility.

Policy Responses

Although many factors contribute to poverty, almost all world leaders agree that reducing fertility is an important step toward increasing the standard of living in the poorer nations of the world. The most successful programs to reduce fertility have combined an aggressive family-planning program, economic and educational development, and improvements in the status of women (Poston & Gu 1987):

1. *Family-planning programs.* These programs are designed to make modern contraceptives and sterilization available inexpensively and conveniently

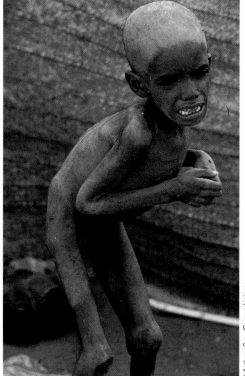

■ Because of war, drought, corrupt governments, and income inequality within and between nations, about half a billion people in the world are seriously malnourished. Some, such as this child, are starving.

A Global Perspective

International Migration

Some are pulled by new opportunities, others are pushed out of their homeland by ethnic conflict, civil war, poverty, or drought. During 2001, approximately 15 million people became refugees and left their homes involuntarily (U.S. Committee for Refugees 2002). About the same number chose voluntarily to seek new lives in other countries. We often hear debate about immigrants and refugees in the United States, but what do we know about international migration? Map 15.1 shows the current migration patterns around the world.

Demographers believe that the political turmoil and violence of the last two decades have substantially increased the numbers of involuntary migrants. At least 160 million people lived outside their country of birth or citizenship in 2000, almost twice the number in 1990 (Martin & Widgren 2002). During 2001, the leading sources of refugees were Sudan, Afghanistan, Pakistan, Colombia, and the Congo, with most of these refugees moving to another developing nation. The result, of course, is that enormous strain is placed on the already limited capacities of host countries to sustain their own growing populations.

Although push factors such as war and famine account for much of the movement between developing

MAP 15.1
Major Migration Patterns in the Early 21st Century
SOURCE: Martin & Widgren (2002).

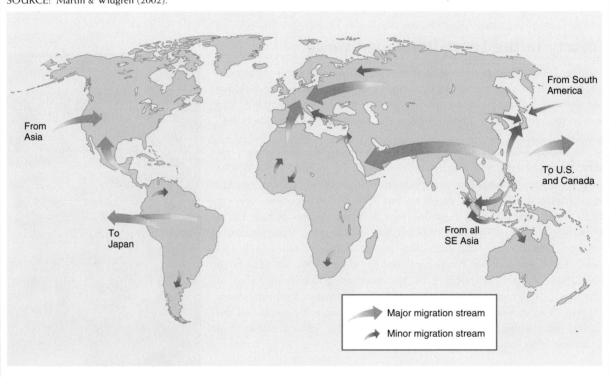

to individuals who desire to limit the number of their children. For example, between 1975 and 1991, an aggressive family-planning program increased contraceptive use in Bangladesh by 500 percent and decreased the average number of children per woman from 7.0 to 4.9 in just 16 years (Kalish 1994).

2. *Economic and educational development.* Experience all over the world shows that fertility declines as education increases and the country undergoes economic development. For example, South Korea's fertility has plummeted from 6.0 children per

countries, some migrants are also pulled by the economic growth and employment opportunities in newly industrializing nations, such as South Korea, Singapore, and Malaysia. Pull factors also account for much of the immigration from less to more developed countries. Strong European economies provide increasing numbers of jobs to a growing non-Western labor force. Migrants traditionally have been young men, but women and girls now make up 40 to 60 percent of the international migrant stream. Many of these are mothers who, in growing numbers, seek employment opportunities in more affluent neighboring countries in order to send money to the family, friends, or neighbors who are raising their children.

The money sent back home by migrants—both men and women—is a large and growing source of revenue for many nations. These funds are Mexico's third-largest source of income, exceeding local and state budgets in rural regions. In the poorest countries of Latin America, money sent home by migrants exceeds all international development assistance and is increasing by an average 11 percent a year (Thompson 2002). Most of that money goes to help relatives, but increasingly migrants are pooling together their funds to build community projects or start industries in their home towns.

It is not yet clear, however, who profits most from the international migrant stream. Although countries such as Germany, France, and Italy do face new challenges stemming from an eth-nically diverse population, workers from developing nations are, in fact, helping to sustain the continued expansion of the European economy. Low birth rates have led to smaller labor forces and aging populations in much of Europe. Thus, migrants from countries such as Turkey and Pakistan fill the demand for more workers, particularly those at the low end of the labor hierarchy. Whether the money that migrants send home will significantly improve the quality of life in less developed nations remains an open question.

© Charles Caratini/Corbis Sygma

 Two-thirds of the world's displaced persons are found in developing nations, where problems are only slightly less severe than those of the refugees' homelands.

For more information, look up the following subjects in InfoTrac College Edition:
Refugees
Emigration and immigration

Or visit the following Web site:
Population Reference Bureau
www.prb.org

woman in 1960 to only 1.3 in the wake of its dramatic economic development (Population Reference Bureau 2003).

3. *Improving the status of women.* In countries where women have low status, having many children—especially sons—is the only way women can increase their social value and guarantee support in their old age. When women have greater education and can earn even a small income on their own, they gain greater power within the family. As a result, they typically marry later and have fewer children. In addition,

they are better able to protect their daughters from being married off while still children. Consequently, the countries that have proven most successful in family planning and in economic growth are those, such as South Korea and Singapore, that have made particular efforts to increase education, economic options, and legal rights for women (United Nations Population Fund 2000).

Population in the United States

The U.S. population picture is similar to that of Western Europe in having low mortality and fertility rates, but there are also several differences. First, although fertility is at near replacement level, it has not dropped significantly below this level as has happened in many European nations. Second, our population is younger than that of Europe. Third, immigration continues to add substantially to the size of our population. In this section, we briefly describe fertility, mortality, and migration issues in the United States.

Fertility

For nearly 20 years, the number of children per woman in the United States has remained at just about 2.1—the level necessary to replace the population. This low fertility has been accompanied by sharp reductions in social-class, racial, and religious differences in fertility. Some women will have their children as teenagers and some when they are 30, but increasingly they will stop at two children.

Mortality

Death is almost a stranger to U.S. families. The average age at death is now in the late 70s, and many people who survive to age 65 live another 20 years. Parents can feel relatively secure that their infants will survive. If they don't divorce, young newlyweds can safely plan on a golden wedding anniversary.

In the last 30 years, we have added 5 years to life expectancy. These increases are due to better diagnosis and treatment of the degenerative diseases (such as heart disease and cancer) that strike elderly people. In addition, increases in life expectancy have been made possible by reducing (although not eliminating) racial and social-class differentials in mortality. At the time of World War II, African American women lived a full 12 years less than white women; by 2000, the gap was down to 6 years.

On the other hand, the AIDS epidemic, first recognized in 1981, has given death a new face. Although death rates from AIDS have fallen in recent years, AIDS remains a leading cause of death for all persons ages 25 to 44, but especially for African Americans and Hispanics. Often spread through intravenous drug use (which has the most appeal for those who have the least to look forward to), AIDS is increasingly becoming a disease of the poor and disadvantaged.

Migration

Immigration is the permanent movement of people into another country.

Although it can safely be ignored as a factor in world population growth, migration often has dramatic effects on the growth of individual nations. The United States is one of the nations for which **immigration,** the permanent movement of people into another country, has had an important impact, particularly in areas such as California, the Southwest, and Florida.

Most U.S. citizens are descended from people who emigrated to the United States to improve their economic prospects, such as many recent migrants from Mexico. Other immigrants, such as those from Cambodia, Bosnia, and the Sudan, are primarily refugees driven from their homes by warfare or the economic destruction that often follows in its wake (see the Focus on a Global Perspective box in this chapter). Patterns of both internal migration and immigration have created a unique set of problems in the United States and have dramatically changed our political landscape.

Immigration

An estimated 1 million people enter the United States each year. Almost all recent immigrants come from Latin America or Asia. Perhaps as many as half are illegal immigrants, most of whom are from Mexico or Central America.

In 2000, immigration accounted for 25 percent of U.S. population growth (U.S. Bureau of the Census 2002b). As a consequence, the United States does not need to fear population decline. The racial and ethnic composition of the nation will change substantially, however. By 2050, it is estimated that the combination of Hispanic immigration and low fertility among whites will reduce the proportion of our population that is white non-Hispanic from 69 to 53 percent (Figure 15.1).

Most immigrants to the United States, both legal and illegal, are pushed from their native lands by poor local economies and are pulled by an unmet demand in the United States for low-skill, low-paid labor. The consequences of current immigration trends are likely to be both economic and cultural. From the standpoint of economics, three generalizations seem to be supported: (1) Immigrants are not taking jobs away from U.S. citizens; but (2) the availability of low-wage illegal immigrants does depress wages in some economic sectors; and (3) because immigrants are often funneled into ethnic conclaves where they compete against each other (Morris & Western 1999), poor Hispanics and other minorities are the ones hardest hit by this situation. From the standpoint of culture, it is likely that the United States will become a more pluralistic society, perhaps one that is multilingual and has no majority ethnic group (white or otherwise).

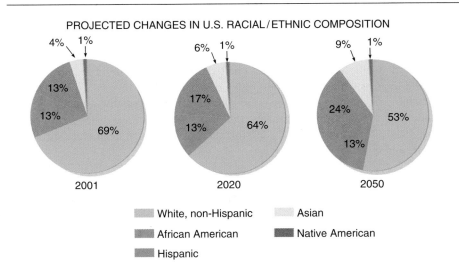

PROJECTED CHANGES IN U.S. RACIAL/ETHNIC COMPOSITION

2001

2020

2050

White, non-Hispanic   Asian
African American   Native American
Hispanic

FIGURE 15.1
Changing Composition of U.S. Population
If annual immigration remains at 1 million and if fertility remains low, the racial and ethnic composition of the U.S. population will change substantially in the decades ahead. The most noticeable effect will be a sharp rise in the proportion who are Hispanic and Asian and a corresponding decrease in the proportion who are non-Hispanic whites.
SOURCE: U.S. Bureau of the Census 2003b.

Hispanics make up a rising proportion of the U.S. population. This family is waiting to apply for legal residency in the United States.

LEGALIZATION INFORMATION & ASSISTANCE

AP/World Photos

Urbanization is the process of population concentration in metropolitan areas.

Internal Migration

Until 1970, the story of the U.S. population was one of progressive **urbanization,** the process of population concentration in cities. For most of our history, urban areas grew faster than rural areas, with the largest urban areas growing the most. During the last 30 years, however, there have been three major variations on this pattern: shrinking central cities, Sunbelt growth, and resurgence of some nonmetropolitan areas.

Because central cities have grown much less rapidly than their suburban rings, the proportion of the U.S. population that lives in central cities has decreased since 1970. Similarly, several large cities have actually decreased in size. In the last three decades, almost all metropolitan growth has occurred in the South and the West (especially in the Sunbelt states of the Southeast and Southwest).

At the same time that central cities have been shrinking, nonmetropolitan areas have experienced some modest growth (Fuguitt et al. 1998; Johnson 1998). Most of this growth, again, has occurred in Sunbelt states, especially in retirement destinations and in areas within a few hours' drive of a big city.

Patterns of internal migration have significantly changed the geographic concentration of the U.S. population. As a result of this relocation, the United States must confront three problems: the urbanization of poverty, a declining central-city tax base, and environmental hazards.

1. *Urbanization of poverty.* Urban poverty has sharply increased over the last 30 years, primarily as a result of the exodus of jobs to the suburbs, lack of public transportation from city to suburb, and lack of affordable housing and child care in cities ("Out of Sight" 2000).

2. *Declining tax base.* As retail businesses, jobs, and more affluent property owners have moved outside the city limits, the loss of tax dollars has left many cities teeter-

■ During the nineteenth and early twentieth centuries, many cities grew up around manufacturing plants. Although the plants are largely gone, lower- and working-class housing is still densely crowded in the blocks surrounding now decaying industrial facilities.

© Library of Congress/Meyers Photo-Art

ing on the edge of bankruptcy. In an attempt to remain solvent, cities have cut some services and eliminated others.

3. *Environmental hazards.* Because suburban life is so dependent on automobile transportation, migration to the suburbs has increased air pollution. In addition, much of the geographic relocation of the U.S. population over the last 30 years has been to those regions of the country that are least able to withstand the ecological impact of a large population. In many areas of Florida, California, and the Southwest, the demand for water already outstrips the supply. As states argue over water rights, political tensions are likely to increase; within states, competition for access to water may increase conflict between agricultural and urban interests.

Fertility, mortality, and migration patterns in the United States provide clear examples of the interrelationships between population and social institutions. Social class, women's roles, and race and ethnic relationships are all intimately connected to changes in population. One additional element of population that is especially important for social relationships is community size, an issue to which we now turn.

Urbanization

Most of our social institutions evolved in agrarian societies, where the vast bulk of the population lived and worked in the countryside. As late as 1850, only 2 percent of the world's population lived in cities of 100,000 or more (Davis 1973). Today, nearly a quarter of the world's population and more than two-thirds of the U.S. population live in cities larger than 100,000. How did these cities develop and what are they like?

Theories of Urban Growth and Decline

Structural-functionalist theorists and conflict theorists hold very different views of the sources, nature, and consequences of urban life. Structural-functionalists emphasize the benefits of urban growth and decline, while conflict theorists emphasize the political struggles that undergird these changes.

Structural-Functional Theory: Urban Ecology

Early structural-functional sociologists, many of whom lived in the booming Chicago of the 1920s and 1930s, assumed that cities grew in predictable ways. Some argued that (like Chicago), cities naturally grew outward in concentric circles from central business districts (Burgess 1925). Others believed that cities grew in wedge-shaped sectors, along transit routes, or in other patterns (Hoyt 1939). All structural functionalists, however, agreed that healthy and natural competition between economic rivals would lead cities to grow in whatever ways offered the most efficient means for producing and distributing goods and services. More recently, structural functionalists have assumed that urban decline and the growth of suburbs similarly reflect natural progress toward superior and more efficient ways of organizing economic and social life.

Conflict Perspectives: White Flight and Government Subsidies

In contrast, conflict theorists note that no patterns of urban growth have yet been discovered that hold across time and across different locations. Thus they conclude that there is nothing natural about urban growth or decline. Rather, they argue, each city grows or declines in its own unique way, depending on the relative power of competing economic and political forces (Feagin & Parker 1990).

These competing forces appear to have played an important role in drawing middle-class Americans from cities during the last half century. Western culture has long held an antiurban bias, assuming that rural life is "purer" than city life. This view garnered strength during the early decades of the twentieth century, as first foreign immigrants and later African Americans moved in large numbers from the South to the cities of the Northeast and Midwest. These changes contributed greatly to white Americans' sense that the city was a dangerous place, and encouraged middle- and upper-class Americans to flee the cities, a process known as "white flight." In contrast, throughout most of the world, the upper classes live in central cities and the poor are relegated to city outskirts and rural areas.

Suburbanization is the growth of suburbs.

The abandonment of American cities was greatly assisted by government subsidies for **suburbanization,** the growth of suburbs (Goddard 1994; Moe and Wilkie 1997). Since the 1930s, federal and local governments have responded to pressure from auto manufacturers and suburban developers by steadily reducing financial support for public transit while tremendously expanding subsidies for auto manufacturing, highways, road maintenance, and the like. As a result, people found it increasingly difficult to live, work, shop, or travel in dense cities with limited parking and decaying transit systems. In addition, since the 1950s the government has provided inexpensive home mortgages (along with tax breaks) to suburbanites while routinely denying mortgages to city dwellers. During the 1960s and 1970s, the government implemented a catastrophic "urban renewal" program that placed highways in the middle of stable, urban neighborhoods (most of which were minority and poor or working class) and moved dislocated residents to poorly constructed, public, high-rise housing. Finally, in the last two decades, local suburban governments have used tax subsidies to entice corporations to relocate to the suburbs.

All these changes pressured middle-class Americans to move to the suburbs, further contributing to the decay of our cities (Jackson 1985; Moe & Wilkie 1997). Of course, many people gratefully left their urban homes for suburbia and relished the freedom automobiles promised. But many others only reluctantly exchanged their close-knit urban neighborhoods, where they could read the newspaper while riding the bus to work, for sprawling suburbs where high walls separate neighbor from neighbor and long, nerve-wracking drives to work are the norm.

The Nature of Modern Cities

From the Industrial Revolution to the present, the modern city has grown in size and changed considerably in character. We look here at the development of industrial and postindustrial cities.

The Industrial City

With the advent of the Industrial Revolution, production moved from the countryside to the urban factory, and industrial cities were born. These cities were mill towns, steel towns, shipbuilding towns, and later, automobile-building towns; they were home to slaughterers, packagers, millers, processors, and fabricators. They were the product of new technologies, new forms of transportation, and vastly increased agricultural productivity that freed most workers from the land.

Fired by a tremendous growth in technology, the new industrial cities grew rapidly during the nineteenth century. In the United States, the urban population grew from 2 to 22 million in the half-century between 1840 and 1890. In 1860, New York was the first U.S. city to reach 1 million in population. The industrial base that provided the impetus for city growth also gave the industrial city its character: tremendous density and a central business district.

Density   Until the middle of the twentieth century, most Americans walked to work—and everywhere else, for that matter. The result was dense crowding of working-class housing around manufacturing plants. Even in 1910, the average New Yorker commuted only two blocks to work. Entire families shared a single room, and in major cities such as New York and London, dozens of people crowded into a single cellar or attic. The crowded conditions, accompanied by a lack of sewage treatment and clean water, fostered tuberculosis, epidemic diseases, and generally high mortality. A glimpse of these conditions is provided by a letter from a poor Londoner that appeared in the London *Times* in 1849:

> Sur,—May we beg and beseach your proteckshion and power. We are Sur, as it may be, livin in a Wilderness, so far as the rest of London knows anything of us, or as the rich and great people care about. We live in muck and filth. We aint got no priviz [i.e., privies], no dust bins [i.e., trash cans], no drains, no water-splies, and no drain or suer in the hole place. The Suer Company, in Greek St., Soho Square, all great, rich powerfool men take no notice whasomdever of our complaints. The Stenche of a Gulley-hole is disguistin. We all of us suffer, and numbers are ill, and if the Cholera comes Lord help us. (As quoted in Thomlinson 1976)

Central Business District   The lack of transportation and communication facilities also contributed to another characteristic of the industrial city, the central business district (CBD). The CBD is a dense concentration of retail trade, banking and finance, and government offices, all clustered close together so messengers could run between offices and businessmen could walk to meet one another. By 1880, most major cities had electric streetcars or railway systems to take traffic into and out of the city. Because most transit routes offered service only into and out of the CBD rather than providing cross-town routes, the earliest improvements over walking enhanced rather than decreased the importance of the CBD as the hub of the city.

The Postindustrial City

The industrial city was a product of a manufacturing economy plus a relatively immobile labor force. Beginning about 1950, these conditions changed and a new type of city began to grow. Among the factors prominent in shaping the character of the postindustrial city are the change from secondary to tertiary production and greater ease of communication and transportation. These changes have led to the rise of urban sprawl and edge cities.

Change from Secondary to Tertiary Production   As we noted in Chapter 14, the last decades have seen a tremendous expansion of jobs in tertiary production and the subsequent decline of jobs in secondary production. The manufacturing plants that shaped the industrial city are disappearing. Many of those that remain have moved to the suburbs where land is cheaper, taking working-class jobs, housing, and trade with them.

Instead of manufacturing, the contemporary central city is dominated by medical and educational complexes, information-processing industries, convention and entertainment centers, and administrative offices. These are the growth industries. They are also white-collar industries. These same industries, plus retail trade, also dominate the suburban economy.

Easier Communication and Transportation   Development of telecommunications and good highways has greatly reduced the importance of physical location. The central business district of the industrial city was held together by the need for physical proximity. Once this need was eliminated, high land values and commuting costs led more and more businesses to locate on the periphery, where land was cheaper and housing more desirable. Many corporate headquarters moved from New York or Chicago all the way to Arizona or Texas.

A key factor in increasing individual mobility was the automobile. Without the automobile, workers and businesses could not have moved to the city periphery, and space-gobbling single-family homes would not have been built. In this sense, the automobile and the automotive industry have been the chief architect of U.S. cities since 1950.

Urban Sprawl and Edge Cities   These changes have led to the collapse of many central business districts. In their stead, urban sprawl and edge cities have emerged.

Postindustrial cities are much larger in geographical area than the industrial cities were. The average city in 1940 was probably less than 15 miles across; now many metropolitan areas are 50 to 75 miles across. No longer are the majority of people bound by subway and railway lines that only go back and forth to downtown. Retail trade is dominated by huge, climate-controlled, suburban malls. A great proportion of the retail and service labor force has also moved out to these suburban centers, and many of the people who live in the suburbs also work in them. Suburban centers that now have an existence largely separate from the cities that spawned them are known as **edge cities** (Garreau 1991).

Urbanization in the United States

What is considered urban in one century or nation is often rural in another. To impose some consistency in usage, the U.S. Bureau of the Census has replaced the common words *urban* and *rural* with two technical terms: *metropolitan* and *nonmetropolitan*.

Edge cities are suburban centers that now have an existence largely separate from the cities that spawned them.

A **metropolitan area** is a county that has a city of 50,000 or more in it *plus* any neighboring counties that are significantly linked, economically or socially, with the core county. Some metropolitan areas have only one county; others, such as New York, San Francisco, or Detroit, include half a dozen neighboring counties. In each case, the metropolitan area goes beyond the city limits and includes what is frequently referred to as, for example, the Greater New York area. A **nonmetropolitan area** is a county that has neither a major city in it nor close ties to such a city.

Currently, 78 percent of the U.S. population lives in metropolitan areas. This metropolitan population is divided between those who live in the central city (within the actual city limits) and those who live in the balance of the county or counties, the suburban ring. More than half of the metropolitan population live in the suburbs rather than in the central city itself. Although these people have access to a metropolitan way of life, they may live as far as 50 miles from the city center.

The nonmetropolitan population of the United States has shrunk to 22 percent of the U.S. population. Although there are nonmetropolitan counties in every state of the Union except New Jersey, the majority of the nonmetropolitan population lives in either the Midwest or the South. Few of these people are farmers; many live in small towns and cities of 10,000 or 30,000.

> A **metropolitan area** is a county that has a city of 50,000 or more in it plus any neighboring counties that are significantly linked, economically or socially, with the core county.

> A **nonmetropolitan area** is a county that has no major city in it and is not closely tied to such a city.

Urbanization in the Less-Developed World

The growth of large cities and an urban way of life has occurred everywhere very recently; in the less- and least-developed nations, this growth is happening almost overnight (Figure 15.2). Mexico City, São Paulo, Bogotá, Seoul, Kinshasa, Karachi, Calcutta and other cities in developing nations continue to grow at a very rapid pace. Their populations are likely to double in about a decade. The roads, the schools, and the sewers that used to be sufficient no longer are; neighborhoods triple their populations and change their character from year to year. These problems are similar to the problems that plagued Western societies at the onset of the industrial revolution, but on a much larger scale.

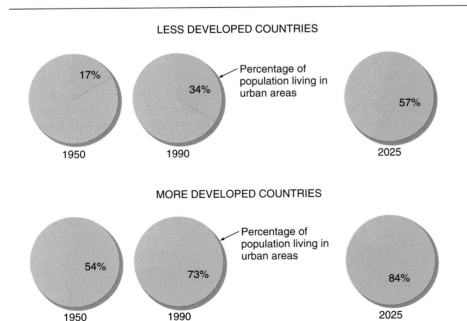

LESS DEVELOPED COUNTRIES

17% — 1950
34% — 1990 (Percentage of population living in urban areas)
57% — 2025

MORE DEVELOPED COUNTRIES

54% — 1950
73% — 1990 (Percentage of population living in urban areas)
84% — 2025

FIGURE 15.2
Urbanization Trends in the Developed and Less-Developed World, 1950–2025
Although the world is still more rural than urban, this is changing within our lifetime. Urbanization is growing particularly quickly in the less-developed world. In 1960, only 3 of the world's 10 largest cities were in developing countries: Shanghai, Buenos Aires, and Calcutta. In 1990, all except three (Tokyo, New York, and Los Angeles) were in the developing world. By the year 2025, three-quarters of the world's urban population will live in less-developed countries.
SOURCE: Haub 1993.

Urbanization in the less-developed world differs from that of the developed world, not only in pace, but also in causal factors. First, more than half of the growth in developing cities arises from a very high excess of births over deaths. Where overall population growth is 3 or even 4 percent annually, cities grow rapidly even without migration from the countryside. Second, many of the large and growing cities in the less-developed world have never been industrial cities. They are government, trade, and administrative centers. More than one-third of the regular full-time jobs in Mexico City are government jobs. These cities offer few working-class jobs, and the growing populations of unskilled men and women become part of the informal economy—artisans, peddlers, bicycle renters, laundrywomen, and beggars.

Place of Residence and Social Relationships

Every year, new films and television shows depict the evils of city life, the boredom of suburbs, and the intolerance of small towns. How realistic are such images? This section explores the pleasures and perils of modern urban, suburban, and small-town life.

Urban Living

Urbanism is a distinctly urban mode of life that is developed in cities but not confined there.

Although it is important to understand the factors that encourage urban growth, more sociologists are interested in **urbanism**—a distinctly urban mode of life that has developed in the city though it is not confined there (Wirth 1938). They are concerned with the extent to which social relationships and the norms that govern them differ between rural and urban places.

Theoretical Views

As we saw earlier, the Western world as a whole has an antiurban bias. Big cities are seen as haunts of iniquity and vice, corruptors of youth and health, and destroyers of family and community ties. City dwellers are characterized as sophisticated but artificial; rural people are characterized as possessing homegrown goodness and warmth. This general antiurban bias (which has been around at least since the time of ancient Rome), coupled with the very real problems of the industrial city, had considerable influence on early sociologists.

The classic statement of the negative consequences of urban life for the individual and for social order was made by Louis Wirth in 1938. In his influential work "Urbanism as a Way of Life," Wirth suggested that the greater size, heterogeneity, and density of urban living necessarily led to a breakdown of the normative and moral fabric of everyday life.

Gemeinschaft refers to society characterized by the personal and permanent ties associated with primary groups.

Gesellschaft refers to society characterized by the impersonal and instrumental ties associated with secondary groups.

Greater size means that many members of the community will be strangers to us. Rural society is characterized by what is called **gemeinschaft,** personal and permanent ties associated with primary groups, and urban society is characterized by **gesellschaft,** the impersonal and instrumental ties associated with secondary groups. Wirth postulated that urban-dwellers would still have primary ties, but would keep their emotional distance from, for example, store clerks or strangers in a crowded elevator by developing a cool and calculating interpersonal style.

Wirth also believed that when faced with a welter of differing norms, the city dweller was apt to conclude that anything goes. Such an attitude, coupled with the lack of informal social control brought on by size, would lead to greater crime and deviance and a greater emphasis on formal controls.

Later theorists have had a more benign view of the city. Sociologists now suggest that individuals experience the city as a mosaic of small worlds that are manageable and knowable. Thus, the person who lives in New York City does not try to cope with 9 million people and 500 square miles of city; rather the individual's private world and primary ties are made up of family, a small neighborhood, and a small work group. In addition, sociologists point out, urban life provides the "critical mass" required for the development of tightknit subcultures, from gays to symphony orchestra aficionados to rugby fans. Wirth might interpret some of these subcultures as evidence of a lack of moral integration of the community, but they can also be seen as private worlds within which individuals find cohesion and primary group support.

Empirical Consequences of Urban Living

Does urban living offer more disadvantages or advantages? This section reviews the evidence about the effects of urban living on social networks, neighborhood integration, and quality of life.

Social Networks The effects of urban living on personal integration are rather slight. Surveys asking about social networks show that urban people have as many intimate ties as rural people. There is a slight tendency for urban people to name fewer kin and more friends than rural people. The kin omitted from the urban lists are not parents, children, and siblings, however, but more distant relatives (Amato 1993; Fischer 1981). There is no evidence that urban people are disproportionately lonely, alienated, or estranged from family and friends.

Many people enjoy the vibrant nature of life in cities, as seen in this Pittsburgh neighborhood.

Neighborhood Integration Empirical research generally reveals the neighborhood to be a very weak group. Most city dwellers, whether central city or suburban, find that city living has freed them from the necessity of liking the people they live next to and has given them the opportunity to select intimates on a basis other than physical proximity. This freedom is something that people in rural areas do not have. There is growing consensus among urban researchers that physical proximity is no longer a primary basis of intimacy (Flanagan 1993). Rather, people form intimate networks on the basis of kin, friendship, and work groups; and they keep in touch by telephone or e-mail rather than relying solely on face-to-face communication. In short, urban people do have intimates, but they are unlikely to live in the same neighborhood with them. When in trouble, they call on their good friends, parents, or adult children for help. In fact, one study of neighborhood interaction in Albany-Schenectady-Troy, New York, found that a substantial share—15 to 25 percent—of all interaction with neighbors was with *family* neighbors—parents or adult children who happen to live in the same neighborhood (Logan & Spitze 1994).

Neighbors are seldom strangers, however, and there are instances in which being nearby is more important than being emotionally close. When we are locked out of the house, need a teaspoon of vanilla, or want someone to accept a United Parcel Service package, we still rely on our neighbors (Wellman & Wortley 1990). Although we generally do not ask large favors of our neighbors and don't want them to rely heavily on us, most of us expect our neighbors to be good people who are willing to help in a pinch. This has much to do with the fact that neighborhoods are often segregated by social class and stage in the family life cycle. We trust our neighbors because we expect them to be people pretty much like us.

Quality of Life Big cities are exciting places to live. People can choose from a wide variety of activities, 24 hours a day, 7 days a week. The bigger the city, the more it offers in the way of entertainment, libraries, museums, zoos, parks, concerts, and galleries. The quality of medical services and police and fire protection also increases with city size. These advantages offer important incentives for big-city living.

On the other hand, there are also disadvantages: more noise, more crowds, more expensive housing, and more crime. The latter is a particularly important problem for many people. Among central city residents, 15 percent believe that crime is a problem in their neighborhood, compared to only 5 percent of suburban residents and 2 percent of rural residents (DeFrances & Smith 1998). Crime is, in fact, more common in cities than in suburbs, and more common in suburbs than in rural areas. This is especially true of violent assaults, the kind of crime that people fear most. During 1999, 40 urban residents per 1,000 were victims of violent crimes, compared to 33 suburban residents and 25 rural residents (Rennison 1999).

Because of these disadvantages, many people would rather live close to a big city than actually in it. For most Americans, the ideal is a large house on a spacious lot in the suburbs, but close enough to a big city that they can spend an evening or afternoon there. Some groups, however, prefer big-city living, in particular, childless people who work downtown. Many of these people are decidedly pro-urban and relish the entertainment and diversity that the city offers. Because of their affluence and childlessness, they can afford to dismiss many of the disadvantages of city living.

Sociological attention has been captured by cities such as Manhattan and San Francisco with their bright lights, ethnic diversity, and crowding. Nevertheless, only one-quarter of our population actually lives in these big-city centers. The rest live in suburbs and small towns. How does their experience differ?

Suburban Living

The classic picture of a suburb is a development of very similar single-family detached homes on individual lots. This low-density housing pattern is the lifestyle to which a majority of people in the United States aspire; it provides room for dogs, children, and barbecues. This is the classic picture of suburbia. How has it changed?

The Growth of the Suburbs

The suburbs are no longer bedroom communities that daily send all their adults elsewhere to work. They are increasingly major manufacturing and retail trade centers. Most people who live in the suburbs work in the suburbs. Thus, many close-in suburban areas have become densely populated and substantially interlaced with retail trade centers, highways, and manufacturing plants.

These changes have altered the character of the suburbs. Suburban lots have become smaller, and neighborhoods of townhouses, duplexes, and apartment buildings have begun to appear. Childless couples, single people, and retired couples are seen in greater numbers. Suburbia has become more crowded and less dominated by the minivan set.

With expansion, suburbia has become more diverse. Although each suburban neighborhood tends to have its own style, stemming in large part from each development including houses of similar size and price, there are a wide variety of styles.

© Peter Menzel

■ The suburbs are the fastest growing part of America, now encompassing almost half of the entire U.S. population.

In addition to the neighborhoods of classic suburbia are spacious mini-estate suburbs where people have horses and riding lawn mowers, as well as dense suburbs of duplexes, townhouses, and apartment buildings. Some of the first suburbs are now 45 years old. Because people tend to age in place, these suburbs are more often characterized by retirees than by young families (Fitzpatrick & Logan 1985). Many of the older suburbs are becoming rundown, and renting is more common than home owning.

Suburban Problems

Many of the people who moved to suburbia did so to escape urban problems: They were looking for lower crime rates, less traffic, less crowding, and lower tax rates. The growth of the suburbs, however, has brought its own problems, including the following (Adler 1995; Langdon 1994):

- *Housing costs.* The increased demand for suburban housing has driven housing costs up to a level that cripples the finances of many families and is simply beyond the reach of people who could have afforded a home 10 years ago.
- *Weak governments.* The county and municipal governments of the small cities in the suburban ring are fragmented and relatively powerless. One result of this is the very haphazard growth associated with weak and inadequate zoning authority.
- *Higher density.* The increased density of the suburbs recreates the urban problems of crowding and traffic congestion.
- *Transportation.* Living in suburbia depends on access to automobiles. People who don't have them are basically excluded from the suburban lifestyle and suburban jobs. If you can't afford a car or can't drive one due to disability, aging, or youth, suburban jobs and opportunities may be simply out of reach.
- *Social isolation and alienation.* Long commutes leave individuals with little time to socialize with coworkers after work or with neighbors and family once they arrive home. In addition, suburban zoning laws that forbid businesses such as cafes, beauty parlors, and taverns in residential neighborhoods deprive people of the natural

Connections

Social Policy

Many urban planners and activists are working to redesign suburbs to reduce the social isolation and alienation that too often characterizes them. New suburbs are being designed, and old ones redesigned, with bike paths, neighborhood parks, and front porches. To stimulate a pedestrian atmosphere and encourage people to get out and interact, some towns have widened sidewalks and encouraged downtown restaurants to add street-side tables. Others have used financial incentives to encourage builders to place interesting stores on the first floor of new downtown buildings with apartments above. Through these actions, planners and activists hope to bring a sense of community to suburbs.

gathering places that foster social relationships. Similarly, suburban houses with high fences and no front porches make it nearly impossible for neighbor to informally meet neighbor (Oldenburg 1997). As a result of living in one community and working in another, individuals may end up feeling alienated from both.

Small-Town and Rural Living

Approximately 25 percent of the nation's population lives in small towns (less than 2,500) or rural areas. Some of these rural and small-town people are included in the metropolitan population count because they live within the orbit of a major metro area, but most live in nonmetropolitan areas—in South Dakota and Alabama, but also in Vermont and Pennsylvania.

The nonmetropolitan population of the United States continues to grow. Although young people often leave to go to school or get jobs elsewhere, enough come back to keep populations growing. In addition, small-town growth is maintained by a small but steady stream of people seeking refuge from the problems of urban and suburban living.

People find small-town living attractive for a number of reasons: It offers lots of open space, low property taxes, affordable housing, and relative freedom from worry about crime. In addition, research supports the popular perception that small towns provide greater opportunities for neighboring and community involvement (Freudenberg 1986). But although knowing your neighbors may promote social integration, it is not an unalloyed blessing. The fact that everybody knows everybody else does help keep crime down, but for some people the lack of privacy and enforced conventionality are oppressive (Johansen & Fuguitt 1984). Indeed, several studies have found small-town and rural residents to be in somewhat poorer mental health than residents of large metropolitan areas (Beeson & Johnson 1987). Another problem with rural life is the dearth of jobs: Both men and women who live outside of metropolitan areas are more likely to be underemployed (Jensen et al. 1999).

■ Although rural residents watch the same television shows and buy their clothes from the same national chains as urban residents, they continue to enjoy distinctive pastimes, such as attending county fairs and frog-jumping contests.

Rubes and Hicks?

According to stereotype, rural people, especially farmers, are hicks, rednecks, and rubes. They use bad grammar and think that if it was good enough for grandpa, it is good enough for me, that a woman's place is in the home, and that children should be seen and not heard. In contrast, the stereotype of the sophisticated city dweller suggests someone who is aware of current events, innovative, and upbeat.

These portraits are very much exaggerated. Although rural and small-town residents are somewhat less tolerant than city dwellers (Wilson 1991), on most social issues from churchgoing to support for welfare, there are almost no size-of-place differentials (Camasso & Moore 1985). Furthermore, there appear to be no size-of-place or metropolitan–nonmetropolitan differences on the importance that people attach to values, such as working hard, achievement, personal freedom, helping others, salvation, or leisure (Christenson 1984). All but the remotest cabin dwellers have access to television, radio, movies, and news magazines. The automobile and a good highway system have also increased the accessibility of urban culture to rural people.

Nevertheless, some important differences remain. Differences in political party affiliation, for instance, were so stark in the 2000 presidential election that one commentator noted, "The presidency essentially came down to a deadlock between urban Democrats and rural Republicans, each suspicious of the other, each protecting its way of life" (Bai 2001, 40). The fundamental issues that divided rural Republicans from urban Democrats appeared to be gun control—in many rural communities almost everyone owns one—and the belief that federal programs have focused on urban transportation, housing, or crime issues while ignoring both rural family values and rural economic problems.

As is true in central cities, well-paying employment opportunities have decreased in many rural communities. In their place have come jobs in low-wage, low-skilled, and often part-time service industries, such as fast-food restaurants and retail sales. In fact, about 27 percent of all rural workers held low-wage jobs in 1999. Furthermore, like many poor, minority, central city residents, rural low-wage earners also receive less training in the technologies and work organization practices that would allow them to successfully compete for better paying jobs (U.S. Department of Agriculture 2000). Despite these economic similarities, continuing differences in availability of government programs and services and in attitudes toward their use (Rank & Hirschl 1993) means that the rural/urban divide is still significant in U.S. political life.

Although they watch the same television shows and shop from the same catalogs as their urban counterparts, rural and urban lifestyles do differ. The city continues to be the major source of innovation and change. New dress styles, music, educational philosophies, and technologies originate in the city and spread to the countryside. Thus, the rural-urban difference is constantly created anew and seems unlikely ever to be totally eliminated (Fischer 1979). Because the speed of cultural diffusion is now much more rapid than before, however, rural-urban differences are far less profound than they were in the past (Flanagan 1993).

Where This Leaves Us

There's no question about it: Numbers matter. As the world's population grows—and, in places, shrinks—there are consequences for all of us. Population growth in the United States has enormous consequences for the environment because of the huge amounts of natural resources Americans use. Population growth in the less-developed nations is especially important because it not only stems from poverty but also produces even more poverty. Meanwhile, population loss in Europe leaves those nations grappling with problems brought on by having too few young people compared to the number of old people.

The problems of population growth are intimately connected to the problems of urbanization—and suburbanization. Cities emerged with the rise in industrialization, a process that is still continuing in the developing nations. In turn, problems with urban life, accentuated by various social policies, have stimulated the growth of suburbs and "edge cities." Each of these environments offers its own dangers and its own rewards.

Summary

1. For most of human history, fertility was about equal to mortality and the population grew slowly or not at all. Childbearing was a lifelong task for most women, and death was a frequent visitor to most households, claiming one-quarter to one-third of all infants in the first year of life.

2. The demographic transition in the West took centuries. Mortality declined because of better nutrition, an improved standard of living, improved public sanitation, and to a much more limited extent, modern medicine. Somewhat later, changes in social structure associated with industrialization caused fertility to decline.

3. Social structure, fertility, and mortality are interdependent; changes in one affect the others. Among the most important causes and consequences of high fertility are the low status of women and a very young age structure.

4. The level of fertility in a society has much to do with the balance of costs and rewards associated with childbearing. In traditional societies, such as that of Ghana, most social structures (the economy and women's roles, for example) support high fertility. In many modern societies, such as those of Europe and the United States, social structure imposes many costs on parents.

5. When a nation's fertility declines, the nation faces several problems. Among these are labor-force shortages, difficulties in funding health and pension benefits for a burgeoning number of older people, and nationalistic fears over growing numbers of foreign workers.

6. Population growth is not the only or even the primary cause of environmental devastation or poverty in the less- and least-developed world, nor is reducing it the primary solution to these problems. Continued population growth does contribute to these problems, however, and reducing growth will make it easier to seek solutions.

7. In the United States, life expectancy is high and continues to increase. Childlessness is increasing and fertility is near replacement level. Because of high immigration rates, however, the U.S. population is unlikely to decline. Because many of the new Americans are Asian and Latino, the racial and ethnic composition of the U.S. population is likely to change substantially.

8. Three-quarters of U.S. residents live in metropolitan areas, but most of these live in the suburban ring rather than the central city. Big cities in the Northeast and Midwest have lost population, and almost all recent growth has occurred in the Sunbelt. Changing patterns of internal migration have led to three problems: the urbanization of poverty, a declining central-city tax base, and environmental hazards.

9. Structural functionalists argue that cities grow and decline in predictable and natural ways, reflecting the most efficient means for producing and distributing goods and services. Conflict theorists, on the other hand, argue that city growth and decline reflect the outcomes of economic and political struggles between competing groups. Government subsidies played a major role in the twentieth-century growth of suburbs and decline of central cities.

10. The industrial city has high density and a central-city business district; the postindustrial city is characterized by lower density and urban sprawl.

11. Urbanization is continuing rapidly in the less-developed world; many of its large cities will double in size in a decade. This urban growth is less the result of industrialization than of high urban fertility.

12. There are competing theories about the consequences of urban living. Wirth's theory suggests that urban living will lead to nonconformity and indifference to others. Other theorists suggest that the size of the city is managed through small groups and allows for the development of unconventional subcultures.

13. Urban living is associated with less reliance on neighbors and kin and more reliance on friends, with greater fear of crime.

14. Suburban living has become more diverse. Retail trade and manufacturing have moved to the suburbs, and the suburbs are now more densely populated, more congested, and less dominated by the minivan set. Subur-

ban living has its own problems, including rising costs of living, weak governments, transportation problems, and social isolation.

15. Small-town and rural living is characterized by more emphasis on family and neighborliness, less crowding,

more informal social control, and somewhat poorer mental health. There are fewer cultural and lifestyle differences between rural and urban areas than there used to be.

Thinking Critically

1. How useful do you think the concept of cultural lag is in explaining population growth in the least-developed world?

2. One alternative to the problems posed by population aging is for the current 20-something generation to have more children. Another solution is to postpone retirement until age 70 or so. Would you rather have more kids or work longer? Which problems of population aging will not be solved by longer-lasting careers?

3. Campuses often bring large numbers of strangers together in crowded circumstances. Which theoretical view best describes the outcomes of this interaction? Can you find some examples to support all three perspectives on the consequences of urban living?

4. We all hold stereotypes about what people are like if they're from New York, California, or North Dakota. Although surveys of actual populations find little difference on many social issues, does this mean that these stereotypes are without foundation? On what kinds of matters do you expect these groups to really differ?

Sociology on the Net

 The Wadsworth Sociology Resource Center: Virtual Society

http://sociology.wadsworth.com/

The companion Web site for this book includes a range of enrichment material. Further your understanding of the chapter by accessing Online Practice Quizzes, Internet Exercises, InfoTrac College Edition Exercises, and many more compelling learning tools.

Chapter-Related Suggested Web Sites

U.S. Bureau of the Census
http://www.census.gov

Population Reference Bureau
http://www.prb.org

Environmental Protection Agency
http://www.epa.gov

U.S. Committee for Refugees
www.refugees.org

 InfoTrac College Edition

http://www.infotrac-college.com/wadsworth

Access the latest news and research articles online—updated daily and spanning four years. InfoTrac College Edition is an easy-to-use online database of reliable, full-length articles from hundreds of top academic journals and popular sources. Conduct an electronic search using the following key search terms:

Demography

World population

Emigration

Urbanization/urbanism

Suburbs

 For more information related to this chapter, visit the Opposing Viewpoints Resource Center. Be sure you check all the options on the toolbar—viewpoints, references, statistics, and so on—and search under the following subjects:

Population growth

Illegal aliens

Emigration and immigration

Suggested Readings

Groth, Paul. 1994. *Living Downtown: The History of Residential Hotels in the United States.* Berkeley: University of California Press. A fascinating social history, augmented with wonderful photographs, about life in the residential hotels and boardinghouses of downtown U.S. cities.

Langdon, Philip. 1994. *A Better Place to Live: Reshaping the American Suburb.* New York: Harper. A cogent discussion of the problems with American suburbs, how these problems developed, and what we can do to solve them.

Riis, Jacob A. 1971. *How the Other Half Lives.* New York: Dover Publications (Original work published in 1901.) A liberally illustrated essay on conditions in U.S. urban slums at the turn of the century. Riis' photographs provide ample documentation of the poverty and filth of the early industrial city.

Martinez, Rubén. 2001. *Crossing Over: A Mexican Family on the Migrant Trail.* New York: Metropolitan. Martinez, a journalist, followed the Chávez family through its illegal journey across the border and into dangerous low-wage jobs in the United States. The book explores how Latino migrants are becoming part of the United States, and how both Latino and Anglo-American culture is changing in the process.

Fertility, Mortality, and Social Structure

The Effects of Social Structure on Fertility

In Ghana, the average woman has four or five children; in Italy, the average woman has only one or two. These differences are the product of values, roles, and statuses in very different societies. The average woman in Ghana *wants* four or five children, and the average woman in Italy *wants* only one or two.

The level of fertility in a society is strongly related to the roles of women. Generally, fertility is higher in places where women marry at younger ages, where their value is measured by the number of children they bear, where they have less access to education and income, where their roles outside the household are limited, and where they are relatively powerless (United Nations Population Fund 2000). Fertility also reflects the development of society's institutions. When the family is the source of security, income, social interaction, and even salvation, fertility is high.

The Effects of Social Structure on Mortality

As noted earlier, the most important factors affecting mortality are public health measures and the standard of living—access to good nutrition, safe drinking water, and protective housing. Differences in these factors almost entirely account for the expectation that the average American will be healthier and will live 33 years longer than the average person living in Sierra Leone; these differences also explain why, in Britain, the United States, and elsewhere, each social class has lower death rates and better health than the social class below it (Marmot, Kogevinas, & Elston 1987; Weitz 2004b). As a result, by focusing on public-health measures and on equalizing income and education, countries such as China, Cuba, Costa Rica, and Sri Lanka with per capita incomes that are far lower have achieved life expectancies that are approaching those of the United States (Caldwell 1993).

More subtly, social structure affects mortality through its structuring of social roles and lifestyle (Weitz 2004b). Race, socioeconomic status, and gender all affect exposure to unhealthy or dangerous lifestyles. People with less education, for example, get less physical exercise and are more likely to smoke than those with more education, and young men are more likely to die in automobile accidents than young women.

The Effects of Fertility on Social Structure

Women's roles: Fertility has powerful effects on the roles of women. The greater the number of children a woman has, the less likely she is to have any involvement in social structures outside the family. When the average woman bears only two children, fertility places much less restriction on her social involvement.

Age structure: In addition to affecting women's roles, fertility has a major impact on the age structure of the population: the higher the fertility, the younger the population. This is graphically shown in the population pyramid in Figure 15.3. When fertility is low, the number of young people is about the same as the number of adults; when fertility is high, there are many more children than adults, and the age structure takes on a pyramidal shape.

This pyramidal age structure has both short- and long-term consequences. In the short term, it means that a large proportion of the population is too young to work, and this reduces society's productivity. In the long run, it translates into potential for explosive population growth. In those parts of sub-Saharan Africa that have not been hard hit by AIDS, the number of girls aged 0 to 4 exceeds the number of women aged 20 to 24 by a ratio of two to one.

Intersections

FIGURE 15.3

A Comparison of Age Structures in Low- and High-Fertility Societies

When fertility is high, the number of children tends to be much larger than the number of parents. When this pattern is repeated for generations, the result is a pyramidal age structure. When fertility is low, however, each generation has a similar size, and a boxier age structure results.

SOURCE: U.S. Census Bureau International Data Base 2003 (http://www.census.gov/ipc/www/idbpyr.html)

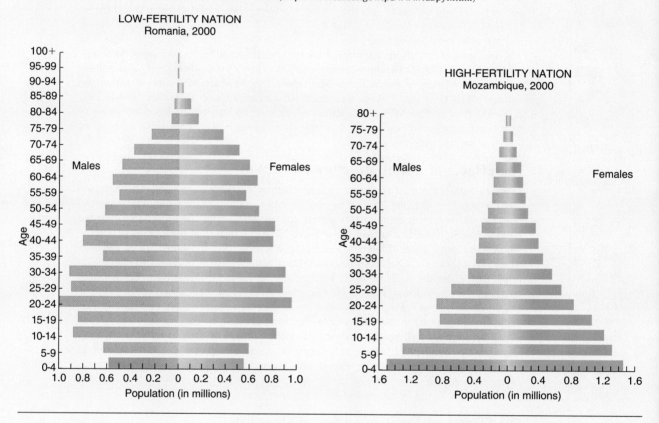

LOW-FERTILITY NATION
Romania, 2000

HIGH-FERTILITY NATION
Mozambique, 2000

This means that in the next generation there will be twice as many mothers as there are today. In those parts of Africa that have been ravaged by AIDS, however, over the next decades the age structure will more closely resemble that of a low- than a high-fertility nation.

The Effects of Mortality on Social Structure

Like fertility, mortality has particularly strong effects on the family. A popular myth about the family in the preindustrial economy is that it was a multiple-generation household, what we call an extended family. A little reflection will demonstrate how unlikely it is that many children lived with their grandparents when life expectancy was only 30 to 35 years and when fertility was seven to eight children per woman. Reconstructions tell us that only a small percentage of all families could have been three-generational. Quite often, if the children lived with their grandparents, it was because their parents were dead. The household of a high-mortality society was probably as fractured, as full of stepmothers, half-sisters, and stepbrothers, as is the current household of the high-divorce society.

MicroCase Online Exercise

For an interactive exercise using MicroCase data sets, go to the text companion website at http://sociology.wadsworth.com/brinkerhoff/essentials6e. Choose Chapter 15 and then select "Intersection Exercise" from the left navigation bar.

CHAPTER 16

Social Movements, Technology, and Social Change

© Owen Franken/Stock, Boston Inc./PictureQuest

How Societies Change

Social institutions do not stand still. Often, things change without our knowing how or why. Immediately after the terrorist attacks of September 11, 2001, thousands of Europeans held candle-lit vigils to express their solidarity with the United States. One and a half years later, after U.S.-led forces invaded Iraq, even more Europeans demonstrated in protest against the United States. Meanwhile, Eastern Europe, South Africa, and Mexico lurch toward new economic and political forms, while the fortunes of all nations increasingly depend upon an international political and economic system.

What is going on? Many Americans shake their heads in confusion. Social change in the last decade has brought wonderful developments such as the fall of the Taliban, new drugs to treat AIDS, and wireless computer and communication technologies. Balanced against these positive changes, however, are civil war and malnutrition in many developing nations, the destruction of the Amazon rainforest, and an epidemic of repetitive stress disorders linked to computer use. The rapid pace of social change and the complexity of late twentieth-century problems lead many individuals to feel a sense of both urgency and helplessness. In this chapter, we describe two sources of social change: social movements and technology.

Social Movements

- In May 2000, a crowd of English and Turkish soccer fans became violent in Copenhagen, Denmark. At least three people were stabbed, and five others were seriously injured. Riot police had to use tear gas to break up the crowd, and at least 26 people were arrested.
- In September 2000, the Yugoslavian government of dictator Slobodan Milosevic was overturned in a popular uprising. During the preceding months, Milosevic's opposition had gradually won the police and army to its side, and so the uprising proved nearly bloodless.

Sociology divides these kinds of activities into two related but distinct topics: collective behavior and social movements. **Collective behavior** is nonroutine action by an emotionally aroused gathering of people who face an ambiguous situation (Lofland 1985, 29). It includes situations such as the impromptu violence in Copenhagen and the candle-lit vigils after 9/11. These are unplanned, relatively spontaneous actions, where individuals and groups improvise some joint response to an unusual or problematic situation (Zygmunt 1986). Collective behavior is usually short-lived, at least in part because participants lack a clearly defined social agenda and the resources and organization needed to affect public policy.

In contrast, a **social movement** is an ongoing, goal-directed effort to fundamentally challenge social institutions, attitudes, or ways of life. Examples include the gay rights and environmentalist movements, as well as the grassroots struggle against drunk driving. A social movement is extraordinarily complex. It may include sit-ins,

Collective behavior is nonroutine action by an emotionally aroused gathering of people who face an ambiguous situation.

A **social movement** is an ongoing, goal-directed effort to fundamentally challenge social institutions, attitudes, or ways of life.

■ Collective behavior, such as mosh pits and crowd-surfing at rock concerts, differs from a social movement in being more spontaneous and relatively unplanned.

© Neal Preston/Corbis

Connections

Example

The word spread rapidly via text messaging and e-mail: Show up in New York City's Grand Central Station at 7 P.M., applaud wildly for 15 seconds, then disperse. Other "flash mobs" have gathered to duck-walk, spin in circles, or ask bookstore clerks for nonexistent titles. Although flash mobs may look like a social movement, they aren't yet. Participants are not motivated by anything other than the desire to have fun, and have no program for social change. Flash mobs could, however, become a technique used by social movements, or could grow into a movement if mobbers develop a more conscious critique of modern-day life.

demonstrations, and even riots, but it also includes meetings, fund-raisers, legislative lobbying, and letter-writing campaigns (Marwell & Oliver 1984).

Both collective behavior and social movements challenge the status quo. The primary distinction between the two is that social movements are organized, relatively broad based, and long term; collective behavior, on the other hand, is unplanned and spontaneous. Although the two are conceptually distinct, in practice they are related in at least two ways. First, social movements need and encourage some instances of collective behavior simply to keep issues in the public eye (Delgado 1986). There is nothing like a riot or police breaking up an illegal demonstration to get people's attention. Second, even though collective behavior is usually limited to a particular place and time, it can be part of a repeated mass response to problematic conditions. When this happens, collective behavior at a grassroots level may be a driving force in mobilizing social movements (Ash 1972; Rule & Tilly 1975; Zygmunt 1986).

In the following section, we focus on the processes through which organized, politicized social movements develop. We adopt the perspective that social movements are best understood as political processes (Tarrow 1988). Some social movements will be closely allied with traditional political groups, such as parties, and others will seek to overthrow or radically change the state. Although all movements have the goal of changing the conventional ways of doing things by affecting public policy, the most successful social movements are those that have been able, through one tactic or another, to mobilize the government, the courts, and the law on their behalf (Burstein 1991). Thus, social movements are a part of the political process, and social movement members tend to be the same kinds of people who vote and write letters to their congressional representatives.

As we documented in Chapter 14, most people in the United States have relatively little interest in politics. Why, then, do some people shake off this lethargy and try to change the system? Under what circumstances do social movements succeed or fail?

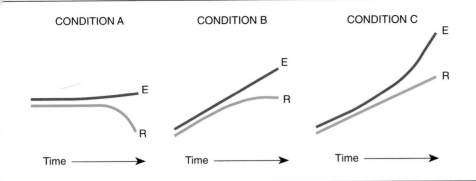

FIGURE 16.1
The Gap between Expectations and Rewards
Relative-deprivation theory suggests that whenever there is a gap between expectations (E) and rewards (R), relative deprivation is created. It may occur when conditions are stable or improving as well as when the real standard of living is declining.

Theories of Social Movements

Three major theories explain the circumstances in which social movements arise: relative-deprivation theory, resource mobilization theory, and political process theory. All three theories suggest that social movements arise out of inequalities and cleavages in society, but they offer somewhat different assessments of the meaning, sources, and tactics of social movements.

Structural-Functional Theory: Relative Deprivation

Poverty and injustice are universal phenomena. Why is it that they so seldom lead to social movements? According to **relative-deprivation theory,** social movements arise when we experience an intolerable gap between our rewards and what we believe we have a right to expect. Our expectations, in turn, are usually determined by comparing ourselves to other groups or to other times. Because the theory refers to deprivation relative to other groups or other times rather than to absolute deprivation, it is called *relative*-deprivation theory.

Figure 16.1 diagrams three conditions for which relative-deprivation theory would predict the development of a social movement. In Condition A, disaster or taxation suddenly reduces the absolute level of living. If there is no parallel drop in what people rightfully expect, they will feel that their deprivation is illegitimate. In Condition B, both expectations and the real standard of living are improving, but expectations continue to rise even after the standard of living has leveled off. Consequently, people feel deprived relative to what they had anticipated. Finally, in Condition C, expectations rise faster than the standard of living, again creating a gap between reality and expectations. Relative-deprivation theory has the merit of providing a plausible explanation for many social movements occurring in times when objective conditions are either improving (Condition C) or at least showing a major improvement over the past (Condition B). Because relative-deprivation theory relies ultimately on the disorganizing effects of social change, it is often referred to as *breakdown theory*.

Relative-deprivation theory is a structural-functional theory. Like other structural-functional theories, it assumes that in normal circumstances society functions smoothly. According to this theory, then, social movements arise only when social change occurs unevenly across different social or cultural institutions or when the pace of change is simply too rapid.

There are two major criticisms of relative-deprivation theory. First, empirical evidence does not bear out the prediction that those who are most deprived, absolutely or relatively, will be the ones most likely to participate in social movement. Often,

Relative-deprivation theory
argues that social movements arise when people experience an intolerable gap between their rewards and what they believe they have a right to expect; it is also known as breakdown theory.

social movement participants are the best off in their groups rather than the worst off. For example, almost all of the 19 terrorists who destroyed the World Trade Center and attacked the Pentagon, as well as Osama Bin-Laden, were well educated and middle class or wealthy. In many other situations, individuals participate in and lead social movements on behalf of groups to which they do not belong, such as South African whites who fought against apartheid and people who fight for animal rights. Second, the theory fails to specify the conditions under which relative deprivation will lead to social movements. Why do some relatively deprived groups form social movements and others don't? Relative deprivation can play a role, but by itself is not a good predictor of the development of social movements (Gurney & Tierney 1982).

Conflict Theory: Resource Mobilization

While structural functionalists assume that society generally works harmoniously, conflict theorists assume that conflict, competition, and, as a result, deprivation are common. From this perspective, it is not surprising that deprivation alone cannot explain why social movements emerge when they do (Oberschall 1973). Rather, conflict theorists argue that social movements emerge when individuals who experience deprivation are able to pull together the resources they need to effectively mobilize for some kind of action. This theory is known as **resource mobilization theory.** According to this theory, social movements develop when organized groups compete for scarce resources; the spark for turning deprivation into a movement is not anger and resentment but rather organization. Similarly, the building blocks of social movements are organized groups, not alienated, discontented individuals.

Resource mobilization theory suggests that social movements develop when individuals who experience deprivation pull together the resources they need to mobilize for action.

Research shows that the most effective social movements emerge from preestablished groups that share two characteristics: relative homogeneity and many overlapping ties (Tilly 1978, 63). This implies that an African American civil-rights group that admits whites will be less cohesive than one consisting only of African Americans. Groups will be stronger if, in addition to homogeneity, their members share a strong network of ties—if they belong to the same clubs, if they work together, or if they live in the same neighborhood. Research on the 1871 Paris Commune revolt shows, for instance, that neighborhood-based insurgent groups were much more cohesive, effective, and long-lived than units that drew volunteers from all parts of the city (Gould 1991).

Symbolic Interaction Theory: Political Process

Resource mobilization theory remains very important within sociology, but it has been criticized for two reasons. First, it downplays the importance of grievances and spontaneity as triggers for social movements (Klandermas 1984; Morris & Mueller 1992;

Theories of Social Movements

	Major Assumption	Causes of Social Movements
Structural-Functional Theory: Relative Deprivation	Social movements are an abnormal part of society	Social change produced disorganization and discontent
Conflict Theory: Resource Mobilization	Social movements are the normal outgrowth of competition between groups	Competition between organized groups
Symbolic Interaction Theory: Political Process	People join social movements because they have developed an "insurgent consciousness"	Political opportunities combined with an individual sense that change is needed and possible

Concept Summary

Zygmunt 1986). Second, it overlooks the importance of the process through which vague individual grievances are transformed into new collective identities and organized political agendas (Jasper & Poulsen 1995; Williams 1995). **Political process theory** has arisen to fill this gap. According to political process theory, a social movement needs two things: political opportunities and an "insurgent consciousness." **Political opportunities** include preexisting organizations that can provide the new movement with leaders, members, phone lines, copying machines, and other resources. Whether or not political opportunities will exist depends on a number of factors, including the level of industrialization in a society, whether a war is going on, and whether other cultural changes are underway.

    Insurgent consciousness is the individual sense that change is both needed and possible. In the same way that symbolic interactionism argues that individuals develop their identities and understanding of the social world through interactions with significant others, political process theory argues that individuals develop their own sense of identity and the possibility of change through interaction with others. For example, until the 1970s, newspapers regularly listed job ads in separate columns for men and for women, top universities refused to admit women as students, and some ministers told battered wives that their husbands would not have abused them if they had behaved as proper wives should. The growth of the women's movement depended upon convincing women that these were not their own private, personal problems but that these were problems they shared with other women *simply because they were women*. This point is neatly summed up in the feminist slogan "the personal is political."

How Do Social Movements Operate?

A social movement is the product of the activities of dozens and even hundreds of groups and organizations, all pursuing, in their own way, the same general goals. For example, there are probably 50 different social movement organizations (SMOs) within the environmental movement, ranging from the relatively conventional Audubon Society and Sierra Club to the radical Greenpeace organization and the

Political process theory suggests that social movements develop when political opportunities are available and when individuals have developed a sense that change is both needed and possible.

Political opportunities are resources that allow a social movement to grow; they include preexisting organizations that can provide the new movement with leaders, members, phone lines, copying machines, and other resources.

Insurgent consciousness is the individual sense that change is both needed and possible.

■ Local chapters of Greenpeace have been particularly successful at keeping the environmental movement in the public eye through direct action tactics.

© David Ulmer/Stock Boston

ecoterrorists of the Earth Liberation Front (ELF). The organizations within a movement may be highly divergent and may compete with each other for participants and supporters. Because this assortment of organizations provides avenues of participation for people with a variety of goals and styles, however, the existence of diverse SMOs is functional for the social movement.

Organization

SMOs can be organized in one of two basic ways: as professional or as volunteer organizations. On the one hand, we have organizations such as the Audubon Society or the Sierra Club that have offices in Washington, D.C., and a relatively large paid staff, many of whom are professional fund-raisers or lobbyists who develop an interest in the environment after being hired. At the other extreme is the SMO staffed on a volunteer basis by people who are personally involved—for example, neighbors who organize in the church basement in order to prevent a nuclear power plant from being built in their neighborhood. These two types of SMO are referred to, respectively, as the *professional* SMO and the *indigenous* SMO.

Evidence suggests that the existence of both types of organizations facilitates a social movement. The professional SMO is usually more effective at soliciting resources

■ Mass demonstrations such as this early women's suffrage march serve to gain publicity for social movements.

© Photoworld/FPG

from foundations, corporations, and government agencies. It appeals to what is called a *conscience constituency*—those who are ideologically or morally committed to the group's cause. On the other hand, because they themselves are not underprivileged and because they work daily with the establishment, professional SMOs sometimes lose the sense of grievance that is necessary to motivate continued, imaginative efforts for change. As a result, a social movement also requires sustained indigenous organizations (Jenkins & Eckert 1986). Indigenous organizations perform two vital functions. First, by keeping the aggrieved group actively supportive of the social movement, they help to maintain the sense of urgency necessary for sustained effort. Second, their anger and grievance propel them to more direct-action tactics (sit-ins, demonstrations, and the like) that publicize the cause and keep it on the national agenda.

The feminist movement is an excellent example of a social movement that combines both professional and indigenous SMOs. Informal networks continue to keep the discussion of equal rights and equal opportunities alive, even in periods when professional SMOs are nonexistent or marginalized. The most successful periods of feminist activism have been when professional SMOs, such as the National Organization for Women (NOW), worked in close cooperation with indigenous SMOs made up of informal networks and conscience constituencies (Buechler 1993). In the absence of direct actions—women's suffrage marches, boycotts of stores that sell violent pornography, or equal rights rallies—pressure from both professional and indigenous SMOs produces modest results, at best.

When a movement is successful in reaching some of its major goals, it often becomes indistinguishable from any other humanitarian or political organization. For example, the League of Women Voters was initially considered a radical group because it fought for women's suffrage, but it now focuses mostly on getting out the vote and is considered middle-of-the-road. As new issues arise, however, the organization is likely to rely again on protests and demonstrations to gain new supporters or to reinvigorate existing rank-and-file members.

Mobilization

Mobilization is the "process by which a unit gains significantly in the control of assets it previously did not control" (Etzioni 1968, 388). These assets may be weapons, technologies, goods, money, or members. The resources available to a social movement depend on two factors: the amount of resources controlled by group members and the proportion of their resources that the members are willing to contribute to the movement. Thus, mobilization can proceed by increasing the size of the membership, increasing the proportion of assets that members are willing to give to the group, or recruiting richer members. Mobilization can also mean getting other organizations to work with a social movement. For example, the Civil Rights movement relied on aid from African American churches, and the antipornography movement has garnered support both from fundamentalist churches and feminist organizations.

Political process theory has pointed to the importance of frame alignment for attracting and mobilizing new members. **Frame alignment** is the process of convincing individuals that their interests, values, and beliefs are complementary to those of the social movement organization (Benford & Snow 2000; Snow et al. 1986). The Sierra Club, for example, might mail pamphlets to members of the Audubon Society in hopes of convincing them to join. It also might hold public meetings in a town plagued by pollution in hopes of convincing parents that their children's illnesses are caused by pollution, not by bad luck or bad genes. Other organizations, like cults and extremist groups, try to gain new members by convincing individuals that the way they have seen things is entirely wrong.

Mobilization is the process by which a unit gains significantly in the control of assets it did not previously control.

Frame alignment is the process of convincing individuals that their interests, values, and beliefs are complementary to those of the social movement organization.

Who is most likely to be affected by these mobilization strategies? Studies of social movement activists show that, although ideology and grievances are important, the key factor is the strength of their personal ties with other movement activists and the absence of ties with significant others who oppose their activism. No matter how deeply committed to the movement they are ideologically, if they are not part of a network of others with similar convictions or if they are embedded in networks that give conflicting messages, people are unlikely to be more than token members (McAdam 1986; McAdam & Paulsen 1993).

Factors Associated with Movement Success

Based on a historical review of 53 diverse SMOs, sociologist William Gamson (1990) identified four possible outcomes of social movement activities. A fully successful SMO is one that wins both *new advantages* for those it aims to help and *acceptance* as a legitimate, reputable organization. Nelson Mandela's African National Congress, for example, now controls the government of the Republic of South Africa and has improved the situation of South Africa's black population enormously. Other SMOs, however, have not been as successful. Some SMOs are co-opted when their rhetoric and ideology gain nominal public approval, but the real social changes they had advocated have not occurred. Other SMOs are preempted when those in power adopt their goals and programs but continue to denigrate the organization and its ideology; many politicians, for example, now support the idea of equal pay for equal work but continue to belittle the feminists who brought the issue to public attention. Table 16.1 outlines the four movement outcomes discussed by Gamson.

Empirical analysis of social movements in the United States and around the world suggests that a number of factors are important for movement success. First, given the resistance they face from the larger society, movements can only succeed if they develop an ideology capable of sustaining continued participation of current members and motivating new members to join. To succeed in this task, SMOs must frame their ideologies in ways that convince potential and current participants that a particular grievance is serious, that there is an urgent need to address the problem, that taking action is proper, and most importantly that such action will be effective (Benford 1993; Benford & Snow 2000). SMOs are more likely to succeed in this task if they focus on single rather than on multiple issues and on reforming rather than revolutionizing the social or political system (Gamson 1990; Marx 1971).

TABLE 16.1
Social Movement Outcomes
Based on his historical review of 53 social movement organizations, Gamson (1990) identified four possible movement outcomes. These four outcomes depend on a combination of two factors: whether or not the movement is able to win advantages for those it seeks to help and whether or not it is able to gain acceptance as a legitimate organization.

	Acceptance	
New Advantages	Full	None
Many	Full response	Preemption
None	Co-optation	Collapse

Organizational factors also affect success of SMOs. SMOs are more likely to be successful when individuals must actively participate in the movement to derive any of the benefits from its victories. SMOs are also more likely to succeed if they have a centralized, bureaucratic structure, are able to avoid in-fighting, and successfully cultivate alliances with other organizations (Gamson 1990; Marx 1971).

Finally, the particular tactics an SMO chooses can affect its chances of success. Innovative tactics often can galvanize public and media attention and sympathies (Marx 1971). For example, although nonviolent sit-ins had a tremendous impact when first used during the Civil Rights movement, these days they have little impact. Police now know how to deal with the tactic, and the media no longer find it newsworthy. Researchers remain divided on whether disruptive strategies such as strikes, boycotts, and violence can actually increase a movement's chances of success (Giugni 1998).

The Role of the Mass Media

A social movement is a deliberate attempt to create change. To do so, it must reach the public and create the appearance that public opinion is on its side. The relationship between the media and social movements is one of mutual need. The movements need publicity, and the media need material. Sometimes both needs can be met satisfactorily. In this mutual exchange, however, most of the power belongs to the media. The media can affect a social movement's success by giving or withholding publicity and by slanting the story positively or negatively. What the media choose to cover and how they cover it also affect how the movement and its ideology are viewed by both participants and the public. In short, the media determine ". . . what the movement actually is" (Molotch 1979, 81; Mulcahy 1995).

A particularly interesting example of the importance of the media can be seen in China's Tiananmen Square uprising. Two major events in the spring of 1989 brought an unusually large contingent of international journalists to Beijing—the historic visit of Mikhail Gorbachev and the Asian Development Bank's first meeting in the People's Republic of China. Aware that the whole world would be watching and that state authorities would feel somewhat constrained from acting repressively, student activists used these journalists to create a global stage. Thus, foreign media played a crucial role in the movement, not only by reporting events to their audiences back home, but also by keeping Chinese citizens informed regarding movement developments (Zuo & Benford 1995).

In both the West and East, then, media coverage appears to be a vital mechanism through which resource-poor organizations can generate public debate over their grievances. What does an organization have to do to get news coverage? Empirical studies show that four factors are critical (Kielbowicz & Scherer 1986):

1. *Dramatic, visible events.* Sit-ins and demonstrations are more newsworthy than news conferences, and both are more newsworthy than a pamphlet. There is no question that part of what motivates suicide bombers is that they are guaranteed news coverage.

2. *Authoritative sources.* Journalists want to save time and gain credibility by going straight to the horse's mouth. This means they are most likely to cover a movement or movement event if it has established, publicly recognized leaders.

3. *Timing.* News is published or aired according to regular deadlines. If you want to be on the evening news, your action should be scheduled in advance (so that the news cameras can be there) to occur before 3 o'clock in the afternoon (so that it

The Environment

Lynx, Minks, and Suburbia: How the Environmental Movement Works

Being in favor of protecting the environment sounds like an innocuous position to take. After all, who is in favor of polluted air, dirty water, and disappearing species? Yet by default, nearly all of us are.

Ruining our environment is part of the status quo; it is part of our accepted way of life, of manufacturing and packaging merchandise, and of dealing with garbage. The average American produces 35 pounds of garbage each week, only a tiny fraction of which is recycled. Environmental protection will entail costs: higher-priced goods, more bother over recycling, more regulation, and fewer consumer goods. It is also likely to result in the loss of some jobs.

The Battle over Environmental Policy

The battle over environmental policy is being fought on many fronts—nuclear power, oil exploration in protected areas, hazardous wastes, forests, and suburban sprawl. Sometimes that battle takes extreme forms. "Mink liberators" in Utah have released animals being raised on fur farms, bombed the fur breeder's cooperative that provides most of the food for the state's $20-million-a-year mink industry, and even set fire to a leather store. The Earth Liberation Front (ELF) announced that it fire-bombed and destroyed a $12-million mountaintop restaurant and ski-lift facility in 1998 to protect the last, best lynx habitat in Colorado (Glick 2001). Between 2001 and 2003, in Colorado, Indiana, Arizona, New York, and California, groups protest-

Ecoterrorists who oppose suburban sprawl and the sale of gas-guzzling vehicles have taken actions such as spray-painting sport-utility vehicles and burning dealerships where SUVs are sold.

can be edited before the news hour) on a day when not much else is going on. Savvy political activists do just that.

4. *Matching news "beats."* Most reporters have a beat, a particular area of the news they cover. If your movement falls in the cracks, you are less likely to get coverage.

Because they rely on the media, social movements find themselves changing to maximize news coverage. The link between action and newsworthiness encourages direct action rather than more quiet forms of activity such as lobbying or letter writing; and because the public tires of watching demonstrators being hauled off, the degree of extremism necessary to get news coverage may escalate. This may lead to inflamed rhetoric, greater conflict within the movement, and disproportionate attention being given to publicity rather than to other movement goals. Of course, if the alternative is no publicity at all, these changes may be a price the movement has to pay. Without free publicity, the cost to a social movement of spreading its message is greatly increased.

ing suburban sprawl have set fire to sport utility vehicles and luxury home construction sites. Although many environmentalists disagree with this kind of illegal sabotage, the spokesperson for one ELF cell says, "We know that the real 'ecoterrorists' are the white male industrial and corporate elite. They must be stopped" (Murr & Morganthau 2001).

Although militants do much to publicize and galvanize the environmental movement, they cannot succeed by themselves. Arson, freeing animals, and bombing may buy time, but permanent victory in protecting forests, wildlife, and the rest of the environment involves court orders and legal battles. Thus, both professional and indigenous, conservative and radical SMOs help to push the movement forward.

The professional SMOs of the environmental movement—the Sierra Club, the National Audubon Society, and the Wilderness Society—write letters to congressional representatives to urge support for clean-air laws or to lobby against dam projects or unrestrained suburban growth; they have a battery of lawyers available to get court injunctions when needed and to push for change in government policies.

The Environmental Movement Assessed

One reason federal agencies have adopted more environmentally friendly policies is that concern for the environment has increased markedly over the last two decades; a large majority of Americans now say they are willing to pay more taxes to clean up the environment. Not all of today's activists are leftover hippies. Many of those involved in campaigns against hazardous waste, environmental contamination from huge corporate-owned hog farms, and nuclear power, for example, are just parents trying to protect their children and citizens trying to protect their communities.

The environmental movement has had some notable successes, such as the rise in recycling, the establishment of new wilderness areas, and the passage of the Endangered Species Act. Nevertheless, there are many controversial issues left to be negotiated. Moreover, in a tight economic climate, Americans may be less willing to sacrifice any economic growth for environmental benefits. If environmental protectionism starts to threaten the lifestyles and livelihoods of people other than mink farmers, resort owners, and housing contractors, we will likely see more controversy rather than less (Cable & Cable 1995).

 For more information, look up the following subjects in InfoTrac College Edition:
Earth Liberation Front
Environmental movement

 Or visit the following Web sites:
Sierra Club
www.sierraclub.org
Environmental Protection Agency
www.epa.gov

Countermovements

Countermovements are social movements that seek to reverse or resist changes advocated by an opposing movement (Lo 1982; Meyer & Staggenborg 1996). Countermovements can arise in response to any movement and can be either left-wing or right-wing.

Countermovements are most likely to develop if three conditions are met (Meyer & Staggenborg 1996). First, the original movement must have achieved moderate success. If the movement appears unsuccessful, then few will feel it worth their while to oppose it. Conversely, if the movement appears totally successful, then opposition will seem futile. Most tobacco smokers, for example, simply accepted the new restrictions on smoking in the workplace rather than trying to resist them. On the other hand, when cities have passed laws banning smoking in restaurants and bars, smokers have realized that they had new allies: bar and

A **countermovement** seeks to reverse or resist change advocated by an opposing social movement.

■ As these antiabortion and pro-choice protesters illustrate, whenever a social movement starts to get a lot of publicity and it appears that it might affect public policy, a countermovement tends to develop to support the status quo.

restaurant owners who feared loss of customers. As a result, a countermovement has appeared to fight these laws.

Second, countermovements only arise when individuals feel that their status, power, or social values are threatened. This is most likely to happen if the original movement frames its goals broadly. The nineteenth-century temperance movement, which opposed all alcohol use, generated a strong countermovement. In contrast, the current movement against drunk driving, which identifies individual drunk drivers as the problem rather than alcohol consumption per se, has met with almost no opposition.

Third, countermovements emerge when individuals who feel their position is threatened by a new movement can find powerful allies. Those allies can come from within political parties, unions, churches, or any other important social group. Again, the alliance between smokers and bar owners is an example.

The conflict over abortion provides an excellent example of the interrelationship between movements and countermovements. The abortion rights movement of the 1960s was a quiet campaign, largely run by political elites—doctors, lawyers, and women active in mainstream political groups. For this reason, perhaps, it received little media coverage (Luker 1985). Its victory in the 1973 *Roe v. Wade* Supreme Court decision caught the country by surprise and galvanized the antiabortion movement (Meyer & Staggenborg 1996). That countermovement drew its supporters from women and men who believed that the legalization of abortion threatened religion, the stability of the family, and traditional ideas regarding women's nature and role. The antiabortion movement gained further support through highly visible "newsworthy" actions that won media coverage for its views. In the years since *Roe v. Wade*, both the movement and the countermovement have sought political allies—the pro-choice movement primarily within the Democratic Party and the antiabortion movement primarily within the Republican Party. Neither group, however, has yet achieved a decisive legal victory.

A Case Study: The Gay and Lesbian Rights Movement

Throughout most of modern Western history, homosexuality was considered a sin or a sickness. As a result, it was—and to some extent still is—often furtive and concealed. An active social movement now exists to change this situation. Like many social movements, it is diverse and fragmented, replete with competing SMOs and threatened by a countermovement.

History of the Movement

As long as those who engaged in homosexual activities feared stigma, discrimination, violence, or even death, most gays and lesbians concealed their sexual preferences. And as long as they did so, there could be no gay and lesbian rights movement. For a movement to exist, there must be a group of people who acknowledge that they are members of the same group, share common goals, and are willing to speak up about it.

The beginning of the gay and lesbian rights movement came when sufficient numbers of prominent individuals were willing to step forward and define themselves as homosexuals. This development began in Germany at the end of the nineteenth century. It was abruptly halted by Hitler, who included gays and lesbians among the undesirables of the world and sent many to their deaths in concentration camps. In the United States, the gay and lesbian rights movement began after World War II with the two founding gay and lesbian SMOs—the Mattachine Society for gay men and the Daughters of Bilitis for lesbians. The movement did not really grow, however, until the late 1960s and early 1970s, when participants began adopting and adapting the ideologies and structures of the civil rights and women's rights movements (Clendinen 1999).

The Current Movement

The gay and lesbian rights movement is not unified. It is divided by sex, class, race, and political ideology (Gamson 1995). Broadly, however, the movement seeks to do five things (Altman 1983, 122):

© Joel Gordon

1. *To define a gay and lesbian community and identity.* The movement seeks to help individuals realize they are not alone. As C. Wright Mills might have put it, the movement wants gays and lesbians to recognize that their problems are not merely personal troubles but are shaped by social structure.
2. *To establish the legitimacy of a gay or lesbian identity.* The movement seeks to reduce the shame, to overcome the internalized self-hatred and doubts, of people who have been socialized to believe they were wicked or sick.
3. *To achieve civil rights for gays and lesbians.* The movement seeks to decriminalize homosexual acts and to establish antidiscrimination laws to protect gays and lesbians.
4. *To challenge the general ascription of gender roles in society.* The movement seeks to give people the right to choose roles rather than being forced to act out a role thrust on them by reason of their sex.
5. *To secure family rights.* The most recent goal of the movement is to obtain for gay and lesbian couples the same legal rights as those other couples enjoy, such as the right to marry, to adopt children, and to obtain health or life insurance for a partner.

Conflict within the Movement

The most important schism within the movement is between men and women. Gay men and lesbians have some goals in common—in particular, civil rights goals. Lesbians, however, face a situation of double jeopardy:

By openly acknowledging their sexual preferences, men and women like the ones in this photograph hope to force society to acknowledge that many normal and decent people are gay or lesbian.

They may be discriminated against on the basis of sex as well as sexual preference. Many lesbian women believe they will make more progress by working with straight women than with gay men; they believe that improvements in the status of women (especially in economic terms) will be more beneficial to them than will general improvements for homosexuals. As a result, there is relatively little cooperation between gay and lesbian groups.

A second major schism involves class and politics. On the one side are the middle-class professionals who wear gender-appropriate suits and insist that gays and lesbians are respectable, decent people—good parents, good credit risks, and good neighbors. On the other side are people who demand the freedom to wear lavender and leather and who wish to dismantle the entire system of gender roles and status politics. These are, respectively, the people who want to tinker with the system—to extend the basic rights package just a little further—and the people who think that the whole system is a sham and want to overthrow it.

Successes, Failures, and Prospects

The gay and lesbian rights movement has achieved some notable successes. The American Psychological Association no longer considers homosexuality per se a sickness. As of 2003, 14 states have passed laws outlawing discrimination on the basis of sexual orientation (National Gay and Lesbian Task Force 2003); and acknowledged homosexuals have been elected to public office, including the U.S. senate. Most importantly, in 2003 the U.S. Supreme Court declared that states could no longer criminalize private, consensual, same-sex activities. Public opinion polls also show substantial increases in support for homosexual rights. In offices and families around the country, it is becoming possible to be open about one's homosexuality without losing the respect of others. Similarly, gay characters now have central roles on television shows like *Will and Grace* and *Queer Eye for the Straight Guy*, the *New York Times* runs wedding announcements from gay and lesbian couples, and in 2003, *Bride* magazine ran its first article on lesbian couples.

In many homes, offices, and neighborhoods, however, the position of an acknowledged lesbian or gay man would still be awkward at best. The military continues to discharge individuals who are found to engage in homosexual activity. "Gay bashing"—violent attacks on gays—is a relatively frequent recreational activity of young toughs (Jenness 1995; Singer & Deschamps 1994). College campuses are not exempt: In a survey conducted in 2003, 51 percent of gay, lesbian, and bisexual students noted that they concealed their sexual identity to avoid harassment, and more than one-third had in fact experienced verbal or physical harassment (Rankin 2003).

The AIDS epidemic has had a complex impact on the gay and lesbian rights movement. On the one hand, the epidemic has become another reason for some to stigmatize gays. On the other hand, the epidemic has brought discussion of homosexuality "out of the closet," and has provided an impetus to organization. AIDS hotlines and information meetings have triggered gay men's networks that cut across class, race, and political cleavages. By increasing the solidarity of the group, these new networks may increase the effectiveness of the gay rights movement.

Not surprisingly, the successes of the gay and lesbian civil rights movement have encouraged the development of a countermovement. Much of that countermovement traces its roots either to the Religious Right or to ultraconservative political groups. Nevertheless, the recent Supreme Court decision, coming from a largely conservative court, suggests the extent to which the gay and lesbian civil rights movement has succeeded.

Technology

The pervasive influence of technology in our daily lives today is obvious. Perhaps you woke up to an alarm this morning to find coffee already brewed in your preset electric coffeemaker, checked your cell-phone for messages, and listened to MP3 files on your laptop, all before you made it to your first class. Not only are our technological tools far more pervasive than they once were, but they are also potentially far more powerful and dangerous. It is vitally important, then, that we understand the relationship between technology and social change.

Technology can be defined as the human application of knowledge to the making of tools and the use of natural resources. It is important to note that the term *technology* refers not only to the tools themselves (material culture) but also to our beliefs, values, and attitudes toward them (nonmaterial culture). While we may be inclined to think of technology in terms of today's "high-tech" advances, it also includes relatively simple tools such as pottery and woven baskets. Thus, technology has been a component of culture from the beginning of human society.

Social change is any significant modification or transformation of social structures or institutions over time; technology is one major cause of social change. Chapter 4 describes how technology helped to transform hunting, fishing, and gathering societies to horticultural, then agricultural, and then industrial societies.

Because technology defines the limits of what a society can do, technological innovation is a major impetus to social change. Meanwhile, new technologies are developed to meet the new needs created by a changing culture and society. The result is a never-ending cycle in which social change both causes and results from new technology. In this section, we briefly review two theories of technologically induced social change and present a case study of how information technology may change society. We then discuss the benefits and costs of two new technologies: information technology and reproductive technology.

> **Technology** involves the human application of knowledge to the making of tools and to the use of natural resources.

> **Social change** is any significant modification or transformation of social structures and sociocultural processes over time.

Two Theories of Technologically Induced Social Change

Social change is a central topic in sociology. As discussed in Chapter 1, the early sociologists were bent on understanding the consequences of the Industrial Revolution, an event that triggered dramatic social change. Auguste Comte, the founder of sociology, argued that any understanding of society required not only an understanding of the sources of order (statics) but also of the process of change (dynamics). Throughout this text, we examine many important aspects of social change such as global inequality (Chapter 8) and population growth (Chapter 15) to name but a few. This section explores how structural functionalism and conflict theory explain social change.

Structural-Functional Theory: Social Change as Evolutionary

While structural-functional theory primarily asks how social organization is maintained in an orderly way, the theory does not ignore the fact that societies and cultures change. As pointed out in Chapter 1, according to the structural-functional perspective, change occurs through evolution: Social structures adapt to new needs and demands in an orderly way while outdated patterns, ideas, and values gradually disappear. Often, the new needs and demands that prompt this evolution are technological advances.

As noted in Chapter 2 and in the Focus on Technology Box in Chapter 11, the sociologist William T. Ogburn (1922) added the concept of cultural lag to the idea of evolutionary change. Because the components of a society are interrelated, Ogburn reasoned, changes in one aspect of the culture invariably affect other aspects. The society will adapt, but only after some time has passed. As an illustration, Ogburn noted that by 1870, large numbers of U.S. industrial workers were being injured in factory accidents, but workers' compensation laws were not passed until the 1920s— a cultural lag of about 50 years. Ogburn pointed out that a society can hardly adapt to a new technology before it is introduced. Hence, cultural lag is a temporary period of maladjustment during which the social structure adapts to new technologies.

Conflict Theory: Power and Social Change

While structural functionalism sees social change as orderly and generally consensual, conflict theorists contend that change results from conflict between competing interests. Furthermore, conflict theorists assert that those with greater power can direct social change to their own advantage. In a process characterized by conflict and disruption, social structure changes (or does not change) as powerful groups act either to alter or to maintain the status quo.

Vested interests are stakes in either maintaining or transforming the status quo.

According to Thorstein Veblen (1919), those for whom the status quo is profitable are said to have a vested interest in maintaining it. **Vested interests** represent stakes in either maintaining or transforming the status quo; people or groups who would suffer from social change have a vested interest in maintaining the status quo, while those who would profit from social change have a vested interest in transforming it. Electric companies have a vested interest in promoting electric cars, oil companies have a vested interest in impeding this. University students have a vested interest in the U.S. government's retaining federally guaranteed student loans, taxpayers have a vested interest in opposing these loans if that will lower their taxes.

Just as the benefits of a particular social innovation are unevenly distributed, so also are the costs. Conflict theorists argue that costs tend to go to the less powerful. Pollution-producing factories, which can earn great profits for corporations, are typically located in poor neighborhoods and never located in places like Beverly Hills or Scarsdale.

Like evolutionary theories, the conflict perspective on social change makes intuitive sense to many, and there is empirical evidence to support it. A general assumption of the conflict perspective is that those with a disproportionate share of society's wealth, status, and power have a vested interest in preserving the status quo. In today's rapidly changing society, this may no longer be the case, as powerful factions may be just as likely to support as to oppose technological innovations. Microsoft, for example, is fully in favor of developing new technologies that it can profit from, like Windows XP, even while it works to impede innovations that others control, like Linux and Apple software and computers. Furthermore, some scholars have argued that technology is virtually "autonomous." That is, once the necessary supporting knowledge is developed, a particular invention—like the personal computer or the atomic bomb—will be created by someone. And once created, it will be used. In other words, technological changes may be put in motion by social forces beyond our effective control.

The Costs and Benefits of New Technologies

Almost all of us are glad that personal computers now exist: their benefits are obvious, and the problems they create seem small by comparison. Far fewer of us are happy that the atomic bomb exists, although most Americans were happy that our government was able to use it during World War II.

As these examples, suggest, new technologies always offer both benefits and costs, many of which are not immediately obvious. This section explores two examples of new technologies: new reproductive technologies and information technology. It then explores two general problems inherent in the expansion of technology: the technological imperative and "normal accidents."

New Reproductive Technologies

New reproductive technologies—some simple, some complex—have substantially expanded the options of women and men who want to raise children genetically related to them. Men whose wives are infertile can have their sperm inseminated into another woman who agrees to serve as a "surrogate mother" (usually for a fee). Women whose husbands are infertile can be inseminated with another man's sperm. Women who cannot conceive can have eggs surgically removed and mixed with sperm in a test tube. If those eggs become fertilized (a process known as in vitro fertilization), doctors can surgically implant them in the woman's uterus in the hopes that at least one egg will develop into a viable pregnancy. The same technology is used to enable women who lack viable eggs, postmenopausal women included, to bear children using another woman's eggs. These technologies are available not only to childless couples but also to single men and women and to gay and lesbian couples who want children genetically related to them. In 1997, 5 percent of women between the ages of 35 and 44 took prescription hormones to stimulate ovulation, 3 percent endured surgical procedures to restore fertility, and another 2 percent used some other technological method for having a child.

Although these reproductive technologies have increased childbearing options, some sociologists have raised concerns about their health, social, and ethical implications (Rothman 2000). The potential health problems are numerous. Prescription hormones are powerful drugs, with multiple short- and long-term effects. Women who take hormones to increase their chances of conceiving may face significantly higher rates of breast cancer or ovarian cancer in the future. Other women face long, difficult, and potentially life-threatening pregnancies when these hormones leave them carrying twins, triplets . . . or even septuplets. The children they give birth to are disproportionately likely to be born prematurely, which carries with it greater risks of a wide variety of lifelong cognitive and health problems. Finally, women who undergo surgical procedures to increase their chances of having a baby face all the dangers inherent in any surgery.

The social and ethical problems implicit in new reproductive technologies are more subtle, but also problematic. Perhaps most important, the more complex of these techniques have low success rates, especially with older women. Even those who eventually give birth typically have to endure several cycles of treatment costing around $10,000, with most of those cycles resulting in no pregnancies, miscarriages, or babies who die quickly. Yet, the constant development of new techniques makes it difficult for childless individuals and couples to either decide to adopt or to come to terms with childlessness. Finally, these technologies raise the question of whether or not we are turning children into commodities that are available to the highest bidder; they also may encourage the far too narrow definition of parenthood as having genetic ties to a child rather than the broader definition that focuses on the process of loving and raising a child.

Information Technology

Consider the college student in 1974 who is assigned the task of writing a term paper on the consequences of parental divorce. She goes to the library and walks through the periodicals section until she stumbles on the *Journal of Marriage and the*

Connections
Historical Note

In the same way that new reproductive technologies are changing society and social attitudes, older reproductive technologies deeply affected social life. Consider "the pill." Prior to the 1960s, having sex meant risking having a baby. The development of birth control pills—inexpensive, reasonably safe, and very effective—meant that couples could now have sexual intercourse without fearing pregnancy. This brought both positive and negative changes: Individuals found it easier to relax and enjoy sex, and found it harder to justify saying no to sex.

■ Because today's computers are better and cheaper than those of even 10 years ago, a very large portion of all college students bring their own to campus with them. In fact, computers are now so widely used and available, that some colleges even require students to have their own notebook computer.

© Bob Daemmrich/Stock Boston

Family, in which she eventually finds five articles—the number her professor requires—on her topic. She takes notes on three-by-five-inch cards (there are no photocopying machines) and goes home to draft her paper on her new electric typewriter. She cuts and tapes her draft copy, moving sections around until it looks good, checks words of dubious spelling in her dictionary, and then retypes a final copy. She uses carbon paper to make a copy for herself. (Ask your mom or dad to explain this to you.) When she makes a mistake, she erases it carefully and tries to type the correction in the original space.

Now consider the student in 2004. This student starts her paper by logging onto Sociological Abstracts, an online bibliography of more than 100,000 sociology articles. When she enters the keywords *divorce* and *parental,* the program prints out full citations and a summary for 41 articles. After identifying, downloading, and/or photocopying the 5 articles she wants, she drafts a report on her computer, edits it to her satisfaction, runs it through her spelling checker, and adjusts the vocabulary a bit by using the built-in thesaurus. She also runs the report through her grammar checker, which will catch errors in punctuation, capitalization, and so forth. Finally, she sends the whole thing to her mother (who lives 2,000 miles away) by electronic mail and asks her to read it for logic and organization. She receives the edited version from her mother in an hour, prints two copies, and hands in the report. Or she may send the paper to her instructor via e-mail or fax.

Information technology comprises computers and telecommunication tools for storing, using, and sending information.

Information technology—computers and telecommunication tools for storing, using, and sending information—has changed many aspects of our daily lives. Over the past few decades, the United States has become an "information society." More and more workers are employed in information acquisition, processing, and communication. More important in its social implications has been the convergence of various information technologies—mass media, telecommunications, and computing—to form a unified system. Aside from enabling us to write term papers more easily, how will information technology change our lives? Will it reduce or increase social class inequality? Will it make life safer and better? Or will it make life more stressful and isolated?

Technology

focus on

Biotechnology's Power, Promise, and Peril

Plants that glow like fireflies, and goats that give milk containing human medicine: This may sound like science fiction, but it's not. These innovations and others like them exist today because of biotechnology—scientists' purposeful and direct manipulation of genetic material in animals and plants. To create the "Flavr Savr" tomato, scientists identified the tomato gene that promotes softening and changed it to allow tomatoes to stay on the vine longer, ripening and gaining flavor without getting too mushy to ship (Shapiro 1994). To invent plants that glow in the dark, scientists inserted firefly genes into the plants' genetic materials.

Biotechnology is a powerful new tool that promises dramatic improvements in human beings' lives. Through biotechnology, we can create longer-lasting foods; plants that yield more food; new plants that can be grown specifically for fuel; animals that secrete medicines; and laboratory animals that allow scientists to better study and perhaps find cures for diseases such as AIDS, cancer, and sickle-cell anemia.

These advantages should not be discounted. Nevertheless, biotechnology also brings serious risks—including many that we will not be aware of for many years. For instance, scientists can now design crops with pesticides in their genes. These plants would kill damaging insects "automatically." But eating the food from such crops would require ingesting the pesticide, a situation possibly hazardous to health. In addition, those pesticides may kill other, needed insects and birds.

Biotechnology also can create plants that are genetically resistant to weed killers. Farmers growing such plants as crops can more effectively spray herbicides over their fields without damaging the crop itself. Meanwhile, environmentalists point out the many dangers inherent in the increased use of pesticides, which can pollute land and groundwater and harm the health of farmworkers and consumers.

Perhaps even more important, what will happen if newly created crops and animals are introduced into the environment? Rabbits are not native to Australia, so there are no native species that prey on them. When rabbits were first introduced into Australia from Britain, they multiplied immediately, began eating up everything in sight, and devastated the landscape. Bioengineered plants and animals pose similar risks. As an example, scientists are now creating a carp with an "antifreeze" gene that can live in very cold water. If introduced into an ecosystem, the carp might displace or destroy some or all of the native fish species.

In addition to these issues, animal rights activists have raised concerns that biotechnology sometimes causes cruel treatment of animals (Varner 1994). About 10 years ago, for instance, scientists genetically implanted a growth hormone into pigs to increase meat production and reduce fat. The pigs experienced serious and very likely painful health complications. In a second example, laboratory mice are genetically engineered to be born with humanlike diseases such as cancer and cystic fibrosis. Producing such mice allows scientists to research new, effective ways to treat such diseases. But the mice suffer.

How society will cope with the benefits and risks of biotechnology remains to be seen. There is currently evidence of cultural lag. In the opinion of many scientists, neither federal law nor the Food and Drug Administration (the federal agency responsible for regulating engineered organisms) is yet equipped to handle all the possible problems that might emerge. Is biotechnology potentially more dangerous than helpful? No, most scientists say, but the new technology must be understood, debated, monitored, and effectively regulated (Donnelly, McCarthy, & Singleton, 1994; Mellon 1993).

For more information, look up the following subjects in InfoTrac College Edition:
Biotechnology, environmental aspects
Bioengineering

Or visit the following Web site:
Center for Genetics and Society
http//www.genetics-and-society.org

The answer is likely to be some of each. As shown in Map 16.1, access to computers is spreading around the world—if unevenly. This means we can link via computer to distant family and friends, to doctors and medical information, to libraries and databanks, and to world events. During the war in Iraq, for example, U.S. soldiers stay in touch with their families via e-mail, and American citizens discuss events

MAP 16.1
Personal Computers per 1,000 People
SOURCE: International Bank for Reconstruction and Development (2002).

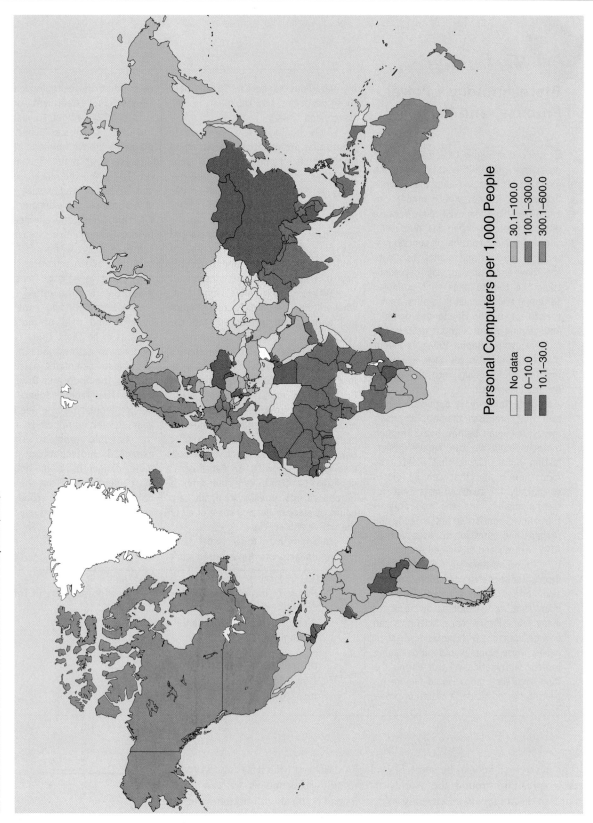

Personal Computers per 1,000 People

No data
0–10.0
10.1–30.0
30.1–100.0
100.1–300.0
300.1–600.0

in Internet chat rooms and follow events on 24-hour satellite and cable news stations (some broadcast from Europe or the Arab world). Information technology may allow us to participate more fully in the political process by making it possible to communicate more effectively and directly with our elected officials. By linking us to distant work sites, computers and electronic mail may allow more employees to work from their homes in small cities, simultaneously reducing the amount of time they need to spend commuting to a job and increasing the flexibility of their working hours so they have more time to spend with their families.

On the downside, advances in information technology have introduced new forms of crime (hacking and electronic theft), new defense worries (breaches of defense data systems and faulty software programs that may inadvertently launch World War III), new health problems (eye-strain and repetitive stress injuries), and new inefficiencies ("I'm sorry; the system is down"). They also have introduced new forms of social control. Information technology has given corporations, the police, lawyers, and government bureaucrats, among others, a greater ability to build databases about you, combining information on the cars you buy, the Web sites you visit, and the type of music you like with whether or not you have recently married, moved, had a child, or gotten a speeding ticket. Similarly, others now can obtain access to your computer files, deleted e-mail messages, and phone logs. A recent survey of 1,000 major corporations showed that almost two-thirds engaged in some form of "electronic surveillance" of their employees (Rosen 2000). Finally, new technologies have lengthened the number of working hours in a day, as notebook computers, e-mail, faxes, and cell phones increasingly invade our homes and even vacations.

The long-term effect of information technology on society will depend as much on social institutions as it does on the technological capacities of computers and telecommunications. Information technology offers us more freedom of residence and more input into local and federal legislative bodies, but we simultaneously lose some privacy and autonomy. Whether the blessings or costs will predominate will depend on how these technologies are implemented in schools, workplaces, and government bureaucracies. To the extent that they affect relationships among work, class, neighborhood, and family, the new technologies are of vital interest to those concerned with social institutions.

Making the best use of advancing technology and helping assure that advances prompt desirable social changes requires social planning—the conscious and deliberate process of investigating, discussing, and coming to some agreement about desirable action based on common values.

The Technological Imperative

As we've already noted, once the knowledge needed to devise a certain technology is available in a society, that new technology is likely to appear and to gain adherents. But we can make an even stronger statement than this: Once that technology is available, it becomes more and more difficult for anyone to choose *not* to use it.

Consider the automobile. In 1925, any city-dweller who had enough money could choose to commute to work by car. But if he chose not to do so, he could rely on a broad network of trolleys running on a frequent schedule to get him to his destination. He almost certainly lived fairly close to where he worked, and could also choose to enjoy the walk instead. These days, the automobile has become completely enmeshed in our way of life. Billions of dollars in public subsidies pay for road-building and parking lots and keep down the price of oil and gas for consumers. Meanwhile, public transportation has been cut to the bone. In many cities, walking

■ Whether traveling through Kenya or Bangladesh or simply relaxing on a park bench, technologies such as Palm Pilots and hand-held computers keep us connected to our office and the world of work.

© Joel Gordon

or bicycling is dangerous or unpleasant because of high-speed traffic or freeways that cut through neighborhoods.

Technological imperative refers to the idea that once a technology is available, it becomes difficult to avoid using it.

This situation is an example of the **technological imperative**: the idea that once a technology is available, it becomes difficult to avoid using it. Think how annoyed people sometimes feel when their friends don't use cell phones, e-mail, or voice-mail, and the pressures on holdouts to get these technologies.

Normal Accidents

As our lives come increasingly to depend on highly complex and interconnected technologies, our vulnerability to technological problems and the consequences of those problems increase exponentially. In the nineteenth century, most people got water from wells and used candles for lighting. If a well dried up or a house burned down, the disaster was limited to no more than a few households. Now we get our water from municipal water systems and our electricity from electric companies. When things go wrong, they go wrong big time.

The blackout of August 2003 provides a perfect example of this vulnerability. Electricity is provided to American households by a network of cooperating utility companies sharing a vast grid of electric cables. This grid is made possible by complex computerized technologies, designed to spread the demand over a broad region and reduce the chance of overloading the system in any one region. But because of its complexity and interconnectedness, a small problem can quickly mushroom to a huge problem and spread to a huge area. This is what happened in 2003, when overloaded circuits in the Midwest resulted in 50 million people throughout the Midwest, Northeast, and even Canada losing electric power. The blackout left people with a new awareness of how dependent we have become on technology, and how vulnerable we are when that technology fails. Because the system for distributing water to consumers runs on electricity, the blackout left thousands without water. Flashlight batteries ran down, leaving people with only candles for lighting. Many found themselves with no way of communicating with friends and relatives. Laptops and PDAs

The blackout of August 2003 illustrated how dependent we have become on technology and how vulnerable we are when technology fails. The blackout shut down most train service in the New York City region, leaving commuters like these waiting for many hours before they could board busses home.

© Reuters/CORBIS

quickly ran out of power, while cell-phone networks either lost power or became overloaded, so people could neither phone nor e-mail. Even those who had working phones could not telephone others if they kept their phone directories on computers.

This process is an example of a "normal accident." **Normal accidents** are accidents that can be expected to happen sooner or later, no matter how many safeguards are built into a system, simply because the system is so complex (Perrow 1984). Normal accidents such as space shuttle crashes, accidental releases of radiation from nuclear power plants, and electric blackouts are the price we pay for the benefits we reap from modern technology.

Normal accidents are accidents that can be expected to happen sooner or later, no matter how many safeguards are built into a system, simply because the system is so complex.

Where This Leaves Us

Whether it originates in a social movement or in a new technology, any social change will have opponents. Every winner potentially produces a loser. This means that change creates a situation of competition and conflict.

In Chapter 1, we discussed the appropriate role of sociologists in studying social issues. Should they be value-free, or should they take a stand? Issues of social change and conflict bring this question into sharp focus. Although most sociologists restrict their work to teaching and researching, a vocal minority argue that sociologists should take a more active role in monitoring and even creating social change. These people believe that sociologists should be actively involved in helping individuals understand and resolve the conflicts that arise from competition, inequality, and social change.

What can sociologists contribute to ensure that social changes enhance social justice? A few of the areas that can be pursued include the following:

1. *The study of conflict resolution.* A growing number of universities have special courses or programs on conflict resolution. These courses are concerned with the development of techniques for handling disputes and negotiating peaceful settlements that

can lead to positive social changes. Sociological research on topics such as small-group decision making and organizational culture are relevant here.

2. *Developing social justice perspectives.* At its core, sociology is concerned with the interaction of social groups, and the role power plays in those interactions. In their research and teaching, sociologists can explore how individuals, groups, and nations obtain and use power, and how that power can be distributed and used more equitably.

3. *Modeling practical social change strategies.* Sociological research may lead to the development of programs that are more effective in improving the well-being of individuals, from Head Start programs to transnational investments.

The involvement of sociologists in issues of conflict resolution, social justice, and social change is not likely to be the crucial factor that creates a better world. We can be sure, however, that scholarly neglect of these issues is both shortsighted and immoral. To the extent that knowledge of the principles of human behavior bears on issues of social conflict, we have an obligation—as scholars, students, and citizens—to seek out knowledge and to apply it. Our future depends on this.

Summary

1. Collective behavior and social movements, although related, are distinct activities. Collective behavior is spontaneous and unplanned; a social movement is organized, goal-oriented, and long term.

2. According to relative-deprivation theory, social movements arise when individuals experience an unacceptable gap between what they have and what they expect to have. Expectations are derived from comparisons to other groups and other points in time.

3. Resource mobilization theory argues that social movements emerge when individuals are able to bring the necessary resources to effectively work together toward social change.

4. Political process theory builds on resource mobilization theory by recognizing that in addition to access to political opportunities and resources, successful movements must build a sense among participants that change is both needed and possible.

5. Each social movement contains a variety of social movement organizations (SMOs). Some are professional SMOs and some are indigenous SMOs. A variety of SMOs tend to enhance a movement's ability to mobilize resources.

6. Mobilization occurs by the recruitment of individual adherents through frame alignment and the recruitment of sympathetic organizations. Mobilization is most likely to occur when people have social ties to movement members and lack ties to those who oppose the movement.

7. To be successful, a movement must convince people that a particular grievance is serious and urgent and that actions to resolve the grievance are proper and likely to succeed. Movements are also more likely to succeed if they focus on single issues, aim to reform rather than revolutionize, develop a central bureaucratic structure, avoid in-fighting, cultivate alliances with other organizations, and provide benefits only to participants.

8. The mass media play an important and growing role in collective behavior and social movements. Because resource-poor social movements depend on free publicity, they are often forced to become more extremist and less democratic to continue attracting media coverage.

9. Countermovements are social movements that seek to resist or reverse changes advocated by another social movement. Countermovements are most likely to develop if the original movement achieves modest success, if some individuals feel that their social position or values are threatened by changes achieved by the original movement, and if potential countermovement participants believe that they will have powerful allies.

10. In its effort to affect public policy, the environmental movement uses a variety of tactics, ranging from courtroom battles to sabotage. Among the reasons for the movement's growing successes are the wide variety of SMOs within the movement.

11. Technology is the human application of knowledge to the making of tools and hence to humans' use of natural resources. The term refers not only to the tools themselves (aspects of material culture) but also to people's beliefs, values, and attitudes regarding those tools (aspects of nonmaterial culture).

12. Social change is any significant modification or transformation of social structures or institutions over time. Technology is one important cause of social change.

13. Structural-functional theory primarily asks how social organization is maintained in an orderly way and sees social change as occurring through gradual evolution. Cultural lag is the time interval between the arrival of a change in society and the completion of the adaptations that this change prompts.

14. Conflict theorists contend that change results from conflict between competing interests. Furthermore, conflict theorists assert that those with greater power actually direct social change to their own advantage. People or groups who would either suffer or profit from social change have vested interests—stakes in either maintaining or transforming the status quo.

15. Information technology has changed many aspects of our daily lives. It links us to people and information but has also created new defense worries, new inefficiencies, new forms of social control, and new illnesses and injuries. Similarly, new reproductive technologies have expanded the options of those who want to raise children genetically related to them. At the same time, they have raised serious questions about their health, social, and ethical implications, such as whether we are turning children into commodities available to the highest bidder.

Thinking Critically

1. What social structural conditions in the larger society do you think helped spark the gay and lesbian rights movement? What countermovements do you know of that may impact the movement's success? Of the five goals identified for the gay and lesbian rights movement, which do you think is most likely to be achieved? Which is least likely? Why?

2. Suppose you were interested in mobilizing public opinion against the death penalty. What kind of activity or event would you try to use to get the media's attention?

3. How would you analyze the current debate over affirmative action policies and programs in terms of various groups' vested interests?

4. Europeans have been much more vocal in their opposition to genetically modified plants and food than have citizens of the United States. How would you explain this difference based on your understanding the factors that make societies more or less likely to adopt new technologies and attitudes (see Chapter 2) and on your understanding of how social movements are able to successfully mobilize?

5. If you were to run for office, how would you use e-mail in your campaign? Which groups of your constituents would you be more likely to hear from via the Internet? How would you know whether they were actually U.S. citizens with the legal right to vote—or would it matter? How might you make sure that other voices, those without high-speed data ports and modems, were heard as well?

Sociology on the Net

The Wadsworth Sociology Resource Center: Virtual Society

http://sociology.wadsworth.com/

The companion Web site for this book includes a range of enrichment material. Further your understanding of the chapter by accessing Online Practice Quizzes, Internet Exercises, InfoTrac College Edition Exercises, and many more compelling learning tools.

Chapter-Related Suggested Web Sites

The Gallup Organization
http://www.gallup.com

National Opinion Research Center
http://www.norc.uchicago.edu

CyberSociology
http://www.cybersociology.com

Social Movement Cultures
http://www.wsu.edu/~amerstu/smc/smcframe.html

 InfoTrac College Edition
http://www.infotrac-college.com/wadsworth

Access the latest news and research articles online—updated daily and spanning four years. InfoTrac College Edition is an easy-to-use online database of reliable, full-length articles from hundreds of top academic journals and popular sources. Conduct an electronic search using the following key search terms:

Social change

Social movements

Collective behavior

Student movements

Cyberspace

For more information related to this chapter, visit the Opposing Viewpoints Resource Center. Be sure you check all the options on the toolbar—viewpoints, references, statistics, and so on—and search under the following subjects:

Internet

Discrimination against gays

Gay and lesbian rights

Suggested Readings

Cable, Sherry, and Cable, Charles. 1995. *Environmental Problems, Grassroots Solutions: The Politics of Grassroots Environmental Conflict.* New York: St. Martin's Press. A description of the environmental movement with details on how various regional, grassroots organizations address environmental issues and concerns.

Marx, Gary T., and McAdam, Douglas. 1994. *Collective Behavior and Social Movements: Structure and Process.* Englewood Cliffs, N.J.: Prentice-Hall. A concise, well-written review of the major ideas and issues in the study of collective behavior and social movements. Written primarily from a resource mobilization perspective, the text pays close attention to the relationship between spontaneous and more sustained forms of collective action.

Morris, Aldon. 1984. *The Origins of the Civil Rights Movement.* New York: Free Press. A powerful account of the modern civil rights movement. Utilizing personal interviews and original documents, Morris draws particular attention to the importance of local African American community groups, in addition to the efforts of national leaders and organizations.

Rothman, Barbara Katz. 1998. *Genetic Maps and Human Imaginations: The Limits of Science in Understanding Who We Are.* New York: W. W. Norton. A thoughtful, provocative, and passionate collection of essays on the social implications of the Human Genome Project.

Stoll, Clifford. 1999. *High Tech Heretic.* New York: Doubleday. An attack by a legendary Internet pioneer and committed computer geek on the idea that computers can take the place of good teaching. Stoll also argues that the emphasis on computers has deprived teachers and students of support and funds for the things they really need.

Glossary

A

Absolute poverty is the inability to provide the minimum requirements of life.

Accommodation occurs when two groups coexist as separate cultures in the same society.

Accounts are explanations of unexpected or untoward behavior. They are of two sorts: excuses and justifications.

Acculturation is the process of acquiring the values and social practices of the dominant group and surrounding society.

An **achieved status** is optional, one that a person can obtain in a lifetime.

Achievement motivation is the continual drive to match oneself against standards of excellence.

Ageism is the belief that age determines individuals' abilities and is a legitimate basis for unequal treatment.

An **aggregate** is people who are temporarily clustered together in the same location.

Alienation occurs when workers have no control over the work process or the product of their labor.

Anglo-conformity is the process of acculturation in which new immigrant groups adopt the English language and English customs.

Anomie is a situation in which the norms of society are unclear or no longer applicable to current conditions.

Anticipatory socialization is role learning that prepares us for roles we are likely to assume in the future.

An **ascribed status** is fixed by birth and inheritance and is unalterable in a person's lifetime.

Assimilation is the full integration of the minority group into the institutions of society and the end of its identity as a subordinate group.

Authoritarian systems are political systems in which the leadership is not selected by the people and legally cannot be changed by them.

Authoritarianism is the tendency to be submissive to those in authority, coupled with an aggressive and negative attitude toward those lower in status.

Authority is power supported by norms and values that legitimate its use.

B

The **bourgeoisie** is the class that owns the tools and materials for their work—the means of production.

Bureaucracy is a special type of complex organization characterized by explicit rules and hierarchical authority structure, all designed to maximize efficiency.

C

Capitalism is the economic system in which most wealth (land, capital, and labor) is private property, to be used by its owners to maximize their own gain; this economic system is based on competition.

Caste systems rely largely on ascribed statuses as the basis for distributing scarce resources.

A **category** is a collection of people who share a common characteristic.

Charisma refers to extraordinary personal qualities that set an individual apart from ordinary mortals.

Charismatic authority is the right to make decisions based on perceived extraordinary personal characteristics.

Churches are religious organizations that have become institutionalized. They have endured for generations, are supported by and support society's norms and values, and have become an active part of society.

Civil religion is the set of institutionalized rituals, beliefs, and symbols sacred to the U.S. nation.

In Marxist theory, **class** refers to a person's relationship to the means of production.

Class consciousness occurs when people are aware of their relationship to the means of production and recognize their true class identity.

Class systems rely largely on achieved statuses as the basis for distributing scarce resources.

Coercion is the exercise of power through force or the threat of force.

Cohesion in a group is characterized by high levels of interaction and by strong feelings of attachment and dependency.

Collective behavior is nonroutine action by an emotionally aroused gathering of people who face an ambiguous situation.

Collective memories are a culture's ideas about its own past.

A **community** is characterized by dense, cross-cutting social networks.

Competition is a struggle over scarce resources that is regulated by shared rules.

Complex organizations are large formal organizations with complex status networks.

In the **concrete operational stage**, usually occurring between the ages of 7 and 12, children develop a basic understanding of cause and effect.

Conflict is a struggle over scarce resources that is not regulated by shared rules; it may include attempts to destroy, injure, or neutralize one's rivals.

Conflict theory addresses the points of stress and conflict in society and the ways in which they contribute to social change.

Consumerism is the philosophy that says "buying is good" because "we are what we buy."

A **control group** is the group in an experiment that does not receive the independent variable.

Cooperation is interaction that occurs when people work together to achieve shared goals.

Core societies are rich, powerful nations that are economically diversified and relatively free of outside control.

Correlation occurs when there is an empirical relationship between two variables.

Countercultures are groups having values, interests, beliefs, and lifestyles that are opposed to those of the larger culture.

A **countermovement** seeks to reverse or resist change advocated by an opposing social movement.

Credentialism is the use of educational credentials to measure an individual's worth.

Crime is behavior that is subject to legal or civil penalties.

A **cross-sectional design** uses a sample (or cross section) of the population at a single point in time.

The **crude birthrate** is the number of live births per 1,000 persons.

The **crude deathrate** is the number of deaths per 1,000 persons.

A **cult** is a sectlike religious organization that is independent of the religious traditions of society.

Cultural capital refers to familiarity and identification with elite culture.

Cultural diffusion is the process by which aspects of one culture or subculture enter and are incorporated into another.

Cultural lag occurs when one part of culture changes more rapidly than another.

Cultural relativity requires that each cultural trait be evaluated in the context of its own culture.

Culture is the total way of life shared by members of a community. It includes not only language, values, and symbolic meanings but also technology and material objects.

The **culture of poverty** is a set of values that emphasizes living for the moment rather than thrift, investment in the future, or hard work.

Culture shock refers to the discomfort that arises from exposure to a different culture.

D

Deduction is the process of moving from theory to data by testing hypotheses drawn from theory.

Democracies are political systems that provide regular, constitutional opportunities for a change in leadership according to the will of the majority.

The **demographic transition** is the process of moving from the traditional balance of high birth- and deathrates to a new balance of low birth- and deathrates.

Demography is the study of population—its size, growth, and composition.

The **dependent variable** is the effect in cause-and-effect relationships. It is dependent on the actions of the independent variable.

Deterrence theories suggest that deviance results when social sanctions, formal and informal, provide insufficient rewards for conformity.

Development refers to the process of increasing the productivity and standard of living of a society—longer life expectancies, more adequate diets, better education, better housing, and more consumer goods.

Deviance refers to norm violations that exceed the tolerance level of the community and result in negative sanctions.

Dialectic philosophy views change as a product of contradictions and conflict between the parts of society.

A **differential** is a difference in the incidence of a phenomenon across subcategories of the population.

Differential association theory argues that people learn to be deviant when more of their associates favor deviance than favor conformity.

Disclaimers are verbal devices employed in advance to ward off doubts and negative reactions that might result from one's conduct.

Discrimination is the unequal treatment of individuals on the basis of their membership in categories.

Disengagement theory, a functionalist theory of aging, argues that older people voluntarily disengage themselves from active social participation.

The **divorce rate** is calculated as the number of divorces each year per 1,000 married women.

Double jeopardy means having low status on two different dimensions of stratification.

Dramaturgy is a version of symbolic interaction that views social situations as scenes manipulated by the actors to convey the desired impression to the audience.

A **dual economy** consists of the complex giants of the industrial core and the small, competitive organizations that form the periphery.

Dysfunctions are consequences of social structures that have negative effects on the stability of society.

E

Economic determinism means that economic relationships provide the foundation on which all other social and political arrangements are built.

Economic institutions are social structures concerned with the production and distribution of goods and services.

Edge cities are suburban centers that now have an existence largely separate from the cities that spawned them.

The **educational institution** is the social structure concerned with the formal transmission of knowledge.

Emotional labor refers to the work of smiling, appearing happy, or in other ways suggesting that one enjoys providing a service.

Endogamy is the practice of choosing a mate from within one's own racial, ethnic, or religious group.

An **established sect** is a sect that has adapted to its institutional environment.

An **ethnic group** is a category whose members are thought to share a common origin and to share important elements of a common culture.

Ethnocentrism is the tendency to view the norms and values of our own culture as standards against which to judge the practices of other cultures.

Exchange is voluntary interaction from which all parties expect some reward.

Excuses are accounts in which one admits that the act in question is wrong or inappropriate but claims one couldn't help it.

The **experiment** is a method in which the researcher manipulates independent variables to test theories of cause and effect.

An **experimental group** is the group in an experiment that experiences the independent variable. Results for this group are compared with those for the control group.

Expressive describes activities or roles that provide integration and emotional support.

An **extended family** is a family in which a couple and their children live with other kin, such as the wife's or husband's parents or siblings.

Extrinsic rewards are tangible benefits such as income and security.

F

False consciousness is a lack of awareness of one's real position in the class structure.

The **family** is a relatively permanent group of persons linked together in social roles by ties of blood, marriage, or adoption who live together and cooperate economically and in the rearing of children.

Felon disenfranchisement is the loss of voting privileges suffered by those who have been convicted of a felony. In some states, felon disenfranchisement applies only to those in prison; in other states, it is lifelong.

Fertility is the incidence of childbearing.

Folkways are norms that are the customary, normal, habitual ways a group does things.

During the **formal operational stage**, usually occurring about age 12, some humans develop the capacity for abstract thought.

Formal social controls are administrative sanctions such as fines, expulsion, or imprisonment.

A **frame** is an answer to the question, What is going on here? It is roughly identical to a definition of the situation.

Frame alignment is the process of convincing individuals that their interests, values, and beliefs are complementary to those of the social movement organization.

Functions are consequences of social structures that have positive effects on the stability of society.

G

Gemeinschaft refers to society characterized by the personal and permanent ties associated with primary groups.

Gender refers to the expected dispositions and behaviors that cultures assign to each sex.

Gender roles refer to the rights and obligations that are normative for men and women in a particular culture.

The **generalized other** is the composite expectations of all the other role players with whom we interact; it is Mead's term for our awareness of social norms.

Gesellschaft refers to society characterized by the impersonal and instrumental ties associated with secondary groups.

Globalization refers to the process through which ideas, resources, practices, and people increasingly operate in a worldwide rather than local framework.

Globalization of culture is the process through which cultural elements (including musical styles, fashion trends, and cultural values) spread around the globe.

A **group** is two or more people who interact on the basis of shared social structure and recognize mutual dependency.

The **guinea-pig effect** occurs when subjects' knowledge that they are participating in an experiment affects their response to the independent variable.

H

The **hidden curriculum** socializes young people into obedience and conformity.

High culture refers to the cultural preferences associated with the upper class.

Homogamy is the tendency to choose a mate similar in status to oneself.

A **hypothesis** is a statement about relationships that we expect to find if our theory is correct.

I

The **I** is the spontaneous, creative part of the self.

The **id** is the natural, unsocialized, biological portion of self, including hunger and sexual urges.

An **identity salience hierarchy** is a ranking of an individual's various role identities in order of their importance to him or her.

Identity work is the process of managing identities to support and sustain our self-esteem.

An **ideology** is a set of norms and values that rationalizes the existing social structure.

Immigration is the permanent movement of people into another country.

Incidence is the frequency with which an attitude or behavior occurs.

The **independent variable** is the cause in cause-and-effect relationships.

The **indirect inheritance model** argues that children have occupations of a status similar to that of their parents because the family's status and income determine children's aspirations and opportunities.

Induction is the process of moving from data to theory by devising theories that account for empirically observed patterns.

The **informal economy** is the part of the economy that escapes the record keeping and regulation of the state; also known as the underground economy.

Informal social control is self-restraint exercised because of fear of what others will think.

Information technology comprises computers and telecommunication tools for storing, using, and sending information.

Institutionalized racism occurs when the normal operation of apparently neutral processes systematically produces unequal results for majority and minority groups.

Institutions are enduring social structures that meet basic human needs.

Instrumental describes activities or roles that are task oriented.

Insurgent consciousness is the individual sense that change is both needed and possible.

The **interaction school of symbolic interaction** focuses on the active role of the individual in creating the self and self-concept.

Intergenerational mobility is the change in social class from one generation to the next.

Internalization occurs when individuals accept the norms and values of their group and make conformity to these norms part of their self-concept.

Intragenerational mobility is the change in social class within an individual's own career.

Intrinsic rewards are rewards that arise from the process of work; they include enjoying the people you work with and pride in your creativity and accomplishments.

J

Justifications are accounts that explain the good reasons the violator had for choosing to break the rule; often they are appeals to some alternate rule.

L

Labeling theory is concerned with the processes by which labels such as *deviant* come to be attached to specific people and specific behaviors.

Latent functions or dysfunctions are consequences of social structures that are neither intended nor recognized.

Laws are rules that are enforced and sanctioned by the authority of government. They may or may not be norms.

Least-developed countries are those nations that are characterized by poverty and political weakness and that are considerably behind on every measure of development.

Less-developed countries are those nations that have lower living standards than the most-developed countries but are substantially better off than the least-developed nations.

Lifetime divorce probability is the estimated probability that a marriage will ever end in divorce.

The **linguistic relativity hypothesis** argues that the grammar, structure, and categories embodied in each language affect how its speakers see reality.

The **looking-glass self** is the process of learning to view ourselves as we think others view us.

M

Macrosociology focuses on social structures and organizations and the relationships between them.

A **majority group** is a group that is culturally, economically, and politically dominant.

Manifest functions or dysfunctions are consequences of social structures that are intended or recognized.

Marriage is an institutionalized social structure that provides an enduring framework for regulating sexual behavior and childbearing.

McDonaldization is the process by which the principles of the fast-food restaurant—efficiency, calculability, predictability, and control—are coming to dominate more sectors of American society.

The **me** represents the self as social object.

The **medical model of mental illness** holds that mental illness can be objectively defined, has identifiable causes, and if left untreated will become worse.

A **metropolitan area** is a county that has a city of 50,000 or more in it plus any neighboring counties that are significantly linked, economically or socially, with the core county.

Microsociology focuses on interactions among individuals.

A **minority group** is a group that is culturally, economically, and politically subordinate.

Mobilization is the process by which a unit gains significantly in the control of assets it did not previously control.

Modernization theory sees development as the natural unfolding of an evolutionary process in which societies go from simple to complex economies and institutional structures.

The **modernization theory of aging** argues that older people have low status in modern societies because the value of their traditional resources has eroded.

Monogamy is marriage in which there is only one wife and one husband.

Moral entrepreneurs are people who are in a position to create and enforce new definitions of morality.

Mores are norms associated with fairly strong ideas of right or wrong; they carry a moral connotation.

Mortality is the incidence of death.

Most-developed countries are those rich nations that have relatively high degrees of economic and political autonomy.

N

Net migration is the number of people who move into an area minus the number who move out.

A **nonmetropolitan area** is a county that has no major city in it and is not closely tied to such a city.

The **norm of reciprocity** is the expectation that people will return favors and strive to maintain a balance of obligation in social relationships.

Normal accidents are accidents that can be expected to happen sooner or later, no matter how many safeguards are built into a system, simply because the system is so complex.

Norms are shared rules of conduct that specify how people ought to think and act.

A **nuclear family** is a family in which the couple and their children form an independent household living apart from other kin.

O

An **operational definition** describes the exact procedure by which a variable is measured.

P

The **panel design** follows a sample over a period of time.

Participant observation includes a variety of research strategies—participating, interviewing, observing—that examine the context and meanings of human behavior.

Peripheral societies are poor and weak, with highly specialized economies over which they have relatively little control.

The **petit bourgeois** are those persons who use their own modest capital to establish small enterprises in which they and their family provide the primary labor.

Political institutions are concerned with the social structure of power; the most prominent political institution is the state.

Political opportunities are resources that allow a social movement to grow; they include preexisting organizations that can provide the new movement with leaders, members, phone lines, copying machines, and other resources.

Political process theory suggests that social movements develop when political opportunities are available and when individuals have developed a sense that change is both needed and possible.

Polyandry is a form of marriage in which one woman may have more than one husband at a time.

Polygamy is any form of marriage in which a person may have more than one spouse at a time.

Polygyny is a form of marriage in which one man may have more than one wife at a time.

Popular culture refers to aspects of culture that are widely accessible and commonly shared by most members of a society, especially those in the middle, working, and lower classes.

Power is the ability to direct others' behavior even against their wishes.

The **power elite** comprises the people who occupy the top positions in three bureaucracies—the military, industry, and the executive branch of government—and who are thought to act together to run the United States in their own interests.

Prejudice is an irrational, negative attitude toward a category of people.

The **preoperational stage** lasts from about age 2 to 7, during which children begin to use language and other symbols, such as numbers.

Primary groups are groups characterized by intimate, face-to-face interaction.

Primary production is extracting raw materials from the environment.

Primary socialization is personality development and role learning that occurs during early childhood.

Privatization is the process of taking goods and services out of governmental control and instead treating them like any other marketable commodity—something to be bought and sold in a competitive market.

The **profane** represents all that is routine and taken for granted in the everyday world, things that are known and familiar and that we can control, understand, and manipulate.

Professional socialization is role learning that provides individuals with both the knowledge and a cultural understanding of their profession.

Professions are occupations that demand specialized skills and creative freedom.

The **proletariat** is the class that does not own the means of production. They must support themselves by selling their labor to those who own the means of production.

Propinquity is spatial nearness.

R

A **race** is a category of people treated as distinct because of *physical* characteristics to which *social* importance has been assigned.

Racism is the belief that inherited physical characteristics associated with racial groups determine individuals' abilities and are a legitimate basis for unequal treatment.

The **rate of natural increase** is the crude birthrate minus the crude deathrate, calculated as a percentage.

Rational-legal authority is the right to make decisions based on rationally established rules.

Relative-deprivation theory argues that social movements arise when people experience an intolerable gap between their rewards and what they believe they have a right to expect; it is also known as breakdown theory.

Relative poverty is the inability to maintain what your society regards as a decent standard of living.

Religion is a system of beliefs and practices related to sacred things that unites believers into a moral community.

Religiosity is an individual's level of commitment to religious beliefs and to acting on those beliefs.

Religious economy refers to the competition between religious organizations to provide better "consumer products," thereby creating greater "market demand" for their own religions.

Replacement level fertility requires that women bear an average of 2.1 children—one to replace themselves, one to replace their partners, and a little extra to cover childhood mortality.

Replication is the repetition of empirical studies with another investigator or a different sample to see if the same results occur.

Reproductive labor refers to traditionally female tasks such as cooking, cleaning, and nurturing that make it possible for a society to continue and possible for others to work and play.

Resocialization occurs when we abandon our self-concept and way of life for a radically different one.

Resource mobilization theory suggests that social movements develop when individuals who experience deprivation pull together the resources they need to mobilize for action.

Rites of passage are formal rituals that mark the end of one age status and the beginning of another.

A **role** is a set of norms specifying the rights and obligations associated with a status.

Role conflict is when incompatible role demands develop because of multiple statuses.

Role exit is the process by which individuals leave important social roles.

Role identity is the image we have of ourselves in a specific social role.

Role strain is when incompatible role demands develop within a single status.

Role taking involves imagining ourselves in the role of the other in order to determine the criteria others will use to judge our behavior.

S

The **sacred** consists of events and things that we hold in awe and reverence—what we can neither understand nor control.

Sampling is the process of systematically selecting representative cases from the larger population.

Sanctions are rewards for conformity and punishments for nonconformity.

Scapegoating occurs when people or groups who are blocked in their own goal attainment blame others for their failures.

School choice refers to a range of options (vouchers, tax credits, magnet and charter schools, home schooling) that enable families to choose where their children go to school.

Science is a way of knowing based on empirical evidence.

Secondary groups are groups that are formal, large, and impersonal.

Secondary production is the processing of raw materials.

Sects are religious organizations that reject the social environment in which they exist.

Secularization is the process of transferring things, ideas, or events from the sacred realm to the profane.

The **segmented labor market** parallels the dual economy. Hiring, advancement, and benefits vary systematically between the industrial core and the periphery.

Segregation refers to the physical separation of minority- and majority-group members.

The **self** is a complex whole that includes unique attributes and normative responses. In sociology, these two parts are called the I and the me.

The **self-concept** is the self we are aware of. It is our thoughts about our personality and social roles.

Self-esteem is the evaluative component of the self-concept; it is our judgment about our worth compared with others' worth.

Self-fulfilling prophecies occur when something is *defined* as real and therefore *becomes* real in its consequences.

A **semi-caste system** is a hierarchical odering of social classes within racial categories that are also hierarchically ordered.

In the **sensorimotor stage**, understanding about the world is limited to what a child can actually see, touch, feel, smell, or hear. It lasts from birth to about age 2.

Sex is a biological characteristic, male or female.

Sexism is a belief that men and women have biologically different capacities and that these form a legitimate basis for unequal treatment.

Sexual harassment consists of unwelcome sexual advances, requests for sexual favors, or other verbal or physical conduct of a sexual nature.

Significant others are the role players with whom we have close personal relationships.

Social change is any significant modification or transformation of social structures and sociocultural processes over time.

Social class is a category of people who share roughly the same class, status, and power and who have a sense of identification with each other.

The **social construction of race and ethnicity** is the process through which a culture (based more on social ideas than on biological facts) defines what constitutes a race or an ethnic group.

Social control consists of the forces and processes that encourage conformity, including self-control, informal control, and formal control.

Social-desirability bias is the tendency of people to color the truth so that they sound nicer, richer, and more desirable than they really are.

Social distance is the degree of intimacy and equality in relationships between two groups.

Social epidemiology is the study of how social statuses relate to the distribution of illness and mortality.

Social mobility is the process of changing one's social class.

A **social movement** is an ongoing, goal-directed effort to fundamentally challenge social institutions, attitudes, or ways of life.

A **social network** is an individual's total set of relationships.

Social processes are the forms of interaction through which people relate to one another; they are the dynamic aspects of society.

A **social structure** is a recurrent pattern of relationships.

Socialism is an economic structure in which productive tools (land, labor, and capital) are owned and managed by the workers and used for the collective good.

Socialization is the process of learning the roles, statuses, and values necessary for participation in social institutions.

A **society** is the population that shares the same territory and is bound together by economic and political ties.

Sociobiology is the study of the biological basis of all forms of human (and non-human) behavior.

Socioeconomic status (SES) is a measure of social class that ranks individuals on income, education, occupation, or some combination of these.

The **sociological imagination** is the ability to see the intimate realities of our own lives in the context of common social structures; it is the ability to see personal troubles as public issues.

Sociology is the systematic study of human social interaction.

The **sociology of everyday life** focuses on the social processes that structure our experience in ordinary face-to-face situations.

The **state** is the social structure that successfully claims a monopoly on the legitimate use of coercion and physical force within a territory.

A **status** is a specialized position within a group.

Status is social honor, expressed in lifestyle.

Strain theory suggests that deviance occurs when culturally approved goals cannot be reached by culturally approved means.

Stratification is an institutionalized pattern of inequality in which social statuses are ranked on the basis of their access to scarce resources.

Strong ties are relationships characterized by intimacy, emotional intensity, and sharing.

Structural-functional theory addresses the question of social organization (structure) and how it is maintained (function).

The **structural school of symbolic interaction** focuses on the self as a product of social roles.

Subcultures are groups that share in the overall culture of society but also maintain a distinctive set of values, norms, lifestyles, and even language.

Suburbanization is the growth of suburbs.

The **superego** is composed of internalized social ideas about right and wrong.

Survey research is a method that involves asking a relatively large number of people the same set of standardized questions.

Sustainable development is development that meets the needs of the present without compromising the ability of future generations to meet their own needs.

Symbolic interaction theory addresses the subjective meanings of human acts and the processes through which people come to develop and communicate shared meanings.

T

Technological imperative refers to the idea that once a technology is available, it becomes difficult to avoid using it.

Technology involves the human application of knowledge to the making of tools and to the use of natural resources.

Tertiary production is the production of services.

A **theory** is an interrelated set of assumptions that explains observed patterns.

Total institutions are facilities in which all aspects of life are strictly controlled for the purpose of radical resocialization.

Tracking occurs when evaluations made relatively early in a child's career determine the educational programs the child will be encouraged to follow.

Traditional authority is the right to make decisions for others that is based on the sanctity of time-honored routines.

Transnational corporations are large corporations that produce and distribute goods internationally.

A **trend** is a change in a variable over time.

U

Urbanism is a distinctly urban mode of life that is developed in cities but not confined there.

Urbanization is the process of population concentration in metropolitan areas.

V

Value-free sociology concerns itself with establishing what is, not what ought to be.

Values are shared ideas about desirable goals.

Variables are measured characteristics that vary from one individual or group to the next.

Vested interests are stakes in either maintaining or transforming the status quo.

Victimless crimes such as drug use, prostitution, gambling, and pornography are voluntary exchanges between persons who desire goods or services from each other.

Voluntary associations are nonprofit organizations designed to allow individuals an opportunity to pursue their shared interests collectively.

W

Weak ties are relationships with coworkers, neighbors, distant relatives, and others that are characterized by low intensity and intimacy.

White-collar crime is crime committed by respectable people of high status in the course of their occupation.

World-systems theory is a conflict perspective of the economic relationships between developed and developing countries, the core and peripheral societies.

References

Abelman, Robert. 1990. "The Selling of Salvation in the Electronic Church." Pp. 173–183 in Robert Abelman and Stewart M. Hoover (eds.), Religious Television: Controversies and Conclusions. Norwood, NJ: Ablex Publishing Company.

Abma, Joyce, Chandra, Anjani, Mosher, William, Peterson, Linda, and Piccinino, Linda. 1997. Fertility, Family Planning, and Women's Health: New Data from the 1995 National Survey of Family Growth. Washington, D.C.: National Center for Health Statistics, Vital and Health Statistics 23 (19).

Achenbaum, W. Andrew. 1985. "Societal Perceptions of Aging and the Aged." In Robert Binstock and Ethel Shanas (eds.), Handbook of Aging and the Social Sciences (2nd ed.). New York: Van Nostrand Reinhold.

Adams, Bert N. 1971. The American Family: A Sociological Interpretation. Chicago: Markham.

———. 1985. "The Family: Problems and Solutions." Journal of Marriage and the Family 47 (August):525–529.

Adams, Carol J. 1994. "Down to Earth: Finding Spirituality in Everyday Acts." Ms. Magazine, May–June:20–21.

Adams, Rebecca G. and Sardiello, Robert. 2000. Deadhead Social Science. Walnut Creek, CA: Alta Mira Press.

Adler, Jerry B. 1995. "Bye-Bye, Suburban Dream." Newsweek, May 15:40–53.

Affleck, Marilyn, Morgan, Carolyn, and Hayes, Maggie. 1989. "The Influence of Gender-Role Attitudes on Life Expectations of College Students." Youth and Society 20:307–319.

Agnew, Robert, and Petersen, David. 1989. "Leisure and Delinquency." Social Problems 36:322–350.

AIDS Horizons. 1983. "The AIDS Front: Good News and Grim News." 1995. U.S. News and World Report, February 13:7.

Alba, Richard D., Logan, John R., and Stults, Brian J. 2000. "How Segregated Are Middle-Class African Americans?" Social Problems 47:543–558.

Albas, Daniel, and Albas, Cheryl. 1988. "Aces and Bombers: The Postexam Impression Management Strategies of Students." Symbolic Interaction 11:289–302.

Albonetti, Celesta A. 1998. "The Avoidance of Punishment: A Legal-Bureaucratic Model of Suspended Sentences in Federal White-Collar Cases Prior to the Federal Sentencing." Social Forces 78:303–330.

Alcock, John. 2001. The Triumph of Sociobiology. New York: Oxford University Press.

Alesci, Nina. 1994. "Can't We Talk? Bertrice Berry Begins National TV Talk Show." Footnotes 21 (December):5.

Alexander, Bobby C. 1994. Televangelism Reconsidered: Ritual in Search of Human Community. Atlanta, GA: Scholars Press.

Alexander, Karl, Entwisle, Doris, Cadigan, Doris, and Pallas, Aaron. 1987. "Getting Ready for First Grade: Standards of Deportment in Home and School." Social Forces 66:57–84.

Alexander, Karl L., Entwisle, Doris, and Thompson, Maxine. 1987. "School Performance, Status Relations, and the Structure of Sentiment: Bringing the Teacher Back In." American Sociological Review 52:665–682.

Ali, Jennifer, and Avison, William R. 1997. "Employment Transitions and Psychological Distress: The Contrasting Experiences of Single and Married Mothers." Journal of Health and Social Behavior 38: 345–362.

Allan, Emilie, and Steffensmeier, Darrell. 1989. "Youth Unemployment and Property Crime." American Sociological Review 54:107–123.

Allen, Katherine R., Blieszner, Rosemary, and Roberto, Karen A. 2000. "Families in the Middle and Later Years: A Review and Critique of Research in the 1990s." Journal of Marriage and the Family 62:911–926.

Altman, Dennis. 1983. The Homosexualization of America. Boston: Beacon Press. (Originally published 1982.)

Amanat, Abbas. 2001. "Empowered Through Violence: The Reinventing of Islamic Extremism." In Strobe Talbott and Nayan Chanda (eds.), The Age of Terror. New York: Basic Books.

Amato, Paul R. 1993. "Urban-Rural Differences in Helping Friends and Family Members." Social Psychology Quarterly 56:249–262.

American Association of Retired Persons. 1999. The AARP Grandparenting Survey: Sharing and Caring between Mature Grandparents and Their Grandchildren. Washington, D.C.: American Association of Retired Persons.

Anderson, David C. 1998. Sensible Justice: Alternatives to Prison. New York: Norton.

Aquilino, William S., and Supple, Khalil R. 1991. "Parent-Child Relations and Parents' Satisfaction with Living Arrangements

When Adult Children Live at Home." Journal of Marriage and the Family 53:13–28.

"Arab American Demographics." 2003. Arab American Institute. http://www.aaiusa.org/demographics.htm

Arana-Ward, Marie. 1997. "As Technology Advances, a Bitter Debate Divides the Deaf." The Washington Post, May 11, p. A1+.

Arrigo, Bruce A. (ed.). 1998. Social Justice/Criminal Justice. Belmont, CA: Wadsworth.

Arrington, Leonard J., and Bitton, Davis. 1992. The Mormon Experience: A History of the Latter-Day Saints. 2nd ed. Urbana, IL: University of Illinois Press.

Asch, Solomon. 1955. "Opinions and Social Pressure." Scientific American 193:31–35.

Ash, Roberta. 1972. Social Movements in America. Chicago: Markham.

Ashford, Lori S. 1995. "New Perspectives on Population: Lessons from Cairo." Population Bulletin 50 (1, March). Washington D.C.: Population Reference Bureau.

Atchley, Robert C. 1982. "Retirement as a Social Institution." American Review of Sociology 8:263–287.

Austin, Roy, and Allen, Mark. 2000. "Racial Disparity in Arrest Rates as an Explanation of Racial Disparity in Commitment to Pennsylvania's Prisons." Journal of Research in Crime and Delinquency 37:200–220.

"Awarded." 1994. Newsweek, August 22:45.

Babbie, Earl. 2004. The Practice of Social Research (10th ed.). Belmont, CA: Wadsworth.

Bai, Matt. 2001. "Red Zone vs. Blue Zone." Newsweek, January 22:38–41.

Barber, Benjamin R. 2001. Jihad vs. McWorld: How Globalism and Tribalism are Reshaping the World (rev. ed.). New York: Ballantine.

Barker, Eileen. 1986. "Religious Movements: Cult and Anticult Since Jonestown." Annual Review of Sociology 12:329–346.

Baron, James N., and Bielby, William T. 1984. "The Organization of a Segmented Economy." American Sociological Review 49:454–473.

Barzansky, Barbara, and Etzel, Sylvia I. 2002. "Educational Programs in U.S. Medical Schools, 2001–2002." Journal of the American Medical Association 288:1067–1072.

Bayer, Ada-Helen, and Harper, Leon. 2000. Fixing to Stay: A National Study on Housing and Home Modification Issues. Washington, D.C.: American Association of Retired Persons.

Beauvais, Fred. 1998. "American Indians and Alcohol." Alcohol Health and Research World 22:253–259.

Becker, Howard S. 1963. Outsiders: Studies in the Sociology of Deviance. New York: Free Press.

Beckwith, Carol. 1983. "Niger's Wodaabe: 'People of the Taboo.'" National Geographic 164 (October):482–509.

Beeson, Peter G., and Johnson, David R. 1987. "A Panel Study of Change (1981–1986) in Rural Mental Health Status: Effects of the Rural Crisis." Paper presented at the National Conference on Mental Health Statistics, Denver, Colorado, May 19.

Bell, Daniel. 1973. The Coming of Post-Industrial Society. New York: Basic Books.

Bell, Derrick. 1992. Race, Racism, and American Law. Boston: Little, Brown.

Bellah, Robert N. 1974. "Civil Religion in America." In Russell B. Richey and Donald G. Jones (eds.), American Civil Religion. New York: Harper & Row.

Bellah, Robert N., Madsen, Richard, Sullivan, William M., Swidler, Ann, and Tipton, Steven M. 1985. Habits of the Heart: Individualism and Commitment in American Life. Berkeley: University of California Press.

_____. 1991. The Good Society. New York: Knopf.

Bell-Fialkoff, Andrew. 1993. "A Brief History of Ethnic Cleansing." Foreign Affairs 72:110–121.

Benford, Robert D. 1993. "'You Could Be the Hundredth Monkey': Collective Action Frames and Vocabularies of Motive within the Nuclear Disarmament Movement." Sociological Quarterly 34:195–216.

Benford, Robert D., and Snow, David A. 2000. "Framing Processes and Social Movements: An Overview and Assessment." Annual Review of Sociology 26:611–639.

Bennett, Neil G., and Lu, Hsien-hen. 2000. "Child Poverty in the States: Levels and Trends from 1979 to 1998." Child Poverty Research Brief 2. New York: National Center for Children in Poverty.

Bensman, Joseph, and Lilienfeld, Robert. 1979. Between Public and Private: The Lost Boundaries of Self. New York: Free Press.

Berezin, Mabel. 1997. "Politics and Culture: A Less Fissured Terrain." Annual Review of Sociology 23:361–383.

Bergen, Raquel Kennedy. 1998. Issues in Intimate Violence. Thousand Oaks, CA: Sage.

Berger, Peter L. 1963. Invitation to Sociology: A Humanistic Perspective. New York: Doubleday.

Berger, Peter L., and Luckmann, Thomas. 1966. The Social Construction of Reality: A Treatise in the Sociology of Knowledge. Garden City, NY: Doubleday.

Berk, Laura. 1989. Child Development. Newton, MA: Allyn & Bacon.

Berliner, David C., and Biddle, Bruce J. 1995. The Manufactured Crisis: Myths, Fraud, and the Attack on America's Public Schools. New York: Longman.

Bian, Yanjie. 1997. "Bringing Strong Ties Back In: Indirect Ties, Network Bridges, and Job Searches in China." American Sociological Review 62:366–385.

Bian, Yanjie, and Soon, Ang. 1997. "Guanxi Networks and Job Mobility in China and Singapore." Social Forces 75:981–1005.

Biddle, B. J. 1986. "Recent Developments in Role Theory." Annual Review of Sociology 12:67–92.

Bielby, William T., and Baron, James N. 1986. "Men and Women at Work: Sex Segregation and Statistical Discrimination." American Journal of Sociology 91:759–798.

Billings, Dwight B. 1990. "Religion as Opposition." American Journal of Sociology 96:1–31.

Binson, D., Michaels, S., Stall, R., Coates, T.J., Gagnon, J.H., and Catania, J.A. 1995. "Prevalence and Social Distribution of Men Who Have Sex with Men: United States and its Urban Centers." Journal of Sex Research 32:245-254.

Blakely, Edward, and Philip Shapiro. 1984. "Industrial Restructuring: Public Policies for Investment in Advanced Industrial Societies." Annals of the American Academy of Political and Social Science 475:96–109.

Blau, Peter M. 1987. "Contrasting Theoretical Perspectives." In J. Alexander, B. Giesen, R. Munch, and N. Smelser (eds.), The Micro-Macro Link. Berkeley: University of California Press.

Blau, Peter M., and Meyer, Marshall W. 1971. Bureaucracy in Modern Society (2nd ed.). New York: Random House.

Blau, Peter M., and Schwartz, Joseph E. 1984. Cross-Cutting Social Circles. Orlando, FL: Academic.

Blumberg, Rae Lesser. 1978. Stratification: Socioeconomic and Sexual Inequality. Dubuque, IA: Wm. C. Brown.

Blumer, H. 1969. Symbolic Interactionism: Perspective and Method. Englewood Cliffs, NJ: Prentice-Hall.

Blumstein, Phillip, and Schwartz, Pepper. 1983. American Couples. New York: William Morrow.

Bobo, Lawrence, and Hutchings, Vincent. 1996. "Perceptions of Racial Group Competition: Extending Blumer's Theory of Group Position to a Multiracial Social Context." American Sociological Review 61:951–972.

Bobo, Lawrence, and Kluegel, James R. 1993. "Opposition to Race-Targeting, Self-Interest, Stratification Ideology, or Racial Attitudes?" American Sociological Review 58:443–464.

Bonacich, Edna. 1972. "A Theory of Ethnic Antagonism: The Split Labor Market." American Sociological Review 37:547–559.

Booth, Alan, and Osgood, D. Wayne. 1993. "The Influence of Testosterone on Deviance in Adulthood: Assessing and Explaining the Relationship." Criminology 31:93–117.

Boritch, Helen, and Hagan, John. 1990. "A Century of Crime in Toronto: Gender, Class, and Patterns of Social Control, 1859 to 1955." Criminology 28:567–599.

Bose, Christine E., and Rossi, Peter H. 1983. "Gender and Jobs: Prestige Standings of Occupations as Affected by Gender." American Sociological Review 48:316–330.

Bourdieu, Pierre. 1984. Distinction: A Social Critique of the Stratification of Taste. Cambridge, MA: Harvard University Press.

Bourgois, Philippe. 1995. In Search of Respect: Selling Crack in El Barrio. New York: Cambridge University Press.

Bowles, Samuel, and Gintis, Herbert. 1976. Schooling in Capitalist America: Educational Reform and the Contradictions of Economic Life. New York: Basic Books.

Boyle, John P., and O'Connor, Liz Clapp. 1996. "Leveraging Technology and Partnerships in Criminal Investigations." Police Chief 63:19–24.

Bozett, Frederick (ed.). 1987. Gay and Lesbian Parents. New York: Praeger.

Bradbury, Thomas N., Fincham, Frank D., and Beach, Steven R. H. 2000. "Research on the Nature and Determinants of Marital Satisfaction: A Decade in Review." Journal of Marriage and the Family 62:964–980.

Bradshaw, York. 1988. "Reassessing Economic Dependency and Uneven Development: The Kenyan Experience." American Sociological Review 53:693–708.

Braithwaite, John. 1981. "The Myth of Social Class and Criminality, Reconsidered." American Sociological Review 46:36–58.

_____. 1985. "White Collar Crime." Annual Review of Sociology 11:1–25.

Bramlett, M. D., and Mosher, W. D. 2002. Cohabitation, Marriage, Divorce, and Remarriage in the United States. Vital Health Statistics 23 (22).

Brannon, Robert L. 1996. "Restructuring Hospital Nursing: Reversing the Trend Toward a Professional Work Force." International Journal of Health Services 26:643–654.

Braverman, Harry. 1974. Labor and Monopoly Capital. New York: Monthly Review Press.

Breuninger, Paul. 1995. "Crime Scene Reconstruction Using 3D Computer-Aided Drafting." Police Chief 62:61–62.

Bright, Chris. 1990. "Shipping Unto Others." E: The Environmental Magazine 1:30–35.

Brinkerhoff, Merlin B., and Kunz, Phillip R. (eds.). 1972. Complex Organizations and Their Environments. Dubuque, IA: Wm. C. Brown.

Bruce, Steve. 1990. Pray TV: Televangelism in America. New York: Routledge.

"Brundtland Report." 2000. Encyclopedia of the Atmospheric Environment. http://www.doc.mmu.ac.uk/aric/eae/Sustainability/Older/Brundtland_Report.html.

Bryson, Bethany. 1996. "Anything but Heavy Metal": Symbolic Exclusion and Musical Dislikes. American Sociological Review 61:884–899.

Buechler, Steven M. 1993. "Beyond Resource Mobilization? Emerging Trends in Social Movement Theory." Sociological Quarterly 34:217–235.

Bulan, Heather Ferguson, Erickson, Rebecca J., and Wharton, Amy S. 1997. "Doing for Others on the Job: The Affective Requirements of Service Work, Gender, and Emotional Well-Being." Social Problems 44:235–255.

Bullard, Robert D. 1993. Confronting Environmental Racism: Voices from the Grassroots. Boston: South End Press.

Bullard, Robert D., Warren, Rueben C., and Johnson, Glenn S. 2001. Pp. 471–488 in Ronald L. Braithwaite and Sandra E. Taylor (eds.). Health Issues in the Black Community (2nd ed.). San Francisco: Jossey-Bass.

Bumpass, Larry L., Raley, R. Karen, and Sweet, James. 1995. "The Changing Character of Stepfamilies: Implications of Cohabitation and Nonmarital Childbearing." Demography 32:425–436.

Bumpass, Larry L., and Sweet, James A. 1991. "The Role of Cohabitation in Declining Rates of Marriage." Journal of Marriage and the Family 53:913–927.

Bunker, John P., Frazier, Howard S., and Mosteller, Frederick. 1994. "Improving Health: Measuring Effects of Medical Care." Milbank Quarterly 72:225–258.

Burgess, Ernest W. 1925. "The Growth of the City: An Introduction to a Research Project." Pp. 47–62 in Robert E. Park, Ernest W. Burgess, and Roderick D. McKenzie (eds.), The City. Chicago: University of Chicago Press.

Burke, Peter J. 1980. "The Self: Measurement Requirements from the Interactionist Perspective." Social Psychological Quarterly 43 (1):18–29.

Burman, Patrick. 1988. Killing Time, Losing Ground: Experiences of Unemployment. Toronto: Thompson Educational Publishing.

Burris, Val, and Salt, James. 1990. "The Politics of Capitalist Class Segments: A Test of Corporate Liberalism Theory." Social Problems 37:341–359.

Burstein, Paul. 1991. "Legal Mobilization as a Social Movement Tactic: The Struggle for Equal Opportunity Employment." American Journal of Sociology 96:1201–1225.

Bush, Corlann Gee. 1993. "Women and the Assessment of Technology." Pp. 192–264 in Albert H. Teich (ed.), Technology and the Future (6th ed.). New York: St. Martin's.

Cable, Sherry, and Cable, Charles. 1995. Environmental Problems, Grassroots Solutions: The Politics of Grassroots Environmental Conflict. New York: St. Martin's.

Cahill, Spencer E. 1983. "Reexamining the Acquisition of Sex Roles: A Social Interactionist Perspective." Sex Roles 9:1–15.

Caldwell, John C. 1993. "Health Transition: The Cultural, Social, and Behavioral Determinants of Health in the Third World." Social Science and Medicine 36:125–135.

Call, Vaughn, Sprecher, Susan, and Schwartz, Pepper. 1995. "The Incidence and Frequency of Marital Sex in a National Sample." Journal of Marriage and the Family 57:639–652.

Callero, Peter L. 1985. "Role Identity Salience." Social Psychology Quarterly 48 (3):203–215.

Camacho, David E. (ed.). 1998. Environmental Injustices, Political Struggles: Race, Class, and the Environment. Durham, NC: Duke University Press.

Camasso, Michael J., and Moore, Dan E. 1985. "Rurality and the Residualist Social Welfare Response." Rural Sociology 50:397–408.

Campbell, Karen E. 1990. "Networks Past: A 1939 Bloomington Neighborhood." Social Forces 69:139–155.

Cancian, Francesca M., and Oliker, Stacey J. 2000. Caring and Gender. Thousand Oaks, CA: Pine Forge Press.

Cancio, A. Silvia, Evans, T. Davic, and Maume, David J., Jr. 1996. "Reconsidering the Declining Significance of Race: Racial Differences in Early Career Wages." American Sociological Review 61:541–556.

Capitman, John. 2002. Defining Diversity: A Primer and a Review. Generations 26 (3):8–14.

Caplow, Theodore, and Chadwick, Bruce. 1979. "Inequality and Life-Style in Middletown, 1920–1978." Social Science Quarterly 60 (December):367–386.

Carson, Rachel. 1962. Silent Spring. Boston: Houghton Mifflin.

Casper, Lynne M. 1997. "My Daddy Takes Care of Me! Fathers as Care Providers." Current Population Reports P-70/2/59. Washington, D.C.: U.S. Census Bureau.

Casper, Lynne M., and Bryson, Ken. 1998. "Household and Family Characteristics: March 1998." Current Population Reports P20–515. Washington, D.C.: U.S. Census Bureau.

Catanzarite, Lisa. 2003. "Race-Gender Composition and Occupational Pay Degradation." Social Problems 50:14-37.

Center for American Women and Politics. 2003. Women in Elective Office 2003. Fact Sheet. New Brunswick, NJ: Rutgers University.

Central Intelligence Agency. 2002. World Factbook 2002. Washington, D.C.: Central Intelligence Agency.

Chafetz, Janet S. 1984. Sex and Advantage. Totowa, NJ: Rowman and Allanheld.

Chambliss, William. 1978. "Toward a Political Economy of Crime." In Charles Reasons and Robert Rich (eds.), The Sociology of Law: A Conflict Perspective. Toronto: Butterworths.

Chapkis, Wendy. 1997. Live Sex Acts: Women Performing Erotic Labor. New York: Routledge.

Chapman, Jane R., and Gates, Margaret (eds.). 1978. The Victimization of Women. Newbury Park, CA: Sage.

Chase-Dunn, Christopher. 1989. Global Formation: Structure of the World Economy. London: Basil Blackwell.

Cherlin, Andrew. 1981. Marriage, Divorce, Remarriage. Cambridge, MA: Harvard University Press.

_____. 1992. Marriage, Divorce, and Remarriage (rev. ed.). Cambridge, MA: Harvard University Press.

Chesney-Lind, Meda, and Shelden, Randall G. 2004. Girls, Delinquency, and Juvenile Justice (3rd ed.). Belmont, CA: Wadsworth.

Chetkovich, Carol. 1998. Real Heat. New Brunswick, NJ: Rutgers University Press.

Chirot, Daniel. 1977. Social Change in the Twentieth Century. San Francisco, CA: Harcourt Brace Jovanovich.

_____. 1986. Social Change in the Modern Era. San Diego, CA: Harcourt Brace Jovanovich.

Chodak, Symon. 1973. Societal Development: Five Approaches with Conclusions from Comparative Analysis. New York: Oxford University Press.

Chodorow, Nancy. 1999. The Power of Feelings: Personal Meaning in Psychoanalysis, Gender, and Culture. New Haven, CT: Yale University Press.

Christenson, James A. 1984. "Gemeinschaft and Gesellschaft: Testing the Spatial and Communal Hypothesis." Social Forces 63:160–168.

Chubb, John E., and Moe, Terry M. 1990. Politics, Markets, and America's Schools. Washington, D.C.: Brookings Institute.

Clarke, Sally C. 1995. "Advance Report of Final Marriage Statistics, 1989 and 1990." Monthly Vital Statistics Report 43 (12).

Clawson, Dan, and Su, Tie-ting. 1990. "Was 1980 Special? A Comparison of 1980 and 1986 Corporate PAC Contributions." Sociological Quarterly 31:371–387.

Clendinen, Dudley. 1999. Out for Good: The Struggle to Build a Gay Rights Movement in America. New York: Simon and Schuster.

Clinard, Marshall B. 1990. Corporate Corruption: The Abuse of Power. New York: Praeger.

Coburn, David, and Willis, Evan. 2000. "The Medical Profession: Knowledge, Power, and Autonomy." Pp. 377–393 in Gary L. Albrecht, Ray Fitzpatrick, and Susan C. Scrimshaw (eds.), Handbook of Social Studies in Health and Medicine. Thousand Oaks, CA: Sage.

Cockerham, William C. 1997. "The Social Determinants of the Decline of Life Expectancy in Russian and Eastern Europe: A Lifestyle Explanation." Journal of Health and Social Behavior 38:117–130.

Cohen, Philip N., and Huffman, Matt L. 2003a. "Individuals, Jobs, and Labor Markets." American Sociological Review 68:443–463.

_____. 2003b. "Occupational Segregation and the Devaluation of Women's Work Across U.S. Labor Markets." Social Forces 81:881–907.

Coleman, James. 1988. "Competition and the Structure of Industrial Society: Reply to Braithwaite." American Journal of Sociology 94: 632–636.

Coleman, Marilyn, Ganong, Lawrence, and Fine, Mark. 2000. "Reinvestigating Remarriage: Another Decade of Progress." Journal of Marriage and the Family 62:1288–1307.

Collins, Patricia Hill. 1991. Black Feminist Thought: Knowledge, Consciousness, and the Politics of Empowerment. New York: Routledge.

Collins, Randall. 1979. The Credential Society. Orlando, FL: Academic Press.

Collins, Sharon M. 1993. "Blacks on the Bubble: The Vulnerability of Black Executives in White Corporations." Sociological Quarterly 34:429–447.

_____. 1997. "Black Mobility in White Corporations: Up the Corporate Ladder but Out on a Limb." Social Problems 44:55–67.

Coltrane, Scott. 2000. "Research on Household Labor: Modeling and Measuring the Social Embeddedness of Routine Family Work." Journal of Marriage and the Family 62:1208–1233.

Conger, Rand D., Lorenz, Frederick O., Elder, Glen H., Jr., Simons, Ronald L., and Ge, Xiaojia. 1993. "Husband and Wife Differences in Response to Undesirable Events." Journal of Health and Social Behavior 34:71–88.

Conrad, John P. 1983. "Deterrence, the Death Penalty, and the Data." In Ernest van den Haag and John P. Conrad (eds.), The Death Penalty: A Debate. New York: Plenum.

Conrad, Peter. 1999. "A Mirage of Genes." Sociology of Health and Illness 21:228–241.

Conrad, Peter, and Schneider, Joseph W. 1992. Deviance and Medicalization: From Badness to Sickness. Philadelphia: Temple University Press

Cook, Philip J., and Laub, John H. 1998. "The Unprecedented Epidemic in Youth Violence." In Michael Tonry and Mark H. Moore (eds.), Youth Violence. Chicago: University of Chicago Press.

Cool, Linda, and McCabe, Justine. 1983. "The 'Scheming Hag' and the 'Dear Old Thing': The Anthropology of Aging Women." In Jay Sokolovsky (ed.), Growing Old in Different Cultures. Belmont, CA: Wadsworth.

Cooley, Charles Horton. 1902. Human Nature and the Social Order. New York: Scribner's.

_____. 1967. "Primary Groups." In A. Paul Hare, Edgar F. Borgotta, and Robert F. Bales (eds.), Small Groups: Studies in Social Interaction (rev. ed.). New York: Knopf. (Originally published 1909.)

Cooney, Mark. 1997. "The Decline of Elite Homicide." Criminology 35:381–407.

Coontz, Stephanie. 1997. The Way We Really Are: Coming to Terms with America's Changing Families. New York: Basic Books.

Corcoran, M. 1995. "Rags to Rags: Poverty and Mobility in the United. States." Annual Review of Sociology 21:237–267.

Coser, Lewis A. 1956. The Functions of Conflict. New York: Free Press.

Council of Europe. 2000. Recent Demographic Developments in Europe, 2000. Strasbourg, France: Council of Europe.

Cousineau, Michael R. 1997. "Health Status of and Access to Health Services by Residents of Urban Encampments in Los Angeles." Journal of Health Care for the Poor and Underserved 8:70–82.

Coverdill, James E. 1988. "The Dual Economy and Sex Differences in Earnings." Social Forces 66:97–993.

Cowan, Ruth Schwartz. 1993. "Less Work for Mother?" Pp. 329–339 in Albert H. Teich (ed.), Technology and the Future (6th ed.). New York: St. Martin's.

Cowgill, Donald O. 1986. Aging Around the World. Belmont, CA: Wadsworth.

Crèvecoeur, J. Hector. 1974. "What Is an American?" In Richard J. Meister (ed.), Race and Ethnicity in Modern America. Lexington, MA: Heath. (Originally published 1782.)

Crimmins, Ellen M., Hayward, Mark D., and Saito, Yashuhiko. 1994. "Changing Mortality and Morbidity Rates and the Health Status and Life Expectancy of the Older Population." Demography 31:168–169.

Crittenden, Ann. 2001. The Price of Motherhood: Why the Most Important Job in the World Is Still the Least Valued. New York: Henry Holt.

Crosnoe, R., Erikson, K.G., and S. Dornbusch. 2002. "Protective Functions of Family Relationships and School Factors on the Deviant Behavior of Adolescent Girls and Boys." Youth and Society 33:515–544.

Crozier, Michael, and Friedberg, Erhard. 1980. Actors and Systems: The Politics of Collective Action. Chicago: University of Chicago Press.

Crutchfield, Robert D. 1989. "Labor Stratification and Violent Crime." Social Forces 68:489–512.

Culver, John H. 1992. "Capital Punishment, 1977–1990: Characteristics of the 143 Executed." Sociology and Social Research 76:59–61.

Cureton, Steven. 2000. "Justifiable Arrests or Discretionary Justice: Predictors of Racial Arrest Differentials." Journal of Black Studies 30:703–719.

Currie, Elliott. 1998. Crime and Punishment in America. New York: Henry Holt.

Curry, Theodore. R. 1996. "Conservative Protestantism and the Perceived Wrongfulness of Crimes: A Research Note." Criminology 34:453–464.

Curtis, James E., Grabb, Edward G., and Baer, Douglas E. 1992. "Voluntary Association Membership in Fifteen Countries: A Comparative Analysis." American Sociological Review 57:139–152.

Dahl, Robert. 1961. Who Governs? New Haven, CT: Yale University Press.

_____. 1971. Polarchy. New Haven, CT: Yale University Press.

Dalby, Andrew. 2003. Language in Danger: The Loss of Linguistic Diversity and the Threat to Our Future. New York: Columbia University Press.

Daly, Martin, and Wilson, Margo. 1983. Sex, Evolution, and Behavior (2nd ed.). Boston: Willard Grant.

Davidman, Lynn. 1991. Tradition in a Rootless World: Women Turn to Orthodox Judaism. Berkeley: University of California Press.

Davies, Annette. 1986. Industrial Relations and New Technology. London: Croom Helm.

Davis, James, and Stasson, Mark. 1988. "Small-Group Performance: Past and Future Research Trends." Advances in Group Processes 5:245–277.

Davis, Kingsley. 1961. "Prostitution." In Robert K. Merton and Robert A. Nisbet (eds.), Contemporary Social Problems. San Francisco: Harcourt Brace Jovanovich.

_____. 1973. "Introduction." In Kingsley Davis (ed.), Cities. New York: W. H. Freeman.

Davis, Kingsley, and Moore, Wilbert E. 1945. "Some Principles of Stratification." American Sociological Review 10:242–249.

Davis, Nancy J., and Robinson, Robert V. 1996. "Are Rumors of War Exaggerated? Religious Orthodoxy and Moral Progressivism in America." American Journal of Sociology 102:756–787.

Deegan, Mary Jo. 1987. Jane Addams and the Men of the Chicago School, 1892–1918. New Brunswick, NJ: Transaction.

DeFrances, Carol J., and Smith, Steven K. 1998. Perceptions of Neighborhood Crime, 1995. Bureau of Justice Statistics Special Report, NCJ-165811. Washington, D.C.: U.S. Department of Justice.

Delgado, Gary. 1986. Organizing the Movement: The Roots and Growth of ACORN. Philadelphia: Temple University Press.

Demo, David H., and Cox, Martha J. 2000. "Families with Young Children: A Review of Research in the 1990s." Journal of Marriage and the Family 62:876–895.

DeNavas-Walt, Carmen, and Cleveland, Robert. 2002. "Money Income in the United States: 2001." Current Population Reports. P60-218. Washington, D.C.: U.S. Census Bureau.

Denzin, Norman K. 1984. "Toward a Phenomenology of Domestic, Family Violence." American Journal of Sociology 90:483–513.

Devine, Joel, Sheley, Joseph, and Smith, M. Dwayne. 1988. "Macroeconomic and Social Control Policy Influences in Crime Rate Changes, 1948–85." American Sociological Review 53:407–420.

DeWitt, J. L. 1943. Japanese in the United States. Final Report: Japanese Evacuation from the West Coast. P. 34 cited in Paul E. Horton and Gerald R. Leslie, Social Problems 1955. East Norwalk, CT: Appleton-Century-Crofts.

Diamond, Irene, and Orenstein, Gloria Feman. 1990. Reweaving the World: The Emergence of Ecofeminism. San Francisco: Sierra Club Books.

DiMaggio, Paul, Evans, John, and Bryson, Bethany. 1996. "Have Americans' Social Attitudes Become More Polarized?" American Journal of Sociology, 102:690–755.

DiMaggio, Paul, Hargittai, Eszter, Neuman, W. Russell, and Robinson, John P. 2001. "Social Implications of the Internet." Annual Review of Sociology 27:307–336.

DiMaggio, Paul, and Mohr, John. 1985. "Cultural Capital, Educational Attainment, and Marital Selection." American Journal of Sociology 90:1231–1261.

DiTomaso, Nancy. 1987. "Symbolic Media and Social Solidarity: The Foundations of Corporate Culture." Sociology of Organizations 5:105–134.

Dixon, William J., and Boswell, Terry. 1996. "Dependency, Disarticulation, and Denominator Effects: Another Look at Foreign Capital Penetration." American Journal of Sociology 102:543–562.

Doane, Ashley W., Jr. 1997. "Dominant Group Ethnic Identity in the United States: The Role of 'Hidden' Ethnicity in Intergroup Relations." Sociological Quarterly 38:375–395.

Dolnick, Edward. 1993. "Deafness as Culture." The Atlantic Monthly, September:37–53.

Domhoff, G. William. 1998. Who Rules America: Power and Politics in the Year 2000 (3rd ed.). Mountain View, CA: Mayfield.

Donnelly, Strachan, McCarthy, Charles R., and Singleton, Rivers, Jr. 1994. "The Brave New World of Animal Biotechnology." Special Supplement. Hastings Center Report 24 (January/February).

Dore, Ronald P. 1973. British Factory, Japanese Factory. Berkeley, CA: University of California Press.

Douglas, Tom. 1983. Groups: Understanding People Gathered Together. London: Tavistock.

Dreze, Jean, and Sen, Amartya. 1989. Hunger and Public Action. Oxford, England: Clarendon Press.

Duncan, Otis Dudley, Featherman, David L., and Duncan, Beverly. 1972. Socioeconomic Background and Achievement. New York: Seminar Books.

Dunlap, Riley E., Gallup, George H., Jr., and Gallup, Alec M. 1993. Health of the Planet: A George H. Gallup Memorial Survey. Princeton: Gallup International Institute.

Dunlap, Riley E., and Van Liere, Kent D. 1984. "Commitment to the Dominant Social Paradigm and Concern for Environmental Quality." Social Science Quarterly 65:1013–1028.

Durkheim, Emile. 1938. The Rules of Sociological Method. New York: Free Press. (Originally published 1895.)

_____. 1951. Suicide: A Study in Sociology. New York: Free Press. (Originally published 1897.)

_____. 1961. The Elementary Forms of the Religious Life. London: Allen & Unwin. (Originally published 1915.)

Dworkin, Andrea. 1981. Pornography: Men Possessing Women. New York: Putnam.

Dworkin, Shari. 2003. "Holding Back: Negotiating a Glass Ceiling on Women's Muscular Strength." In Rose Weitz (ed.), Politics of Women's Bodies: Sexuality, Appearance, and Behavior. New York: Oxford University Press.

Ebaugh, Helen Rose Fuchs. 1988. Becoming an Ex: The Process of Role Exit. Chicago: University of Chicago Press.

Eisenstadt, S. N. 1985. "Macrosocietal Analysis—Background, Development, and Indications." In S. N. Eisenstadt and H. J. Helle (eds.), Macrosociological Theory: Perspectives on Sociological Theory. Newbury Park, CA: Sage.

Elliott, Delbert S., and Ageton, Suzanne S. 1980. "Reconciling Race and Class Differences in Self-Reported Official Estimates of Delinquency." American Sociological Review 45:95–110.

Ellison, Christopher. 1991. "Religious Involvement and Subjective Well-Being." Journal of Health and Social Behavior 32:80–99.

———. 1992. "Are Religious People Nice People? Evidence from the National Survey of Black Americans." Social Forces 71:411–430.

Ellison, Christopher G., Bartkowski, John P., and Segal, Michelle L. 1996. "Conservative Protestantism and the Parental Use of Corporal Punishment." Social Forces 74:1003–1028.

Elmer-Dewitt, Philip. 1994. "A Royal Pain in the Wrist." Time, October 24:60–61.

———. 1995. "Snuff Porn on the Net." Time, February 20:69.

Ember, Lois. 1994. "Minorities Still More Likely to Live Near Toxic Sites." Chemical and Engineering News 72 (36, September 5):19.

Emerson, Richard. 1962. "Power-Dependence Relations." American Sociological Review 27:31–41.

Engels, Friedrich. 1965. "Socialism: Utopian and Scientific." In Arthur P. Mendel (ed.), The Essential Works of Marxism. New York: Bantam Books. (Originally published 1880.)

Entman, Robert M., and Rojecki, Andrew. 2000. The Black Image in the White Mind: Media and Race in America. Chicago: University of Chicago Press.

Entwisle, Doris E., and Alexander, Karl L. 1992. "Summer Setback: Race, Poverty, School Composition and Mathematics Achievement in the First Two Years of School." American Sociological Review 57:72–84.

Erickson, Kai. 1986. "On Work and Alienation." American Sociological Review 51:1–8.

Etzioni, Amitai. 1968. The Active Society. New York: Free Press.

Evans, Ellis D., Rutberg, Judith, Sather, Carmela, and Turner, Chari. 1991. "Content Analysis of Contemporary Teen Magazines for Adolescent Females." Youth and Society 23:99–120.

Evans, Sara. 2003. Tidal Wave: How Women Changed America at Century's End. New York: Free Press.

Farkas, George, Grobe, Robert P., Sheehan, Daniel, and Shuan, Yuan. 1990. "Cultural Resources and School Success." American Sociological Review 55:127–142.

Farmer, Paul. 1999. Infections and Inequalities: The Modern Plagues. Berkeley: University of California Press.

———. 2003. The Uses of Haiti. Monroe, ME: Common Courage Press.

Feagin, Joe R., and Parker, Robert. 1990. Building American Cities: The Urban Real Estate Game. Englewood Cliffs, NJ: Prentice-Hall.

Feagin, Joe R., and Sikes, Melvin P. 1994. Living with Racism: The Black Middle-Class Experience. Boston: Beacon.

Federal Interagency Forum on Aging Related Statistics. 2000. Older Americans 2000: Key Indicators of Well-Being. Washington, D.C.

Feeley, Malcolm M., and Simon, Jonathan. 1992. "The New Penology: Notes on the Emerging Strategy of Corrections and Its Implications." Criminology 30:449–474.

Felson, Richard B. 1996. "Mass Media Effects on Violent Behavior." Annual Review of Sociology 22:103–128.

Felson, Richard B., and Trudeau, Lisa. 1991. "Gender Differences on Mathematics Performance." Social Psychology Quarterly 54:113–126.

Feshbach, Morris. 1999. "Dead Souls." Atlantic Monthly 283 (1):26–27.

Festinger, Leon, Schacter, Stanley, and Back, Kurt. 1950. Social Pressure in Informal Groups. New York: Harper & Row.

Fine, Gary Alan. 1984. "Negotiated Orders and Organizational Cultures." Annual Review of Sociology 10:239–262.

———. 1996. Kitchens: The Culture of Restaurant Work. Berkeley: University of California Press.

Finke, Roger, and Stark, Rodney. 1992. The Churching of America: Winners and Losers in Our Religious Economy. New Brunswick, NJ: Rutgers University Press.

Finz, Stacy. 1999. "Race a Factor in Slaying of Mail Carrier." San Francisco Chronicle, August 13. http://www.sfgate.com/cgi-bin/article.cgi?file=/chronicle/archive/1999/08/13/MN84143.DTL.

Firebaugh, Glenn. 1996. "Does Foreign Capital Harm Poor Nations? New Estimates Based on Dixon and Boswell's Measures of Capital Penetration." American Journal of Sociology 102:563–575.

Fischer, Claude S. 1979. "Urban-to-Rural Diffusion of Opinion in Contemporary America." American Journal of Sociology 84:151–159.

———. 1981. "The Public and Private Worlds of City Life." American Sociological Review 46:306–317.

———. 1982. To Dwell Among Friends: Personal Networks in Town and City. Chicago: University of Chicago Press.

Fisher, A. D. 1987. "Alcoholism and Race: The Misapplication of Both Concepts to North American Indians." Canadian Sociological and Anthropological Review 24:80–95.

Fitzgerald, Bridget. 1999. "Children of Lesbian and Gay Parents: A Review of the Literature." Marriage and Family Review 29:57–75.

Fitzpatrick, Kevin M., and Logan, John. 1985. "The Aging of the Suburbs, 1960–1980." American Sociological Review 50:106–117.

Flanagan, William G. 1993. Contemporary Urban Sociology. New York: Cambridge University Press.

Form, William. 1985. Divided We Stand: Working Class Stratification in America. Urbana: University of Illinois Press.

"The Forbes 400." 2003. Forbes.com. http://www.forbes.com/richlist2003/rich400land.html.

Foster, John Bellamy. 1993. "Let Them Eat Pollution: Capitalism and the World Environment." Monthly Review 44:10–20.

Foster, John L. 1990. Bureaucratic Rigidity Revisited. Social Science Quarterly 71:223–238.

Freedman, Estelle B. 2002. No Turning Back: The History of Feminism and the Future of Women. New York: Ballantine.

Freeman, Sue J. M. 1990. Managing Lives: Corporate Women and Social Change. Amherst: University of Massachusetts Press.

Freud, Sigmund. 1925. The Standard Edition of the Complete Psychological Works of Sigmund Freud. Volume 19. London: Hogarth Press. (Republished in 1971.)

Freudenberg, William R. 1986. "The Density of Acquaintanceship: An Overlooked Variable in Community Research." American Journal of Sociology. 92:27–63.

Friedlander, D., and Okun, B. S. 1996. "Fertility Transition in England and Wales: Continuity and Change." Health Transition Review Supplement 1–18.

Friedsam, H. J. 1965. Competition. In Julius Gould and William L. Kolb (eds.), A Dictionary of the Social Sciences. New York: Free Press.

Fuguitt, Glenn V., Beale, Calvin L., Fulton, John A., and Gibson, Richard M. 1998. "Recent Population Trends in Nonmetropolitan Cities and Villages: From the Turnaround, Through Reversal, to the Rebound. Research in Rural Sociology and Development 7:1–21.

Funk, Richard, and Willits, Fern. 1987. "College Attendance and Attitudinal Change: A Panel Study, 1970–81." Sociology of Education 60:224–231.

Gaes, Gerald G., and McGuire, William J. 1985. "Prison Violence: The Contribution of Crowding Versus Other Determinants of Prison Assault Rates." Journal of Research in Crime and Delinquency 22 (February):41–65.

Gallup Poll. 2001. Religion after 9/11. Dec. 21.

_____. 2002. Health Care Coverage: Who's Responsible? Nov. 14.

Gamoran, Adam. 1992. "The Variable Effects of High School Tracking." American Sociological Review 57:812–828.

Gamoran, Adam, and Mare, Robert D. 1989. "Secondary School Tracking and Educational Inequality: Compensation, Reinforcement, or Neutrality?" American Journal of Sociology 94:1146–1183.

Gamson, Joshua. 1995. "Must Identity Movements Self Destruct? A Queer Dilemma." Social Problems 42:390–407.

Gamson, William. 1990. The Strategy of Social Protest (2nd ed.). Belmont, CA: Wadsworth.

Gardner, L. I. 1972. "Deprivation Dwarfism." Scientific American 227 (July):76–82.

Gardner, LeGrande, and Shoemaker, Donald. 1989. "Social Bonds and Delinquency: A Comparative Analysis." Sociological Quarterly 30:481–500.

Garfinkel, H. 1967. Studies in Ethnomethodology. Englewood Cliffs, NJ: Prentice-Hall.

Garreau, Joel. 1991. Edge City: Life on the New Frontier. New York: Doubleday.

Gatto, John Taylor. 2002. Dumbing Us Down: The Hidden Curriculum of Compulsory Schooling (10th ed.). New York: New Society.

Gautier, Ann H. 1999. The State and the Family: A Comparative Analysis of Family Policies in Industrial Countries. New York: Oxford University Press.

Gautier, Ann H., and Hatzius, Jan. 1997. "Family Benefits and Fertility: An Econometric Analysis." Population Studies 51:295–306.

Gecas, Viktor. 1981. "Contents of Socialization." In Morris Rosenberg and Ralph H. Turner (eds.), Social Psychology: Sociological Perspectives. New York: Basic Books.

_____. 1989. "The Social Psychology of Self-Efficiency." Annual Review of Sociology 15:291–316.

Gecas, Viktor, and Schwalbe, Michael. 1983. "Beyond the Looking-Glass Self: Structure and Efficacy-Based Self-Esteem." Social Psychological Quarterly 46 (2):77–88.

General Social Survey. 2002. University of California–Berkeley. http://sda.berkeley.edu:7502/archive.htm.

Genovese, Frank. 1988. "An Examination of Proposals for a U.S. Industrial Policy." American Journal of Economics and Sociology 47:441–453.

Ghana Statistical Service. 1999. Ghana Demographic and Health Survey 1998. Accra, Ghana: Ghana Statistical Service.

Giddens, Anthony. 1984. The Constitution of Society. Cambridge, England: Polity Press.

Giugni, Marco G. 1998a. "Frontiers in Social Movement Theory." Sociological Forum 13:365–375.

_____. 1998b. "Was It Worth the Effort? The Outcomes and Consequences of Social Movements." Annual Review of Sociology 24:371–393.

Glassner, Barry. 1999. The Culture of Fear: Why Americans Are Afraid of the Wrong Things. New York: Basic Books.

Glick, Daniel. 2001. "Web-Exclusive Excerpt: 'Powder Burn.'" http://www.msnbc.com/news/512636.asp?cp1=1.

Goddard, Stephen B. 1994. Getting There: The Epic Struggle between Road and Rail in the American Century. Chicago: University of Chicago Press.

Goffman, Erving. 1959. The Presentation of Self in Everyday Life. New York: Doubleday.

_____. 1961a. Asylums: Essays on the Social Situation of Mental Patients and Other Inmates. New York: Doubleday.

_____. 1961b. Encounters: Two Studies in the Sociology of Interaction. Indianapolis, IN: Bobbs-Merrill.

_____. 1963. Behavior in Public Places: Notes on the Social Organization of Gatherings. New York: Free Press.

_____. 1974. Gender Advertisements. New York: Harper & Row.

Goldin, Claudia. 1992. Understanding the Gender Gap: An Economic History of American Women. New York: Oxford University Press.

Goldscheider, Frances, and Goldscheider, Calvin. 1994. "Leaving and Returning Home in 20th Century America." Population Bulletin 48:1–50.

Gordon, Myles. 1993. "Is There an Islamic Threat?" Scholastic Update, October 22:11.

Gorman, Christine. 1992. "Sizing Up the Sexes." Time, January 20:42–51.

Goudy, Willis J., Powers, Edward A., Keith, Patricia, and Reger, Richard A. 1980. "Changes in Attitude Toward Retirement: Evidence from a Panel Study of Older Males." Journal of Gerontology, 35:942–948.

Gould, Roger V. 1991. "Multiple Networks and Mobilization in the Paris Commune, 1871." American Sociological Review 56:716–729.

Gouldner, Alvin. 1960. "The Norm of Reciprocity." American Sociological Review 25:161–178.

Granovetter, Mark. 1973. "The Strength of Weak Ties." American Journal of Sociology 78:1360–1380.

_____. 1974. Getting a Job: A Study of Contacts and Careers. Cambridge, MA: Harvard University Press.

Grant, Don S., and Martinez-Ramiro, J. R. 1997. "Crime and restructuring of the U.S. economy: A Reconsideration of the Class Linkages." Social Forces 75:769-798.

Greenberg, David F. 1985. "Age, Crime, and Social Explanation." American Journal of Sociology, 91 (July):1–21.

Greenhouse, Steven. 2003. "I.B.M. Explores Shift of Some Jobs Overseas." New York Times, July 22, P. C1+.

Grimes, Michael D. 1989. "Class and Attitudes Toward Structural Inequalities: An Empirical Comparison of Key Variables in Neo- and Post-Marxist Scholarship." Sociological Quarterly 30:441–463.

Gross, Edward. 1958. Work and Society. New York: Crowell.

Gullotta, Thomas P., Adams, Gerald R., and Markstrom, Carol A. 2000. The Adolescent Experience (4th ed.). San Diego, CA: Academic.

Gurney, Joan N., and Tierney, Kathleen, J. 1982. "Relative Deprivation and Social Movements: A Critical Look at Twenty Years of Theory and Research." Sociological Quarterly 23:33–47.

Guterbock, Thomas M., and London, Bruce. 1983. "Race, Political Orientation, and Participation: An Empirical Test of Four Competing Theories." American Sociological Review 48:439–453.

Hafferty, Frederic W. 1991. Into the Valley: Death and the Socialization of Medical Students. New Haven, CT: Yale University Press.

Hagan, Frank. 2002. Introduction to Criminology. Belmont, CA: Wadsworth.

Hagan, John, Gillis, A. R., and Simpson, John. 1985. "The Class Structure of Gender and Delinquency: Toward a Power-Control Theory of Common Delinquent Behavior." American Journal of Sociology 90:1151–1178.

Halle, David. 1993. Inside Culture: Art and Class in the American Home. Chicago: University of Chicago Press.

Hallinan, Maureen T. 1994. "School Differences in Tracking Effects on Achievement." Social Forces 72:799–820.

Hallinan, Maureen T., and Sorenson, Aage B. 1986. "Student Characteristics and Assignment to Ability Groups: Two Conceptual Formulations." Sociological Quarterly 27 (1):1–13.

Hamilton, Brady E., Martin, Joyce A., and Sutton, Paul D. 2003. "Births: Preliminary Data for 2002." National Vital Statistics Reports 51(11).

Harlow, H. F., and Harlow, M. K. 1966. "Learning to Live." Scientific American 1:244–272.

Harris, Irving B. 1996. Children in Jeopardy. New Haven, CT: Yale University Press.

Harris, Judith Rich. 1998. The Nurture Assumption: Why Children Turn Out the Way They Do. New York: Free Press.

Harris Poll. 2000. "The Public Tends to Blame the Poor, the Unemployed, and Those on Welfare for Their Problems." 24 (May 3).

Harrison, Paige M., and Karberg, Jennifer C. 2003. Prison and Jail Inmates at Midyear 2002. Bureau of Justice Statistics Bulletin. April.

Haub, Carl. 1993. "Tokyo Now Recognized as World's Largest City." Population Today 21 (March):1–2.

_____. 1994. "Population Change in the Former Soviet Republics." Population Bulletin, Vol. 49. Washington, D.C.: Population Reference Bureau.

Hayes-Bautista, David, Hsu, Paul, and Perez, Aide. 2002. The Browning of the Graying in America: Diversity in the Elderly Population and Policy Implications. Generations 26 (3):15–24.

Hechter, Michael. 1987. Principles of Group Solidarity. Berkeley: University of California Press.

Heidensohn, Frances. 1985. Women and Crime: The Life of the Female Offender. New York: New York University Press.

Hendricks, Jon, and C. Davis Hendricks. 1981. Aging in Mass Society: Myths and Realities (2nd ed.). Cambridge, MA: Winthrop.

Henley, Nancy M. 1985. "Psychology and Gender." Signs 11:101–119.

Herbert, Bob. 1999. "Children in Crisis." New York Times, June 10, p. A31.

Herdt, Gilbert (ed.). 1994. Third Sex, Third Gender: Beyond Sexual Dimorphism in Culture and History. New York: Zone Books.

Hertzman, Clyde, Frank, J., and Evans, Robert G. 1994. "Heterogeneities in Health Status and the Determinants of Population Health." Pp. 67–92 in R. Evans, M. Barer, and T. Marmor (eds.), Why Are Some People Healthy and Others Not? The Determinants of Health of Populations. New York: Aldine de Gruyter.

Hesse-Biber, Sharlene, and Carter, Gregg Lee. 2000. Working Women in America: Split Dreams. New York: Oxford University Press.

Hewitt, John, and Stokes, Randall. 1975. "Disclaimers." American Sociological Review 40:1–11.

Higley, Stephen R. 1995. Privilege, Power, and Place: The Geography of the American Upper Class. Lanham, MD: Rowman & Littlefield.

Hirschi, Travis, and Gottfredson, Michael. 1983. "Age and the Explanation of Crime." American Journal of Sociology 89:552–584.

Hochschild, Arlie R. 1985. The Managed Heart: The Commercialization of Human Feeling. Berkeley: University of California Press.

_____. 1997. The Time Bind: When Work Becomes Home and Home Becomes Work. New York: Holt.

Hodge, Robert W., Siegel, Paul, and Rossi, Peter. 1964. "Occupational Prestige in the United States, 1925–63." American Journal of Sociology 70:286–302.

Hodge, Robert W., Treiman, Donald J., and Rossi, Peter. 1966. "A Comparative Study of Occupational Prestige." In Reinhard Bendix and Seymour Martin Lipset (eds.), Class, Status, and Power (2nd ed.). New York: Free Press.

Hogan, Dennis P., and Astone, Nan Marie. 1986. "The Transition to Adulthood." Annual Review of Sociology 12:109–130.

Hogan, Dennis P., Eggebeen, David J., and Clogg, Clifford C. 1993. "The Structure of Intergenerational Exchanges in American Families." American Journal of Sociology 98:1428–1458.

Holden, Karen, Burkhauser, Richard, and Feaster, Daniel. 1988. "The Timing of Falls into Poverty After Retirement and Widowhood." Demography 25:405–414.

Holloway, Marguerite. 1994. "Trends in Women's Health: A Global View." Scientific American, August:76–83.

Homans, George. 1950. The Human Group. San Diego, CA: Harcourt Brace Jovanovich.

Hondagneu-Sotelo, Pierrette. 2001. Domestica: Immigrant Workers Cleaning and Caring in the Shadows of Affluence. Berkeley: University of California Press.

Hooks, Gregory. 1984. "The Policy Response to Factory Closings: A Comparison of the United States, Sweden, and France." Annals of the American Academy of Political and Social Science 475:110–124.

———. 1990. "The Rise of the Pentagon and U.S. State Building: The Defense Program as Industry Policy." American Journal of Sociology 96:358–404.

Hoover, Stewart M. 1990. "Ten Myths about Religious Broadcasting." Pp. 23–39 in Robert Abelman and Stewart M. Hoover (eds.), Religious Television: Controversies and Conclusions. Norwood, NJ: Ablex.

Horowitz, Allan, and White, Helene Raskin. 1987. "Gender Role Orientations and Styles of Pathology Among Adolescents." Journal of Health and Social Behavior 28:158–170.

Hout, Michael. 1986. "Opportunity and the Minority Middle Class: A Comparison of Blacks in the United States and Catholics in Northern Ireland." American Sociological Review 51:214–223.

"How Many People Experience Homelessness?" 2002. NCH Fact Sheet #2. National Coalition for the Homeless. http://www.nationalhomeless.org/numbers.html.

Hoyt, Homer. 1939. The Structure and Growth of Residential Neighborhoods in American Cities. Washington, D.C.: Federal Housing Administration.

Huber, Joan, and Form, William H. 1973. Income and Ideology: An Analysis of the American Political Formula. New York: Free Press.

Human Rights Watch. 2004. Women's Rights Division. http://www.hrw.org/women/.

Hummert, Mary Lee, Garstka, Teri A., Shaner, Jaye L., and Strahm, Sharon. 1994. "Stereotypes of the Elderly Held by Young, Middle-aged, and Elderly Adults." Journal of Gerontology Psychological Sciences 49:240–249.

———. 1995. "Judgments about Stereotypes of the Elderly." Research on Aging 17:168–169.

Hunt, Darnell M. 1997. Screening the Los Angeles "Riots": Race, Seeing, and Resistance. New York: Cambridge University Press.

Hunter, James Davison. 1991. Culture Wars: The Struggle to Redefine America. New York: Basic Books.

Hyde, Janet S., Fennema, Elizabeth H., and Lamon, Susan J. 1990. "Gender Differences in Mathematics Performance: A Meta-Analysis." Psychological Bulletin 107:139–155.

Inglehart, Ronald. 1997. Modernization and Postmodernization: Cultural, Economic, and Political Change in 43 Societies. Princeton, NJ: Princeton University Press.

Inglehart, Ronald, and Baker, Wayne E. 2000. "Modernization, Cultural Change, and the Persistence of Traditional Values." American Sociological Review 65:19–51.

Ingoldsby, Bron B., and Smith, Suzanne. 1995. Families in Multicultural Perspective. New York: Guildford Press.

International Bank for Reconstruction and Development. 2002. World Development Indicators 2002. Washington, D.C.: World Bank.

"The International Campaign for Justice in Bhopal: Background." n.d. http://www.bhopal.net.

International Centre for Prison Studies. 2002. "World Prison Brief." http://www.prisonstudies.org/.

"Islam." 2001. Ontario Consultants on Religious Tolerance. http://www.religioustolerance.org/islam.htm.

Jacobs, David. 1988. "Corporate Economic Power and the State: A Longitudinal Assessment of Two Explanations." American Journal of Sociology 93:852–881.

Jacobs, David, and Helms, Ronald E. 1997. "Testing Coercive Explanations for Order: The Determinants of Law Enforcement Strength over Time." Social Forces 75:1361–1392.

Jacobs, Jerry. 1989. Revolving Doors: Sex Segregation and Women's Careers. Stanford, CA: Stanford University Press.

Jacobson, Matthew Frye. 1998. Whiteness of a Different Color: European Immigrants and the Alchemy of Race. Cambridge, MA: Harvard University Press.

Jacquard, Roland. 2002. In the Name of Osama Bin Laden. Durham, NC: Duke University Press.

James, Angela D., Grant, David M., and Cranford, Cynthia. 2000. "Moving Up, but How Far? African American Women and Economic Restructuring in Los Angeles, 1970–1990." Sociological Perspective 43:399–420.

Janis, Irving. 1982. Groupthink: Psychological Studies of Policy Decisions and Fiascoes. Boston: Houghton Mifflin.

Jasper, James M., and Poulsen, Jane D. 1995. "Recruiting Strangers and Friends: Moral Shocks and Social Networks in Animal Rights and Anti-Nuclear Protests." Social Problems 42:493–512.

Jenkins, J. Craig, and Brent, Barbara. 1989. "Social Protest, Hegemonic Competition, and Social Reform." American Sociological Review 54:891–909.

Jenkins, J. Craig, and Eckert, Craig M. 1986. "Channeling Black Insurgency: Elite Patronage and Professional Social Movement Organizations in the Development of the Black Movement." American Sociological Review 51:812–829.

Jenness, Valerie. 1990. "From Sex to Sin to Sex as Work: COYOTE and the Reorganization of Prostitution as a Social Problem." Social Problems 37:403–420.

———. 1995. "Social Movement Growth, Domain Expansion, and Framing Processes: The Gay/Lesbian Movement and Violence Against Gays and Lesbians as a Social Problem." Social Problems 42:145–170.

Jensen, Holger. 1990. "The Cost of Neglect." Maclean's, May 7:54–55.

Jensen, Leif, Findeis, Jill L., Hsu, Wan Ling, and Schachter, Jason P. 1999. "Slipping Into and Out of Underemployment: Another Disadvantage for Nonmetropolitan Workers?" Rural Sociology 64:417–438.

Johansen, Harley, and Fuguitt, Glenn. 1984. The Changing Rural Village in America: Demographic and Economic Trends Since 1950. Cambridge, MA: Ballinger.

Johnson, Kenneth M. 1998. "Renewed Population Growth in Rural America." Research in Rural Sociology and Development 7:23–45.

Johnson, Michael P., and Ferraro, Kathleen J. 2000. "Research on Domestic Violence in the 1990s: Making Distinctions." Journal of Marriage and the Family 62:948–963.

Johnson, Richard E. 1980. "Social Class and Delinquent Behavior: A New Test." Criminology 18 (1):86–93.

Johnson, Robert C. 1993. "Science, Technology, and Black Community Development." Pp. 265–282 in Albert H. Teich (ed.), Technology and the Future (6th ed.). New York: St. Martin's.

Jones, Jeffrey M. 2001. "Americans Felt Uneasy Toward Arabs Even Before September 11." Gallup Poll Monthly, September:52–53.

Jones, Steve. 2003. Let the Games Begin: Gaming Technology and Entertainment among College Students. Washington, D.C.: Pew Internet and American Life Project.

Joseph, Brian D., DeStephano, Johanna, Jacobs, Neil G., and Lehiste, Ilse. 2003. When Languages Collide: Perspectives on Language Conflict, Language Competition, and Language Coexistence. Columbus, OH: Ohio State University Press.

Joynt, Jen, and Poe, Marshall. 2003. "Waterworld." Atlantic Monthly 292 (1):42–43.

Juster, Susan, and Vinovskis, Maris. 1987. "Changing Perspectives on the American Family in the Past." Annual Review of Sociology 13:193–216.

Kalab, Kathleen. 1987. "Student Vocabularies of Motive Accounts for Absence." Symbolic Interaction 10:71–83.

Kalish, Susan. 1994. "Culturally Sensitive Family Planning: Bangladesh Story Suggests It Can Reduce Family Size." Population Today 22 (February):5.

_____. 1995. "Multiracial Births Increase as U.S. Ponders Racial Definition." Population Today 23 (4):1–2.

Kalmijn, Matthijs. 1998. "Intermarriage and Homogamy: Causes, Patterns, and Trends." Annual Review of Sociology 24:395–421.

Kalmijn, Matthijs, and Kraaykamp, Gerbert. 1996. "Race, Cultural Capital, and Schooling: An Analysis of Trends in the United States." Sociology of Education 69:22–34.

Kanter, Rosabeth Moss. 1981. Women and the Structure of Organizations: Explorations in Theory and Behavior. Pp. 395–424 in O. Grusky and G. Miller (eds.), The Sociology of Organizations (2nd ed.). New York: Free Press.

Kaplan, Howard B., Martin, Steven S., and Johnson, Robert J. 1986. "Self-Rejection and the Explanation of Deviance: Specification of the Structure Among Latent Constructs." American Journal of Sociology 92:384–411.

Katel, Peter, Liu, Meerglinda, and Cohn, Bob. 1994. "The Bust in Boot Camps." Newsweek, February 21:26.

Katz, Jack. 1988. The Seductions of Crime: Moral and Sensual Attractions of Doing Evil. New York: Basic Books.

Kay, Jane. 1994. "Still Trying to Repair Nature." San Francisco Examiner, September 18, p. A4.

Keister, Lisa A., and Moller, Stephanie. 2000. "Wealth Inequality in the United States." Annual Review of Sociology 26:63–81.

Kemper, Peter. 1992. "Use of Formal and Informal Home Care by the Disabled Elderly." Health Services Research 27:421–451.

Kerbo, Harold R. 1991. Social Stratification and Inequality: Class Conflict in Historical and Comparative Perspective (2nd ed.). New York: McGraw-Hill.

Kessler, Ronald C., McGonagle, Katherine A., Zhao, Shanyang, Nelson, Christopher B., Hughes, Michael, Eshleman, Suzanne, Wittchen, Hans-Urlich, and Kendler, Kenneth S. 1994. "Lifetime and 12-Month Prevalence of DSMIII-R Psychiatric Disorders in the United States: Results from the National Comorbidity Study." Archives of General Psychiatry 51:8–19.

Kessler, Ronald C., and McLeod, Jane. 1984. "Sex Differences in Vulnerability to Undesirable Life Events." American Sociological Review 49:620–631.

Kielbowicz, Richard B., and Scherer, Clifford. 1986. "The Role of the Press in the Dynamics of Social Movements." Research in Social Movements, Conflicts, and Change 9:71–96.

Kimmel, Michael S. 2000. The Gendered Society. New York: Oxford University Press.

Kinzer, Stephen. 2003. All the Shah's Men: An American Coup and the Roots of Middle East Terror. New York: Wiley.

Kiple, Kenneth F. 1993. Cambridge World History of Human Disease. New York: Cambridge University Press.

Klandermas, Bert. 1984. "Mobilization and Participation: Social-Psychological Expansion of Resource Mobilization Theory." American Sociological Review 49:583–600.

Kleinman, Lawrence C., Freeman, Howard, Perlman, Judy, and Gelberg, Lillian. 1996. "Homing In on the Homeless: Assessing the Physical Health of Homeless Adults in Los Angeles County Using an Original Method to Obtain Physical Examination Data in a Survey." Health Services Research 31:533–549.

Kluegel, James R., and Smith, Eliot R. 1983. "Affirmative Action Attitudes: Effects of Self-Interest, Racial Affect, and Stratification Beliefs on Whites' Views." Social Forces 61:170–181.

Koblik, Steven. 1975. Sweden's Development from Poverty to Affluence 1750–1970. Minneapolis: University of Minnesota Press.

Kohn, Melvin, Schooler, Carmi, and Associates. 1983. Work and Personality: An Inquiry into the Impact of Social Stratification. Norwood, NJ: Ablex.

Kolko, Gabriel. 1999. "Ravaging the Poor: The International Monetary Fund Indicted by Its Own Data." International Journal of Health Services 29:51–57.

Kollock, Peter, Blumstein, Phillip, and Schwartz, Pepper. 1985. "Sex and Power in Conversation: Conversational Privileges and Duties." American Sociological Review 50:34–46.

Konig, René. 1968. "Auguste Comte." In David J. Sills (ed.), International Encyclopedia of the Social Sciences. Vol. 3. New York: Macmillan and Free Press.

Korpi, Walter. 1989. "Power, Politics, and State Autonomy in the Development of Social Citizenship." American Sociological Review 54:309–328.

Kraska, Peter B., and Kappeler, Victor E. 1997. "Militarizing American Police: The Rise and Normalization of Paramilitary Unites." Social Problems 44:1–18.

Krohn, Marvin D., Akers, Ronald L., Radosevich, Marcia J., and Lanza-Kaduce, Lonn. 1980. "Social Status and Deviance." Criminology 18:303–318.

Ku, Leighton, Sonenstein, Freya, Lindberg, Laura, Bradner, Carolyn H., Boggess, Scott, and Pleck, Joseph. 1998. "Understanding Changes in Sexual Activity among Young Metropolitan Men: 1979–1995." Family Planning Perspectives 30 (6):256–262.

Kunstler, James Howard. 1994. The Geography of Nowhere: The Rise and Decline of America's Man-Made Landscape. New York: Simon & Schuster.

Lamanna, Mary Ann, and Riedmann, Agnew. 2000. Marriages and Families: Making Choices in a Diverse Society (7th ed.). Belmont, CA: Wadsworth.

Lamont, Michelle, and Fournier, Marcel (eds.). 1992. Cultivating Differences: Symbolic Boundaries and the Making of Inequality. Chicago: University of Chicago Press.

Lane, Sandra D., and Cibula, Donald A. 2000. "Gender and Health." Pp. 136–153 in Gary L. Albrecht, Ray Fitzpatrick, and Susan C. Scrimshaw (eds.), Handbook of Social Studies in Health and Medicine. Thousand Oaks, CA: Sage Publications.

Langdon, Philip. 1994. A Better Place to Live: Reshaping the American Suburb. New York: Harper.

Lappé, Frances Moore, Collins Joseph, and Rosset, Peter. 1998. World Hunger: Twelve Myths. New York: Grove.

Laumann, Edward O., Gagnon, John H., Michael, Robert T., and Michael, Stuart. 1994. The Social Organization of Sexuality: Sexual Practices in the United States. Chicago: University of Chicago Press.

Laumann, Edward O., Knoke, David, and Kim, Yong-hat. 1985. "An Organizational Approach to State Policy Formation: A Comparative Study of Energy and Health Domains." American Sociological Review 50:1–19.

Lavee, Yoar, McCubbin, Hamilton I., and Patterson, Joan M. 1985. "The Double ABCX Model of Family Stress and Adaptation: An Empirical Test by Analysis of Structural Equations with Latent Variables." Journal of Marriage and the Family 47:811–825.

Lawrence, Bruce B. 1998. Shattering the Myth: Islam Beyond Violence. Princeton, NJ: Princeton University Press.

Laz, Cheryl. 1998. "Act Your Age." Sociological Forum 13:85–113.

Lebergott, Stanley. 1975. Wealth and Want. Princeton, NJ: Princeton University Press.

Leblanc, Lauraine. 1999. Pretty in Punk: Girls' Gender Resistance in a Boys' Subculture. New Brunswick, NJ: Rutgers University Press.

Lefkowitz, Bernard. 1997. Our Guys: The Glen Ridge Rape and the Secret Life of the Perfect Suburb. Berkeley: University of California Press.

Lemert, Edwin. 1981. "Issues in the Study of Deviance." Sociological Quarterly 22:285–305.

Lenski, Gerhard. 1966. Power and Privilege: A Theory of Social Stratification. New York: McGraw-Hill.

Levin, William. 1988. "Age Stereotyping: College Student Evaluations." Research on Aging 10:134–148.

Levy, Emanuel. 1990. "Stage, Sex and Suffering: Images of Women in American Films." Empirical Studies of the Arts 8 (1):53–76.

Lewin, Tamar. 2000. "Boom in Gene Testing Raises Questions on Sharing Results." New York Times, July 21, p. A1+.

Lewin, Tamar, and Medina, Jennifer. 2003. "To Cut Failure Rate, Schools Shed Students." New York Times July 31, p. A1+.

Lewis, Oscar. 1969. "The Culture of Poverty." In Daniel P. Moynihan (ed.), On Understanding Poverty. New York: Basic Books.

Lichter, Daniel T., LeClere, Felicia B., and McLaughlin, Diane K. 1991. "Local Marriage Markets and the Marital Behavior of Black and White Women." American Journal of Sociology 96:843–867.

Lichter, Daniel T., McLaughlin, Diane K., Kephart, George, and Landry, David J. 1992. "Race and the Retreat from Marriage: A Shortage of Marriageable Men?" American Sociological Review 57:781–799.

Lieberson, Stanley, and Waters, Mary C. 1993. "The Ethnic Responses of Whites: What Causes Their Instability, Simplification, and Inconsistency?" Social Forces 72:421–450.

Lin, Chien, and Liu, William T. 1993. "Intergenerational Relationships Among Chinese Immigrant Families from Taiwan." Pp. 271–286 in Harriett Pipes McAdoo (ed.), Family Ethnicity: Strength in Diversity. Newbury Park, CA: Sage.

Lin, Nan. 1990. Social Resources and Social Mobility: A Structured Theory of Status Attainment. Pp. 247–271 in R. Breiger (ed.), Social Mobility and Social Structure. New York: Cambridge University Press.

Lincoln, James R., and McBride, Kerry. 1987. "Japanese Industrial Organization in Comparative Perspective." Annual Review of Sociology 13:289–312.

Link, Bruce G., Cullen, Francis T., Frank, James, and Wozniak, John F. 1987. "The Social Rejection of Former Mental Patients: Understanding Why Labels Matter." American Journal of Sociology 92:1461–1500.

Link, Bruce G, Dohrenwend, Bruce P., and Skodol, Andrew E. 1986. "Socio-economic Status and Schizophrenia: Noisome Occupational Characteristics as a Risk Factor." American Sociological Review 51: 242–258.

Link, Bruce G., Struening, Elmer L., Rahav, Michael, Phelan, Jo C., and Nuttbrock, Larry. 1997. "On Stigma and Its Consequences: Evidence from a Longitudinal Study of Men with Dual Diagnoses of Mental Illness and Substance Abuse." Journal of Health and Social Behavior 38:177–190.

Liska, Allen E., Chamlin, Mitchell B., and Reed, Mark. 1985. "Testing the Economic Production and Conflict Models of Crime Control." Social Forces 64:119–138.

Little, Ruth E. 1998. "Public Health in Central and Eastern Europe and the Role of Environmental Pollution." Annual Review of Public Health 19:153–172.

Litwak, Eugene. 1960. "Geographic Mobility and Extended Family Cohesion." American Sociological Review 25:385–394.

Litwak, Eugene, and Kulis, Stephen. 1988. "Technology, Proximity, and Measures of Kin Support." Journal of Marriage and the Family 49:649–661.

Lo, Clarence, Y. H. 1982. "Countermovements and Conservative Movements in the Contemporary U.S." Annual Review of Sociology 8:10–34.

Lofland, John. 1985. Protest: Studies of Collective Behavior and Social Movements. New Brunswick, NJ: Transaction.

Logan, John R., and Spitze, Glenna D. 1994. "Family Neighbors." American Journal of Sociology 100:453–476.

London, Bruce, and Robinson, Thomas. 1989. "The Effect of International Dependence on Income Inequality and Political Violence." American Sociological Review 54:305–308.

Lorber, Judith. 1994. Paradoxes of Gender. New Haven, CT: Yale University Press.

Love, Douglas, and Torrence, William. 1989. "The Impact of Worker Age on Unemployment and Earnings After Plant Closings." Journal of Gerontology 44:S190–S195.

Love, John M., Harrison, Linda, Sagi-Schwartz, Abraham, van Ijzendoorn, Marinus, Ross, Christine, Ungerer, Judy A., Raikes, Helen, Brady-Smith, Christy, Boller, Kimberly, Brooks-Gunn, Jeanne, Constantine, Jill, Eliason Kisker, Ellen, Paulsell, Diane, and Chazan-Cohen, Rachel. 2003. "Child Care Quality Matters: How Conclusions May Vary With Context." Child Development 74:1021–1033.

Luker, Kristen. 1985. Abortion and the Politics of Motherhood. Berkeley: University of California Press.

———. 1996. Dubious Conceptions: The Politics of Teenage Pregnancy. Cambridge, MA: Harvard University Press.

Lurigio, Arthur. 1990. "Introduction." Crime and Delinquency 36:3–5.

Luxembourg Income Study. 2002. http://www.lisproject.org/.

Lydall, Harold. 1989. Yugoslavia in Crisis. Oxford: Clarendon Press.

Lye, Diane N. 1996. "Adult Child-Parent Relationships." Annual Review of Sociology 22: 79–102.

Lynch, J. J. 1979. The Broken Heart: The Medical Consequences of Loneliness. New York: Basic Books.

MacKenzie, Doris Layton, Brame, Robert, McDowall, David, and Souryal, Claire. 1995. "Boot Camps and Recidivism in Eight States." Criminology 33:327–357.

MacKenzie, Doris Layton, Shaw, James W., and Gowdy, Voncile B. 1993. "An Evaluation of Shock Incarceration in Louisiana." National Institute of Justice Research in Brief (June). Washington, D.C.: National Institute of Justice.

Maher, Lisa, and Daly, Kathleen. 1996. "Women in the Street-Level Drug Economy: Continuity or Change?" Criminology 34:465–491.

Mannheim, Karl. 1929. Ideology and Utopia: An Introduction to the Sociology of Knowledge. San Diego, CA: Harcourt Brace Jovanovich.

Marger, Martin. 2003. Race and Ethnic Relations (6th ed.). Belmont, CA: Wadsworth.

Marini, Margaret. 1989. "Sex Differences in Earnings in the United States." Annual Review of Sociology 15:343–380.

Marmot, M. G., Kogevinas, M., and Elston, M. 1987. "Social/Economic Status and Disease." Annual Review of Public Health 8:111–135.

Marsden, Peter V. 1987. "Core Discussion Networks of Americans." American Sociological Review 52:122–131.

Marsh, Robert M., and Mannari, Niroshi. 1976. Modernization and the Japanese Factory. Princeton, NJ: Princeton University Press.

Martin, Karin A. 1998. "Becoming a Gendered Body: Practices of Preschools." American Sociological Review 63:494–511.

Martin, Philip, and Widgren, Jonas. 2002. International Migration: Facing the Challenge. Population Bulletin 57(1).

Martin, Teresa, and Bumpass, Larry. 1989. "Recent Trends in Marital Disruption." Demography 26:37–51.

Martindale, Diane. 2001. "Pink Slip in Your Genes." Scientific American, January:19–20.

Marwell, Gerald, and Oliver, Pamela. 1984. "Collective Action Theory and Social Movement Research." Research in Social Movements, Conflicts, and Change 7:1–27.

Marx, Gary T. (ed.). 1971. Racial Conflict. Boston: Little, Brown.

Marx, Karl, and Engels, Friedrich. 1967. The Communist Manifesto. London: Penguin.

Masatsugu, Mitsuyuki. 1982. The Modern Samurai Society: Duty and Dependence in Contemporary Japan. New York: American Management Association.

Massey, Douglas S. 1990. "American Apartheid: Segregation and the Making of the Underclass." American Journal of Sociology 96:329–357.

Massey, Douglas S., and Bitterman, Brooks. 1985. "Explaining the Paradox of Puerto Rican Segregation." Social Forces 64:306–331.

Massey, Douglas S., and Denton, Nancy. 1988. "Suburbanization and Segregation in U.S. Metropolitan Areas." American Journal of Sociology 94:592–626.

Massey, Douglas S., and Mullan, Brendan P. 1984. "Processes of Hispanic and Black Spatial Assimilation." American Journal of Sociology 89:836–873.

Massey, Garth, Hodson, Randy, and Sekulic, Dusko. 1999. "Ethnic Enclaves and Intolerance: The Case of Yugoslavia." Social Forces: 78:669–691.

Matsueda, Ross, and Heimer, Karen. 1987. "Race, Family Structure, and Delinquency: A Test of Differential Association and Social Control Theories." American Sociological Review 52:826–840.

McAdam, Doug. 1986. "Recruitment to High-Risk Activism." American Journal of Sociology 92:64–90.

McAdam, Doug, and Paulsen, Ronnelle. 1993. "Specifying the Relationship between Social Ties and Activism." American Journal of Sociology 99:640–667.

McBrier, Debra Branch. 2003. "Gender and Career Dynamics Within a Segmented Professional Labor Market: the Case of Law Academia." Social Forces 81:1201–1266.

McCarthy, Bill. 2002. "New Economics of Sociological Criminology." Annual Review of Sociology 28:417-42.

McDill, Edward L., Natriello, Gary, and Pallas, Aaron. 1986. "A Population at Risk: Potential Consequences of Tougher School Standards for School Dropouts." American Journal of Education 94:135–181.

McGhee, Jerrie L. 1985. "The Effect of Siblings on the Life Satisfaction of the Rural Elderly." Journal of Marriage and the Family 47:85–90.

McGinn, Robert E. 1991. Science, Technology, and Society. Englewood Cliffs, NJ: Prentice Hall.

McGuffey, C. Shawn, and Rich, B. Lindsay. 1999. "Playing in the Gender Transgression Zone: Race, Class, and Hegemonic Masculinity in Middle Childhood." Gender and Society 13:608–627.

McKinlay, John B., and McKinlay, Sonja J. 1977. "The Questionable Effect of Medical Measures on the Decline of Mortality in the United States in the Twentieth Century." Milbank Memorial Fund Quarterly 55:405–428.

McPherson, J. Miller. 1983. "The Size of Voluntary Organizations." Social Forces 61:1044–1064.

McPherson, J. Miller, Popielarz, Pamela A., and Drobnic, Sonja. 1992. "Social Networks and Organizational Dynamics." American Sociological Review 57:153–170.

McPherson, J. Miller, and Smith-Lovin, Lynn. 1982. "Women and Weak Ties: Differences by Sex in the Size of Voluntary Organizations." American Journal of Sociology 87:883–904.

Mead, George Herbert. 1934. Mind, Self, and Society: From the Standpoint of a Social Behaviorist (Charles W. Morris, ed.). Chicago: University of Chicago Press.

Mead, Lawrence M. 1986. Beyond Entitlement: The Social Obligations of Citizenship. New York: Free Press.

_____. 1992. The New Politics of Poverty: The Nonworking Poor in America. New York: Basic Books.

Mellon, Margaret. 1993. "Altered Traits." Nucleus (Fall):4–6, 12.

Merton, Robert. 1949. "Discrimination and the American Creed." In Robert MacIver (ed.), Discrimination and National Welfare. New York: Harper & Row.

_____. 1957. Social Theory and Social Structure (2nd ed.). New York: Free Press.

Messerschmidt, James W. 1993. Masculinities and Crime: Critique and Reconceptualization of Theory. Lanhan, MN: Rowman & Littlefield.

Messner, Steven F. 1989. "Economic Discrimination and Societal Homicide Rates: Further Evidence on the Cost of Inequality." American Sociological Review 54:597–611.

Messner, Steven F., and Krohn, Marvin D. 1990. "Class, Compliance Structures, and Delinquency: Assessing Integrated Structural-Marxist Theory." American Journal of Sociology 96:300–328.

Meyer, David S., and Staggenborg, Suzanne. 1996. "Movements, Countermovements, and the Structure of Political Opportunity." American Journal of Sociology 101:1628–1660.

Meyerowitz, Joanne. 2002. How Sex Changed: A History of Transsexuality in the United States. Cambridge, MA: Harvard University Press.

Michalowski, Raymond J., and Kramer, Ronald C. 1987. "The Space Between Laws: The Problem of Corporate Crime in a Transnational Context." Social Problems 34:34–53.

Milkie, Melissa A. 1999. "Social Comparisons, Reflected Appraisals, and Mass Media: The Impact of Pervasive Beauty Images on Black and White Girls' Self Concepts." Social Psychology Quarterly 62:190–210.

Miller, Alan S., and Hoffmann, John P. 1999. "The Growing Divisiveness: Culture Wars or a War of Words." Social Forces 78:721–752.

Miller, Karen A., Kohn, Melvin L., and Schooler, Carmi. 1985. "Educational Self-Direction and the Cognitive Functioning of Students." Social Forces 63:923–944.

Miller, Karen, Kohn, Melvin, and Schooler, Carmi. 1986. "Educational Self-Direction and Personality." American Sociological Review 51:372–390.

Mills, C. Wright. 1940. "Situated Actions and Vocabularies of Motives." American Sociological Review 5:904–913.

_____. 1956. The Power Elite. New York: Oxford University Press.

_____. 1959. The Sociological Imagination. Oxford, England: Oxford University Press.

Mills, Robert J. 2001. "Health Insurance Coverage, 2001." Current Population Reports P60-215. Washington, D.C.: U.S. Census Bureau.

Minino, Arialdi M. 2002. "Deaths: Final Statistics for 2000." National Vital Statistics Report 50(15).

Mirowsky, John, and Ross, Catherine E. 1995. "Sex Differences in Distress: Real or Artifact?" American Sociological Review 60:449–468.

Mizrahi, Terry. 1986. Getting Rid of Patients: Contradictions in the Socialization of Physicians. New Brunswick, NJ: Rutgers University Press.

Mizruchi, Mark. 1989. "Similarity of Political Behavior Among Large American Corporations." American Journal of Sociology 95:401–424.

_____. 1990. "Determinants of Political Opposition among Large American Corporations." Social Forces 68:1065–1088.

Moe, Richard, and Wilkie, Carter. 1997. Changing Places: Rebuilding Community in the Age of Sprawl. New York: Henry Holt.

Moen, Phyllis, Dempster-McClain, Donna, and Williams, Robin. 1989. "Social Integration and Longevity: An Event-History Analysis of Women's Roles and Resilience." American Sociological Review 54:635–647.

Molm, Linda D. 2003. "Theoretical Comparisons of Forms of Exchange." Sociological Theory 21:1–17.

Molm, Linda D., and Cook, Karen S. 1995. "Social Exchange Theory." Pp. 209–235 in Karen S. Cook, Gary A. Fine, and James S. House (eds.), Sociological Perspectives on Social Psychology. New York: Allyn & Bacon.

Molotch, Harvy. 1979. "Media and Movements." In M. Zald and J. McCarthy (eds.), The Dynamics of Social Movements. Cambridge, MA: Winthrop.

Monroe, Charles R. 1995. World Religions: An Introduction. Amherst, NY: Prometheus.

Moore, Gwen S. 1990. "Structural Determinants of Men's and Women's Personal Networks." American Sociological Review 55:726–735.

Moore, Helen A., and Whitt, Hugh P. 1986. "Multiple Dimensions of the Moral Majority Platform: Shifting Interest Group Coalitions." Sociological Quarterly 27 (3):423–439.

Morgeanthau, Tom. 1995. "What Color Is Black?" Newsweek, February 13:63–70.

Morris, Aldon D., and Mueller, Carol. 1992. Frontiers in Social Movement Theory. New Haven, CT: Yale University Press.

Morris, Martina, and Western, Bruce. 1999. "Inequality in Earnings at the Close of the Twentieth Century." Annual Review of Sociology 25:623–657.

Mortimer, Jeylan, T., and Simmons, R. G. 1978. "Adult Socialization." Annual Review of Sociology 4:421–454.

Moshoeshoe II. 1993. "Return to Self-Reliance: Balancing the African Condition and the Environment." Pp. 158–170 in Pablo Piacetini (ed.), Story Earth: Native Voices on the Environment. San Francisco, CA: Mercury House.

Mulcahy, Aogan. 1995. "Claims-Making and the Construction of Legitimacy: Press Coverage of the 1981 Northern Irish Hunger Strike." Social Problems 42:449–467.

Munch, Allison, McPherson, J. Miller, and Smith-Lovin, Lynn. 1997. "Gender, Children, and Social Contact: The Effects of Childrearing for Men and Women." American Sociological Review 62:509–520.

Murdock, George Peter. 1949. Social Structure. New York: Free Press.

_____. 1957. "World Ethnographic Sample." American Anthropologist 59: 664–697.

Murphy, Caryle. 1994. "Egypt: An Uneasy Portent of Change." Current History 93:78–82.

Murr, Andrew, and Morganthau, Tom. 2001. "Burning Suburbia." Newsweek, January 15:32–33.

Murray, Charles A. 1984. Losing Ground: American Social Policy 1950–1980. New York: Basic Books.

Mutran, Elizabeth, and Reitzes, Donald C. 1984. "Intergenerational Support Activities and Well-Being Among the Elderly: A Convergence of Exchange and Symbolic Interaction Perspectives." American Sociological Review 49:117–130.

"NAFTA and Workers' Rights and Jobs." 2003. Public Citizen: Global Trade Watch. http://www.citizen.org/trade/nafta/jobs/.

Nagel, Joane. 1994. "Constructing Ethnicity: Creating and Recreating Ethnic Identity and Culture." Social Problems 41:152–176.

Nardi, Peter. 1992. Men's Friendships: Research on Men and Masculinities. Newbury Park, CA: Sage.

National Alliance for Caregiving. 1997. Family Caregiving in the United States: Findings from a National Survey. Bethesda, MD: National Alliance for Caregiving.

National Gay and Lesbian Task Force. 2003. "GLBT Civil Rights Laws in the United States." http://www.ngltf.org/downloads/civilrightsmap.pdf.

National Institute of Child Health and Human Development, Early Child Care Research Network. 2003. Does Amount of Time Spent in Child Care Predict Socioemotional Adjustment During the Transition to Kindergarten? Child Development 74:976–1005.

New York Times. 2000. "Reservations with Casinos Gain Ground on Poverty." September 2, p. A19.

Newcomb, Chad C. 1994. "Fearful and Punitive Responses to Concern about Crime." Unpublished manuscript. University of Nebraska–Lincoln.

Newman, Katherine S. 1999a. Falling from Grace: Downward Mobility in the Age of Affluence (rev. ed.). Berkeley: University of California Press.

_____. 1999b. No Shame in My Game: The Working Poor in the Inner City. New York: Knopf.

Nie, Norman H., Verba, Sidney, and Petrocik, John R. 1976. The Changing American Voter. Cambridge, MA: Harvard University Press.

Niebuhr, H. Richard. 1957. The Social Sources of Denominationalism. New York: Holt, Rinehart & Winston. (Originally published 1929.)

Norrish, Barbara R., and Rundall, Thomas G. 2001. "Hospital Restructuring and the Work of Registered Nurses." Milbank Quarterly 79:55–79.

Oakes, Jeannie. 1985. Keeping Track: How Schools Structure Inequality. New Haven, CT: Yale University Press.

Oberschall, A. 1973. Social Conflict and Social Movements. Englewood Cliffs, NJ: Prentice-Hall.

Oberschall, Anthony, and Leifer, Eric J. 1986. "Efficiency and Social Institutions: Uses and Misuses of Economic Reasoning in Sociology." Annual Review of Sociology 12:233–253.

Ogburn, William F. 1922. Social Change with Respect to Culture and Original Nature. New York: Huebsch. Reprinted 1966 Dell.

Oldenburg, Ray. 1997. The Great Good Place: Cafés, Coffee Shops, Community Centers, Beauty Parlors, General Stores, Bars, Hangouts, and How They Get You Through the Day. New York: Marlowe.

Olsen, Gregg M. 1996. "Re-modeling Sweden: The Rise and Demise of the Compromise in a Global Economy." Social Problems 43:1–20.

Orenstein, Peggy. 1994. School Girls: Young Women, Self-Esteem, and the Confidence Gap. New York: Doubleday.

Ortega, Suzanne T., and Corzine, Jay. 1990. "Socioeconomic Status and Mental Disorders." Research in Community and Mental Health 6:149–182.

Osgood, D. Wayne, and Wilson, Janet. 1989. "Role Transitions and Mundane Activities in Late Adolescence and Early Adulthood." Paper read at the 1989 meetings of the Midwest Sociological Society, St. Louis.

Ouichi, William G., and Wilkins, Alan L. 1985. "Organizational Culture." Annual Review of Sociology 11:457–483.

"Out of Sight, Out of Mind." 2000. Economist. May 20:27–28.

Parreñas, Rhacel Salazar. 2000. "Migrant Filipina Domestic Workers and the International Division of Reproductive Labor." Gender and Society 14:560–581.

Parsons, Talcott. 1964. "The School Class as a Social System: Some of Its Functions in American Society." In Talcott Parsons (ed.), Social Structure and Personality. New York: Free Press.

Paternoster, Raymond. 1989. "Absolute and Restrictive Deterrence in a Panel of Youth: Explaining the Onset, Persistence/Desistance, and Frequency of Delinquent Offending." Social Problems 36:289–309.

Patterson, Charlotte J. 2000. "Family Relationships of Lesbians and Gay Men." Journal of Marriage and the Family 62:1052–1069.

Paules, Greta Foff. 1991. Dishing It Out: Power and Resistance among Waitresses in a New Jersey Restaurant. Philadelphia: Temple University Press.

Paz, Juan J. 1993. "Support of Hispanic Elderly." Pp. 177–183 in Harriett Pipes McAdoo (ed.), Family Ethnicity: Strength in Diversity. Newbury Park, CA: Sage.

Peabody, John W. 1996. "Economic Reform and Health Sector Policy: Lessons from Structural Adjustment Programs." Social Science and Medicine 43:823–835.

Perrow, Charles. 1984. Normal Accidents: Living with High-Risk Technologies. New York: Basic Books.

———. 1986. Complex Organizations: A Critical Essay (3rd ed.). New York: Random House.

Petersilia, Joan. 1999. "A Decade of Experimenting with Intermediate Sanctions: What Have We Learned?" Justice Research and Policy 1:9–23.

Peterson, Nicolas. 1993. "Demand Sharing: Reciprocity and the Pressure for Generosity among Foragers." American Anthropologist 95:860–874.

Peterson, Peter G. 1999. Gray Dawn: How the Coming Age Wave Will Transform America—and the World. New York: Times Books.

Pettigrew, Thomas F. 1982. "Prejudice." In Thomas F. Pettigrew, George M. Fredrickson, Dale T. Knobel, Nathan Glazer, and Reed Ueda (eds.), Prejudice: Dimensions of Ethnicity. Cambridge, MA: Harvard University Press.

Pew Research Center. 2003a. Views of a Changing World 2003. Washington, D.C.: Pew Research Center.

———. 2003b. Religion and Politics: Contention and Consensus. Washington, D.C.: Pew Research Center.

Piaget, Jean. 1954. The Construction of Reality in the Child. New York: Basic Books.

Piliavin, Irving, Gartner, Rosemary, Thornton, Craig, and Matsueda, Ross. 1986. "Crime, Deterrence, and Rational Choice." American Sociological Review 51:101–119.

Piven, Frances Fox, and Cloward, Richard A. 1988. Why Americans Don't Vote. New York: Pantheon.

———. 2000. Why Americans Still Don't Vote: And Why Politicians Want It That Way. Boston: Beacon.

Plummer, Gayle. 1985. "Haitian Migrants and Backyard Imperialism." Race and Class 26:35–43.

Pollock, Philip H., III. 1982. "Organizations and Alienation. The Mediation Hypothesis Revisited." Sociological Quarterly 23:143–155.

Population Bulletin. 1999. "World Population Beyond 6 Billion." Vol. 54, No. 1.

Population Reference Bureau. 2003. World Population Data Sheet. Washington D.C.: Population Reference Bureau.

Portes, Alejandro, and Sassen-Koob, Saskia. 1987. "Making It Underground: Comparative Material on the Informal Sector in Western Market Economies." American Journal of Sociology 93:30–61.

Portes, Alejandro, and Truelove, Cynthia. 1987. "Making Sense of Diversity: Recent Research on Hispanic Minorities in the United States." Annual Review of Sociology 13:359–385.

Post, Tom. 1993. "Sailing into Big Trouble." Newsweek, November 1:34–35.

Poston, Dudley, and Gu, Baochang. 1987. "Socioeconomic Development, Family Planning, and Fertility in China." Demography 24:531–551.

Preimsberger, Duane. 1996. "Cops and Space Scientists: New Crime-Fighting Partners." Police Chief 63:108–114.

Prestby, J. E., Wandersman, A., Florin, P., Rich, R., and Chavis, D. M. 1990. "Benefits, Costs, Incentive Management and Participation in Voluntary Associations: A Means to Understanding and Promoting Empowerment." American Journal of Community Psychology 18:117–150.

Proctor, Bernadette, and Dalaker, Joseph. 2002. "Poverty in the United States, 2001." Current Population Reports P60-219. Washington, D.C.: U.S. Census Bureau.

Provence, Sally, and Lipton, Rose. 1962. Infants in Institutions: A Comparison of Their Development with Family-Reared Infants During the First Year of Life. New York: International Universities Press.

Putnam, Robert D. 2000. Bowling Alone: The Collapse and Revival of American Community. New York: Simon & Schuster.

Qian, Zhenchao. 1999. "Who Intermarries? Education, Nativity, Region, and Interracial Marriage, 1980 and 1990." Journal of Comparative Family Studies 30:579–597.

Quadagno, Jill. 2002. Aging and the Life Course: An Introduction to Social Gerontology. (2nd ed.). New York: McGraw Hill.

Quart, Alissa. 2003. Branded: The Buying and Selling of Teenagers. Cambridge, MA: Perseus.

Quillian, Lincoln. 1996. "Group Threat and Regional Change in Attitudes toward African-Americans." American Journal of Sociology 102:816–860.

Quinney, Richard. 1980. Class, State, and Crime (2nd ed.). New York: Longman.

Radelet, Michael L. 1981. "Racial Characteristics and the Imposition of the Death Penalty." American Sociological Review 46:918–927.

Raley, J. Kelly. 1996. "A Shortage of Marriageable Men? A Note on the Role of Cohabitation in Black-White Differences in Marriage Rates." American Sociological Review 61:973–983.

Rank, Mark R., and Hirschl, Thomas A. 1993. "The Link between Population Density and Welfare Participation." Demography 30:607–622.

Rankin, Susan R. 2003. Campus Climate for Gay, Lesbian, Bisexual, and Transgender People: A National Perspective. New York: National Gay and Lesbian Taskforce.

Reichman, Nancy. 1989. "Breaking Confidences: Organizational Influences on Insider Trading." Sociological Quarterly 30:185–204.

Reid, Lori L. 2002. "Occupational Segregation, Human Capital, and Motherhood: Black Women's Higher Exit Rates from Full-Time Employment." Gender and Society 16:728-747.

Reid, Robert. 1986. Land of Lost Content: The Luddite Revolt, 1812. London: Heinemann.

Reiman, Jeffrey. 1998. The Rich Get Richer and the Poor Get Prison: Ideology, Class, and Criminal Justice (5th ed.). Boston: Allyn & Bacon.

Rennison, Callie M. 1999. Criminal Victimization 1999: Changes 1998–99, with Trends 1993–99. Washington, D.C.: U.S. Department of Justice, Bureau of Justice Statistics.

———. 2002. Criminal Victimization 2001: Changes 2000–01, with Trends 1993–2001. Washington, D.C.: U.S. Department of Justice, Bureau of Justice Statistics.

Reskin, Barbara. 1989. "Women Taking 'Male' Jobs Because Men Leave Them." IlliniWeek, July 20:7.

Rhea, Joseph Tilden. 1997. Race Pride and the American Identity. Cambridge, MA: Harvard University Press.

Ricento, Thomas, and Burnaby, Barbara (eds.). 1998. Language and Politics in the United States and Canada: Myths and Realities. Mahwah, NJ: Erlbaum.

Rich, Paul. 1999. "American Voluntarism, Social Capital, and Political Culture." Annals of the American Academy of Political and Social Science 565:15–34.

Rich, Robert. 1977. The Sociology of Law. Washington, D.C.: University Press of America.

Ridgeway, Cecilia L., and Smith-Lovin, Lynn. 1999. "The Gender System and Interaction." Annual Review of Sociology 25:191–216.

Riedmann, Agnes. 1987. "Ex-Wife at the Funeral: Keyed Anti-Structure." Free Inquiry in Sociology 16:123–129.

———. 1993. Science That Colonizes. Philadelphia: Temple University Press.

Rietschlin, John. 1998. "Voluntary Association Membership and Psychological Distress." Journal of Health and Social Behavior 39:348–355.

Rindfuss, Ronald R., Swicogood, C. Gray, and Rosenfeld, Rachel A. 1987. "Disorder in the Life Course: How Common and Does it Matter?" American Sociological Review 52:785–801.

Risman, Barbara J. 1998. Gender Vertigo: American Families in Transition. New Haven, CT: Yale University Press.

Ritzer, George. 1996. The McDonaldization of Society (rev. ed.). Thousand Oaks, CA: Pine Forge Press.

Robert, Christopher, and Carnevale, Peter J. 1997. "Group Choice in Ultimatum Bargaining." Organizational Behavior and Human Decision Processes 72:256–279.

Robert, Stephanie A., and House, James S. 2000. "Socioeconomic Inequalities in Health: Integrating Individual-, Community-, and Societal-Level Theory and Research." Pp. 115–135 in Gary L. Albrecht, Ray Fitzpatrick, and Susan C. Scrimshaw (eds.), Handbook of Social Studies in Health and Medicine. Thousand Oaks, CA: Sage.

Robinson, J. Gregg, and McIlwee, Judith. 1989. "Women in Engineering: A Promise Unfulfilled." Social Problems 36:455–472.

Robinson, Robert V. 1984. "Reproducing Class Relations in Industrial Capitalism." American Sociological Review 49:182–196.

Romaine, Suzanne. 2000. Language in Society: An Introduction to Sociolinguistics (2nd ed.). Oxford, England: Oxford University Press.

Romero, Simon, and Elder, Janet. 2003. "Hispanics in U.S. Report Optimism." New York Times, August 6, p. 1+.

Rose, Peter. 1981. They and We: Racial and Ethnic Relations in the United States (3rd ed.). New York: Random House.

Rosen, Jeffrey. 2000. The Unwanted Gaze: The Destruction of Privacy in America. New York: Random House.

Rosenbaum, Emily. 1996. "Racial/Ethnic Differences in Home Ownership and Housing Quality, 1991." Social Problems 43:403–426.

Rosenberg, Debra. 1994. "Men, Women, Computers." Newsweek, May 16:48–55.

Rosenberg, Morris, Schooler, Carmi, and Schoenbach, Carrie, 1989. "Self-Esteem and Adolescent Problems Modeling Reciprocal Effects." American Sociological Review 54:1004–1018.

Rosenthal, Carolyn J. 1985. "Kinkeeping in the Familial Division of Labor." Journal of Marriage and the Family 47:965–974.

Ross, Catherine E., and Mirowsky, John. 1999. "Refining the Association between Education and Health: The Effects of Quantity, Credential, and Selectivity." Demography 36:445–460.

Rothman, Barbara Katz. 2000. Recreating Motherhood: Ideology and Technology in a Patriarchal Society. New Brunswick, NJ: Rutgers University Press.

Rothman, David J. 1993. "A Century of Failure: Health Care Reform in America." Journal of Health, Politics, Policy and Law 18:271–286.

Rubel, Maxmilien. 1968. "Karl Marx." In David Sills (ed.), International Encyclopedia of the Social Sciences. Vol. 10. New York: Macmillan and Free Press.

Rubin, Barry. 2002. Islamic Fundamentalism in Egyptian Politics, (2nd ed.). New York: Macmillan.

Rule, James, and Tilly, Charles. 1975. "Political Process in Revolutionary France 1830–1932." Pp. 41–85 in J. Merriman (ed.), 1830 in France. New York: New Viewpoints.

Rutter, Michael, Anderson-Wood, Lucie, Beckett, Celia, Bredenkamp, Diana, Castle, Jenny, Groothues, Christine, Kreppner, Jana, Keaveney, Lisa, Lord, Catherine, O'Conner, Thomas G., and the English and Romanian Adoptees (ERA) Study Team. 1999. "Quasi-Autistic Patterns Following Severe Early Global Privation." Journal of Child Psychology. 40:537–549.

Ryan, William. 1981. Equality. New York: Pantheon Books.

Sadker, Myra, and Sadker, David. 1994. Failing at Fairness: How Our Schools Cheat Girls. New York: Simon & Schuster.

Saltman, Juliet. 1991. "Maintaining Racially Diverse Neighborhoods." Urban Affairs Quarterly 26:416–441.

Sampson, Robert. 1987. "Urban Black Violence: The Effect of Male Joblessness and Family Disruption." American Journal of Sociology 93:348–382.

———. 1988. "Local Friendship Ties and Community Attachment in Mass Society: A Multilevel Systemic Model." American Sociological Review 53:766–779.

Sampson, Robert, and Groves, W. Byron. 1989. "Community Structure and Crime: Testing Social-Disorganization Theory." American Journal of Sociology 94:774–802.

Samuelson, Robert J. 1999. "The PC Boom—And Now Bust?" Newsweek, April 5:52.

Sanday, Peggy Reeves. 1990. Fraternity Gang Rape: Sex, Brotherhood, and Privilege on Campus. New York: New York University Press.

Saporito, Salvatore. 2003. "Private Choices, Public Consequences: Magnet School Choice and Segregation by Race and Poverty." Social Problems 50:181–203.

Sardon, Jean-Paul. 2002. "Recent Demographic Trends in the Developed Countries." Population 57:111–156.

Schachter, Jason. 2001. "Geographic Mobility—Population Characteristics March 1999–March 2000." Current Population Reports, March. Washington, D.C.: U.S. Census Bureau.

Schaefer, Kristin D., Hennessy, James J. and Ponterotto, Joseph G. 1999. "Race as a Variable in Imposing and Carrying out the Death Penalty in the U.S." Journal of Offender Rehabilitation 30:35–45.

Schaefer, Richard R. 1990. Racial and Ethnic Groups (4th ed.). New York: HarperCollins.

Scharff, Virginia. 1991. Taking the Wheel: Women and the Coming of the Motor Age. New York: Free Press.

Scheck, Barry, Neufeld, Peter, and Dwyer, Jim. 2000. Actual Innocence: Five Days to Execution and Other Dispatches from the Wrongfully Convicted. New York: Doubleday.

Scheff, Thomas. 1966. Being Mentally Ill: A Sociological Theory. Chicago: Aldine.

Schmidt, Hans. 1971. The U.S. Occupation of Haiti, 1915–1934. New Brunswick, NJ: Rutgers University Press.

Schneider, David J. 1981. "Tactical Self-Presentations: Toward a Broader Conceptualization." In J. T. Tedeschi (ed.), Impression Management Theory and Social Psychological Research. Orlando, FL: Academic.

Schneider, Mark, Teske, Paul, and Marschall, Melissa. 2000. Choosing Schools: Consumer Choice and the Quality of American Schools. Princeton, NJ: Princeton University Press.

Schor, Juliet B. 1998. The Overspent American: Upscaling, Downshifting, and the New Consumer. New York: Basic Books.

Schudson, Michael. 1995. The Power of News. Cambridge, MA: Harvard University Press.

Schultz, T. Paul. 1993. "Investments in the Schooling and Health of Men and Women." The Journal of Human Resources 28:694–734.

Schur, Edwin M. 1979. Interpreting Deviance: A Sociological Introduction. New York: Harper & Row.

Schwartz, Barry. 1983. "George Washington and the Whig Conception of Heroic Leadership." American Sociological Review 48:18–33

———. 1996. "Memory as a Cultural System: Abraham Lincoln in World War II." American Sociological Review 61:908–927.

———. 1998. "Postmodernity and Historical Reputation: Abraham Lincoln in Late Twentieth-Century American Memory." Social Forces 77:63–103.

Scott, Marvin B., and Lyman, Stafford M. 1968. "Accounts." American Sociological Review 33:46–62.

Seccombe, Karen, and Warner, Rebecca L. 2004. Marriage and Families: Relationships in Context. Belmont, CA: Wadsworth.

Sedlak, Andrea, and Broadhurst, Diane D. 1996. Third National Incidence Study of Child Abuse and Neglect. Washington, D.C.: U.S. Department of Health and Human Services.

Sekulic, Dusko, Massey, Garth, and Hodson, Randy. 1994. "Who Were the Yugoslavs? Failed Sources of Common Identity in the Former Yugoslavia." American Sociological Review 59:83–97.

Seltzer, Judith A. 1994. "Consequences of Marital Dissolution for Children." Annual Review of Sociology 20: 235–266.

Sen, Amartya. 1999. Development as Freedom. New York: Knopf.

Sen, Gita, and Grown, Caren. 1987. Development, Crises, and Alternative Visions. New York: Monthly Review Press.

Serpe, Richard. 1987. "Stability and Change in Self: A Structural Symbolic Interactionist Explanation." Social Psychology Quarterly 50 (1):44–55.

Shalin, Dmitri. 1986. "Pragmatism and Social Interaction." American Sociological Review 51:9–29.

Shapiro, Laura. 1994. "A Tomato with a Body that Just Won't Quit." Newsweek, June 6:80–82.

Shavit, Yossi. 1984. "Tracking and Ethnicity in Israeli Secondary Education." American Sociological Review 49:210–220.

Sheler, Jeffrey L. 1995. "Keeping Faith in His Time." U.S. News and World Report, October 9:72–77.

Shenon, Philip. 2003. "Report on U.S. Antiterrorism Law Identifies Accusations of Abuses." New York Times July 20, p. A1+.

Sherif, Muzafer. 1936. The Psychology of Social Norms. New York: Harper & Row.

Sherkat, Darren E., and Ellison, Christopher G. 1999. "Recent Developments and Current Controversies in the Sociology of Religion." Annual Review of Sociology 25:363–394.

Shkilnyk, Anastasia M. 1985. A Poison Stronger Than Love: The Destruction of an Ojibwa Community. New Haven, CT: Yale University Press.

Shortt, Samuel. 1996. "Is Unemployment Pathogenic? A Review of Current Concepts with Lessons for Policy Planners." International Journal of Health Services 26:569–589.

Siegel, Larry J. 1995. Criminology (5th ed.). Minneapolis, MN: West.

Simons, Ronald, and Gray, Phyllis. 1989. "Perceived Blocked Opportunity as an Explanation of Delinquency Among Lower-Class Black Males." Journal of Research on Crime and Delinquency 26:90–101.

Simpson, Miles. 1990. "Political Rights and Income Inequality: A Cross-National Test." American Sociological Review 55:682–693.

Simpson, Richard L. 1985. "Social Control of Occupations and Work." Annual Review of Sociology 11:415–436.

Singer, Bennet L., and Deschamps, David. 1994. Gay and Lesbian Stats: A Pocket Guide of Facts and Figures. New York: New Press.

Singh, Karan. 1993. "Let No Enemy Ever Wish Us Ill: The Hindu Vision of the Environment." Pp. 146–156 in Pablo Piacentini (ed.), Story Earth: Native Voices on the Environment. San Francisco: Mercury House.

Sjoberg, Gideon. 1960. The Preindustrial City. New York: Free Press.

Sluka, Jeffrey A. (ed.). 2000. Death Squad: The Anthropology of State Terror. Philadelphia: University of Pennsylvania Press.

Smaje, Chris. 2000. Natural Hierarchies: the Historical Sociology of Race and Caste. Malden, MA: Blackwell.

Small, Mario Luis, and Newman, Katherine. 2002. "Urban Poverty After 'The Truly Disadvantaged': The Rediscovery of the Family, the Neighborhood, and Culture." Annual Review of Sociology 27:23–45.

Smith, Christian. 1991. The Emergence of Liberation Theology: Radical Religion and Social Movement Theory. Chicago: University of Chicago Press.

Smith, Jane I. 1999. Islam in America. New York: Columbia University Press.

Smith, Page. 1995. Democracy on Trial: The Japanese American Evacuation and Relocation in World War II. New York: Simon & Schuster.

Smith, Ryan A. 1997. "Race, Income, and Authority at Work: A Cross-Temporal Analysis of Black and White Men (1972–1994)." Social Problems 44:19–37.

Smock, Pamela J. 2000. "Cohabitation in the United States: An Appraisal of Research Themes, Findings, and Implications." Annual Review of Sociology 26:1–20.

Snow, David A., and Anderson, Leon. 1987. "Identity Work Among the Homeless: The Verbal Construction and Avowal of Personal Identities." American Journal of Sociology 92:1336–1371.

_____. 1993. Down on Their Luck: A Study of Homeless Street People. Berkeley: University of California Press.

Snow, David A., Rochford, E. Burke, Jr., Worden, Steven K., and Benford, Robert D. 1986. "Frame Alignment Processes, Micromobilization, and Movement Participation." American Sociological Review 51:464–481.

Sohoni, Neera Kuckreja. 1994. "Where Are the Girls?" Ms. Magazine, July/August:96.

South, Scott, and Crowder, Kyle. 1997. "Escaping Distressed Neighborhoods: Individual, Community, and Metropolitan Influences." American Journal of Sociology 102:1040–1084.

Southwell, Priscilla Lewis, and Everest, Marcy Jean. 1998. "The Electoral Consequences of Alienation: Nonvoting and Protest Voting in the 1992 Presidential Race." Social Science Journal 35:53–51.

Spilka, Bernard, Shaver, Phillip, and Kirkpatrick, Lee A. 1985. "A General Attribution Theory for the Psychology of Religion." The Journal for the Scientific Study of Religion 24 (1):1–20.

Spring, Joel. 1997. Deculturalization and the Struggle for Equality: A Brief History of the Education of Dominated Cultures in the United States. New York: McGraw-Hill.

Stanglin, Douglas. 1992. "Toxic Wasteland." U.S. News and World Report, April 13:40–46.

Stark, Rodney, and Bainbridge, William Sims. 1979. "Of Churches, Sects, and Cults: Preliminary Concepts for a Theory of Religious Movements." Journal for the Scientific Study of Religion 18 (2):117–133.

_____. 1985. The Future of Religion: Secularization, Revival, and Cult Formation. Berkeley: University of California Press.

_____. 1987. A Theory of Religion. Toronto: Lang.

Stark, Rodney, and Finke, Roger. 2000. Acts of Faith. Berkeley: University of California Press.

Starr, Paul. 1982. The Social Transformation of American Medicine. New York: Basic Books.

Stearn, Peter N. 1976. "The Evolution of Traditional Culture Toward Aging." In Jon Hendricks and C. Davis Hendricks (eds.), Dimensions of Aging: Readings. Cambridge, MA: Winthrop.

Steffensmeier, Darrell, and Allan, Emilie. 1996. "Gender and Crime: Toward a Gendered Theory of Female Offending." Annual Review of Sociology 22:459–487.

Steffensmeier, Darrell J., Allan, Emilie, Harer, Miles, and Streifel, Cathy. 1989. "Age and the Distribution of Crime." American Journal of Sociology 94:803–831.

Stern, Jessica. 2003. Terror in the Name of God: Why Religious Militants Kill. New York: HarperCollins.

Stiglitz, Joseph E. 2003. Globalization and Its Discontents. New York: Norton.

Stokes, Randall, and Anderson, Andy. 1990. "Disarticulation and Human Welfare in Less Developed Countries." American Sociological Review 55:63–74.

Stolte, John F., Fine, Gary Alan, and Cook, Karen S. 2001. "Sociological Miniaturism: Seeing the Big Through the Small in Social Psychology." Annual Review of Sociology 27:387–412.

Straus, Murray, and Gelles, Richard. 1986. "Societal Change and Change in Family Violence from 1975 to 1985 as Revealed by Two National Surveys." Journal of Marriage and the Family 48:465–479.

Stretesky, Paul, and Hogan, Michael J. 1998. "Environmental Justice: An Analysis of Superfund Sites in Florida." Social Problems 45:268–287.

Stroebe, Margaret S., and Stroebe, Wolfgang. 1983. "Who Suffers More? Sex Differences in Health Risks of the Widowed." Psychological Bulletin 93:279–301.

Stryker, Sheldon. 1981. "Symbolic Interactionism: Themes and Variations." In Morris Rosenberg and Ralph H. Turner (eds.), Social Psychology: Sociological Perspectives. New York: Basic Books.

Sturken, Marita. 1997. Tangled Memories: The Vietnam War, the AIDS Epidemic, and the Politics of Remembering. Berkeley: University of California Press.

Sullivan, Deborah. 2001. Cosmetic Surgery: The Cutting Edge of Commercial Medicine in America. New Brunswick, NJ: Rutgers University Press.

Sullivan, Teresa A., Warren, Elizabeth, and Westbrook, Jay Lawrence. 2000. The Fragile Middle Class: Americans in Debt. New Haven, CT: Yale University Press.

Suomi, S. J., Harlow, H. H., and McKinney, W. T. 1972. "Monkey Psychiatrists." American Journal of Psychiatry 128 (February): 927–932.

Sutherland, Edwin H. 1961. White-Collar Crime. New York: Holt, Reinhart & Winston.

Suzuki, Bob. 1989. "Asian Americans as the Model Minority." Change, November–December:12–20.

Swidler, Ann. 1986. "Culture in Action: Symbols and Strategies." American Sociological Review 51:273–286.

Takagi, Dana Y. 1990. "From Discrimination to Affirmative Action: Facts in the Asian American Admissions Controversy." Social Problems 37:578–592.

Takamura, Jeanette. 2002. Social Policy Issues and Concerns in a Diverse Aging Society. Generations 26 (3):33–38.

Tang, Joyce 1993. "The Career Attainment of Caucasian and Asian Engineers." Sociological Quarterly 34:467–496.

Tannen, Deborah. 1990. You Just Don't Understand. New York: Morrow.

———. 1994. Talking from 9 to 5: How Women's and Men's Conversational Styles Affect Who Gets Heard, Who Gets Credit, and What Gets Done at Work. New York: Morrow.

Tarrow, Sidney. 1988. "National Politics and Collective Action: Recent Theory and Research in Western Europe and the United States." Annual Review of Sociology 14:421–440.

Teachman, Jay. 1987. "Family Background, Educational Resources, and Educational Attainment." American Sociological Review 52:548–557.

———. 2002. Stability Across Cohorts in Divorce Risk Factors. Demography 39:331–351.

Teachman, Jay D., Tedrow, Lucky M., and Crowder, Kyle D. 2000. "The Changing Demography of America's Families." Journal of Marriage and the Family 62:1234–1246.

Tedeschi, James T., and Riess, Marc. 1981. "Identities, the Phenomenal Self, and Laboratory Research." In J. T. Tedeschi (ed.), Impression Management Theory and Social Psychological Research. Orlando, FL: Academic.

Teitelbaum, Michael. 1975. "Relevance of Demographic Transition Theory to Developing Countries." Science 188 (May 2):420–425.

Thomas, Darwin L., and Cornwall, Marie. 1990. "Religion and Family in the 1980's." Pp. 265–274 in Alan Booth (ed.), Contemporary Families. Minneapolis, MN: National Council on Family Relations.

Thomas, W. I., and Thomas, Dorothy. 1928. The Child in America: Behavior Problems and Programs. New York: Knopf.

Thomis, Malcolm I. 1970. The Luddites: Machine-Breaking in Regency England. Hamden, CT: Archon.

Thomlinson, Ralph. 1976. Population Dynamics: Causes and Consequences of World Demographic Change (2nd ed.). New York: Random House.

Thompson, Ginger. 2002. Big Mexican Breadwinner: The Migrant Worker. New York Times, March 25, p. 3A.

Thompson, Kevin. 1989. "Gender and Adolescent Drinking Problems: The Effects of Occupational Structure." Social Problems 36:30–47.

Thompson, Linda, and Walker, Alexis J. 1989. "Gender in Families: Women and Men in Marriage, Work and Parenthood." Journal of Marriage and the Family 51:845–872.

Thornberry, Terence P., and Farnworth, Margaret. 1982. "Social Correlates of Criminal Involvement: Further Evidence on the Relationship Between Social Status and Criminal Behavior." American Sociological Review 47:505–518.

Thorson, James A. 1995. Aging in a Changing Society. Belmont, CA: Wadsworth.

Tillman, Robert, and Pontell, Henry N. 1992. "Is Justice Collar-Blind?: Punishing Medicaid Provider Fraud." Criminology 30:547–574.

Tilly, Charles. 1978. From Mobilization to Revolution. Reading, MA: Addison-Wesley.

———. 1998. Durable Inequality. Berkeley: University of California Press.

Tittle, Charles R., and Meier, Robert F. 1990. "Specifying the SES/Delinquency Relationship." Criminology 28:271–299.

Tjaden, Patricia, and Thoennes, Nancy. 1998. "Prevalence, Incidence, and Consequences of Violence against Women: Findings from the National Violence against Women Survey." National Institute of Justice Research in Brief. November.

———. 2000. Nature and Consequences of Intimate Partner Violence. Research Report 181867. Washington, D.C.: U.S. Department of Justice, National Institute of Justice.

Tomaskovich-Devey, Donald, and Skaggs, Sheryl. 2002. Sex Segregation, Labor Process Organization, and Gender Earnings Inequality. American Journal of Sociology. 108:102–128.

Treas, Judith. 1995. "Older Americans in the 1990s and Beyond." Population Bulletin. Vol. 50. Washington, D.C.: Population Reference Bureau.

Troeltsch, Ernst. 1931. The Social Teaching of the Christian Churches. New York: Macmillan.

Trudgill, Peter. 2000. Sociolinguistics: A Introduction to Language and Society. New York: Penguin.

Turner, Jonathan, and Beeghley, Leonard. 1981. The Emergence of Sociological Theory. Homewood, IL: Dorsey.

Turner, Jonathan, and Musick, David. 1985. American Dilemmas. New York: Columbia University Press.

Turner, R. Jay, Wheaton, Blair, and Lloyd, Donald A. 1995. "The Epidemiology of Social Stress." American Sociological Review 60: 104–125.

Turner, Ralph H. 1985. "Unanswered Questions in the Convergence Between Structuralist and Interactionist Role Theories." In S. N. Eisenstadt and H.J. Helle (eds.), Microsociological Theory: Perspectives on Sociological Theory. Vol. 2. Newbury Park, CA: Sage.

Uehara, Edwina S. 1995. Reciprocity Reconsidered: Gouldner's Moral Norm of Reciprocity and Social Support. Journal of Social and Personal Relationships 12:483–502.

Ugger, Christopher, and Jeff Manza. 2002. "Democratic Contraction?: Political Consequences of Felon Disenfranchisement in the United States." American Sociological Review 67:777–803.

Ulbrich, Patricia, Warheit, George, and Zimmerman, Rick. 1989. "Race, Socioeconomic Status, and Psychological Distress: An Examination of Differential Vulnerability." Journal of Health and Social Behavior 30:131–146.

UNAIDS/WHO. 2002. AIDS Epidemic Update: December 2002. Geneva, Switzerland: World Health Organization.

United Nations Development Programme. 2003. Human Development Report 2003. New York: United Nations.

United Nations Environment Programme. 2002a. Global Environmental Outlook 3. New York: United Nations Environment Programme.

———. 2002b. Vital Water Graphics: Problems Related to Freshwater Resources. http://www.unep.org/vitalwater/21.htm.

United Nations Population Fund. 1991. Population and the Environment: The Challenges Ahead. New York: United Nations Population Fund.

United Nations Population Fund. 1992. State of World Population 1992. New York: United Nations Population Fund.

United Nations Population Fund. 2000. State of World Population 2000. New York: United Nations Population Fund.

United Nations Statistics Division. 2003. The World's Women 2000: Trends and Statistics. http://unstats.un.org/unsd/demographic/ww2000/table6a.htm.

U.S. Bureau of the Census. 1975a. Historical Statistics of the United States: Colonial Times to 1970 (Bicentennial ed., Part 1). Washington, D.C.: U.S. Government Printing Office.

_____. 1975b. Statistical Abstract of the United States, 1975. Washington, D.C.: U.S. Government Printing Office.

_____. 1989a. "The Black Population of the United States: March 1988." Current Population Reports P20-442. Washington, D.C.: U.S. Government Printing Office.

_____. 1989b. "The Hispanic Population of the United States: March 1988." Current Population Reports P20-438. Washington, D.C.: U.S. Government Printing Office.

_____. 1989c. "Household and Family Characteristics: March 1988." Current Population Reports P20-437. Washington, D.C.: U.S. Government Printing Office.

_____. 1993a. Statistical Abstract of the United States: 1993. Washington, D.C.: U.S. Government Printing Office.

_____. 1993b. "The Hispanic Population in the U.S.: March 1992." Current Population Reports, Series P-20, no. 465RV. Washington, D.C.: U.S. Government Printing Office.

_____. 1995. Statistical Abstract of the United States: 1995. Washington, D.C.: U.S. Government Printing Office.

_____. 1998. Statistical Abstract of the United States: 1998. Washington, D.C.: U.S. Government Printing Office.

_____. 1999. Statistical Abstract of the United States: 1999. Washington, D.C.: U.S. Government Printing Office:

_____. 2000a. Social and Demographic Characteristics of the U.S. Population, 1998. Issued July 2000 (CD-ROM).

_____. 2000b. Statistical Abstract of the United States: 2000. Washington, D.C.: U.S. Government Printing Office.

_____. 2001. Statistical Abstract of the United States: 2001. http://www.census.gov/prod/2001pubs/statab/sec25.pdf.

_____. 2002a. The Population Profile of the United States: 2000 (Internet Release). http://www.census.gov/population/www/pop-profile/profile2000.html#cont.

_____. 2002b. Statistical Abstract of the United States: 2002. Washington, D.C.: U.S. Government Printing Office.

_____. 2002c. Table 8: Income in 2001 by Educational Attainment for People 18 Years Old and Over, by Age, Sex, Race, and Hispanic Origin: March 2002. http://www.census.gov/population/socdemo/education/ppl-169/tab08.pdf

_____. 2003a. Asset Ownership of Households: 2000. http://www.census.gov/hhes/www/wealth/1998_2000/wlth00-1.html.

_____. 2003b. Statistical Abstract of the United States: 2003. Washington, D.C.: U.S. Government Printing Office.

U.S. Bureau of Labor Statistics. 2001. Highlights of Women's Earnings in 2000. Report 952. Washington, D.C.: U.S. Government Printing Office.

_____. 2002. Occupational Outlook Quarterly. Winter 2001–02.

_____. 2003. Occupational Outlook Handbook 2002–03. Washington D.C.: U.S. Government Printing Office.

U.S. Committee for Refugees. 2002. World Refugee Survey 2001. Washington, D.C.: U.S. Committee for Refugees.

U.S. Congressional Budget Office. 2004. CBO's Current Budget Projections. www.cbo.gov/showdoc.cfm?index=1944&sequence=0#table4.

U.S. Department of Agriculture. 2000. Rural Conditions and Trends: Socioeconomic Conditions. Economic Research Service. Vol. 11, No. 2. http://www.ers.usda.gov/Publications/rcat/rcat112/.

U.S. Department of Education. 2000. Digest of Education Statistics 1999. NCES2000-031. Washington D.C.: U.S. Government Printing Office.

U.S. Department of Health and Human Services. 2002. Health, United States, 2002. Washington D.C.: U.S. Government Printing Office.

_____. 2003. Health, United States, 2003. Washington D.C.: U.S. Government Printing Office.

U.S. Department of Justice. 1995. Crime in the United States: Uniform Crime Reports, 1995. Bureau of Justice Statistics. Washington D.C.: U.S. Government Printing Office.

_____. 1998. Violence by Intimates: Analysis of Data on Crimes by Current or Former Spouses, Boyfriends, and Girlfriends. Report no. NCJ-167-237. Washington D.C.: U.S. Government Printing Office.

_____. 2001a. Crime in the United States: Uniform Crime Reports, 2001. Bureau of Justice Statistics. Washington D.C.: U.S. Government Printing Office.

_____. 2001b. Sourcebook of Criminal Justice Statistics, 2001. Bureau of Justice Statistics. Washington D.C.: U.S. Government Printing Office.

_____. 2002. Sourcebook of Criminal Justice Statistics, 2002. Bureau of Justice Statistics. Washington D.C.: U.S. Government Printing Office.

U.S. Department of Labor. 1985. The Impact of Technology on Labor in Four Industries. Bulletin 2263. Washington, D.C.: U.S. Government Printing Office.

_____. 1986. The Impact of Technology on Labor in Four Industries. Bulletin 2263. Washington, D.C.: U.S. Government Printing Office.

_____. 2002. Current Population Survey. Bureau of Labor Statistics. http://www.bls.gov/cps/cpsa2002.pdf.

U.S. Department of State. 1999. Country Report on Human Rights Practices. Washington, D.C.: U.S. Government Printing Office.

U.S. General Accounting Office. 1996. "Death Penalty Sentencing: Research Indicates Pattern of Racial Disparities." Pp. 268–272 in H. A. Bedau (ed.), The Death Penalty in America: Current Controversies. New York: Oxford University Press.

U.S. Internet Council and ITTA. 2000. State of the Internet 2000. http//www.digitaldividenetwork.org/res_stateofnet.adp.

Vago, Steven. 1989. Law and Society (2nd ed.). Englewood Cliffs, NJ: Prentice-Hall.

Vallas, Steven P. 1987. "White Collar Proletarians? The Structure of Clerical Work and Levels of Class Consciousness." Sociological Quarterly 28:523–540.

Vallas, Steven P., and Yarrow, Michael. 1987. "Advanced Technology and Worker Alienation." Working and Occupations 14 (February):126–142.

van de Walle, Etienne, and Knodel, John. 1980. "Europe's Fertility Transition." Population Bulletin 34 (6):1–43.

van den Berghe, Pierre L. 1978. Man in Society. New York: Elsevier North-Holland.

Varner, Gary E. 1994. "The Prospects for Consensus and Convergence in the Animal Rights Debate." Hastings Center Report 24:24–28.

Vaughan, Diane. 1996. The Challenger Launch Decision: Risky Technology, Culture, and Deviance at NASA. Chicago: University of Chicago Press.

Veblen, Thorstein. 1919. The Vested Interests and the State of the Industrial Arts. New York: Huebsch.

Vega, W.A., and Amero, H. 1994. "Latino Outlook: Good Health, Uncertain Prognosis." American Review of Public Health 10: 333–361.

Villemez, Wayne, and Bridges, William. 1988. "When Bigger Is Better: Differences in the Individual-Level Effect of Firm and Establishment Size." American Sociological Review 53:237–255.

Vosters, Helene. 2003. "Partial Chronology of Union Carbide's Bhopal Disaster." CorpWatch, May 15, 2003. http://www.corpwatch.org/search/PSR.jsp?term=bhopal&first=1.

Wald, Kenneth D. 1987. Religion and Politics in the United States. New York: St. Martin's.

Waldman, Steven. 1992. "Benefits 'R' Us." Newsweek, August 10:56–58.

Waldron, Ingrid. 1994. "What Do We Know About Causes of Sex Differences in Mortality? A Review of the Literature." Pp. 42–54 in Peter Conrad and Rochelle Kern (eds.), The Sociology of Health and Illness: Critical Perspectives. New York: St. Martin's.

Wallace, Walter. 1969. Sociological Theory. Hawthorne, NY: Aldine.

Wallerstein, Judith, and Kelly, J. 1980. Surviving the Breakup. New York: Basic Books.

Walshok, Mary Lindenstein. 1993. "Blue Collar Women." Pp. 256–264 in Albert H. Teich (ed.), Technology and the Future (6th ed.). New York: St. Martin's.

Watamura, Sarah E., Donzella, Bonny, Alwin, Jan, and Gunnar, Megan R. 2003. Morning-to-Afternoon Increases in Cortisol Concentrations for Infants and Toddlers at Child Care: Age Differences and Behavioral Correlates. Child Development 74:1006–1020.

Weakliem David L. and Biggert, Robert. 1999. "Region and Political Opinion in the Contemporary United States." Social Forces 77:863–886.

Weber, Max. 1954. Law in Economy and Society. (Max Rheinstein, ed., Edward Shils and Max Reinstein, trans.) Cambridge, MA: Harvard University Press. (Originally published 1914.)

_____. 1958. The Protestant Ethic and the Spirit of Capitalism. (Talcott Parsons, trans.) New York: Scribner's. (Originally published 1904–1905.)

_____. 1970a. "Bureaucracy." In H. H. Gerth and C. Wright Mills (trans.), From Max Weber: Essays in Sociology. New York: Oxford University Press. (Originally published 1910.)

_____. 1970b. "Class, Status, and Party." In H. H. Gerth and C. Wright Mills (trans.), From Max Weber: Essays in Sociology. New York: Oxford University Press. (Originally published 1910.)

_____. 1970c. "Religion." In H. H. Gerth and C. Wright Mills (trans.), From Max Weber: Essays in Sociology. New York: Oxford University Press. (Originally published 1910).

Wechsler, David. 1958. The Measurement and Appraisal of Adult Intelligence (4th ed.). Baltimore, MD: Williams & Wilkins.

Weil, Frederick. 1985. "The Variable Effects of Education on Liberal Attitudes." American Sociological Review 50:458–474.

_____. 1989. "The Sources and Structure of Legitimation in Western Democracies." American Sociological Review 54:682–706.

Weiss, Michael J. 1994. Latitudes and Attitudes: From Abilene, Texas to Zanesville, Ohio. Boston: Little, Brown.

Weitz, Rose. 2004a. Rapunzel's Daughters: What Women's Hair Tells Us About Women's Lives. New York: Farrar, Straus, and Giroux.

_____. 2004b. The Sociology of Health, Illness, and Health Care: A Critical Approach (3rd ed.). Belmont, CA: Wadsworth.

Wellman, Barry (ed.). 1999. Networks in the Global Village: Life in Contemporary Communities. Boulder, CO: Westview.

Wellman, Barry, and Berkowitz, S. D. (eds.) 1988. Social Structures: A Network Approach. New York: Cambridge University Press.

Wellman, Barry, Salaff, Janet, Dimitrova, Dimitrina, Garton, Laura, and Gulia, Milena. 1996. "Computer Networks as Social Networks: Collaborative Work, Telework, and Virtual Community." Annual Review of Sociology 22:213–238.

Wellman, Barry, and Wortley, Scot. 1990. "Different Strokes from Different Folks: Community Ties and Social Support." American Journal of Sociology 96:558–588.

Welsh, Sandy. 1998. "Gender and Sexual Harassment." Annual Review of Sociology 25:169–190.

Wessells, Michael G. 1990. Computer, Self, and Society. Englewood Cliffs, NJ: Prentice Hall.

West, Candace. 1984. Routine Complications: Troubles with Talk Between Doctors and Patients. Bloomington: Indiana University Press.

West, Candace, and Zimmerman, Don H. 1987. "Doing Gender." Gender and Society 1:125–151.

Weston, Kath. 1991. Families We Choose: Lesbians, Gays, Kinship. New York: Columbia University Press.

White, Lynn, Jr. 1967. "The Historical Roots of Our Ecologic Crisis." Science 155:1203–1207.

White, Lynn K. 1994. "Coresidence and Leaving Home: Young Adults and Their Parents." Annual Review of Sociology 20:81–102.

White, Lynn K., and Edwards, John. 1990. "Emptying the Nest and Parental Well-Being." American Sociological Review 55:235–242.

Whorf, Benjamin L. 1956. Language, Thought, and Reality. Cambridge, MA: MIT Press.

Wilentz, Amy. 1993. "Love and Haiti." New Republic 209:18–19.

Wilkinson, Richard G. 1996. Unhealthy Societies: The Afflictions of Inequality. London: Routledge.

Williams, Christine L. 1992. "The Glass Escalator: Hidden Advantages for Men in the 'Female' Professions." Social Problems 39:253–267.

Williams, J. Allen, Jr., and Ortega, Suzanne T. 1986. "The Multidimensionality of Joining." Journal of Voluntary Action Research 15:35–44.

Williams, Kirk, and Drake, Susan. 1980. "Social Structure, Crime, and Criminalization: An Empirical Examination of the Conflict Perspective." Sociological Quarterly 21:563–575.

Williams, Marian R., and Holcolm, Jefferson E. 2001. "Racial Disparity and Death Sentences in Ohio." Journal of Criminal Justice 29:207–218.

Williams, Rhys H. 1995. "Constructing the Public Good: Social Movements and Cultural Resources." Social Problems 42:124–144.

Wilson, Edward O. 1978. "Introduction: What Is Sociobiology?" In Michael S. Gregory, Anita Silvers, and Diane Sutch (eds.), Sociobiology and Human Nature. San Francisco: Jossey-Bass.

Wilson, George. 1997. "Pathways to Power: Racial Differences in the Determinants of Job Authority." Social Problems 44: 38–52.

Wilson, James Q. 1992. "Crime, Race, and Values." Society 30: 90–93.

Wilson, Thomas C. 1986. "Interregional Migration and Racial Attitudes." Social Forces 65:177–186.

———. 1991. "Urbanism, Migration, and Tolerance: A Reassessment." American Sociological Review 56:117–123.

Wilson, William J. 1978. The Declining Significance of Race. Chicago: University of Chicago Press.

———. 1987. The Truly Disadvantaged. Chicago: University of Chicago Press.

———. 1996. When Work Disappears: the World of the New Urban Poor. New York: Knopf.

Wimberly, Dale. 1990. "Investment Dependence and Alternative Explanations of Third World Mortality: A Cross-National Study." American Sociological Review 55:75–91.

Winders, Bill. 1999. "The Roller Coaster of Class Conflict: Class Segments, Mass Mobilization, and Voter Turnout in the U.S., 1840–1996." Social Forces 77:833–860.

Winerip, Michael. 2003. "Rigidity in Florida and its Consequences." New York Times, July 23, p. A15.

Wirth, Louis. 1938. "Urbanism as a Way of Life." American Journal of Sociology 44 (1):1–24.

Wiseman, Claire V., Gray, James J., Mosimann, James E., and Ahrens, Anthony H. 1992. "Cultural Expectations of Thinness in Women: An Update." International Journal of Eating Disorders 11:85–89.

Woodberry, Robert D., and Smith, Christian S. 1998. "Fundamentalism et al: Conservative Protestants in America." Annual Review of Sociology 24:25–56.

World Bank. 2003. World Development Indicators 2003. Washington, D.C.: The World Bank.

World Health Organization. 2000. Fact Sheet No. 241: Female Genital Mutilation. Geneva, Switzerland: World Health Organization.

Wright, Erik O. 1985. Classes. London: Verso.

Wrigley, Julia. 1995. Other People's Children. New York: Basic Books.

Wrong, Dennis. 1961. "The Oversocialized Conception of Man in Modern Sociology." American Sociological Review 26 (April):183–193.

———. 1979. Power. New York: Harper & Row.

Wuthnow, Robert. 1988. The Restructuring of American Religion. Princeton, NJ: Princeton University Press.

Wuthnow, Robert, and Witten, Marsha. 1988. "New Directions in the Sociology of Culture." Annual Review of Sociology 8:49–67.

Yamane, David. 1994. "Professional Socialization for What?" Footnotes 22 (March):7.

Yount, Kristin R. 1991. "Ladies, Flirts, and Tomboys: Strategies for Managing Sexual Harassment in an Underground Coal Mine." Contemporary Journal of Ethnography 19:396–422.

Zhang, Lening, and Messner, Steven F. 1995. "Family Deviance and Delinquency in China." Criminology 33:359–387.

Zuo, JiPing, and Benford, Robert. 1995. "Mobilization Processes and the 1989 Chinese Democracy Movement." Sociological Quarterly 36:131–156.

Zweigenhaft, Richard L., and Domhoff, G. William. 1998. Diversity in the Power Elite: Have Women and Minorities Reached the Top? New Haven, CT: Yale University Press.

Zygmunt, Joseph E. 1986. "Collective Behavior as a Phase of Societal Life: Blumer's Emergent Views and Their Implications." Research in Social Movements, Conflicts, and Change 9:25–46.

Photo Credits

Author Index

Subject Index